D1342728

LIBREX–

ANDERSONIAN LIBRARY
WITHDRAWN
FROM
LIBRARY
STOCK
UNIVERSITY OF STRATHCLYDE

VOLUME FOUR HUNDRED AND SEVENTY-SIX

METHODS IN ENZYMOLOGY

Guide to Techniques in Mouse Development, Part A

Mice, Embryos, and Cells, 2nd Edition

METHODS IN ENZYMOLOGY

Editors-in-Chief

JOHN N. ABELSON AND MELVIN I. SIMON

Division of Biology
California Institute of Technology
Pasadena, California, USA

Founding Editors

SIDNEY P. COLOWICK AND NATHAN O. KAPLAN

VOLUME FOUR HUNDRED AND SEVENTY-SIX

METHODS IN ENZYMOLOGY

Guide to Techniques in Mouse Development, Part A

Mice, Embryos, and Cells, 2nd Edition

EDITED BY

PAUL M. WASSARMAN AND PHILIPPE M. SORIANO
Department of Developmental and Regenerative Biology
Mount Sinai School of Medicine
New York, USA

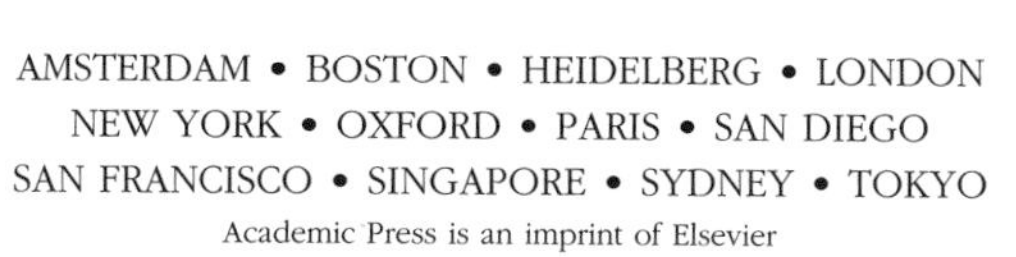

Academic Press is an imprint of Elsevier
525 B Street, Suite 1900, San Diego, CA 92101-4495, USA
30 Corporate Drive, Suite 400, Burlington, MA 01803, USA
32 Jamestown Road, London NW1 7BY, UK

First edition 2010

Notice
No responsibility is assumed by the publisher for any injury and/or damage to persons or property as a matter of products liability, negligence or otherwise, or from any use or operation of any methods, products, instructions or ideas contained in the material herein. Because of rapid advances in the medical sciences, in particular, independent verification of diagnoses and drug dosages should be made

For information on all Academic Press publications
visit our website at elsevierdirect.com

ISBN: 978-0-12-384883-3 (Paperback)
ISBN: 978-0-12-374775-4 (Hardback)
ISSN: 0076-6879

Printed and bound in United States of America
10 11 12 10 9 8 7 6 5 4 3 2 1

Contents

CONTRIBUTORS

Gokhan Akkoyunlu
Department of Histology and Embryology, Faculty of Medicine, Akdeniz University, Antalya, Turkey

David F. Albertini
KU Cancer Center, Kansas University School of Medicine, Kansas City, Kansas, USA

Yacine M. Amrani
Department of Gene and Cell Medicine, and Black Family Stem Cell Institute, Mount Sinai of School of Medicine, New York, USA

Wojtek Auerbach
VelociGene Division, Regeneron Pharmaceutical, Inc., Tarrytown, New York

Serine Avagyan
Department of Gene and Cell Medicine, and Black Family Stem Cell Institute, Mount Sinai of School of Medicine, New York, USA

Kenneth H. Ban
Institute of Molecular and Cell Biology, Proteos, Singapore

Margaret H. Baron
Division of Hematology and Medical Oncology, Department of Medicine; Tisch Cancer Institute, and Black Family Stem Cell Institute, Mount Sinai School of Medicine, New York, USA

Yong Cheng
The Fels Institute for Cancer Research and Molecular Biology, Temple University School of Medicine, Philadelphia, Pennsylvania, USA

Rostislav Chernomorsky
VelociGene Division, Regeneron Pharmaceutical, Inc., Tarrytown, New York

Neal G. Copeland
Institute of Molecular and Cell Biology, Proteos, Singapore

Thomas M. DeChiara
VelociGene Division, Regeneron Pharmaceutical, Inc., Tarrytown, New York

Mary E. Dickinson
Department of Molecular Physiology and Biophysics, and Program in Development Biology, Baylor College of Medicine, Houston, Texas, USA

Yubin Du
Transgenic Core Facility, Division of Intramural Research, National Heart, Lung, and Blood Institute, National Institutes of Health, Bethesda, Maryland, USA

Lakeisha Esau
VelociGene Division, Regeneron Pharmaceutical, Inc., Tarrytown, New York

Anna Ferrer-Vaquer
Developmental Biology Program, Sloan-Kettering Institute, New York, USA

Stuart T. Fraser
Division of Hematology and Medical Oncology, Department of Medicine; Tisch Cancer Institute, and Black Family Stem Cell Institute, Mount Sinai School of Medicine, New York, USA

David Frendewey
VelociGene Division, Regeneron Pharmaceutical, Inc., Tarrytown, New York

Marina Gertsenstein
Mount Sinai Hospital, Samuel Lunenfeld Research Institute, Toronto, Ontario, Canada

Anna-Katerina Hadjantonakis
Developmental Biology Program, Sloan-Kettering Institute, New York, USA

Zhiming Han
The Fels Institute for Cancer Research and Molecular Biology, Temple University School of Medicine, Philadelphia, Pennsylvania, USA

Joan Isern
Division of Hematology and Medical Oncology, Department of Medicine, and Tisch Cancer Institute, Mount Sinai School of Medicine, New York, USA

Nancy A. Jenkins
Institute of Molecular and Cell Biology, Proteos, Singapore

Vanessa Jones
Embryology Unit, Children's Medical Research Institute, Sydney Medical School, University of Sydney, Westmead, New South Wales, Australia

Kevin A. Kelley
Department of Developmental and Regenerative Biology, Mouse Genetics Shared Resource Facility, Mount Sinai School of Medicine, New York, New York, USA

Jeong Beom Kim
Department of Cell and Developmental Biology, Max Planck Institute for Molecular Biomedicine, Münster, NRW, Germany, and UNIST, School of Nano-Biotechnology and Chemical Engineering, Ulsan, Korea

Carlisle P. Landel
Department of Microbiology and Immunology, and Kimmel Cancer Center, Thomas Jefferson University, Philadelphia, Pennsylvania, USA

Keith E. Latham
The Fels Institute for Cancer Research and Molecular Biology, and Department of Biochemistry, Temple University School of Medicine, Philadelphia, Pennsylvania, USA

Cheng-Guang Liang
The Fels Institute for Cancer Research and Molecular Biology, Temple University School of Medicine, Philadelphia, Pennsylvania, USA

Eveline S. Litscher
Department of Developmental and Regenerative Biology, Mount Sinai School of Medicine, New York, New York, USA

Chengyu Liu
Transgenic Core Facility, Division of Intramural Research, National Heart, Lung, and Blood Institute, National Institutes of Health, Bethesda, Maryland, USA

Eiji Mizutani
Center for Developmental Biology, RIKEN, Kobe, Japan

Ken Muneoka
Division of Developmental Biology, Department of Cell and Molecular Biology, Tulane University, New Orleans, Louisiana, USA

Andrew J. Murphy
VelociGene Division, Regeneron Pharmaceutical, Inc., Tarrytown, New York

Rafidah Mutalif
Institute of Medical Biology, Immunos, Singapore

Andras Nagy
Mount Sinai Hospital, Samuel Lunenfeld Research Institute, Toronto, Ontario, Canada

Kristina Nagy
Mount Sinai Hospital, Samuel Lunenfeld Research Institute, Toronto, Ontario, Canada

Valérie Ngô-Muller
CNRS EAC4413, Functional and Adaptative Biology, Physiology of the Gonadotrope Axis, Université Paris Diderot, Paris, France

Brian J. Nieman
Mouse Imaging Centre, Hospital for Sick Children, and Department of Medical Biophysics, University of Toronto, Toronto, Canada

Sonja Nowotschin
Developmental Biology Program, Sloan-Kettering Institute, New York, USA

Jinsop Om
VelociGene Division, Regeneron Pharmaceutical, Inc., Tarrytown, New York

William T. Poueymirou
VelociGene Division, Regeneron Pharmaceutical, Inc., Tarrytown, New York

Jaime A. Rivera-Pérez
Department of Cell Biology, University of Massachusetts Medical School, Worcester, Massachusetts, USA

Thomas L. Saunders
Transgenic Animal Model Core, and Division of Molecular Medicine and Genetics, Department of Internal Medicine, University of Michigan Medical School, Ann Arbor, Michigan, USA

Hans R. Schöler
Department of Cell and Developmental Biology, Max Planck Institute for Molecular Biomedicine, Münster, NRW, Germany

Richard M. Schultz
Department of Biology, University of Pennsylvania, Philadelphia, Pennsylvania, USA

Hans-Willem Snoeck
Department of Gene and Cell Medicine, and Black Family Stem Cell Institute, Mount Sinai of School of Medicine, New York, USA

Paula Stein
Department of Biology, University of Pennsylvania, Philadelphia, Pennsylvania, USA

Colin L. Stewart
Institute of Medical Biology, Immunos, Singapore

Hideko Takahashi
Veterinary Research & Resource Section, and Genetic Engineering Facility, National Eye Institute, National Institutes of Health, Bethesda, Maryland, USA

Patrick P. L. Tam
Embryology Unit, Children's Medical Research Institute, Sydney Medical School, University of Sydney, Westmead, New South Wales, Australia

Daniel H. Turnbull
Kimmel Center for Biology and Medicine at the Skirball Institute of Biomolecular Medicine, and Departments of Radiology and Pathology, New York University School of Medicine, New York, USA

Ryan S. Udan
Department of Molecular Physiology and Biophysics, and Program in Development Biology, Baylor College of Medicine, Houston, Texas, USA

David M. Valenzuela
VelociGene Division, Regeneron Pharmaceutical, Inc., Tarrytown, New York

Sayaka Wakayama
Center for Developmental Biology, RIKEN, Kobe, Japan

Teruhiko Wakayama
Center for Developmental Biology, RIKEN, Kobe, Japan

Paul M. Wassarman
Department of Developmental and Regenerative Biology, Mount Sinai School of Medicine, New York, New York, USA

Esther S. M. Wong
Institute of Medical Biology, Immunos, Singapore.

Wen Xie
Transgenic Core Facility, Division of Intramural Research, National Heart, Lung, and Blood Institute, National Institutes of Health, Bethesda, Maryland, USA

Yingzi Xue
VelociGene Division, Regeneron Pharmaceutical, Inc., Tarrytown, New York

George D. Yancopoulos
VelociGene Division, Regeneron Pharmaceutical, Inc., Tarrytown, New York

Holm Zaehres
Department of Cell and Developmental Biology, Max Planck Institute for Molecular Biomedicine, Münster, NRW, Germany

PREFACE

It has been 17 years since the first edition of the "Guide to Techniques in Mouse Development," Volume 225 of *Methods in Enzymology*, was published by Academic Press. Needless to say, the development of technology used to investigate mouse development has not stood still during the interim. Enormous advances have occurred in genomics, transgenic and ES cell methodology, and reprogramming, culminating in the development of iPS cells. At both the cellular and molecular levels, a great many technological advances have been made that permit investigators to probe ever more deeply into all aspects of mouse development. Consequently, it appeared to be an appropriate time to publish a completely new version of the Guide, highlighting the technological advances used to study mouse development.

As in the first edition of the Guide in 1993, "Our purpose in assembling this volume is to create a source of state-of-the-art experimental approaches in mouse development useful at the laboratory bench to a diverse group of investigators. The aim is to provide investigators with reliable experimental protocols and recipes that are described in sufficient detail by leaders in the field." We believe that these goals have been achieved with publication of the new version of the Guide in 2010.

It is notable that the new version of the Guide is divided into two volumes, whereas the 1993 version was published as a single volume. This change reflects the increased number of topics covered and the increased sophistication of the methodology. In addition, we have not shied away from including articles on the same topic written by authors from different laboratories. Users of the first edition found this aspect of the Guide particularly helpful, since it enabled them to compare protocols on the same or similar topics and choose the methodology just right for them.

We sincerely hope that the Guide will find its way into many laboratories and prove useful at the bench. We are grateful to the authors for both their contributions and forbearance in dealing with publication schedules. Finally, one of us (PMW) thanks the other (PS) for agreeing to co-edit the volume, which made the job infinitely more enjoyable.

PAUL M. WASSARMAN AND PHILIPPE M. SORIANO

Methods in Enzymology

Volume I. Preparation and Assay of Enzymes
Edited by Sidney P. Colowick and Nathan O. Kaplan

Volume II. Preparation and Assay of Enzymes
Edited by Sidney P. Colowick and Nathan O. Kaplan

Volume III. Preparation and Assay of Substrates
Edited by Sidney P. Colowick and Nathan O. Kaplan

Volume IV. Special Techniques for the Enzymologist
Edited by Sidney P. Colowick and Nathan O. Kaplan

Volume V. Preparation and Assay of Enzymes
Edited by Sidney P. Colowick and Nathan O. Kaplan

Volume VI. Preparation and Assay of Enzymes *(Continued)*
Preparation and Assay of Substrates
Special Techniques
Edited by Sidney P. Colowick and Nathan O. Kaplan

Volume VII. Cumulative Subject Index
Edited by Sidney P. Colowick and Nathan O. Kaplan

Volume VIII. Complex Carbohydrates
Edited by Elizabeth F. Neufeld and Victor Ginsburg

Volume IX. Carbohydrate Metabolism
Edited by Willis A. Wood

Volume X. Oxidation and Phosphorylation
Edited by Ronald W. Estabrook and Maynard E. Pullman

Volume XI. Enzyme Structure
Edited by C. H. W. Hirs

Volume XII. Nucleic Acids (Parts A and B)
Edited by Lawrence Grossman and Kivie Moldave

Volume XIII. Citric Acid Cycle
Edited by J. M. Lowenstein

Volume XIV. Lipids
Edited by J. M. Lowenstein

Volume XV. Steroids and Terpenoids
Edited by Raymond B. Clayton

Volume XVI. Fast Reactions
Edited by Kenneth Kustin

Volume XVII. Metabolism of Amino Acids and Amines (Parts A and B)
Edited by Herbert Tabor and Celia White Tabor

Volume XVIII. Vitamins and Coenzymes (Parts A, B, and C)
Edited by Donald B. McCormick and Lemuel D. Wright

Volume XIX. Proteolytic Enzymes
Edited by Gertrude E. Perlmann and Laszlo Lorand

Volume XX. Nucleic Acids and Protein Synthesis (Part C)
Edited by Kivie Moldave and Lawrence Grossman

Volume XXI. Nucleic Acids (Part D)
Edited by Lawrence Grossman and Kivie Moldave

Volume XXII. Enzyme Purification and Related Techniques
Edited by William B. Jakoby

Volume XXIII. Photosynthesis (Part A)
Edited by Anthony San Pietro

Volume XXIV. Photosynthesis and Nitrogen Fixation (Part B)
Edited by Anthony San Pietro

Volume XXV. Enzyme Structure (Part B)
Edited by C. H. W. Hirs and Serge N. Timasheff

Volume XXVI. Enzyme Structure (Part C)
Edited by C. H. W. Hirs and Serge N. Timasheff

Volume XXVII. Enzyme Structure (Part D)
Edited by C. H. W. Hirs and Serge N. Timasheff

Volume XXVIII. Complex Carbohydrates (Part B)
Edited by Victor Ginsburg

Volume XXIX. Nucleic Acids and Protein Synthesis (Part E)
Edited by Lawrence Grossman and Kivie Moldave

Volume XXX. Nucleic Acids and Protein Synthesis (Part F)
Edited by Kivie Moldave and Lawrence Grossman

Volume XXXI. Biomembranes (Part A)
Edited by Sidney Fleischer and Lester Packer

Volume XXXII. Biomembranes (Part B)
Edited by Sidney Fleischer and Lester Packer

Volume XXXIII. Cumulative Subject Index Volumes I-XXX
Edited by Martha G. Dennis and Edward A. Dennis

Volume XXXIV. Affinity Techniques (Enzyme Purification: Part B)
Edited by William B. Jakoby and Meir Wilchek

VOLUME XXXV. Lipids (Part B)
Edited by JOHN M. LOWENSTEIN

VOLUME XXXVI. Hormone Action (Part A: Steroid Hormones)
Edited by BERT W. O'MALLEY AND JOEL G. HARDMAN

VOLUME XXXVII. Hormone Action (Part B: Peptide Hormones)
Edited by BERT W. O'MALLEY AND JOEL G. HARDMAN

VOLUME XXXVIII. Hormone Action (Part C: Cyclic Nucleotides)
Edited by JOEL G. HARDMAN AND BERT W. O'MALLEY

VOLUME XXXIX. Hormone Action (Part D: Isolated Cells, Tissues, and Organ Systems)
Edited by JOEL G. HARDMAN AND BERT W. O'MALLEY

VOLUME XL. Hormone Action (Part E: Nuclear Structure and Function)
Edited by BERT W. O'MALLEY AND JOEL G. HARDMAN

VOLUME XLI. Carbohydrate Metabolism (Part B)
Edited by W. A. WOOD

VOLUME XLII. Carbohydrate Metabolism (Part C)
Edited by W. A. WOOD

VOLUME XLIII. Antibiotics
Edited by JOHN H. HASH

VOLUME XLIV. Immobilized Enzymes
Edited by KLAUS MOSBACH

VOLUME XLV. Proteolytic Enzymes (Part B)
Edited by LASZLO LORAND

VOLUME XLVI. Affinity Labeling
Edited by WILLIAM B. JAKOBY AND MEIR WILCHEK

VOLUME XLVII. Enzyme Structure (Part E)
Edited by C. H. W. HIRS AND SERGE N. TIMASHEFF

VOLUME XLVIII. Enzyme Structure (Part F)
Edited by C. H. W. HIRS AND SERGE N. TIMASHEFF

VOLUME XLIX. Enzyme Structure (Part G)
Edited by C. H. W. HIRS AND SERGE N. TIMASHEFF

VOLUME L. Complex Carbohydrates (Part C)
Edited by VICTOR GINSBURG

VOLUME LI. Purine and Pyrimidine Nucleotide Metabolism
Edited by PATRICIA A. HOFFEE AND MARY ELLEN JONES

VOLUME LII. Biomembranes (Part C: Biological Oxidations)
Edited by SIDNEY FLEISCHER AND LESTER PACKER

Volume 71. Lipids (Part C)
Edited by John M. Lowenstein

Volume 72. Lipids (Part D)
Edited by John M. Lowenstein

Volume 73. Immunochemical Techniques (Part B)
Edited by John J. Langone and Helen Van Vunakis

Volume 74. Immunochemical Techniques (Part C)
Edited by John J. Langone and Helen Van Vunakis

Volume 75. Cumulative Subject Index Volumes XXXI, XXXII, XXXIV–LX
Edited by Edward A. Dennis and Martha G. Dennis

Volume 76. Hemoglobins
Edited by Eraldo Antonini, Luigi Rossi-Bernardi, and Emilia Chiancone

Volume 77. Detoxication and Drug Metabolism
Edited by William B. Jakoby

Volume 78. Interferons (Part A)
Edited by Sidney Pestka

Volume 79. Interferons (Part B)
Edited by Sidney Pestka

Volume 80. Proteolytic Enzymes (Part C)
Edited by Laszlo Lorand

Volume 81. Biomembranes (Part H: Visual Pigments and Purple Membranes, I)
Edited by Lester Packer

Volume 82. Structural and Contractile Proteins (Part A: Extracellular Matrix)
Edited by Leon W. Cunningham and Dixie W. Frederiksen

Volume 83. Complex Carbohydrates (Part D)
Edited by Victor Ginsburg

Volume 84. Immunochemical Techniques (Part D: Selected Immunoassays)
Edited by John J. Langone and Helen Van Vunakis

Volume 85. Structural and Contractile Proteins (Part B: The Contractile Apparatus and the Cytoskeleton)
Edited by Dixie W. Frederiksen and Leon W. Cunningham

Volume 86. Prostaglandins and Arachidonate Metabolites
Edited by William E. M. Lands and William L. Smith

Volume 87. Enzyme Kinetics and Mechanism (Part C: Intermediates, Stereo-chemistry, and Rate Studies)
Edited by Daniel L. Purich

Volume 88. Biomembranes (Part I: Visual Pigments and Purple Membranes, II)
Edited by Lester Packer

Volume 123. Vitamins and Coenzymes (Part H)
Edited by Frank Chytil and Donald B. McCormick

Volume 124. Hormone Action (Part J: Neuroendocrine Peptides)
Edited by P. Michael Conn

Volume 125. Biomembranes (Part M: Transport in Bacteria, Mitochondria, and Chloroplasts: General Approaches and Transport Systems)
Edited by Sidney Fleischer and Becca Fleischer

Volume 126. Biomembranes (Part N: Transport in Bacteria, Mitochondria, and Chloroplasts: Protonmotive Force)
Edited by Sidney Fleischer and Becca Fleischer

Volume 127. Biomembranes (Part O: Protons and Water: Structure and Translocation)
Edited by Lester Packer

Volume 128. Plasma Lipoproteins (Part A: Preparation, Structure, and Molecular Biology)
Edited by Jere P. Segrest and John J. Albers

Volume 129. Plasma Lipoproteins (Part B: Characterization, Cell Biology, and Metabolism)
Edited by John J. Albers and Jere P. Segrest

Volume 130. Enzyme Structure (Part K)
Edited by C. H. W. Hirs and Serge N. Timasheff

Volume 131. Enzyme Structure (Part L)
Edited by C. H. W. Hirs and Serge N. Timasheff

Volume 132. Immunochemical Techniques (Part J: Phagocytosis and Cell-Mediated Cytotoxicity)
Edited by Giovanni Di Sabato and Johannes Everse

Volume 133. Bioluminescence and Chemiluminescence (Part B)
Edited by Marlene DeLuca and William D. McElroy

Volume 134. Structural and Contractile Proteins (Part C: The Contractile Apparatus and the Cytoskeleton)
Edited by Richard B. Vallee

Volume 135. Immobilized Enzymes and Cells (Part B)
Edited by Klaus Mosbach

Volume 136. Immobilized Enzymes and Cells (Part C)
Edited by Klaus Mosbach

Volume 137. Immobilized Enzymes and Cells (Part D)
Edited by Klaus Mosbach

Volume 138. Complex Carbohydrates (Part E)
Edited by Victor Ginsburg

Volume 139. Cellular Regulators (Part A: Calcium- and Calmodulin-Binding Proteins)
Edited by Anthony R. Means and P. Michael Conn

Volume 140. Cumulative Subject Index Volumes 102–119, 121–134

Volume 141. Cellular Regulators (Part B: Calcium and Lipids)
Edited by P. Michael Conn and Anthony R. Means

Volume 142. Metabolism of Aromatic Amino Acids and Amines
Edited by Seymour Kaufman

Volume 143. Sulfur and Sulfur Amino Acids
Edited by William B. Jakoby and Owen Griffith

Volume 144. Structural and Contractile Proteins (Part D: Extracellular Matrix)
Edited by Leon W. Cunningham

Volume 145. Structural and Contractile Proteins (Part E: Extracellular Matrix)
Edited by Leon W. Cunningham

Volume 146. Peptide Growth Factors (Part A)
Edited by David Barnes and David A. Sirbasku

Volume 147. Peptide Growth Factors (Part B)
Edited by David Barnes and David A. Sirbasku

Volume 148. Plant Cell Membranes
Edited by Lester Packer and Roland Douce

Volume 149. Drug and Enzyme Targeting (Part B)
Edited by Ralph Green and Kenneth J. Widder

Volume 150. Immunochemical Techniques (Part K: *In Vitro* Models of B and T Cell Functions and Lymphoid Cell Receptors)
Edited by Giovanni Di Sabato

Volume 151. Molecular Genetics of Mammalian Cells
Edited by Michael M. Gottesman

Volume 152. Guide to Molecular Cloning Techniques
Edited by Shelby L. Berger and Alan R. Kimmel

Volume 153. Recombinant DNA (Part D)
Edited by Ray Wu and Lawrence Grossman

Volume 154. Recombinant DNA (Part E)
Edited by Ray Wu and Lawrence Grossman

Volume 155. Recombinant DNA (Part F)
Edited by Ray Wu

Volume 156. Biomembranes (Part P: ATP-Driven Pumps and Related Transport: The Na, K-Pump)
Edited by Sidney Fleischer and Becca Fleischer

Volume 157. Biomembranes (Part Q: ATP-Driven Pumps and Related Transport: Calcium, Proton, and Potassium Pumps)
Edited by Sidney Fleischer and Becca Fleischer

Volume 158. Metalloproteins (Part A)
Edited by James F. Riordan and Bert L. Vallee

Volume 159. Initiation and Termination of Cyclic Nucleotide Action
Edited by Jackie D. Corbin and Roger A. Johnson

Volume 160. Biomass (Part A: Cellulose and Hemicellulose)
Edited by Willis A. Wood and Scott T. Kellogg

Volume 161. Biomass (Part B: Lignin, Pectin, and Chitin)
Edited by Willis A. Wood and Scott T. Kellogg

Volume 162. Immunochemical Techniques (Part L: Chemotaxis and Inflammation)
Edited by Giovanni Di Sabato

Volume 163. Immunochemical Techniques (Part M: Chemotaxis and Inflammation)
Edited by Giovanni Di Sabato

Volume 164. Ribosomes
Edited by Harry F. Noller, Jr., and Kivie Moldave

Volume 165. Microbial Toxins: Tools for Enzymology
Edited by Sidney Harshman

Volume 166. Branched-Chain Amino Acids
Edited by Robert Harris and John R. Sokatch

Volume 167. Cyanobacteria
Edited by Lester Packer and Alexander N. Glazer

Volume 168. Hormone Action (Part K: Neuroendocrine Peptides)
Edited by P. Michael Conn

Volume 169. Platelets: Receptors, Adhesion, Secretion (Part A)
Edited by Jacek Hawiger

Volume 170. Nucleosomes
Edited by Paul M. Wassarman and Roger D. Kornberg

Volume 171. Biomembranes (Part R: Transport Theory: Cells and Model Membranes)
Edited by Sidney Fleischer and Becca Fleischer

Volume 172. Biomembranes (Part S: Transport: Membrane Isolation and Characterization)
Edited by Sidney Fleischer and Becca Fleischer

Volume 190. Retinoids (Part B: Cell Differentiation and Clinical Applications)
Edited by Lester Packer

Volume 191. Biomembranes (Part V: Cellular and Subcellular Transport: Epithelial Cells)
Edited by Sidney Fleischer and Becca Fleischer

Volume 192. Biomembranes (Part W: Cellular and Subcellular Transport: Epithelial Cells)
Edited by Sidney Fleischer and Becca Fleischer

Volume 193. Mass Spectrometry
Edited by James A. McCloskey

Volume 194. Guide to Yeast Genetics and Molecular Biology
Edited by Christine Guthrie and Gerald R. Fink

Volume 195. Adenylyl Cyclase, G Proteins, and Guanylyl Cyclase
Edited by Roger A. Johnson and Jackie D. Corbin

Volume 196. Molecular Motors and the Cytoskeleton
Edited by Richard B. Vallee

Volume 197. Phospholipases
Edited by Edward A. Dennis

Volume 198. Peptide Growth Factors (Part C)
Edited by David Barnes, J. P. Mather, and Gordon H. Sato

Volume 199. Cumulative Subject Index Volumes 168–174, 176–194

Volume 200. Protein Phosphorylation (Part A: Protein Kinases: Assays, Purification, Antibodies, Functional Analysis, Cloning, and Expression)
Edited by Tony Hunter and Bartholomew M. Sefton

Volume 201. Protein Phosphorylation (Part B: Analysis of Protein Phosphorylation, Protein Kinase Inhibitors, and Protein Phosphatases)
Edited by Tony Hunter and Bartholomew M. Sefton

Volume 202. Molecular Design and Modeling: Concepts and Applications (Part A: Proteins, Peptides, and Enzymes)
Edited by John J. Langone

Volume 203. Molecular Design and Modeling: Concepts and Applications (Part B: Antibodies and Antigens, Nucleic Acids, Polysaccharides, and Drugs)
Edited by John J. Langone

Volume 204. Bacterial Genetic Systems
Edited by Jeffrey H. Miller

Volume 205. Metallobiochemistry (Part B: Metallothionein and Related Molecules)
Edited by James F. Riordan and Bert L. Vallee

Volume 223. Proteolytic Enzymes in Coagulation, Fibrinolysis, and Complement Activation (Part B: Complement Activation, Fibrinolysis, and Nonmammalian Blood Coagulation Factors)
Edited by Laszlo Lorand and Kenneth G. Mann

Volume 224. Molecular Evolution: Producing the Biochemical Data
Edited by Elizabeth Anne Zimmer, Thomas J. White, Rebecca L. Cann, and Allan C. Wilson

Volume 225. Guide to Techniques in Mouse Development
Edited by Paul M. Wassarman and Melvin L. DePamphilis

Volume 226. Metallobiochemistry (Part C: Spectroscopic and Physical Methods for Probing Metal Ion Environments in Metalloenzymes and Metalloproteins)
Edited by James F. Riordan and Bert L. Vallee

Volume 227. Metallobiochemistry (Part D: Physical and Spectroscopic Methods for Probing Metal Ion Environments in Metalloproteins)
Edited by James F. Riordan and Bert L. Vallee

Volume 228. Aqueous Two-Phase Systems
Edited by Harry Walter and Göte Johansson

Volume 229. Cumulative Subject Index Volumes 195–198, 200–227

Volume 230. Guide to Techniques in Glycobiology
Edited by William J. Lennarz and Gerald W. Hart

Volume 231. Hemoglobins (Part B: Biochemical and Analytical Methods)
Edited by Johannes Everse, Kim D. Vandegriff, and Robert M. Winslow

Volume 232. Hemoglobins (Part C: Biophysical Methods)
Edited by Johannes Everse, Kim D. Vandegriff, and Robert M. Winslow

Volume 233. Oxygen Radicals in Biological Systems (Part C)
Edited by Lester Packer

Volume 234. Oxygen Radicals in Biological Systems (Part D)
Edited by Lester Packer

Volume 235. Bacterial Pathogenesis (Part A: Identification and Regulation of Virulence Factors)
Edited by Virginia L. Clark and Patrik M. Bavoil

Volume 236. Bacterial Pathogenesis (Part B: Integration of Pathogenic Bacteria with Host Cells)
Edited by Virginia L. Clark and Patrik M. Bavoil

Volume 237. Heterotrimeric G Proteins
Edited by Ravi Iyengar

Volume 238. Heterotrimeric G-Protein Effectors
Edited by Ravi Iyengar

VOLUME 256. Small GTPases and Their Regulators (Part B: Rho Family)
Edited by W. E. BALCH, CHANNING J. DER, AND ALAN HALL

VOLUME 257. Small GTPases and Their Regulators (Part C: Proteins Involved in Transport)
Edited by W. E. BALCH, CHANNING J. DER, AND ALAN HALL

VOLUME 258. Redox-Active Amino Acids in Biology
Edited by JUDITH P. KLINMAN

VOLUME 259. Energetics of Biological Macromolecules
Edited by MICHAEL L. JOHNSON AND GARY K. ACKERS

VOLUME 260. Mitochondrial Biogenesis and Genetics (Part A)
Edited by GIUSEPPE M. ATTARDI AND ANNE CHOMYN

VOLUME 261. Nuclear Magnetic Resonance and Nucleic Acids
Edited by THOMAS L. JAMES

VOLUME 262. DNA Replication
Edited by JUDITH L. CAMPBELL

VOLUME 263. Plasma Lipoproteins (Part C: Quantitation)
Edited by WILLIAM A. BRADLEY, SANDRA H. GIANTURCO, AND JERE P. SEGREST

VOLUME 264. Mitochondrial Biogenesis and Genetics (Part B)
Edited by GIUSEPPE M. ATTARDI AND ANNE CHOMYN

VOLUME 265. Cumulative Subject Index Volumes 228, 230–262

VOLUME 266. Computer Methods for Macromolecular Sequence Analysis
Edited by RUSSELL F. DOOLITTLE

VOLUME 267. Combinatorial Chemistry
Edited by JOHN N. ABELSON

VOLUME 268. Nitric Oxide (Part A: Sources and Detection of NO; NO Synthase)
Edited by LESTER PACKER

VOLUME 269. Nitric Oxide (Part B: Physiological and Pathological Processes)
Edited by LESTER PACKER

VOLUME 270. High Resolution Separation and Analysis of Biological Macromolecules (Part A: Fundamentals)
Edited by BARRY L. KARGER AND WILLIAM S. HANCOCK

VOLUME 271. High Resolution Separation and Analysis of Biological Macromolecules (Part B: Applications)
Edited by BARRY L. KARGER AND WILLIAM S. HANCOCK

VOLUME 272. Cytochrome P450 (Part B)
Edited by ERIC F. JOHNSON AND MICHAEL R. WATERMAN

VOLUME 273. RNA Polymerase and Associated Factors (Part A)
Edited by SANKAR ADHYA

Volume 293. Ion Channels (Part B)
Edited by P. Michael Conn

Volume 294. Ion Channels (Part C)
Edited by P. Michael Conn

Volume 295. Energetics of Biological Macromolecules (Part B)
Edited by Gary K. Ackers and Michael L. Johnson

Volume 296. Neurotransmitter Transporters
Edited by Susan G. Amara

Volume 297. Photosynthesis: Molecular Biology of Energy Capture
Edited by Lee McIntosh

Volume 298. Molecular Motors and the Cytoskeleton (Part B)
Edited by Richard B. Vallee

Volume 299. Oxidants and Antioxidants (Part A)
Edited by Lester Packer

Volume 300. Oxidants and Antioxidants (Part B)
Edited by Lester Packer

Volume 301. Nitric Oxide: Biological and Antioxidant Activities (Part C)
Edited by Lester Packer

Volume 302. Green Fluorescent Protein
Edited by P. Michael Conn

Volume 303. cDNA Preparation and Display
Edited by Sherman M. Weissman

Volume 304. Chromatin
Edited by Paul M. Wassarman and Alan P. Wolffe

Volume 305. Bioluminescence and Chemiluminescence (Part C)
Edited by Thomas O. Baldwin and Miriam M. Ziegler

Volume 306. Expression of Recombinant Genes in Eukaryotic Systems
Edited by Joseph C. Glorioso and Martin C. Schmidt

Volume 307. Confocal Microscopy
Edited by P. Michael Conn

Volume 308. Enzyme Kinetics and Mechanism (Part E: Energetics of Enzyme Catalysis)
Edited by Daniel L. Purich and Vern L. Schramm

Volume 309. Amyloid, Prions, and Other Protein Aggregates
Edited by Ronald Wetzel

Volume 310. Biofilms
Edited by Ron J. Doyle

Volume 328. Applications of Chimeric Genes and Hybrid Proteins (Part C: Protein–Protein Interactions and Genomics)
Edited by Jeremy Thorner, Scott D. Emr, and John N. Abelson

Volume 329. Regulators and Effectors of Small GTPases (Part E: GTPases Involved in Vesicular Traffic)
Edited by W. E. Balch, Channing J. Der, and Alan Hall

Volume 330. Hyperthermophilic Enzymes (Part A)
Edited by Michael W. W. Adams and Robert M. Kelly

Volume 331. Hyperthermophilic Enzymes (Part B)
Edited by Michael W. W. Adams and Robert M. Kelly

Volume 332. Regulators and Effectors of Small GTPases (Part F: Ras Family I)
Edited by W. E. Balch, Channing J. Der, and Alan Hall

Volume 333. Regulators and Effectors of Small GTPases (Part G: Ras Family II)
Edited by W. E. Balch, Channing J. Der, and Alan Hall

Volume 334. Hyperthermophilic Enzymes (Part C)
Edited by Michael W. W. Adams and Robert M. Kelly

Volume 335. Flavonoids and Other Polyphenols
Edited by Lester Packer

Volume 336. Microbial Growth in Biofilms (Part A: Developmental and Molecular Biological Aspects)
Edited by Ron J. Doyle

Volume 337. Microbial Growth in Biofilms (Part B: Special Environments and Physicochemical Aspects)
Edited by Ron J. Doyle

Volume 338. Nuclear Magnetic Resonance of Biological Macromolecules (Part A)
Edited by Thomas L. James, Volker Dötsch, and Uli Schmitz

Volume 339. Nuclear Magnetic Resonance of Biological Macromolecules (Part B)
Edited by Thomas L. James, Volker Dötsch, and Uli Schmitz

Volume 340. Drug–Nucleic Acid Interactions
Edited by Jonathan B. Chaires and Michael J. Waring

Volume 341. Ribonucleases (Part A)
Edited by Allen W. Nicholson

Volume 342. Ribonucleases (Part B)
Edited by Allen W. Nicholson

Volume 343. G Protein Pathways (Part A: Receptors)
Edited by Ravi Iyengar and John D. Hildebrandt

Volume 344. G Protein Pathways (Part B: G Proteins and Their Regulators)
Edited by Ravi Iyengar and John D. Hildebrandt

Volume 381. Oxygen Sensing
Edited by Chandan K. Sen and Gregg L. Semenza

Volume 382. Quinones and Quinone Enzymes (Part B)
Edited by Helmut Sies and Lester Packer

Volume 383. Numerical Computer Methods (Part D)
Edited by Ludwig Brand and Michael L. Johnson

Volume 384. Numerical Computer Methods (Part E)
Edited by Ludwig Brand and Michael L. Johnson

Volume 385. Imaging in Biological Research (Part A)
Edited by P. Michael Conn

Volume 386. Imaging in Biological Research (Part B)
Edited by P. Michael Conn

Volume 387. Liposomes (Part D)
Edited by Nejat Düzgüneş

Volume 388. Protein Engineering
Edited by Dan E. Robertson and Joseph P. Noel

Volume 389. Regulators of G-Protein Signaling (Part A)
Edited by David P. Siderovski

Volume 390. Regulators of G-Protein Signaling (Part B)
Edited by David P. Siderovski

Volume 391. Liposomes (Part E)
Edited by Nejat Düzgüneş

Volume 392. RNA Interference
Edited by Engelke Rossi

Volume 393. Circadian Rhythms
Edited by Michael W. Young

Volume 394. Nuclear Magnetic Resonance of Biological Macromolecules (Part C)
Edited by Thomas L. James

Volume 395. Producing the Biochemical Data (Part B)
Edited by Elizabeth A. Zimmer and Eric H. Roalson

Volume 396. Nitric Oxide (Part E)
Edited by Lester Packer and Enrique Cadenas

Volume 397. Environmental Microbiology
Edited by Jared R. Leadbetter

Volume 398. Ubiquitin and Protein Degradation (Part A)
Edited by Raymond J. Deshaies

Volume 399. Ubiquitin and Protein Degradation (Part B)
Edited by Raymond J. Deshaies

VOLUME 418. Embryonic Stem Cells
Edited by IRINA KLIMANSKAYA AND ROBERT LANZA

VOLUME 419. Adult Stem Cells
Edited by IRINA KLIMANSKAYA AND ROBERT LANZA

VOLUME 420. Stem Cell Tools and Other Experimental Protocols
Edited by IRINA KLIMANSKAYA AND ROBERT LANZA

VOLUME 421. Advanced Bacterial Genetics: Use of Transposons and Phage for Genomic Engineering
Edited by KELLY T. HUGHES

VOLUME 422. Two-Component Signaling Systems, Part A
Edited by MELVIN I. SIMON, BRIAN R. CRANE, AND ALEXANDRINE CRANE

VOLUME 423. Two-Component Signaling Systems, Part B
Edited by MELVIN I. SIMON, BRIAN R. CRANE, AND ALEXANDRINE CRANE

VOLUME 424. RNA Editing
Edited by JONATHA M. GOTT

VOLUME 425. RNA Modification
Edited by JONATHA M. GOTT

VOLUME 426. Integrins
Edited by DAVID CHERESH

VOLUME 427. MicroRNA Methods
Edited by JOHN J. ROSSI

VOLUME 428. Osmosensing and Osmosignaling
Edited by HELMUT SIES AND DIETER HAUSSINGER

VOLUME 429. Translation Initiation: Extract Systems and Molecular Genetics
Edited by JON LORSCH

VOLUME 430. Translation Initiation: Reconstituted Systems and Biophysical Methods
Edited by JON LORSCH

VOLUME 431. Translation Initiation: Cell Biology, High-Throughput and Chemical-Based Approaches
Edited by JON LORSCH

VOLUME 432. Lipidomics and Bioactive Lipids: Mass-Spectrometry–Based Lipid Analysis
Edited by H. ALEX BROWN

VOLUME 433. Lipidomics and Bioactive Lipids: Specialized Analytical Methods and Lipids in Disease
Edited by H. ALEX BROWN

VOLUME 434. Lipidomics and Bioactive Lipids: Lipids and Cell Signaling
Edited by H. ALEX BROWN

VOLUME 435. Oxygen Biology and Hypoxia
Edited by HELMUT SIES AND BERNHARD BRÜNE

VOLUME 436. Globins and Other Nitric Oxide-Reactive Protiens (Part A)
Edited by ROBERT K. POOLE

VOLUME 437. Globins and Other Nitric Oxide-Reactive Protiens (Part B)
Edited by ROBERT K. POOLE

VOLUME 438. Small GTPases in Disease (Part A)
Edited by WILLIAM E. BALCH, CHANNING J. DER, AND ALAN HALL

VOLUME 439. Small GTPases in Disease (Part B)
Edited by WILLIAM E. BALCH, CHANNING J. DER, AND ALAN HALL

VOLUME 440. Nitric Oxide, Part F Oxidative and Nitrosative Stress in Redox Regulation of Cell Signaling
Edited by ENRIQUE CADENAS AND LESTER PACKER

VOLUME 441. Nitric Oxide, Part G Oxidative and Nitrosative Stress in Redox Regulation of Cell Signaling
Edited by ENRIQUE CADENAS AND LESTER PACKER

VOLUME 442. Programmed Cell Death, General Principles for Studying Cell Death (Part A)
Edited by ROYA KHOSRAVI-FAR, ZAHRA ZAKERI, RICHARD A. LOCKSHIN, AND MAURO PIACENTINI

VOLUME 443. Angiogenesis: *In Vitro* Systems
Edited by DAVID A. CHERESH

VOLUME 444. Angiogenesis: *In Vivo* Systems (Part A)
Edited by DAVID A. CHERESH

VOLUME 445. Angiogenesis: *In Vivo* Systems (Part B)
Edited by DAVID A. CHERESH

VOLUME 446. Programmed Cell Death, The Biology and Therapeutic Implications of Cell Death (Part B)
Edited by ROYA KHOSRAVI-FAR, ZAHRA ZAKERI, RICHARD A. LOCKSHIN, AND MAURO PIACENTINI

VOLUME 447. RNA Turnover in Bacteria, Archaea and Organelles
Edited by LYNNE E. MAQUAT AND CECILIA M. ARRAIANO

VOLUME 448. RNA Turnover in Eukaryotes: Nucleases, Pathways and Analysis of mRNA Decay
Edited by LYNNE E. MAQUAT AND MEGERDITCH KILEDJIAN

Volume 449. RNA Turnover in Eukaryotes: Analysis of Specialized and Quality Control RNA Decay Pathways
Edited by Lynne E. Maquat and Megerditch Kiledjian

Volume 450. Fluorescence Spectroscopy
Edited by Ludwig Brand and Michael L. Johnson

Volume 451. Autophagy: Lower Eukaryotes and Non-Mammalian Systems (Part A)
Edited by Daniel J. Klionsky

Volume 452. Autophagy in Mammalian Systems (Part B)
Edited by Daniel J. Klionsky

Volume 453. Autophagy in Disease and Clinical Applications (Part C)
Edited by Daniel J. Klionsky

Volume 454. Computer Methods (Part A)
Edited by Michael L. Johnson and Ludwig Brand

Volume 455. Biothermodynamics (Part A)
Edited by Michael L. Johnson, Jo M. Holt, and Gary K. Ackers (Retired)

Volume 456. Mitochondrial Function, Part A: Mitochondrial Electron Transport Complexes and Reactive Oxygen Species
Edited by William S. Allison and Immo E. Scheffler

Volume 457. Mitochondrial Function, Part B: Mitochondrial Protein Kinases, Protein Phosphatases and Mitochondrial Diseases
Edited by William S. Allison and Anne N. Murphy

Volume 458. Complex Enzymes in Microbial Natural Product Biosynthesis, Part A: Overview Articles and Peptides
Edited by David A. Hopwood

Volume 459. Complex Enzymes in Microbial Natural Product Biosynthesis, Part B: Polyketides, Aminocoumarins and Carbohydrates
Edited by David A. Hopwood

Volume 460. Chemokines, Part A
Edited by Tracy M. Handel and Damon J. Hamel

Volume 461. Chemokines, Part B
Edited by Tracy M. Handel and Damon J. Hamel

Volume 462. Non-Natural Amino Acids
Edited by Tom W. Muir and John N. Abelson

Volume 463. Guide to Protein Purification, 2nd Edition
Edited by Richard R. Burgess and Murray P. Deutscher

Volume 464. Liposomes, Part F
Edited by Nejat Düzgüneş

SECTION ONE

GENERAL RESOURCES

CHAPTER ONE

A Survey of Internet Resources for Mouse Development

Thomas L. Saunders*,†

Contents

Abstract

The Internet contains many sources of information useful for mouse developmental biology research. These include anatomical atlases, gene expression atlases, and indexes to genetically engineered mouse strains and embryonic stem (ES) cells. Online atlases supersede earlier printed atlases in the quantity of available online images and the depth of specialized anatomical terminology and gene ontologies. Atlases annotated with gene expression data have increased value for comparisons with mouse models designed to study genetic perturbations of developmental processes. Gene expression libraries and microarray analyses of developmental stages are also available in Internet repositories. Bioinformatic interrogation of this data can identify regulatory gene networks and suggest putative transcriptional regulators that control cell fate. *In silico* formulated hypotheses about regulatory genes may be tested *in vivo* with mouse models obtained from the international Knockout Mouse

* Transgenic Animal Model Core, University of Michigan Medical School, Ann Arbor, Michigan, USA
† Division of Molecular Medicine and Genetics, Department of Internal Medicine, University of Michigan Medical School, Ann Arbor, Michigan, USA

Methods in Enzymology, Volume 476
ISSN 0076-6879, DOI: 10.1016/S0076-6879(10)76001-8

Project (KOMP). Many of the genes targeted ES cells in KOMP are designed to mark cells with reporter molecules and produce conditional alleles. In combination with Cre recombinase mouse strains, genes can be inactivated at different developmental stages in specific cell types to study their function in embryogenesis. Atlases of mouse development, gene expression atlases, transcriptome databases, and repositories of genetically engineered mouse strains and ES cells are discussed.

1. Introduction

Internet access vitalizes academic research, opening windows into worldwide resources that provide a standardized, common vocabulary for embryonic structures, developmental stages, gene nomenclature, and phenotypes that give context to biological questions. The large datasets in online repositories provide benchmarks of normal mouse development that can be used to discover regulatory networks in embryonic processes and against which mouse mutant phenotypes can be contrasted. The capacity of computers to store and deliver images far exceeds the scope of information in the classical mouse embryo anatomy books. For example, Karl Theiler's atlas of mouse development (Theiler, 1972) is available as a download from the Edinburgh Mouse Atlas Project (EMAP) but only represents a minuscule fraction of Internet accessible data.

One outcome of the efforts to develop comprehensive anatomical atlases and gene expression databases for the mouse is the establishment of sophisticated ontologies that describe embryonic structures and gene function. The precision and descriptive power of gene ontologies can be used to gain insight into gene regulatory networks active in embryogenesis (Hill *et al.*, 2010; Huang *et al.*, 2009; Martinez-Morales *et al.*, 2007; Wang *et al.*, 2009). Databases that join normal gene expression patterns with anatomical localization can be combined with new observations to find coexpressed genes in the same regulatory pathway. *In silico* hypotheses can be tested *in vivo* with mouse models from the international Knockout Mouse Project (KOMP) collaboration to knockout every gene in the mouse (Adams and van der Weyden, 2008; Austin *et al.*, 2004; Collins *et al.*, 2007; Raymond and Soriano, 2006; Schnütgen *et al.*, 2006). The number of KOMP mouse models for fate mapping and conditional gene ablation studies grows larger every year. In many instances, it is possible to acquire genetically engineered mice from repositories to build mouse models to ask developmental questions. When investigators sidestep the time-consuming process of engineering mouse models, they can focus on important biological questions instead of building research tools.

2. Mouse Embryo Atlases

Internet atlases of the developing mouse use anatomical nomenclature based on the classical published mouse developmental atlases (Kaufman, 1992; Rugh 1968; Theiler, 1972). The most comprehensive web site for the anatomy of the developing mouse embryo is the EMAP (Table 1.1; Baldock *et al.*, 2003). Histological sections with resolution down to 2 μm are combined with optical projection tomography (OPT) to generate both

Table 1.1 Atlases of mouse development

Edinburgh Mouse Atlas Project (EMAP)
http://genex.hgu.mrc.ac.uk
EMAP Anatomy Browser
http://genex.hgu.mrc.ac.uk/Databases/Anatomy
EMAP 3D Digital Atlas
http://genex.hgu.mrc.ac.uk/Atlas/
Duke Center for *In Vivo* Microscopy
http://www.civm.duhs.duke.edu/devatlas/index.html
Mouse Biomedical Research Informatics Network (BIRN) Atlasing Toolkit
http://mbat.loni.ucla.edu
Caltech μMRI Atlas of Mouse Development
http://mouseatlas.caltech.edu/
NIEHS Developing Heart Images[a]
https://niehsimages.epl-inc.com
Username: ToxPath /Password: embryohearts
Mouse Limb Anatomy Atlas
http://www.nimr.mrc.ac.uk/3dlimb
Allen Institute for Brain Science: Developing Mouse Brain Atlas
http://developingmouse.brain-map.org/atlas/index.html
Mouse Diffusion Tensor Imaging Atlas of Developing Mouse Brains at BIRN
http://www.birncommunity.org/data-catalog/
Laboratory of Neuro Imaging at UCLA: Mouse Atlas Project
http://map.loni.ucla.edu/atlas/
The Electronic Prenatal Mouse Brain Atlas
http://www.epmba.org/
Vascular Atlas of the Developing Mouse Embryo
http://www.mouseimaging.ca/research/mouse_atlas.html
The Mouse Anatomical Dictionary
http://www.informatics.jax.org/mgihome/GXD/GEN/AD/

[a] Login information is available to the public, see Savolainen *et al.* (2009).

two- and three-dimensional images and animations. Embryo stages are given as Theiler stages (TS), based on the criteria established by Theiler (1972). A systematized nomenclature for structures in the developing mouse embryo is provided to assist in the interpretation of the information presented in the images. EMAP combines atlas images with standardized nomenclature to precisely describe anatomical regions in the three-dimensional models of the mouse embryo. The EMAP Anatomy Browser (Table 1.1) provides two-dimensional sections of mouse embryos through TS 26. This Java application allows the user to specify the section through the embryo at each TS. Structures within each section are annotated. Hovering a cursor on an embryo section will produce descriptive text for each structure. Selecting a region will colorize the structure and produce annotation. The Anatomy Browser shows sections through the embryo, the original whole-mount embryo, and the anatomical nomenclature in three separate windows. A second EMAP resource is the three-dimensional digital atlas of mouse development. Users can download data and software for three-dimensional embryo sections and manipulate the images locally. When structures within images are selected, associated anatomical terms can be used to link out to gene expression data in the Edinburgh Mouse Atlas of Gene Expression (EMAGE) database or Gene Expression Database (GXD) (see below). EMAGE maps gene expression to anatomical domains and provides visual localization of embryonic gene expression. GXD provides text-based description of gene expression domains and links out to published papers.

An atlas of the developing mouse obtained by magnetic resonance microscopy (MRI) is available from the Duke Center for *In Vivo* Microscopy (Table 1.1; Petiet *et al.*, 2008). The atlas delivers dorsal, transverse, and sagittal sections of mouse embryos from embryonic day 10.5 to 19.5 at resolutions of 19.5 μm. Section magnification can be changed at will. Registered users can view datasets of sections through embryos in VoxStation software, which is provided by the center as a Java application. Alternatively, data can be downloaded from the web site to a local desktop computer. Sections can be viewed serially or played continuously as an animation. When the data is viewed in this format structural annotations are absent except for datasets that describe the developing heart. In order to view sections with structural annotation, it is necessary to download data configured for the Mouse Biomedical Informatics Research Network (BIRN) Atlasing Toolkit (MBAT; Table 1.1). The Caltech μMRI Atlas of Mouse Development offers interactive transverse, sagittal, and frontal sections through mouse embryos from embryonic day 11.5 to 18.5 (Table 1.1; Dhenain *et al.*, 2001). Although the Caltech atlas lacks structural annotation, it is a good introductory resource for mouse development and can be easily accessed for teaching.

2.1. Specialized atlases of mouse development

An atlas for the anatomy of the developing mouse heart is found in the digital histological atlas at the National Institute of Environmental Health Sciences (Table 1.1). This web site serves up a series of hematoxylin- and eosin-stained images of the developing heart from embryonic day 11 through embryonic day 18. Gene expression data is not annotated on the histological images. Examples of commonly observed abnormalities are included in addition to views of normal heart. Savolainen *et al.* (2009) provide a username and password that can be used to access this public database (see Table 1.1).

The Mouse Limb Anatomy Atlas is available for e14.5 embryos (Table 1.1; DeLaurier *et al.*, 2008). Three-dimensional views of limbs can be manipulated with a Java application downloaded from the web site. Data for bones, muscles, and tendons of the hindlimb and forelimb can be loaded. These animated datasets are an excellent teaching tool and provide detailed structural annotation.

The Developing Mouse Brain Atlas (Table 1.1; Jones *et al.*, 2009) is available at the Allen Institute for Brain Science (AIBS). Essential anatomical information about structures with annotated sagittal sections is provided through the Brain Explorer 2 desktop application. The AIBS atlas provides three-dimensional and sagittal sections through developing mouse brains from embryos staged at e11.5, e13.5, e15.5, and e18.5. The images can be compared to an annotated color-coded schematic of the section that identifies brain structures. Postnatal day 4, 14, and 28 mouse brain images are also available for viewing and study. This software tool is an excellent teaching tool and provides a detailed interactive environment to explore brain anatomy.

The Electronic Prenatal Mouse Brain Atlas (Table 1.1; Schambra, 2008) provides a section-by-section view through the mouse brain at e12 and a second series at e16. The 0.5 mm sections are stained with hematoxylin and eosin and can be browsed sequentially. Anatomical structures are abundantly annotated and accessible through the web interface. This resource is an excellent tool for finding major landmarks and teaching.

Another developing mouse brain atlas, the Mouse Diffusion Tensor Imaging Atlas, can be found in the catalog of atlases at BIRN (Table 1.1; Zhang *et al.*, 2003). This atlas is available as a download that includes both image data and Windows 2000/XP compatible viewer software. Images of e14, e15, e16, d17, e18, and adult mouse brains are available. Other datasets of possible interest in the BIRN catalog include a P0 mouse brain atlas and images from a mouse model of Parkinson's disease.

The Laboratory of Neuro Imaging (LONI) at UCLA provides a digital atlas of the P0 mouse brain with images that include structural annotations (Table 1.1; Lee *et al.*, 2007). The mouse P0 brain atlas includes color or gray scale Nissl-stained sections 50 μm thick. Anatomical structures in the sections are color coded and labeled. Viewing the atlas at the web site can be done with the LONI Visualization Tool software that runs on Windows and Linux platforms. Alternatively, the atlas can be viewed with Multitracer software, a Java application. In its default configuration, Multitracer may have insufficient memory to view the atlas. In this case it will be necessary to change the maximum Java heap size on the computer used to view the atlas. A third option is to view the atlas with the MBAT, which is available for Windows, Mac, and Unix operating systems. Simply viewing the data online in a web browser, as can be done for EMAP, is not possible. The effort required to download the data files and find compatible viewing software may be an obstacle to the casual user.

The Vascular Atlas of the Developing Mouse Embryo (Table 1.1; Walls *et al.*, 2008) uses PECAM-1 immunostained embryos to specifically mark endothelial cells lining blood vessels in combination with OPT to acquire high-resolution images. Images from embryos between e8 and e10 (5–30 somites) are available as sections proceeding through the embryos in the viewer's choice of sagittal, coronal, or transverse planes.

Standardized nomenclature for anatomical structures in the developing embryo is essential for investigators who wish to compare their observed phenotypes to normal mouse development. The Jackson Laboratory and EMAP collaborated to develop a systematic vocabulary to describe cells and tissues in the embryo. This information is available as the Mouse Anatomical Dictionary (Table 1.1; Hayamizu *et al.*, 2005). Ontologies for genes, anatomy, and phenotypes establish a common language for independent research groups to use as they annotate embryonic stages and structures with gene expression information (Hayamizu *et al.*, 2005; Hill *et al.*, 2010; Sam *et al.*, 2009).

3. Gene Expression Atlases and Databases

Understanding how a fertilized mouse egg establishes an embryo and develops into a newborn pup requires knowledge of developmental gene expression patterns and regulatory gene networks. The embryo initiates organogenesis from a handful of precursor cells that proliferate and differentiate into multiple specialized lineages. As cells commit to specific differentiation pathways they activate genes critical to their specialized function and shut down genes expressed in multipotent precursor cells. Annotation of gene expression in the developing mouse embryo is especially challenging because of the temporal nature of gene expression compared to the relative stability of gene expression in the cells and tissues of adult animals. It is in the growing embryo that processes of rapid proliferation and differentiation occur. Some transcription factors necessary for cellular commitment are expressed only briefly during the embryogenesis. As a result, gene expression profiles need to be collected over many time points as each developmental stage presents a unique gene expression profile.

3.1. Mouse embryo *in situ* hybridization databases

Internet resources for gene expression can be divided into those which collect data on the basis of *in situ* hybridization (ISH) with gene-specific probes and those which collect data from microarrays or transcriptome libraries of gene expression. The EMAGE database, part of EMAP, collects and curates data from a variety of public sources (Table 1.2; Richardson *et al.*, 2010). Gene expression patterns during the mouse development revealed by ISH, transgenic reporter mice, and immunohistochemical staining are assembled and assigned to EMAP images. In a typical view, the original image is shown side by side with the expression pattern mapped to the whole-mount embryo and also to two-dimensional embryo sections.

Table 1.2 *In situ* hybridization gene expression databases

Edinburgh Mouse Atlas of Gene Expression (EMAGE)
http://www.emouseatlas.org/emage/home.php
Mouse Genome Informatics (MGI) and Mouse Genome Database (MGD)
http://www.informatics.jax.org
Gene Expression Database (GXD)
http://www.informatics.jax.org/expression.shtml
Gene Expression Notebook
http://www.informatics.jax.org/mgihome/GXD/GEN
GenePaint
http://www.genepaint.org
EURExpress Transcriptome Atlas
http://www.eurexpress.org/ee
The Allen Developing Mouse Brain Atlas
http://developingmouse.brain-map.org
Brain Gene Expression Map
http://www.stjudebgem.org
GENSAT Brain Atlas of Gene Expression
http://www.gensat.org/index.html
Preimplantation Embryo Whole Mount *In Situ* Hybridization
http://lgsun.grc.nia.nih.gov/data/supplemental/WISH/images.html
GenitoUrinary Development Molecular Anatomy Project (GUDMAP)
http://www.gudmap.org/index.html
European Renal Genome Project
http://www.euregene.org

The user can search for expression patterns by entering the gene name, the name of the anatomical structure (see Mouse Anatomical Dictionary, Table 1.1), the region in the mouse embryo according to its TS, or review all gene expression patterns for a specific embryo stage to see which expression domains might match the user's experimental results. The cluster analysis and heat maps of expression domains are especially informative when identifying genes with similar expression patterns. Researchers can submit their data to EMAGE and use the resources there to interpret their gene expression patterns. By mapping their results onto a common framework they can compare their data with other laboratories from around the world.

The Mouse Genome Informatics (MGI) group at The Jackson Laboratory (Table 1.2; Bult *et al.*, 2010) and EMAGE share data. When an interesting expression pattern is found in the EMAGE database, the user can link out to the Mouse Genome Database (MGD) maintained by MGI and obtain comprehensive information about the gene. In addition to the MGD, MGI maintains the mouse GXD (Table 1.2; Smith *et al.*, 2007) that complements the information at EMAGE. It is advisable to check both EMAGE and GXD for expression information since the datasets are independent and do not necessarily duplicate each other. For example, a query for the gene Nanog at EMAGE produces the message "No results found" while a query for Nanog at GXD matches 38 "Gene Expression Literature Records." GXD and EMAGE use the same anatomic ontology so that the same terms can be searched in both databases. For example, a search for the term "Rathke's pouch" returns 111 results from GXD and over 5000 results at EMAGE. GXD differs from EMAGE in that gene expression search terms can also consist of genetic intervals on mouse chromosomes, genome coordinates, or intervals bounded by genetic markers in addition to gene names. A useful resource available at the GXD is the Gene Expression Notebook (Table 1.1). This spreadsheet organizes data from expression studies and simplifies the process of submitting data from completed studies to the GXD for public access.

The GenePaint database includes ISH analysis of genes on sections from e10.5 to e14.5 embryos along with e15.5 head sections (Table 1.2; Visel *et al.*, 2004). Gene names, nucleotide sequences or anatomical structures can be used as search terms. JPG images of sections through stained embryos are provided. Images can be viewed in browsers, plugins, or a Java applet. EURExpress (Table 1.2) is an ISH-based database for gene expression in the e14.5 mouse embryo. Data is viewed as stained sagittal sections of embryos on the web site. Strength of expression is indicated by color-coded bars and links out to GenePaint. Both of these databases present the user with viewable ISH images at limited development stages without labeled structural landmarks, data submission from users is not supported.

3.2. Specialized atlases—*In situ* localization

Specialized atlases include the AIBS mouse developing brain atlas (Table 1.2; Jones *et al.*, 2009). ISH data is viewed with Brain Explorer 2 software. Gene expression data imported into the software can be toggled on or off in different brain structures. Data for brains at e11.5, e13.6, e15.5, e18.5, P4, P14, and P28 are available. Alternatively, ISH of brain sagittal sections can be viewed with the web browser interface provided at AIBS. A second atlas providing gene expression data for the developing mouse brain is the Brain Gene Expression Map (BGEM) (Table 1.2; Magdaleno *et al.*, 2006). This atlas includes data from the ISH of selected genes on sagittal sections of brains collected from e11.5, e15.5, P7, and P42 mice. Images are mapped to Nissl-stained sections for comparison and can be downloaded for reference. The BGEM group is collaborating with GENSAT (Table 1.2; Heintz, 2004) to identify candidates for the preparation of bacterial artificial chromosome transgenic reporter mice. Data on e10.4 and e15.5 gene expression patterns obtained from transgenic reporters are available at GENSAT. Should the reporter mice be of interest, GENSAT links out to mouse repositories to facilitate the acquisition of transgenic mouse lines for further study.

Other datasets that explore gene expression by ISH in specific developmental stages and tissues are available. For example, a panel of whole-mount ISH gene expression patterns in preimplantation mouse embryos is available for 91 selected genes (Table 1.2; Yoshikawa *et al.*, 2006). Analysis of gene expression in the developing genitourinary tract by whole-mount ISH of 3000+ genes is available for e10.5 embryos through adult mice from GUDMAP (Table 1.2; McMahon *et al.*, 2008). ISH images for hundreds of genes on mouse kidney sections from TS 23 to 28 imaged by OPT are available for viewing at the European Renal Genome Project (Table 1.2; Willnow *et al.*, 2005). If a specialized database does not exist for the tissue or developmental stage(s) of interest, it is possible to construct an expression database with ISH from published papers (Wittmann *et al.*, 2009). This group mined spatial gene expression patterns from publications to predict a signaling network in the developing brain and established a mathematical model to validate the gene expression data.

3.3. Genomic expression databases

Gene expression databases at the genomic level include the Mouse Atlas of Gene Expression (Table 1.3; Siddiqui *et al.*, 2005). This atlas provides access to Serial Analysis of Gene Expression (SAGE) libraries prepared from mouse embryos throughout every stage of development, including fertilized eggs and embryonic stem (ES) cells. The approach was developed to identify

Table 1.3 Genomic libraries of gene expression

Mouse Atlas of Gene Expression http://www.mouseatlas.org/data/mouse/devstages
GermSAGE: Male Germ Cell Transcriptome http://germsage.nichd.nih.gov
GonadSAGE: Male Embryonic Gonad Transcriptome http://gonadsage.nichd.nih.gov
ArrayExpress Gene Expression Atlas http://www.ebi.ac.uk/microarray-as/atlas
Gene Expression Omnibus http://www.ncbi.nlm.nih.gov/geo
Gene Expression Database in 4D http://4dx.embl.de/4DXpress

differentially expressed genes throughout embryogenesis. The SAGE libraries and DiscoverySpace software for analysis are publicly available and include links to other public databases. Genes of interest in the Mouse Atlas of Gene Expression will need to be localized by reference to expression databases that include anatomical annotation.

GermSAGE and GonadSAGE are dedicated to the transcriptome of male germ cells and the developing mouse male gonad, respectively (Table 1.3; Lee *et al.*, 2009, 2010). For GermSAGE, libraries were prepared from type A spermatogonia, pachytene spermatocytes, or round spermatids. For GonadSage, libraries were constructed from male mouse gonads collected at e10.5, e11.5, e12.5, e13.5, e15.5, and e17.5. A database searchable by gene name is available for each developmental stage. Localization of genes to germ cells is available in GermSAGE, but cell-specific expression in gonads is not annotated.

Gene expression data based on microarray experiments have proliferated and can be accessed through different web interfaces. These include the ArrayExpress Gene Expression Atlas maintained at the European Bioinformatics Institute, the Gene Expression Omnibus (GEO) maintained by the National Center for Biotechnology Information (NIH), and the Gene Expression Database in 4D maintained at the European Molecular Biology Laboratory (Table 1.3; Barrett *et al.*, 2009; Haudry *et al.*, 2008; Kapushesky *et al.*, 2010). Specific subsets of data from microarrays that may be of interest include data from a number of mouse ES cell lines deposited at GEO. Researchers can download datasets of genes expressed in mouse ES cell lines including R1, J1, V6.5, homozygous mutant ES cells, differentiated ES cells, etc. Public access to gene expression data is vital for research and can lead to important discoveries. For example, digital differential display was used to analyze expressed sequence tag libraries in GEO to identify genes expressed specifically in mouse ES cells (Tokuzawa *et al.*, 2003, 2006). This study led to the generation of the *Fbx15* knockout mouse which was instrumental in the identification of the four transcription factors (*Pou5f1*, *Sox2*, *c-myc*, and *Klf4*) necessary for the generation of induced pluripotent stem cells from mouse fibroblasts (Takahashi and Yamanaka, 2006).

4. Locating and Acquiring Mutant Mice and Embryonic Stem Cells

When *in silico* analysis of an embryonic stage of interest identifies a potentially important gene for the cell type or tissue, the function of the gene needs to be validated *in vivo*. Often the model of choice for organogenesis or cellular differentiation is a genetically engineered mouse model that carries a null allele for the gene in question or a conditional allele that can be inactivated by the expression of Cre recombinase in an informative cell type. Mice with null alleles, conditional alleles, and Cre transgenes are available from repositories located around the world (Table 1.4). The most direct method to find a mouse with a spontaneous mutation, transgenic overexpression, or gene-targeted modification of the gene of interest is to use the name of the gene as the search term to query the databases. The International Mouse Strain Resource (IMSR, Table 1.4) and the Federation of International Mouse Resources (FIMRe, Table 1.4) (Davisson, 2006; Eppig and Strivens, 1999) are portals to information from multiple resource centers, including the Canadian Mouse Mutant Repository, the Center for Animal Resources and Development Database, the European Mouse Mutant Archive, the Japan Mouse/Rat Strain Resources Database, the RIKEN BioResource Center, and the Taconic Knockout Repository (Table 1.4; Nakagata and Yamamura, 2009; Wilkinson *et al.*, 2010; Yoshiki *et al.*, 2009). Searches with the correct genetic nomenclature are essential. If gene synonyms or outdated gene names are used they may return incorrect results. For example, the mouse synonym of the human eukaryotic translation elongation factor 1a1 gene, *PTI1*, is *Eef1a1*. A search of the IMSR for mice with mutations in *PTI1* or *Pti1* finds no matching results while a search for *Eef1a1* returns 50 matches, showing availability of mice, ES cells, and cryopreserved mouse sperm and embryos with *Eef1a1* mutations. Current and accurate mouse gene names can be located online at the MGI web site (Table 1.2).

A number of repositories submit data to IMSR and FIMRe; however, submissions may not always be up-to-date. If a mutant mouse strain is not located in IMSR or FIMRe, then an MGI search for the gene name may identify mouse strains that have not yet been uploaded to a public repository. For example, examination of the Alleles and Phenotypes panel on the

Table 1.4 Sources of genetically engineered mice and ES cells

International Mouse Strain Resource (IMSR)
http://www.findmice.org
Federation of International Mouse Resources (FIMRe)
http://www.fimre.org
Mutant Mouse Regional Resource Centers (MMRRC)
http://www.mmrrc.org
German Gene Trap Consortium (GGTC)
http://genetrap.helmholtz-muenchen.de
Database of Genome Survey Sequences (dbGSS)
http://www.ncbi.nlm.nih.gov/nucgss
Knockout Mouse Project (KOMP) Repository
http://www.komp.org
International Knockout Mouse Consortium (IKMC)
http://www.knockoutmouse.org
European Conditional Mouse Mutagenesis Program (EUCOMM)
http://www.eucomm.org/
Texas A&M Institute of Genomic Medicine (TIGM)
http://www.tigm.org
International Gene Trap Consortium (IGTC)
http://www.igtc.org
Cre-X-Mice: A Database of Cre Transgenic Lines
http://nagy.mshri.on.ca/cre
Cre Portal at The Jackson Laboratory
http://www.creportal.org/
Canadian Mouse Mutant Repository (CMMR)
http://www.cmmr.ca/
Center for Animal Resources and Development Database (CARD)
http://cardb.cc.kumamoto-u.ac.jp/transgenic/index.jsp
European Mouse Mutant Archive (EMMA)
http://www.emmanet.org/
Japan Mouse/Rat Strain Resources Database
http://www.shigen.nig.ac.jp/mouse/jmsr/top.jsp
RIKEN BioResource Center
http://www2.brc.riken.jp/lab/animal/search.php
Taconic Knockout Repository
http://kodatabase.taconic.com/database.php

MGI gene detail page for *Eef1a1* shows that one mouse strain carrying a targeted allele has been produced for this gene and that 51 gene trapped ES cell clones are available. Drilling down into the gene detail reveals that: (1) the targeted mouse allele is named $Eef1a1^{tm1(Kras*)Arge}$ (Klinakis *et al.*,

2009) and (2) this mouse strain is not listed as available from a repository. In this example it would be necessary to contact the originating laboratory to inquire about obtaining the mice.

If the desired mouse strain is extinct or are not available from a provider, the MGI gene detail page links out to ES cell clones with mutations in *Eef1a1*. A gene trapped ES cell clone with a null allele for *Eef1a1* that interrupts the gene transcript with a lacZ reporter, such as gene trap clone DC0383 could be ordered from IMSR or the Mutant Mouse Regional Resource Centers (MMRRC, Table 1.4). Alternatively, an ES cell carrying the conditional FLEX gene trap vector (Schnütgen *et al.*, 2006), such as clone D134H06, can be ordered from the German Gene Trap Consortium (Table 1.4). The ES cells are then used to generate a suitable mouse model. Details on the genomic structure of gene trap clones can be found by entering their identifying cell line ID number in the Database of Genome Survey Sequences (dbGSS, Table 1.4). It is not always possible to navigate to an ES cell repository from MGI or dbGSS to order clones. Searches for clone IDs such as DC0383 and D134H06 will not find ES cells at the major repositories (Table 1.4): KOMP Repository, International Knockout Mouse Consortium, MMRRC, European Conditional Mouse Mutagenesis Program (Friedel *et al.*, 2007), Texas A&M Institute of Genomic Medicine (Hansen *et al.*, 2008), although searches by clone ID will produce results at MGI and the International Gene Trap Consortium database (Table 1.4; Nord *et al.*, 2006). The value of a preliminary search for the gene of interest at MGI is to determine whether a mutant mouse is available. If mice are not available but ES cell clones are identified then the most effective way to find and order ES cell clones is to go through a gene name search at KOMP or IKMC.

Growing numbers of mouse strains and ES cell clones are engineered to use Cre/loxP technology to produce conditional alleles. Cre recombinase mouse lines can then be used to ablate genes to mark cells for lineage studies or to observe effects on developing tissues (reviewed in Lewandoski, 2007). To take the fullest advantage of conditional alleles, it is advantageous to use mouse lines that express the tamoxifen inducible Cre-ERT2 recombinase under the control of cell-specific promoters (Ellisor *et al.*, 2009; Kwon and Hadjantonakis, 2009; Savory *et al.*, 2009). Substantial progress in the accumulation and characterization of mouse strains that express Cre recombinase in specific cell types has been made. Databases to identify Cre strains are available at the Cre-X-Database (Table 1.4; Nagy *et al.*, 2009) and the Cre Portal at The Jackson Laboratory (Table 1.4). Depending on the alleles and transgenes needed for the study, it may be possible to obtain mice with conditional alleles and Cre recombinase mice from repositories to test gene function. In this case, the lengthy process of generating and characterizing new genetically modified mouse strains is bypassed.

5. Discussion

The sequencing of the mouse genome has moved research from gene discovery to the era of determining gene function. Digital atlases of developing mouse embryos provide interactive three-dimensional models that can be manipulated in ways that could only be imagined in the time of print atlases. The same is true for expression atlases that assign gene expression domains to specific locations in embryos at different developmental stages. Microarray experiments designed to capture global gene expression patterns across development are available online. Large datasets from expression libraries and microarrays can be accessed effortlessly and used in customized bioinformatic analyses to formulate hypotheses concerning the functions of genes in networks that affect developmental programs. The international consortium to knockout every gene in the mouse makes ES cell and mouse resources available to validate *in silico* formulated models of gene function with *in vivo* mouse models.

Internet tools to educate researchers about the anatomy of developing mouse embryos and to provide systematized vocabularies to harmonize findings across research groups are essential as digital atlases supplant printed atlases (Hill *et al.*, 2010). Gene expression atlases for the mouse at EMAGE and GXD databases are designed to incorporate user submissions. Numerous other groups are developing specialized atlases of gene expression (Table 1.2). The value of centralized repositories for gene expression data is that they can curate information from narrow studies of embryonic gene expression. Submitting data to a centralized location increases its visibility in comparison to maintaining data on a local server. Images generated on a small scale for an in-depth analysis of the Wnt signaling pathway are a case in point: images are accessible both locally (http://www.tcd.ie/Zoology/research/WntPathway) and through EMAP (Summerhurst *et al.*, 2008).

Gene expression data is accumulating at a rapid rate, including libraries designed to capture transcribed genes from all embryonic stages. Data can be mined to identify gene regulatory networks involved in specification of cell types or organogenesis (Zeitlinger and Stark, 2010). Bioinformatic comparison of ISH datasets for 1030 genes in wild type and mutant *Pax6* e14.5 mouse cortex led to the identification of 16 coexpressed genes with changes in expression levels that were further validated (Visel *et al.*, 2007). Transcriptome approaches to differential gene expression are also valuable for the identification of developmentally programmed genes. Pituitary glands isolated from e12.5, e14.5, and e14.5 $Prop1^{df/df}$ mutant embryos were used to prepare cDNA libraries that were then subtracted to identify differentially expressed genes (Brinkmeier *et al.*, 2009). In this process, 45 homeobox genes were identified, several of which were not previously thought to play a role in pituitary development and were subsequently shown to be expressed in the developing pituitary. Bioinformatic analysis of expression data played a role in identifying the four transcription factors that opened the door to the rapidly growing field of induced pluripotent stem cells (see above).

Without an information management system, the myriad database on the Internet can be overwhelming. One example of a portal tool is BioGPS (http://biogps.gnf.org; Wu *et al.*, 2009). BioGPS aggregates results from a multitude of databases and summarize the results in a single browser window. This portal serves up *in situ* images and curated data from EMAGE, GXD, GenePaint, EURExpress, and other sources. Users can easily add their favorite web-based databases and annotation tools to the collection of BioGPS plugins and customize the list of databases that are searched. BioGPS will simultaneously query selected databases for a gene and present results from ISH experiments, microarray expression levels, biological pathways, ontology descriptions, genomic and transcript structures, availability of mutant ES cells and mice, and even results from commercial vendors offering relevant RNAi or antibody reagents. To maximize the value of online resources, a systematic approach prepared in consultation with an expert in bioinformatics is highly recommended.

6. Further Reading

In addition to the articles in the reference list, each of the Internet sites in Tables 1.1–1.4 includes associated help files and tutorials that can be used to gain proficiency in navigating the sites. To take full advantage of genome-wide expression datasets, a working knowledge of bioinformatics software and statistical analysis of results is required. These topics are beyond the scope of this chapter. See Lacroix and Critchlow (2003) for an introduction to field of bioinformatics. Examples and reviews of bioinformatic approaches to discovering new genes controlling development processes are available (Bard *et al.*, 2008; Busser *et al.*, 2008; Rodriguez-Zas *et al.*, 2008; Sharova *et al.*, 2007; Tokuzawa *et al.*, 2006; Vokes *et al.*, 2008; Zeitlinger and Stark, 2010). The rapid proliferation of Internet resources and bioinformatic tools ensures that any survey of Internet resources will be incomplete. A continually updated list of links, including those described in this chapter, can be found at http://www.med.umich.edu/tamc/links.html. Readers with suggestions for additional links are encouraged to contact the author.

References

Adams, D. J., and van der Weyden, L. (2008). Contemporary approaches for modifying the mouse genome. *Physiol. Genomics* **34,** 225–338.

Austin, C. P., Battey, J. F., Bradley, A., Bucan, M., Capecchi, M., Collins, F. S., Dove, W. F., Duyk, G., Dymecki, S., Eppig, J. T., Grieder, F. B., Heintz, N., *et al.* (2004). The knockout mouse project. *Nat. Genet.* **36,** 921–924.

Baldock, R. A., Bard, J. B., Burger, A., Burton, N., Christiansen, J., Feng, G., Hill, B., Houghton, D., Kaufman, M., Rao, J., Sharpe, J., Ross, A., *et al.* (2003). EMAP and EMAGE: A framework for understanding spatially organized data. *Neuroinformatics* **4,** 309–325.

Bard, J. B., Lam, M. S., and Aitken, S. (2008). A bioinformatics approach for identifying candidate transcriptional regulators of mesenchyme-to-epithelium transitions in mouse embryos. *Dev. Dyn.* **237,** 2748–2754.

Barrett, T., Troup, D. B., Wilhite, S. E., Ledoux, P., Rudnev, D., Evangelista, C., Kim, I. F., Soboleva, A., Tomashevsky, M., Marshall, K. A., Phillippy, K. H., Sherman, P. M., *et al.* (2009). NCBI GEO: Archive for high-throughput functional genomic data. *Nucleic Acids Res.* **37,** D885–D890.

Brinkmeier, M. L., Davis, S. W., Carninci, P., MacDonald, J. W., Kawai, J., Ghosh, D., Hayashizaki, Y., Lyons, R. H., and Camper, S. A. (2009). Discovery of transcriptional regulators and signaling pathways in the developing pituitary gland by bioinformatic and genomic approaches. *Genomics* **93,** 449–460.

Bult, C. J., Kadin, J. Z., Richardson, J. E., Blake, J. A., and Eppig, J. T., Mouse Genome Database Group (2010). The Mouse Genome Database: Enhancements and updates. *Nucleic Acids Res.* **38,** D586–D592.

Busser, B. W., Bulyk, M. L., and Michelson, A. M. (2008). Toward a systems-level understanding of developmental regulatory networks. *Curr. Opin. Genet. Dev.* **18,** 521–529.

Collins, F. S., Rossant, J., and Wurst, W. (2007). A mouse for all reasons. *Cell* **128,** 9–13.

Davisson, M. (2006). FIMRe: Federation of International Mouse Resources: Global networking of resource centers. *Mamm. Genome* **17,** 363–364.

DeLaurier, A., Burton, N., Bennett, M., Baldock, R., Davidson, D., Mohun, T. J., and Logan, M. P. (2008). The Mouse Limb Anatomy Atlas: An interactive 3D tool for studying embryonic limb patterning. *BMC Dev. Biol.* **8,** 83.

Dhenain, M., Ruffins, S. W., and Jacobs, R. E. (2001). Three-dimensional digital mouse atlas using high-resolution MRI. *Dev. Biol.* **232,** 458–470.

Ellisor, D., Koveal, D., Hagan, N., Brown, A., and Zervas, M. (2009). Comparative analysis of conditional reporter alleles in the developing embryo and embryonic nervous system. *Gene Expr. Patterns* **9,** 475–489.

Eppig, J. T., and Strivens, M. (1999). Finding a mouse: The International Mouse Strain Resource (IMSR). *Trends Genet.* **15,** 81–82.

Friedel, R. H., Seisenberger, C., Kaloff, C., and Wurst, W. (2007). EUCOMM—The European conditional mouse mutagenesis program. *Brief. Funct. Genomics Proteomics* **6,** 180–185.

Hansen, G. M., Markesich, D. C., Burnett, M. B., Zhu, Q., Dionne, K. M., Richter, L. J., Finnell, R. H., Sands, A. T., Zambrowicz, B. P., and Abuin, A. (2008). Large-scale gene trapping in C57BL/6 N mouse embryonic stem cells. *Genome Res.* **18,** 1670–1679.

Haudry, Y., Berube, H., Letunic, I., Weeber, P. D., Gagneur, J., Girardot, C., Kapushesky, M., Arendt, D., Bork, P., Brazma, A., Furlong, E. E., Wittbrodt, J., *et al.* (2008). 4DXpress: A database for cross-species expression pattern comparisons. *Nucleic Acids Res.* **36,** D847–D853.

Hayamizu, T. F., Mangan, M., Corradi, J. P., Kadin, J. A., and Ringwald, M. (2005). The Adult Mouse Anatomical Dictionary: A tool for annotating and integrating data. *Genome Biol.* **6,** R29.

Heintz, N. (2004). Gene expression nervous system atlas (GENSAT). *Nat. Neurosci.* **7,** 483.

Hill, D. P., Berardini, T. Z., Howe, D. G., and Van Auken, K. M. (2010). The Gene Ontology Consortium. Representing ontogeny through ontology: A developmental biologist's guide to the gene ontology. *Mol. Reprod. Dev.* **77,** 314–329.

Huang, D. W., Sherman, B. T., and Lempicki, R. A. (2009). Systematic and integrative analysis of large gene lists using DAVID bioinformatics resources. *Nat. Protoc.* **4,** 44–57.

Jones, A. R., Overly, C. C., and Sunkin, S. M. (2009). The Allen Brain Atlas: 5 years and beyond. *Nat. Rev. Neurosci.* **10,** 821–828.

Kapushesky, M., Emam, I., Holloway, E., Kurnosov, P., Zorin, A., Malone, J., Rustici, G., Williams, E., Parkinson, H., and Brazma, A. (2010). Gene expression atlas at the European bioinformatics institute. *Nucleic Acids Res.* **38,** D690–D698.

Kaufman, M. H. (1992). The Atlas of Mouse Development. Academic Press, London.

Klinakis, A., Szabolcs, M., Chen, G., Xuan, S., Hibshoosh, H., and Efstratiadis, A. (2009). Igf1r as a therapeutic target in a mouse model of basal-like breast cancer. *Proc. Natl. Acad. Sci. USA* **106,** 2359–2364.

Kwon, G. S., and Hadjantonakis, A. K. (2009). Transthyretin mouse transgenes direct RFP expression or Cre-mediated recombination throughout the visceral endoderm. *Genesis* **47,** 447–455.

Lacroix, Z., and Critchlow, T. (2003). Bioinformatics: Managing Scientific Data. Morgan Kaufmann Publishers, San Francisco, CA.

Lee, E. F., Boline, J., and Toga, A. W. (2007). A high-resolution anatomical framework of the neonatal mouse brain for managing gene expression data. *Front. Neuroinformatics* **1,** 6.

Lee, T. L., Cheung, H. H., Claus, J., Sastry, C., Singh, S., Vu, L., Rennert, O., and Chan, W. Y. (2009). GermSAGE: A comprehensive SAGE database for transcript discovery on male germ cell development. *Nucleic Acids Res.* **37,** D891–D897.

Lee, T. L., Cheung, H. H., Claus, J., Sastry, C., Singh, S., Vu, L., Rennert, O., and Chan, W. Y. (2010). GonadSAGE: A comprehensive SAGE database for transcript discovery on male embryonic gonad development. *Bioinformatics* **26,** 585–586.

Lewandoski, M. (2007). Analysis of mouse development with conditional mutagenesis. *Handb. Exp. Pharmacol.* **178,** 235–262.

Magdaleno, S., Jensen, P., Brumwell, C. L., Seal, A., Lehman, K., Asbury, A., Cheung, T., Cornelius, T., Batten, D. M., Eden, C., Norland, S. M., Rice, D. S., *et al.* (2006). BGEM: An in situ hybridization database of gene expression in the embryonic and adult mouse nervous system. *PLoS Biol.* **4,** e86.

Martinez-Morales, J. R., Henrich, T., Ramialison, M., and Wittbrodt, J. (2007). New genes in the evolution of the neural crest differentiation program. *Genome Biol.* **8,** R36.

McMahon, A. P., Aronow, B. J., Davidson, D. R., Davies, J. A., Gaido, K. W., Grimmond, S., Lessard, J. L., Little, M. H., Potter, S. S., Wilder, E. L., and Zhang, P., GUDMAP Project (2008). GUDMAP: The genitourinary developmental molecular anatomy project. *J. Am. Soc. Nephrol.* **19,** 667–671.

Nagy, A., Mar, L., and Watts, G. (2009). Creation and use of a Cre recombinase transgenic database. *Methods Mol. Biol.* **530,** 365–378.

Nakagata, N., and Yamamura, K. (2009). Current activities of CARD as an international core center for mouse resources. *Exp. Anim.* **58,** 343–350.

Nord, A. S., Chang, P. J., Conklin, B. R., Cox, A. V., Harper, C. A., Hicks, G. G., Huang, C. C., Johns, S. J., Kawamoto, M., Liu, S., Meng, E. C., Morris, J. H., *et al.* (2006). The International Gene Trap Consortium Website: A portal to all publicly available gene trap cell lines in mouse. *Nucleic Acids Res.* **34,** D642–D648.

Petiet, A. E., Kaufman, M. H., Goddeeris, M. M., Brandenburg, J., Elmore, S. A., and Johnson, G. A. (2008). High-resolution magnetic resonance histology of the embryonic and neonatal mouse: A 4D atlas and morphologic database. *Proc. Natl. Acad. Sci. USA* **105,** 12331–12336.

Raymond, C. S., and Soriano, P. (2006). Engineering mutations: Deconstructing the mouse gene by gene. *Dev. Dyn.* **235,** 2424–2436.

Richardson, L., Venkataraman, S., Stevenson, P., Yang, Y., Burton, N., Rao, J., Fisher, M., Baldock, R. A., Davidson, D. R., and Christiansen, J. H. (2010). EMAGE mouse embryo spatial gene expression database: 2010 update. *Nucleic Acids Res.* **38,** D703–D709.

Rodriguez-Zas, S. L., Ko, Y., Adams, H. A., and Southey, B. R. (2008). Advancing the understanding of the embryo transcriptome co-regulation using meta-, functional, and gene network analysis tools. *Reproduction* **135,** 213–224.

Rugh, R. (1968). The Mouse: Its Reproduction and Development. Burgess Pub. Co., Minneapolis, MN.

Sam, L. T., Mendonça, E. A., Li, J., Blake, J., Friedman, C., and Lussier, Y. A. (2009). PhenoGO: An integrated resource for the multiscale mining of clinical and biological data. *BMC Bioinformatics* **10**(Suppl. 2), S8.

Savolainen, S. M., Foley, J. F., and Elmore, S. A. (2009). Histology atlas of the developing mouse heart with emphasis on E11.5 to E18.5. *Toxicol. Pathol.* **37,** 395–414.

Savory, J. G., Bouchard, N., Pierre, V., Rijli, F. M., De Repentigny, Y., Kothary, R., and Lohnes, D. (2009). Cdx2 regulation of posterior development through non-Hox targets. *Development* **136,** 4099–4110.

Schambra, U. (2008). Prenatal Mouse Brain Atlas. Springer, New York, NY.

Schnütgen, F., Stewart, A. F., von Melchner, H., and Anastassiadis, K. (2006). Engineering embryonic stem cells with recombinase systems. *Methods Enzymol.* **420,** 100–136.

Sharova, L. V., Sharov, A. A., Piao, Y., Shaik, N., Sullivan, T., Stewart, C. L., Hogan, B. L., and Ko, M. S. (2007). Global gene expression profiling reveals similarities and differences among mouse pluripotent stem cells of different origins and strains. *Dev. Biol.* **307,** 446–459.

Siddiqui, A. S., Khattra, J., Delaney, A. D., Zhao, Y., Astell, C., Asano, J., Babakaiff, R., Barber, S., Beland, J., Bohacec, S., Brown-John, M., Chand, S., *et al.* (2005). A mouse atlas of gene expression: Large-scale digital gene-expression profiles from precisely defined developing C57BL/6J mouse tissues and cells. *Proc. Natl. Acad. Sci. USA* **102,** 18485–18490.

Smith, C. M., Finger, J. H., Hayamizu, T. F., McCright, I. J., Eppig, J. T., Kadin, J. A., Richardson, J. E., and Ringwald, M. (2007). The mouse Gene Expression Database (GXD): 2007 update. *Nucleic Acids Res.* **35,** D618–D623.

Summerhurst, K., Stark, M., Sharpe, J., Davidson, D., and Murphy, P. (2008). 3D representation of Wnt and Frizzled gene expression patterns in the mouse embryo at embryonic day 11.5 (Ts19). *Gene Expr. Patterns* **8,** 331–348.

Takahashi, K., and Yamanaka, S. (2006). Induction of pluripotent stem cells from mouse embryonic and adult fibroblast cultures by defined factors. *Cell* **126,** 663–676.

Theiler, K. (1972). The House Mouse: Development and Normal Stages from Fertilization to 4 Weeks of Age. Springer-Verlag, Berlin.

Tokuzawa, Y., Kaiho, E., Maruyama, M., Takahashi, K., Mitsui, K., Maeda, M., Niwa, H., and Yamanaka, S. (2003). Fbx15 is a novel target of Oct3/4 but is dispensable for embryonic stem cell self-renewal and mouse development. *Mol. Cell. Biol.* **23,** 2699–2708.

Tokuzawa, Y., Maruyama, M., and Yamanaka, S. (2006). Utilization of digital differential display to identify novel targets of Oct3/4. *Methods Mol. Biol.* **329,** 223–231.

Visel, A., Thaller, C., and Eichele, G. (2004). GenePaint.org: An atlas of gene expression patterns in the mouse embryo. *Nucleic Acids Res.* **32,** D552–D556.

Visel, A., Carson, J., Oldekamp, J., Warnecke, M., Jakubcakova, V., Zhou, X., Shaw, C. A., Alvarez-Bolado, G., and Eichele, G. (2007). Regulatory pathway analysis by high-throughput in situ hybridization. *PLoS Genet.* **3,** 1867–1883.

Vokes, S. A., Ji, H., Wong, W. H., and McMahon, A. P. (2008). A genome-scale analysis of the cis-regulatory circuitry underlying sonic hedgehog-mediated patterning of the mammalian limb. *Genes Dev.* **22,** 2651–2663.

Walls, J. R., Coultas, L., Rossant, J., and Henkelman, R. M. (2008). Three-dimensional analysis of vascular development in the mouse embryo. *PLoS One* **3,** e2853.

Wang, N., Xue, L., Yuan, A., and Xu, D. (2009). Identification of mouse 8-cell embryo stage-specific genes by Digital Differential Display. *Exp. Anim.* **58,** 547–556.

Wilkinson, P., Sengerova, J., Matteoni, R., Chen, C. K., Soulat, G., Ureta-Vidal, A., Fessele, S., Hagn, M., Massimi, M., Pickford, K., Butler, R. H., Marschall, S., *et al.* (2010). EMMA—Mouse mutant resources for the international scientific community. *Nucleic Acids Res.* **38,** D570–D576.

Willnow, T., Antignac, C., Brändli, A., Christensen, E., Cox, R., Davidson, D., Davies, J., Devuyst, O., Eichele, G., Hastie, N., Verroust, P., Schedl, A., *et al.* (2005). The European renal genome project: An integrated approach towards understanding the genetics of kidney development and disease. *Organogenesis* **2,** 42–47.

Wittmann, D. M., Blöchl, F., Trümbach, D., Wurst, W., Prakash, N., and Theis, F. J. (2009). Spatial analysis of expression patterns predicts genetic interactions at the mid-hindbrain boundary. *PLoS Comput. Biol.* **5,** e1000569.

Wu, C., Orozco, C., Boyer, J., Leglise, M., Goodale, J., Batalov, S., Hodge, C. L., Haase, J., Janes, J., Huss, III, J. W., and Su, A. I. (2009). BioGPS: An extensible and customizable portal for querying and organizing gene annotation resources. *Genome Biol.* **10,** R130.

Yoshikawa, T., Piao, Y., Zhong, J., Matoba, R., Carter, M. G., Wang, Y., Goldberg, I., and Ko, M. S. (2006). High-throughput screen for genes predominantly expressed in the ICM of mouse blastocysts by whole mount in situ hybridization. *Gene Expr. Patterns* **6,** 213–224.

Yoshiki, A., Ike, F., Mekada, K., Kitaura, Y., Nakata, H., Hiraiwa, N., Mochida, K., Ijuin, M., Kadota, M., Murakami, A., Ogura, A., Abe, K., *et al.* (2009). The mouse resources at the RIKEN BioResource center. *Exp. Anim.* **58,** 85–96.

Zeitlinger, J., and Stark, A. (2010). Developmental gene regulation in the era of genomics. *Dev. Biol.* **339,** 230–239.

Zhang, J., Richards, L. J., Yarowsky, P., Huang, H., van Zijl, P. C., and Mori, S. (2003). Three-dimensional anatomical characterization of the developing mouse brain by diffusion tensor microimaging. *Neuroimage* **20,** 1639–1648.

SECTION TWO

HANDLING MOUSE LINES

CHAPTER TWO

Transport of Mouse Lines by Shipment of Live Embryos

Kevin A. Kelley

Contents

Abstract

Advances in techniques for the genetic manipulation of the laboratory mouse have resulted in a vast array of novel mouse lines for research. One challenge facing researchers is the ability to rapidly share genetically modified mouse lines with collaborators at other institutions. The standard method of shipping live animals has its share of problems, including the acceptability of the mice at the receiving institution based on health status, as well as the length of time that mice are maintained in quarantine at the receiving institution. Transfer of mouse lines between institutions can also be accomplished by shipment of cryopreserved embryos or sperm. This option, however, is limited by the availability of properly trained staff at the shipping institution who can prepare the cryopreserved materials, as well as staff at the receiving institution who can recover live animals from the transferred samples. Overnight shipment of live, preimplantation mouse embryos circumvents many of the issues involved with shipping live animals or cryopreserved samples. The technique described in

Department of Developmental and Regenerative Biology, Mouse Genetics Shared Resource Facility, Mount Sinai School of Medicine, New York, New York, USA

Methods in Enzymology, Volume 476

ISSN 0076-6879, DOI: 10.1016/S0076-6879(10)76002-X

this chapter for shipping live embryos provides a simple method for transferring mouse lines between institutions.

1. Introduction

The commonly accepted method for transferring mouse lines between investigators at different institutions is the shipment of live animals. There are several distinct issues with the shipment of live mice, however, that have made this method less desirable in recent years. First, the health status of the animals to be shipped must be acceptable to the receiving institution. Even if health reports provided by the shipping institution are found to be acceptable, most receiving institutions consider imported mice to be carriers of microorganisms until on-site health tests prove otherwise. The risk of contaminating existing mouse colonies by the introduction of animals from other nonvendor sources is a major concern (Mahabir *et al.*, 2008). Receiving institutions generally require some form of quarantine for newly arrived animals, which can be several months in duration. During this quarantine period, breeding of the mice is generally prohibited, preventing the researcher from expanding the line for maintenance or experimental purposes. In addition, the amount of quarantine space may be limited, resulting in additional delays before the animals can be received. Second, transport of live laboratory animals has become subject to increasing regulatory guidelines which generally necessitates the use of specialized courier companies that are familiar with the country- and species-specific requirements for shipping live animals within a country as well as internationally. All of these considerations generally make for lengthy delays in a researcher's access to available mouse lines due to the time it takes between requesting a line and its acceptance by the receiving institution, availability of quarantine space, and the length of quarantine itself once space becomes available. In some cases, receiving institutions will admit a new mouse line into the general mouse colony only after rederivation of the line by transfer of zona-intact embryos into pathogen-free hosts (embryo rederivation). This process also results in significant delays in the transfer of a mouse line between investigators, since mice generally have to be embryo rederived out of quarantine at the receiving institution, or with a commercial vendor that offers this service. Finally, the shipment of live mice, whether destined for quarantine or rederivation, is a relatively expensive method for transferring mice between institutions, especially when a courier company is used for the shipment.

In recent years, transferring mouse lines using cryopreserved embryos or sperm has become more widespread. The major advantage of this method is that live animals do not need to be shipped, alleviating some, but not all, of the concerns about transmission of pathogens to existing mouse colonies. Zona-intact embryos are generally considered to be pathogen free, which is the basis for embryo rederivations to remove pathogens from a mouse colony (Carthew *et al.*, 1985; Mohanty and Bachman, 1974; Morrell, 1999; Reetz *et al.*, 1988; Van Keuren and Saunders, 2004). Pathogen transmission through sperm used for *in vitro* fertilization (IVF), however, has not been studied as extensively. Suzuki *et al.* (1996) have shown that in a small number of mouse lines (four strains) embryo transfer following IVF with sperm from males that were known to be infected with mouse hepatitis virus (MHV) and *Pasteurella pneumotropica* resulted in offspring that were pathogen free. While this suggests that disease transmission through sperm from males infected with these two pathogens might not be a major concern, additional studies need to be undertaken to examine a wider range of murine pathogens. Since sperm pass through the protective zona pellucida of oocytes during fertilization, there are some concerns that pathogen-contaminated sperm can result in transmission of the pathogen into the embryo during IVF. While it is known that some murine pathogens can infect testes and spermatozoa (Mahabir *et al.*, 2008), further studies are needed to determine the risk associated with IVF recovery of lines using cryopreserved sperm from pathogen-infected males. Therefore, transfer of mouse lines between institutions using cryopreserved sperm should be done in conjunction with routine quarantine of any resulting offspring to alleviate potential issues with pathogen transmission.

In addition to the concerns about potential pathogen transmission, there are other distinct disadvantages for the transfer of mouse lines through shipment of cryopreserved embryos or sperm. Both institutions need to have staff available who can either prepare the cryopreserved samples (shipping institution) or recover the mouse lines after thawing the sperm or embryos (receiving institution). Both procedures can generally be performed by trained staff in existing transgenic production facilities, provided both the shipping and receiving sites have established transgenic groups, which is not always the case. The yields of live mice from cryopreserved samples can also be quite variable, and depend on multiple factors, including the methods of cryopreservation or thawing, and the skill of the staff involved with the cryopreservation or thawing, and recovery surgeries. Finally, shipping cryopreserved samples requires the use of a specialized liquid nitrogen shipper. These "dry" liquid nitrogen shippers use an absorbent material that can contain liquid nitrogen so that it does not leak from the container during transit. Generally, shipping cryopreserved embryos or sperm is a relatively costly process due to the weight of the shipping container after it is charged with liquid nitrogen, the requirement for a specialized courier company to handle the shipping details, and the need to return the container to the shipping institution.

Alternative methods have been developed to transfer mouse lines between institutions by the shipment of live, preimplantation stage embryos.

Shipment of mouse morula at 0 °C in a sucrose-containing media has been shown to be an effective means of transferring mouse lines (Miyoshi *et al.*, 1992). Low-temperature shipment of two-cell stage mouse embryos within oviducts has also been shown to be useful for transferring mouse lines between institutions (Kamimura *et al.*, 2003). The shipment of live embryos circumvents many of the issues surrounding transfer of pathogens since the zona-intact embryos are essentially being used for embryo rederivation surgeries upon arrival at the receiving institution. The risks of pathogen transfer as well as the complications of shipping animals or cryopreserved samples are essentially eliminated. Live embryos can also be transported by standard overnight shipping companies at a fraction of the cost of shipping animals or cryopreserved samples. As with recovery of lines from cryopreserved embryos or sperm, successful transfer of live embryos requires staff at both the shipping and receiving institutions that can isolate the embryos and transfer them to pseudopregnant hosts, respectively. The following protocol details the isolation, shipment, and surgical transfer of two-cell stage mouse embryos as a method for transporting mouse lines between institutions.

2. Materials

2.1. Mice

Five- to six-week old B6C3 F1 hybrid mice (stock number 100010) used as embryo donors were purchased from Jackson Laboratories. Swiss-Webster (SW) outbred mice used as embryo hosts were purchased from Taconic Farms. Pseudopregnant hosts are prepared by mating naturally ovulating SW females with vasectomized SW males. All mice are maintained in microisolator cages (Allentown Caging) on a 14 h:10 h light:dark cycle with access to an automatic watering system, and *ad libitum* access to PicoLab Rodent Diet 20 (product code 5053, LabDiet). All mice are euthanized by CO_2 inhalation.

2.2. Hormones

Embryo donors are superovulated as described in Section 3 with pregnant mare's serum (PMS; National Hormone and Peptide Program, http://www.humc.edu/hormones/material.html) and human chorionic gonadotropin (HCG; Sigma catalog number CG10). PMS and HCG are reconstituted to 1 International Unit (IU)/μl from lyophilized stocks with sterile saline (0.9% NaCl). Stocks containing 50 IU of each hormone are prepared and stored at

– 80 °C. Dilutions of each hormone to 5 IU/0.1 ml are prepared immediately before injection by the addition of sterile saline to a final volume of 1 ml.

2.3. Embryo isolation and surgical transfer

All embryo isolations and washes are performed using a Nikon SMZ-2 dissecting microscope to visualize the embryos at low magnification (10–63×). Embryos are transferred between washes using a mouth aspiration pipette system (Sigma, catalog number A-5177) described in Nagy *et al.* (2003). Surgeries to transfer two-cell stage embryos into the oviduct of pseudopregnant SW female mice are performed in a laminar flow hood, using a Nikon SMZ10 stereomicroscope on a boom stand.

2.4. Culture media

FHM (Millipore/Specialty Media, catalog number MR-024-D) is a specially formulated HEPES-buffered media that is used when embryos are to be isolated or manipulated outside of a 5% CO_2 atmosphere. FHM with hyaluronidase (Millipore/Specialty Media, catalog number MR-056-F) is used to dissociate embryos from cumulus cells in isolated cumulus masses at the pronuclear stage. KSOM + AA (Millipore/Specialty Media, catalog number MR-121-D) is bicarbonate-buffered, and used for the culture of embryos in a 5% CO_2 incubator at 37 °C. All incubations in a 5% CO_2 incubator at 37 °C are performed with the KSOM + AA media covered with embryo-tested mineral oil (Sigma, catalog number M8410) in 35 mm tissue culture dishes for *in vitro* embryo development.

2.5. Shipping materials

Mouse embryos are shipped in Nalgene cryogenic vials with external threads (Nalge Nunc International, catalog number 5000-0020). To minimize temperature variations during the shipment, small (outside dimensions: 8 in. L × 6 in. W × 5.25 in. H) styrofoam shipping boxes with 1 in. thick walls are used (ThermoSafe #440, distributed by Fisher Scientific, catalog number 03-528-10). Two pieces of 1.5 in. thick convoluted polyurethane foam (also known as egg carton or egg crate foam) are cut to fit the interior dimensions of the shipping box. Convoluted polyurethane foam can be purchased from numerous Internet suppliers of packaging material. American Foam Group (http://www.foampackagingcompany.com/egg-crate-convoluted-foam.html) supplies an inexpensive (~$20 USD) 1.5 in. × 2 ft × 6 ft sheet of convoluted polyurethane foam that can be cut to size to fit the shipping boxes.

2.6. Anesthetic

Avertin anesthesia is used for all surgical transfer of embryos to the reproductive tract of pseudopregnant SW female mice. Avertin is prepared from 2,2,2-tribromoethanol (Sigma-Aldrich, catalog number T48042) and *tert*-amyl alcohol (Sigma-Aldrich catalog number 249486) as described by Nagy *et al.* (2003). Briefly, a stock solution (100%) is prepared by dissolving 10 g of 2,2,2-tribromoethanol in 10 ml of *tert*-amyl alcohol in a 50 °C water bath until it is fully dissolved. A working solution of 2.5% avertin is prepared by a 1:40 dilution of the 100% stock with warm (37 °C) water. The diluted solution is sterilized through a 0.2-μm filter and stored at 4 °C for 2–3 weeks before being replaced. The avertin is administered intraperitoneally at 0.016 ml/g of body weight.

3. Methods

The successful transfer of mouse lines through shipping of live embryos is dependent upon close interaction between staff at the shipping institution who are isolating the embryos and staff at the receiving institution who are transferring the embryos upon arrival into pseudopregnant hosts. The timing is critical, and it is important that day 0.5 postcoital (pc) pseudopregnant hosts are available at the receiving institution on the expected day of arrival of the shipped embryos. It is also equally important that the males to be used for the matings to generate donor embryos are unmated and individually housed for at least 1 week prior to mating with superovulated donor females at the shipping institution. Finally, there may be special requirements for the receiving institution, such as health reports of the colony that the embryos are isolated from, as well as country-specific regulations that must be followed by the shipping company for international shipments. The shipping institution should work out these details in advance.

3.1. Shipping institution

3.1.1. Superovulation of donor females

(1) *Day 1*. Superovulation is initiated by intraperitoneal injection of 5 IU of PMS per donor female at 4:00 p.m. Typically, 5–10 immature (5–6-week-old) wild-type females (actual strain depends on the background of the line to be transferred) are used for an embryo transfer procedure. It is also possible to use immature females from an investigator's colony if it is essential to send animals of a particular genotype, such as knockouts.
(2) *Day 3*. Donor females receive intraperitoneal injection of 5 IU of HCG per female 44–46 h (12 noon to 2 p.m.) after PMS injection. After

injection of HCG, donor females are mated with males (2–10 months of age) from the line to be transferred. The matings are set up in a 1:1 ratio (one superovulated female with one male).

3.1.2. Embryo isolation and culture

(3) *Day 4.* The day after mating, the females are checked in the morning (8–9 a.m.) for the presence of a vaginal copulation plug. The presence of the plug indicates which females successfully mated. These females are euthanized, and their oviducts are removed and collected into a 35-mm tissue culture dish containing FHM media at room temperature. Individual oviducts are sequentially transferred to a depression slide or 35-mm tissue culture dish containing FHM media with hyaluronidase (prewarmed to 37 °C), and observed on a dissecting microscope (Nikon SMZ2). Each oviduct is immobilized behind the ampulla (the swelling containing the cumulus mass of cells) with a #3 forcep (Fine Science Tools), and the outer wall of the ampulla is opened by tearing with a second #3 forcep to release the cumulus mass. After all cumulus masses have been released into the hyaluronidase media, the dissolution of the pronuclear stage embryos from the cumulus cells is carefully monitored. Individual embryos are recovered, washed through three changes of FHM media, transferred to 10–20 μl microdrops of KSOM + AA media under mineral oil (prepared in a 35-mm tissue culture dish), and incubated overnight for development of two-cell stage embryos.

3.1.3. Shipping two-cell stage embryos

(4) *Day 5.* One milliliter of warm (37 °C) FHM media is added to a Nalgene cryogenic vial (with external threads). Using a mouth pipette transfer assembly, two-cell stage embryos that developed from overnight culture of pronuclear stage embryos are pipetted into the cryogenic vial containing FHM. After transfer of the embryos, the vial is filled almost to the top with warm (37 °C) FHM media and closed and sealed with parafilm. The vial is placed into a 50 ml centrifuge tube (Falcon, catalog number 352098) using Kimwipes (Fisher Scientific, catalog number 06-666) or a suitable packing material to keep it firmly in place within the tube (Fig. 2.1A and B). The 50 ml tube is then placed into a small ThermoSafe styrofoam shipper with packing material to keep the tube firmly in place for the shipment (Fig. 2.1C and D). Convoluted polyurethane foam works well as the packing material, and provides additional insulation to the embryos. Other packing materials, such as bubble wrap, would also work. The box is sealed with packing tape, and shipped to the receiving institution using an overnight

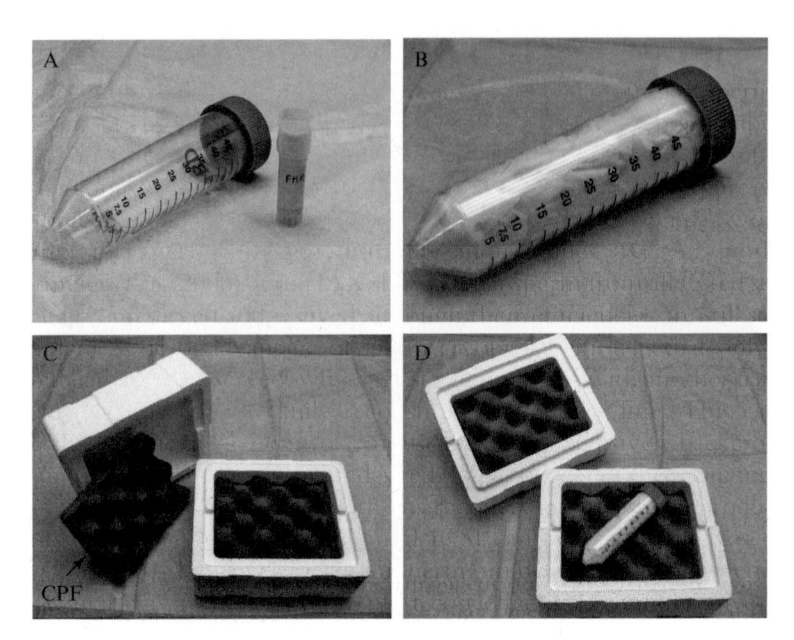

Figure 2.1 Preparation of embryos for shipping. (A) Nalgene cryogenic vial with FHM media and 50 ml Falcon centrifuge tube. (B) Vial of FHM media and Kimwipes packed inside 50 ml Falcon tube. (C) Stryofoam shipping container with convoluted polyurethane foam (CPF) cut to fit the interior dimensions of the box. (D) 50 ml Falcon tube containing the vial with FHM media loaded into the shipping container with CPF packing material.

delivery company (such as Federal Express), with delivery specified as early as possible the following morning. To minimize the length of time that the embryos are in transit, the embryos and shipping container should not be prepared until late in the afternoon, immediately prior to being turned over to the delivery company.

3.2. Receiving institution

3.2.1. Embryo recovery and surgical transfers

(5) *Day 6.* Embryos that arrive at the receiving institution are retrieved from the cryogenic vial with a 0.1–1.0 ml Selectapette pipette (Clay-Adams, model number 4690), using Selectapette pipette tips (Clay-Adams, catalog number 4696). The media in the tube is collected and transferred to a depression slide or 35-mm culture dish. The tube is rinsed one time with 1 ml of room-temperature FHM media, and this rinse is added to the contents of the tube that were removed previously. Embryos are identified using a low-power stereomicroscope, isolated

with a mouth-pipette transfer assembly, and transferred into 10–20 μl microdrops of KSOM + AA media under mineral oil. The recovered embryos are generally still at the two-cell stage as the shipment at lower temperatures delays development to the four-cell stage. The embryos are maintained at 37 °C in a 5% CO_2 incubator until the surgeries are performed. The embryos are transferred to the oviduct of day 0.5 pseudopregnant SW female mice on the same day that they are received (oviduct surgeries are described in Nagy *et al.*, 2003).

(6) Pups should be born 19 days after the surgical transfer of embryos into pseudopregnant hosts. If they are not born by the morning of day 20 postsurgery, the hosts are euthanized, and any pups are removed by caesarian section and transferred to foster mothers. At weaning, host females are used for health analyses to test the pathogen-free status of the litter. Additional health testing of wild-type littermates can be performed several weeks after weaning to confirm the pathogen-free status of the line.

4. Results and Discussion

The shipment of live embryos between institutions is generally subject to the same temperature variations that are experienced during the shipment of live animals. These variations include both ground and air transport conditions. Syversen *et al.* (2008) found that temperature monitoring during 103 shipments of live animals revealed approximately 50% of the shipments were subject to temperatures greater than 29.4 °C, while approximately 15% of shipments were exposed to low temperatures (<7.2 °C). In addition, they observed that approximately 60% of all shipments were subjected to temperature variations of 11 °C or more (Syversen *et al.*, 2008). Based on these results, and the observation that mouse preimplantation embryos can survive incubation at 4 °C for extended periods of time (Kamimura *et al.*, 2003; Miyoshi *et al.*, 1992), the ability of pronuclear stage embryos to survive overnight incubation at various temperatures that would mimic overnight shipment was tested. We found that in KSOM + AA media both ambient temperature (22 °C) and 37 °C incubations yielded live embryos at embryonic day 14.5 (E14.5), whereas incubation in KSOM + AA at 4 °C yielded no E14.5 embryos (Table 2.1). In contrast, live E14.5 embryos were obtained at all tested incubation temperatures (4, 22, and 37 °C) when pronuclear stage embryos were incubated in FHM media. These results have been confirmed for incubation of two-cell stage embryos (data not shown). As a result, the protocol described in this chapter has been designed using FHM media for the shipment of live two-cell stage embryos at ambient temperature. FHM is HEPES buffered, and is therefore a better

Table 2.1 Effect of temperature and media on survival of B6C3 embryos after overnight incubation

	FHM			KSOM + AA		
	4 °C	22 °C	37 °C	4 °C	22 °C	37 °C
Preimplantation embryos[a]	28	28	28	28	28	28
E14.5 embryos (%)[b]	10 (36%)	6 (21%)	17 (61%)	0 (0%)	12 (43%)	16(57%)

[a] Number of embryos transferred to the oviduct of an SW pseudopregnant host after overnight incubation at the pronuclear stage.

[b] Number of embryos recovered after euthanasia of the host on embryonic day 14.5 (E14.5). Values in the parentheses indicate the percentage of transferred preimplantation embryos that developed to viable E14.5 embryos.

choice for shipping live embryos, since it is not as subject to pH variations as media like KSOM + AA that are bicarbonate buffered. In addition, KSOM + AA media did not work as well as FHM at lower temperatures, which is an important consideration in light of the fact that the embryos may be exposed to low temperatures during the shipment (Syversen *et al.*, 2008).

We have designed the protocol described in this chapter to use the shipment of two-cell stage embryos rather than pronuclear stage embryos since the development of isolated embryos to the two-cell stage is an easy visual observation of viable, fertilized eggs. There is no need for a microscope with high magnification for the identification of embryos at the pronuclear stage. The protocol also involves the use of a small, insulated styrofoam mailer (1 in. thick walls) for shipping the embryos. The styrofoam mailer, in combination with the convoluted polyurethane foam packing material, is used to help minimize temperature variations that the embryos will be subjected to during the shipment. The insulated shipping container will minimize the effects of low temperatures during winter shipments, or high temperatures during summer shipments.

Given the disadvantages of shipping live animals, and some of the inconveniences of transferring lines between institutions using cryopreserved embryos or sperm, a method for shipping live embryos is a suitable alternative. The overnight transfer of live preimplantation stage embryos is a quick, convenient, and inexpensive method for transferring novel mouse lines between institutions. In essence, this method is an embryo rederivation that is performed with the assistance of an express overnight mail company, and will provide the receiving investigator with a pathogen-free mouse line from the shipping institution in a shorter time-frame than the more conventional methods of transferring mice.

REFERENCES

Carthew, P., Wood, M. J., and Kirby, C. (1985). Pathogenicity of mouse hepatitis virus for preimplantation mouse embryos. *J. Reprod. Fertil.* **73,** 207–213.

Kamimura, E., Nakashima, T., Ogawa, M., Ohwada, K., and Nakagata, N. (2003). Study of low-temperature (4 degrees C) transport of mouse two-cell embryos enclosed in oviducts. *Comp. Med.* **53,** 393–396.

Mahabir, E., Bauer, B., and Schmidt, J. (2008). Rodent and germplasm trafficking: Risks of microbial contamination in a high-tech biomedical world. *ILAR J.* **49,** 347–355.

Miyoshi, I., Ishikawa, K., Kasai, M., and Kasai, N. (1992). Useful short-range transport of mouse embryos by means of a nonfreezing technique. *Lab. Anim. Sci.* **42,** 198–201.

Mohanty, S. B., and Bachman, P. A. (1974). Susceptibility of fertilized mouse eggs to minute virus of mice. *Infect. Immun.* **9,** 762–763.

Morrell, J. M. (1999). Techniques of embryo transfer and facility decontamination used to improve the health and welfare of transgenic mice. *Lab. Anim.* **33,** 201–216.

Nagy, A., Gertsenstein, M., Vintersten, K., and Behringer, R. (2003). Manipulating the Mouse Embryo: A Laboratory Manual, 3rd edn. Cold Spring Harbor Laboratory Press, Cold Spring Harbor, NY.

Reetz, I. C., Wullenweber-Schmidt, M., Kraft, V., and Hedrich, H. J. (1988). Rederivation of inbred strains of mice by means of embryo transfer. *Lab. Anim. Sci.* **38,** 696–701.

Suzuki, H., Yorozu, K., Watanabe, T., Nakura, M., and Adachi, J. (1996). Rederivation of mice by means of in vitro fertilization and embryo transfer. *Exp. Anim.* **45,** 33–38.

Syversen, E., Pineda, F. J., and Watson, J. (2008). Temperature variations recorded during interinstitutional air shipments of laboratory mice. *J. Am. Assoc. Lab. Anim. Sci.* **47,** 31–36.

Van Keuren, M. L., and Saunders, T. L. (2004). Rederivation of transgenic and gene-targeted mice by embryo transfer. *Transgenic Res.* **13,** 363–371.

CHAPTER THREE

Strategies and Considerations for Distributing and Recovering Mouse Lines

Yubin Du, Wen Xie, *and* Chengyu Liu

Contents

Abstract

As more and more genetically modified mouse lines are being generated, it becomes increasingly common to share animal models among different research institutions. Live mice are routinely transferred between animal facilities. Due to various issues concerning animal welfare, intellectual property rights, colony health status and biohazard, significant paperwork and coordination are required before any animal travel can take place. Shipping fresh or frozen preimplantation embryos, gametes, or reproductive organs can bypass some of the issues associated with live animal transfer, but it requires the receiving facilities to be able to

Transgenic Core Facility, Division of Intramural Research, National Heart, Lung, and Blood Institute, National Institutes of Health, Bethesda, Maryland, USA

Methods in Enzymology, Volume 476
ISSN 0076-6879, DOI: 10.1016/S0076-6879(10)76003-1

perform delicate and sometimes intricate procedures such as embryo transfer, *in vitro* fertilization (IVF), or ovary transplantation. Here, we summarize the general requirements for live animal transport and review some of the assisted reproductive technologies (ART) that can be applied to shipping and reviving mouse lines. Intended users of these methods should consult their institution's responsible official to find out whether each specific method is legal or appropriate in their own animal facilities.

1. Introduction

The laboratory mouse has remained a favored model organism in biomedical research for over a century. Especially since the development of transgenic mouse technology in the late 1970s and the knockout mouse technology in late 1980s, the number of useful mouse models has increased exponentially. Majority of these genetically modified mouse lines are being created by individual research laboratories and transgenic core facilities throughout the world, and large-scale production facilities now exist which aim at generating at least one null mutant mouse line for every gene. It is critically important that the research community can share these valuable genetic resources by acquiring mouse lines through transportation. Thus far, majority of the mouse line transfers occur as shipment of live animals, which is technically simple and reliable, but requires substantial paperwork and coordination. Mouse lines can also be transferred in the form of embryos, gametes, or dissected reproductive organs. These forms of transfers are generally more technically challenging because they involve the use of sometimes complicated micromanipulative and surgical procedures. But they are often easier to arrange for the actual transportation. With recent improvements in mouse sperm cryopreservation and *in vitro* fertilization (IVF) methods, archiving and delivering frozen sperm is bound to play an important role in future mouse line distribution (Ostermeier *et al.*, 2008). In fact, systematic archiving and distributing genetic modified mouse lines in the form of frozen sperm have already begun at large centers, such as the Jackson Laboratory (JAX) and Mutant Mouse Regional Resource Centers (MMRRC). Individual researchers are also encouraged to deposit their strains to the repositories to facilitate such efforts.

This chapter is intended to summarize the general considerations for animal transport, and to review some of the technologically proven methods that have been used to transfer and revive mouse lines. Different countries, states, or institutions may have specific requirements and policies for animal transport, and the rules should therefore be conformed to when arranging for the actual transportation. We advise intended users of these methods to consult with their institutional animal care and use committee (ACUC) and animal facility manager about whether these methods are legal or appropriate for their specific situations.

2. Shipping Live Mice

Live animals can be transported domestically and internationally following proper procedures and regulations. Transferring live mice can avoid the technical challenging and sometimes problematic procedures, such as IVF and embryo transfer. Usually, a few breeding pairs of live mice can ensure the receiving institution to establish a sizeable colony within 3–6 months. A large group of ready-to-use mice can also be shipped to the receiving side for immediate study. Great majority of animal facilities should already be familiar with shipping and receiving live animals, and many of them have specially designated personnel to handle the paperwork and coordinate the shipment. The cost for the actual shipment is low (about 100–200 US dollars/crate) and many freight companies can provide the transportation. However, before the shipment can take place, substantial paperwork and careful coordination are required. Generally, the following issues need to be considered for each animal shipment.

2.1. Animal welfare issues

In the United States, transportation of most animal species is regulated by the United States Department of Agriculture (USDA) and individual state (http://www.aphis.usda.gov/import_export and State requirements therein). However, research mice and rats are generally exonerated but are governed by the Public Health Service (PHS) policy, the Humane Care and Use of Laboratory Animals (http://grants.nih.gov/grants/olaw/references/phspol.htm), if the research is funded by the PHS. Under the Health Research Extension Act of 1985, the PHS Policy mandates adherence to the guidelines in the *Guide for the Care and Use of Laboratory Animals* (the *Guide*) (NRC, 1996, http://www.nap.edu/openbook.php?record_id=5140&page=57). Specific policies regarding to rodent transportation on institutional level vary under this general guideline, for example, NIH has policy manual 3043-1 to regulate the introduction of rodents into NIH facilities.

Internationally, for countries within the World Organization for Animal Health, formerly known as the Office International des Epizooties (OIE; http://www.oie.int/eng/OIE/en_about.htm?e1d1), arrangements can be made to send and receive mice with health certificates. Unless they are known to carry zoonotic disease or diseases of risk to domestic or wild animals, rodents can be transported internationally with specific permits, bill of lading for transportation of animals, and commercial invoice (e.g., NIH1884-1). For a comprehensive review of live animal transport, see *Guidelines for the Humane Transportation of Research Animals* (2006) prepared by the Committee on Guidelines for the Humane Transportation of Laboratory Animals (http://books.nap.edu/openbook.php?record_id=11557&page=11).

The *Guide* provides performance standards on the transportation of research animals. It states that all transportation of animals should be planned to minimize transit time and the risk of zoonoses, protect against environmental extremes (below 45 °F or above 85 °F), avoid overcrowding, provide food and water when indicated, and protect against physical trauma. Sick mice are not advised to travel, because of the stress involved during transportation. Animal distributors such as Harlan, Taconic, and Charles River Laboratories all provide shipment services. Other established ground and air shippers can be found at http://laboratoryanimalsciencebuyersguide.com/results.php?category=Transportation&heading=202&category_id=2813. If traveling by air, the International Air Transport Association (IATA) Live Animals Regulations (LAR) provides guidelines for the packaging and documentations needed for the transport (http://www.iata.org/ps/publications/live-animals.htm). Ground transportation between airport and destinations should also be arranged to avoid any delay.

Use of laboratory animals for biomedical research is a privilege, not a right. Therefore, all major research institutions have specifically designated committee or office to evaluate and approve investigators' proposals for using animals. The mouse line to be sent should already be listed in the sender's animal study proposal before requesting the shipment. Most, if not all, receiving institutions also require the receiving investigators to add the mouse lines to be received in their animal protocols before they can access and use the animals for experimental studies. Therefore, it is advisable that the receiving investigators should start early to amend their animal protocols in order to avoid delays in receiving and using the new mouse lines.

2.2. Animal health status

One major challenge for operating an animal facility is to keep the animal colonies healthy and free of microorganism contamination. Keeping the mice disease-free is important not only for the welfare of the animals, but also for the validity of the research findings. Experimental results obtained from sick or contaminated animals should not be trusted. There are many mouse pathogens (http://www.radil.missouri.edu/info/dora/mousepag/mouse.htm), and some are tolerated in certain research colonies (e.g., http://jaxmice.jax.org/health/agents_list.html). Most institutions have specific pathogen free (SPF) environment, while their pathogen lists may be different. During transportation, it is important to separate animals from different animal rooms to avoid cross-contamination. Improper care and monitoring during animal shipment can lead to contamination of mouse colonies and lost strains.

Most animal facilities have strict policies against the introduction of diseases or unwanted microorganisms into their animal rooms. Therefore, the receiving facilities usually require the sending animal facilities to provide direct or sentinel serological testing (http://www.radil.missouri.edu/) results for the past several months. Based on these health reports (e.g., NIH Form 1192), the receiving facility veterinarian can decide whether the animals should be imported directly, imported after certain period of quarantine, or they need to be cleaned up through rederivation before entering the facility. If quarantine is required, the receiving investigator must make sure there is enough quarantine space available on the day of receiving animals. Some strict animal facilities require that all animals, except from the proven venders, to be rederived regardless of the health report.

2.3. Intellectual property rights

A good animal model may lead to scientific discoveries with significant commercial values, such as the development of new drugs or new diagnostic methods. Because of the unpredictability of scientific research, it is impossible to foresee which mouse lines are economically important and which ones are not. Therefore, most institutions require the receiving institutions to sign Material Transfer Agreements (MTA) on all mouse lines transferred to protect their intellectual properties. Often an Animal Transfer Agreement (ATA) is also required. The MTAs among most governmental agencies, universities, and nonprofit entities are quite generous, without too much limitation on the use of the animal models. However, almost all MTAs prohibit the receiving investigator from further distributing the mouse lines.

2.4. Biosafety concerns

Mice of different biosafety levels (http://www.cdc.gov/OD/ohs/biosfty/bmbl4/bmbl4s3.htm) should be sent and received with clear labels. Infectious disease carrying mice is subject to additional regulations imposed by the Centers for Disease Control and Prevention (CDC). CDC is part of the US Department of Health and Human Services (DHHS) and enforces regulations to prevent the introduction, transmission, or spread of communicable diseases in the United States. CDC regulates the importation of any animal or animal product capable of carrying a zoonotic disease (regulation 42CFR71.54) through permitting, registration, and quarantine. CDC may implement prohibitions on the importation of animals when new health risks to humans arise. The CDC Etiologic Agent Import Permit Program regulates the importation and subsequent transfer of animals or animal tissues within the US that contain etiologic agents. CDC defines etiologic agent as a viable microorganism or its toxin that may cause human disease, but does not specifically identify them. Contact information and permit applications for importation of etiologic agents are available at: http://www.cdc.gov/od/eaipp/.

Mice carrying diseases of risk to domestic or wild animals are also subject to regulations by the US Fish and Wildlife Service (http://www.fws.gov/le/ImpExp/CommWildlifeImportExport.htm), as they may be considered injurious species (http://www.fws.gov/contaminants/ANS/pdf_files/50CF_16_10-05.pdf). Mice carrying infectious agent are also subject to Department of Transportation (DOT) Hazardous Material (HAZMAT) regulation (http://www.phmsa.dot.gov/hazmat/regs), and if traveling by air, by the IATA Dangerous Goods Regulations (DGR) (http://www.iata.org/ps/publications/elist.htm).

3. Shipping Nonfrozen Embryos or Gametes

Freshly collected preimplantation stage embryos or sperms in epididymis can be shipped without cryopreservation. Some issues discussed for live animal transfer still apply to embryo and gamete shipment, such as ATA and MTA. However, the animal welfare issue associated with transportation is alleviated. It could translate into significant savings in both time and cost, particularly for international shipment. For contaminated mouse lines, the transportation of embryos or gametes offers an opportunity to clean them up through rederivation (Van Keuren and Saunders, 2004). Embryos from contaminated mouse lines can be washed at the sender and/or receiver's facility to be sanitized before transferring into clean surrogate mothers. Another advantage of shipping embryos is that the receiving institution does not have to have a "dirty" facility to house the infected mice. Many animal facilities actually require rederivation regardless of the health status to avoid bring in common pathogens such as mouse parvovirus virus (mpv), mouse hepatitis virus (mhv), helicobacter into their colonies.

The disadvantage of shipping embryos or gametes is that the receiving facilities must be able to revive the mouse line through embryo transfer for preimplantation embryos, or IVF (intracytoplasmic sperm injection (ICSI) if necessary) followed by embryo transfer for the sperm. Time coordination is critical for shipping fresh embryos or gametes.

3.1. Shipping fresh embryos

Preimplantation stage embryos ranging from two-cell stage to blastocyst can be shipped without freezing (Miyoshi *et al.*, 1992a), with blastocysts appear to be the most robust and commonly used. In fact, oviducts isolated from mated females, which contains two-cell embryos, can be shipped directly at low temperatures (Miyoshi *et al.*, 1992b). Shipping oviducts is particular useful if the sender's laboratory lacks the necessary skills to isolate and handle the tiny early embryos. For mouse embryos, 0.75 *M* sucrose was found to improve the survival rate stored at 0 °C (Kasai *et al.*, 1983). Shipments can be made in regular test tubes with an appropriate thermal package, and a clear label so that it would not go through the X-rays. At the receiving institution, mouse line can be rederived through embryo transfer.

3.1.1. Brief protocols

- At sender's laboratory:
  1. Set up mating and check plug the next morning. Euthanize the plugged females on an appropriate day depending on the stage of embryos wanted as listed below:
     Day 1 (the same day as checking plug): one-cell stage (fertilized eggs)
     Day 2: two-cell stage embryos
     Day 3: eight-cell and morula stage embryos
     Day 4: blastocysts
  2. For one-cell stage embryos, dissect out oviducts and tear open the ampullae to release the cumulus mass into M2 medium (Millipore/Specialty Media, MR-015-D). Briefly treat the cumulus mass with hyaluronidase (Sigma, H4272) to isolate individual fertilized eggs.
     For two-cell to morula stage embryos, flush oviducts with M2 medium to obtain embryos.
     For blastocysts, dissect out uteri and flush the embryos out with M2 medium.
     Two-cell through blastocyst stage embryos can also be obtained by culturing one-cell stage fertilized eggs in KSOM medium (Millipore/Specialty Media, MR-106-D) for 1–3 days in a 37 °C incubator with 5% CO_2.
  3. Wash the embryos in M2 medium with 0.75 *M* sucrose (Sigma, S7903), and then transfer them into a 1.5–2.0 ml cryovial containing M2. Cover the medium with mineral oil (Sigma, M8410).
  4. Early embryos inside oviducts can be shipped directly. In this case, the dissected oviducts from Day 2 or 3 are placed in a cryovial directly. Simply cover the tissue with mineral oil and the tube is ready to be shipped.
  5. Ship the tubes in a thermal package with refrigerated ice gel packs through express mail.
- At receiver's laboratory:
  1. Prepare pseudopregnant foster mothers 1–3 days before the anticipated embryo arrival.
  2. After receiving the tubes containing embryos, transfer all contents into a 35 mm tissue culture dish (NUNC 153066) using a plastic transfer pipette (Thomas Scientific, 7760M25). Wash the tube with M2 and transfer all wash media into the same 35 mm dish. Gently swirl the dish to concentrate the embryos into the center region of the dish. Under a stereomicroscope, collect the embryos using a mouth pipette

(Sigma, A5177, modified) and wash by passing through new M2 drops under mineral oil.
3. If whole oviducts were shipped, wash the oviducts in M2 to remove mineral oil. Then flush out embryos with M2 medium.
4. Implant the embryos into pseudopregnant foster mothers through embryo transfer. Blastocysts should be transferred into the uteri of 2.5 dpc (day post coitus) foster mothers, while two-cell, eight-cell, or morula stage embryos should be transferred into the oviducts of 0.5 dpc foster mothers. It is recommended to culture the embryos at 37 °C for 1–6 h before transferring into the foster mothers. Alternatively, two-cell through morula stage embryos can be cultured in KSOM till they become blastocysts, and then they are transferred into the uteri of 2.5 dpc foster mothers.

3.2. Shipping fresh sperm in epididymis

Intact epididymis was tested as a short-term storage and transportation method for mouse spermatozoa (Mochida *et al.*, 2005; Sankai *et al.*, 2001). Certain strains have reduced survival rate due to cold temperature. For C57BL/6, it was determined that 7 °C is the optimal low temperature for nonfrozen sperm storage (Mochida *et al.*, 2005). Strains therefore should be tested at low temperature conditions for survival rate before arranging for such transport. When high concentrations of B6C3F1 sperm (4×10^5 ml^{-1}) were stored at different temperatures, it was found that sperms retained motility at 22 °C better than at 4 or 37 °C, and can fertilized eggs after storage up to 3 days albeit at a very low rate (Sato *et al.*, 2001). We have stored epididymides from 129S6 and C57BL/6 hybrid stains at 4 °C overnight, and were still able to achieve acceptable IVF rate. Sending mouse embryos will entail less work for the receiving end, while nonfrozen transport of epididymis seems to be the simplest for the sender.

3.2.1. Brief protocols

- At sender's laboratory:
  1. Euthanize two 3- to 6-month-old male mice, and quickly dissect out epididymides and vasa deferentia. Place epididymides and vasa deferentia from each mouse into a separate 1.5–2.0 ml sterile tube. Add mineral oil into the tube to prevent the tissues from drying out.
  2. Place the tubes in a thermal package with refrigerated ice gel packs and send through express mail.
- At receiver's laboratory:
  1. Three days before the scheduled IVF, choose 5–10 young donor females (3–6 weeks). At 8 p.m., inject 5 IU pregnant mare's serum gonadotropin (PMSG; Sigma G4527) intraperitoneally per mouse.

2. After receiving the package containing the epididymal tissue, store the tubes at 4 °C until they are used for IVF the next morning.
3. One day before the scheduled IVF, 48 h after PMSG injection, inject 5 IU human chorionic gonadotropin (hCG; Sigma C8554) per mouse intraperitoneally into the same female mice.
4. On the day of IVF, at 8 a.m., ~12 h after hCG injection, take the epididymides and vasa deferentia out of the tubes and place them in a 35 mm tissue culture dish containing 1 ml of IVF medium (MVF medium, Cook's Medical, K-RVFE-50).
5. Tear open the epididymis with Dumont tweezers and squeeze the sperms out of vas deferens.
6. Incubate at 37 °C (Cook's Medical, K-MINC-1000-US) equilibrated with 5% O_2, 5% CO_2, 90% N_2 (Robert Oxygen premixed gas, R0133502001) for ~60 min to allow sperm swim out of the tissue and to be capacitated.
7. Euthanize the superovulated females, and quickly dissect out ampullae. Try to keep the eggs at 37 °C at all time, and keep the time between egg collection and fertilization as short as possible.
8. Tear open the ampulla with a 29½ G needle and release the egg clutches.
9. Use a wide mouth pipette, transfer 10 μl capacitated sperm suspension into 0.5 ml IVF medium covered under mineral oil. Transfer egg clutches from about three females into the drop.
10. Incubate the dish containing sperm and eggs for 4–6 h at 37 °C with 5% O_2, 5% CO_2, and 90% N_2.
11. Wash the eggs in three drops of M2 medium, and then culture them in KSOM medium overnight in a 37 °C incubator with 5% CO_2.
12. On the same day as IVF, set up recipient females with vasectomized males.
13. On the day after IVF, check plug and transfer the embryos that have been fertilized (two-cell) into the plugged pseudopregnant females through oviduct transfer.

While cold temperature can induce sperm capacitation (Fuller and Whittingham, 1997), it can cause the disassembly of meiotic spindle of the oocytes (Pickering and Johnson, 1987). The oocytes are not normally used as a form of animal transfer, also because they only remain competent for IFV for a short period of time after ovulation (~13 h after hCG injection for most mouse strains) due to zona hardening. However, with the use of chemically- (Gordon and Talansky, 1986), laser- (Tadir *et al.*, 1991) or piezo-assisted (Kawase *et al.*, 2002) partial zona dissection (PZD) method or with ICSI (Markert, 1983; Yoshida and Perry, 2007), one can overcome the zona hardening and increase the efficiency of IVF. Fresh oocytes beyond the 13-h window can presumably be shipped and used to recover mouse lines for successful IVF, providing they do not undergo apoptosis. In fact, oocytes rescued from postmortem mouse (Schroeder *et al.*, 1991) and equine ovaries that had been store at room temperature (Ribeiro *et al.*, 2008) can still give rise to normal embryos. Not a suggested way to ship animals, but nonetheless useful in desperate situations to rescue important mouse lines.

4. Archiving and Distributing Mouse Lines Through Cryopreservation

Mouse lines should be cryopreserved as soon as possible to prevent accidental loss, genetic drift, and to reduce cost associate with the maintaining of live animals (Landel, 2005; Shaw and Nakagata, 2002). They can be frozen as sperm (Ostermeier *et al.*, 2008; Polge *et al.*, 1949; Tada *et al.*, 1990), oocytes (Nakagata, 1989; Tsunoda *et al.*, 1976; Whittingham, 1971, 1977; Whittingham *et al.*, 1972), embryos (Miyoshi *et al.*, 1992b; Whittingham, 1971; Whittingham *et al.*, 1972), ovary (Agca, 2000; Candy *et al.*, 2000; Cox *et al.*, 1996; Harp *et al.*, 1994; Migishima *et al.*, 2003; Sztein *et al.*, 1998), and testis (Ogonuki *et al.*, 2006; Ohta *et al.*, 2008). Many large animal resource centers, such as the JAX (http://cryo.jax.org/index.html; http://jaxservices.jax.org/cryopreservation/index.html), MMRRC (http://www.mmrrc.org/), and European Mouse Mutant Cell Repository (EuMMCR; http://www.eummcr.org/), are systematically archiving mutant mouse lines through cryopreservation. Many public and commercial repositories also offer custom cryopreservation services for individual investigators.

The cryopreserved embryos, gametes, or gonads can be conveniently shipped worldwide for distribution and reconstitution of mouse lines. Although sperm was successfully shipped at −80 °C (Okamoto *et al.*, 2001), it is recommended that cryopreserved material be shipped in liquid nitrogen dry shippers, such as Cryo Express (CX) series from Taylor-Wharton (http://www.taylorwharton.com/default.aspx?pageid=72) and Vapor Shipper containers from MVE (http://www.chartbiomed.com/p_vapor_shippers.cfm; http://www.cryodepot.com/shippers.htm). Dry shippers are designed to maintain the liquid nitrogen temperature at approximately −190 °C, yet to prevent the spill of liquid nitrogen if tipped over. It is important to completely saturate the absorbent material in the dry shipper with liquid nitrogen before put in the samples.

4.1. Freezing and shipping oocytes and preimplantation embryos

Preimplantation stage embryos from thousands of mouse lines have been cryopreserved, which can be shipped around the world to reconstitute live mice through embryo transfer. A major drawback for embryo cryopreservation is that a relatively large breeding colony is required in order to collect enough embryos (usually 200–500) to safely store and distribute the line. The cryopreservation methods can generally be categorized into two groups, the slow equilibration method and the rapid nonequilibration method, with corresponding thawing procedures. The slow method relies on the use of a controlled rate freezer to cool the embryos at a slow rate of ~1 °C/min. It is very reliable and the cryoprotectant used in this method is relatively nontoxic. On the other hand, the rapid nonequilibration method, commonly called vitrification, uses high concentration (>5 *M*) of cryoprotectant and an extremely fast (>1000 °C/min) cooling speed. It solidifies the solution inside the embryos into a glass state without forming ice crystals. The vitrification method is fast and does not require sophisticated instruments, and thus easy to adopt for laboratories not commonly doing cryopreservations.

4.1.1. Brief vitrification protocol

- At sender's laboratory:
  1. Wash freshly isolated embryos in PBS containing 1 *M* DMSO (Sigma D2650).
  2. Transfer embryos in 5 μl of 1 *M* DMSO solution to a freezing tube (NUNC 366656) and incubate at 0 °C (Nalgene Labtop Cooler 5115-0012) for 5 min.
  3. Add 45 μl cold cryoprotective solution DAP213 (PBS containing 2 *M* DMSO, 1 *M* acetamide, 3 *M* propylene glycerol) into the freezing tube and further incubate for 5 min on ice.
  4. Quickly fix the tube to a cane and plunge samples directly into liquid nitrogen in a Dewar container. Ship the tube containing the embryos in liquid nitrogen dry shipper through express mail.
- At receiver's laboratory:
  1. Take the tube out of the dry shipper and leave it at room temperature for 30 s.
  2. Add 0.9 ml PBS containing 0.25 *M* sucrose prewarmed at 37 °C into the freezing tube and thaw the samples quickly by pipetting. Transfer the contents into a 35 mm culture dish. For morullae before the blastocyst stage, pipet gently so that the blastomeres will not be dislodged.
  3. Wash the tube with the same solution and combine. Recover the embryos and transfer to M2 or KSOM medium. Wash after 10 min of incubation or 60 min for blastocysts.
  4. Transfer to recipient females according to protocols in Section 3.1.1.

Oocytes can be vitrified similarly by first free them from cumulus cells by hyaluronidase treatment, and incubate in Human Tubal Fluid (HTF) medium containing 20% FCS for 10–20 min to prevent zona hardening (George *et al.*, 1992). After thawing, follow protocols described in Section 3.2.1 for IVF.

4.2. Freezing and shipping sperm

With recent improvements in methodology, cryopreservation of spermatozoa is quickly becoming the favored method for archiving and distributing mouse lines. Unlike the cryopreservation of preimplantation embryos, which requires many donor mice, sperm cryopreservation requires only two breeding aged males. Therefore, much less "front-end" work is required for safely store and distribute mouse lines. Recent work by Ostermeier *et al.* (2008) at the JAX showed that sperm from all tested strains can be cryopreserved and revived although their efficiencies are still strain-dependent. One disadvantage of sperm cryopreservation is that only one haploid gamete is preserved, which must be combined with another haploid gamete (egg) in order to fully reconstitute the mouse line. This can be problematic if the full genome of a mouse line with extensively backcrossing or complicated breeding history is desired. IVF is needed at the receiving end to recover the strain. For strains with low recovery rate, assisted reproductive technologies (ART) such as PZD and ICSI can be used to increase efficiency. With ICSI, even freeze-dried sperm can be used to achieve fertilization, which may prove to be the most cost effective way to archive large number of mouse lines (Wakayama and Yanagimachi, 1998).

4.2.1. Brief protocols

- At sender's laboratory:
  1. Prepare cryoprotective medium (CPM) with distilled water (Invitrogen #15230-162) containing 18% raffinose (Sigma R7630), 3% skim milk (BD Diagnostics #232100), and 478 μM monothioglycerol (MTG; Sigma M6145). Warm the water to 40 °C to facilitate the dissolving of chemicals. Spin at 13,000 × g for 15 min to sediment debris, and filter sterilize supernatant with a 0.22 μm filter cup (Millipore, SCGPU02RE). CPM can be aliquoted and stored at −80 °C for up to 6 months.
  2. Euthanize two males, preferably 3–6 months old, that have been single housed for more than 48 h.
  3. Dissect out vasa deferentia and epididymides. Tear open epididymis and squeeze out semen from vas deferens in 1 ml CPM in a 3.5 cm petri dish.
  4. Incubated the dish at 37 °C for 10 min to allow sperm to swim out.
  5. Load the sperm suspension into labeled straws (IMV, AAA201, 005565) as 4 × 0.5 cm segments spaced by air with straw attached to a 1 ml Monoject syringe.
  6. Seal the straw with a heat sealer (American International Electric, AIE-305HD) and place them in labeled straw cassettes (Zanders Medical Supplies, 16980/0601).

7. Place the cassettes on top of a Styrofoam raft floating on liquid nitrogen (6–9 cm deep) in a ThermoSafe (Thermo Fisher, 11-676-13) for 10 min.
8. Plunge the cassettes in liquid nitrogen. Ship the straw in liquid nitrogen dry shipper through express mail.

- At receiver's laboratory:

The straw containing frozen sperm should be thawed quickly at 37 °C about 1 h prior to the scheduled IVF procedure. Perform IVF and embryo transfer according to the procedures described in Section 3.2.1 to revive the mouse lines.

4.3. Reviving mouse lines through transplantation of fresh or cryopreserved gonads

Orthotopic transplant of ovary had been done to produce live offspring (Robertson, 1940). Ovaries have been successfully frozen either by slow freezing (Candy *et al.*, 2000; Cox *et al.*, 1996; Gunasena *et al.*, 1997; Parrott, 1960; Sztein *et al.*, 1998) or vitrification (Kagabu and Umezu, 2000; Migishima *et al.*, 2003; Takahashi *et al.*, 2001), and transplanted after thawing into immunologically compatible recipient to generate live pups. Even postmortem ovaries were successfully cryopreserved and transplanted to rescue the mouse line (Takahashi *et al.*, 2001).

Spermatogonia were successfully transferred to produce mature spermatozoa (Brinster and Zimmermann, 1994) and transmitted through germ line (Brinster and Avarbock, 1994). Cryopreserved germline cells were later demonstrated to do the same (Avarbock *et al.*, 1996). If immature germ-cells have to be used for transfer, this spermatogonial transplant technique can be deployed. Recently, spermatozoa and spermatids retrieved from frozen testis and epididymis were used successfully in ICSI to produce normal offspring (Ogonuki *et al.*, 2006). The organs or even the whole body were simply frozen without any cryoprotectant. The authors found that potassium-rich Ca^{2+}- and Mg^{2+}-free nucleus isolation medium (NIM) medium always gave better results for ICSI. Similarly, spermatozoa retrieved from frozen testicular sections were also successfully used in ICSI to produce live born mice (Ohta *et al.*, 2008), confirming frozen testis as a valid source for spermatozoa to be used in ICSI, and providing another way to archive large number of mouse lines.

5. Conclusion

Sharing genetically modified mouse lines is becoming increasingly important for biomedical research. So far, shipping live animals is the most widely used method for sharing mouse models. Breeding pairs can be transported among different animal facilities after necessary paperwork and coordination. As more and more animal facilities are becoming capable of performing embryo transfers and IVF procedures, shipping fresh or frozen embryos/gametes is becoming an important alternative method, particularly for international transfers where shipment of live animals can be cumbersome. Coordinated large-scale efforts are currently underway to cryopreserve and bank essentially all genetically modified mouse lines. With the recent improvements in methods, sperm cryopreservation is becoming the favored method for archiving mouse lines. Therefore, it is anticipated that shipping cryopreserved sperm will become more popular for distributing mouse lines in the future. At the present time, the efficiency and reliability of mouse sperm cryopreservation and IVF are generally acceptable, but they still depend upon the genetic background of the mice, and demand the involved personnel to strictly follow the specified experimental conditions and timelines. Consequently, experienced staff with proven equipment setup is required in order to reliably recover the mouse line. Hopefully, further improvement in mouse sperm cryopreservation and IVF methods will reduce the technical challenge and make this method more robust and dependable.

ACKNOWLEDGMENT

This work was supported by the Intramural Research Program of NIH, NHLBI.

REFERENCES

Agca, Y. (2000). Cryopreservation of oocyte and ovarian tissue. *ILAR J.* **41,** 207–220.

Avarbock, M. R., Brinster, C. J., and Brinster, R. L. (1996). Reconstitution of spermatogenesis from frozen spermatogonial stem cells. *Nat. Med.* **2,** 693–696.

Brinster, R. L., and Avarbock, M. R. (1994). Germline transmission of donor haplotype following spermatogonial transplantation. *Proc. Natl. Acad. Sci. USA* **91,** 11303–11307.

Brinster, R. L., and Zimmermann, J. W. (1994). Spermatogenesis following male germ-cell transplantation. *Proc. Natl. Acad. Sci. USA* **91,** 11298–11302.

Candy, C. J., Wood, M. J., and Whittingham, D. G. (2000). Restoration of a normal reproductive lifespan after grafting of cryopreserved mouse ovaries. *Hum. Reprod.* **15,** 1300–1304.

Cox, S. L., Shaw, J., and Jenkin, G. (1996). Transplantation of cryopreserved fetal ovarian tissue to adult recipients in mice. *J. Reprod. Fertil.* **107,** 315–322.

Fuller, S. J., and Whittingham, D. G. (1997). Capacitation-like changes occur in mouse spermatozoa cooled to low temperatures. *Mol. Reprod. Dev.* **46,** 318–324.

George, M. A., Johnson, M. H., and Vincent, C. (1992). Use of fetal bovine serum to protect against zona hardening during preparation of mouse oocytes for cryopreservation. *Hum. Reprod.* **7,** 408–412.

Gordon, J. W., and Talansky, B. E. (1986). Assisted fertilization by zona drilling: A mouse model for correction of oligospermia. *J. Exp. Zool.* **239,** 347–354.

Gunasena, K. T., Villines, P. M., Critser, E. S., and Critser, J. K. (1997). Live births after autologous transplant of cryopreserved mouse ovaries. *Hum. Reprod.* **12,** 101–106.

Harp, R., Leibach, J., Black, J., Keldahl, C., and Karow, A. (1994). Cryopreservation of murine ovarian tissue. *Cryobiology* **31,** 336–343.

Kagabu, S., and Umezu, M. (2000). Transplantation of cryopreserved mouse, Chinese hamster, rabbit, Japanese monkey and rat ovaries into rat recipients. *Exp. Anim.* **49,** 17–21.

Kasai, M., Niwa, K., and Iritani, A. (1983). Protective effect of sucrose on the survival of mouse and rat embryos stored at 0 degree C. *J. Reprod. Fertil.* **68,** 377–380.

Kawase, Y., Iwata, T., Ueda, O., Kamada, N., Tachibe, T., Aoki, Y., Jishage, K., and Suzuki, H. (2002). Effect of partial incision of the zona pellucida by piezo-micromanipulator for in vitro fertilization using frozen-thawed mouse spermatozoa on the developmental rate of embryos transferred at the 2-cell stage. *Biol. Reprod.* **66,** 381–385.

Landel, C. P. (2005). Archiving mouse strains by cryopreservation. *Lab. Anim. (NY)* **34,** 50–57.

Markert, C. L. (1983). Fertilization of mammalian eggs by sperm injection. *J. Exp. Zool.* **228,** 195–201.

Migishima, F., Suzuki-Migishima, R., Song, S. Y., Kuramochi, T., Azuma, S., Nishijima, M., and Yokoyama, M. (2003). Successful cryopreservation of mouse ovaries by vitrification. *Biol. Reprod.* **68,** 881–887.

Miyoshi, I., Ishikawa, K., Kasai, M., and Kasai, N. (1992a). A practical transport system for mouse embryos cryopreserved by simple vitrification. *Lab. Anim. Sci.* **42,** 323–325.

Miyoshi, I., Ishikawa, K., Kasai, M., and Kasai, N. (1992b). Useful short-range transport of mouse embryos by means of a nonfreezing technique. *Lab. Anim. Sci.* **42,** 198–201.

Mochida, K., Ohkawa, M., Inoue, K., Valdez, D. M., Jr., Kasai, M., and Ogura, A. (2005). Birth of mice after in vitro fertilization using C57BL/6 sperm transported within epididymides at refrigerated temperatures. *Theriogenology* **64,** 135–143.

Nakagata, N. (1989). High survival rate of unfertilized mouse oocytes after vitrification. *J. Reprod. Fertil.* **87,** 479–483.

Ogonuki, N., Mochida, K., Miki, H., Inoue, K., Fray, M., Iwaki, T., Moriwaki, K., Obata, Y., Morozumi, K., Yanagimachi, R., and Ogura, A. (2006). Spermatozoa and spermatids retrieved from frozen reproductive organs or frozen whole bodies of male mice can produce normal offspring. *Proc. Natl. Acad. Sci. USA* **103,** 13098–13103.

Ohta, H., Sakaide, Y., and Wakayama, T. (2008). The birth of mice from testicular spermatozoa retrieved from frozen testicular sections. *Biol. Reprod.* **78,** 807–811.

Okamoto, M., Nakagata, N., and Toyoda, Y. (2001). Cryopreservation and transport of mouse spermatozoa at −79 degrees C. *Exp. Anim.* **50,** 83–86.

Ostermeier, G. C., Wiles, M. V., Farley, J. S., and Taft, R. A. (2008). Conserving, distributing and managing genetically modified mouse lines by sperm cryopreservation. *PLoS ONE* **3,** e2792.

Parrott, D. M. V. (1960). The fertility of mice with orthotopic ovarian grafts derived from frozen tissue. *J. Reprod. Fertil.* **1,** 230–241.

Pickering, S. J., and Johnson, M. H. (1987). The influence of cooling on the organization of the meiotic spindle of the mouse oocyte. *Hum. Reprod.* **2,** 207–216.

Polge, C., Smith, A. U., and Parkes, A. S. (1949). Revival of spermatozoa after vitrification and dehydration at low temperatures. *Nature* **164,** 666.

Ribeiro, B. I., Love, L. B., Choi, Y. H., and Hinrichs, K. (2008). Transport of equine ovaries for assisted reproduction. *Anim. Reprod. Sci.* **108,** 171–179.

Robertson, G. (1940). Ovarian transplantation in the house mouse. *Proc. Soc. Exp. Biol. Med.* **44,** 302–304.

Sankai, T., Tsuchiya, H., and Ogonuki, N. (2001). Short-term nonfrozen storage of mouse epididymal spermatozoa. *Theriogenology* **55,** 1759–1768.

Sato, M., Ishikawa, A., Nagashima, A., Watanabe, T., Tada, N., and Kimura, M. (2001). Prolonged survival of mouse epididymal spermatozoa stored at room temperature. *Genesis* **31,** 147–155.

Schroeder, A. C., Johnston, D., and Eppig, J. J. (1991). Reversal of postmortem degeneration of mouse oocytes during meiotic maturation in vitro. *J. Exp. Zool.* **258,** 240–245.

Shaw, J. M., and Nakagata, N. (2002). Cryopreservation of transgenic mouse lines. *Methods Mol. Biol.* **180,** 207–228.

Sztein, J., Sweet, H., Farley, J., and Mobraaten, L. (1998). Cryopreservation and orthotopic transplantation of mouse ovaries: New approach in gamete banking. *Biol. Reprod.* **58,** 1071–1074.

Tada, N., Sato, M., Yamanoi, J., Mizorogi, T., Kasai, K., and Ogawa, S. (1990). Cryopreservation of mouse spermatozoa in the presence of raffinose and glycerol. *J. Reprod. Fertil.* **89,** 511–516.

Tadir, Y., Wright, W. H., Vafa, O., Liaw, L. H., Asch, R., and Berns, M. W. (1991). Micromanipulation of gametes using laser microbeams. *Hum. Reprod.* **6,** 1011–1016.

Takahashi, E., Miyoshi, I., and Nagasu, T. (2001). Rescue of a transgenic mouse line by transplantation of a frozen-thawed ovary obtained postmortem. *Contemp. Top. Lab. Anim. Sci.* **40,** 28–31.

Tsunoda, Y., Parkening, T. A., and Chang, M. C. (1976). In vitro fertilization of mouse and hamster eggs after freezing and thawing. *Experientia* **32,** 223–224.

Van Keuren, M. L., and Saunders, T. L. (2004). Rederivation of transgenic and gene-targeted mice by embryo transfer. *Transgenic Res.* **13,** 363–371.

Wakayama, T., and Yanagimachi, R. (1998). Development of normal mice from oocytes injected with freeze-dried spermatozoa. *Nat. Biotechnol.* **16,** 639–641.

Whittingham, D. G. (1971). Survival of mouse embryos after freezing and thawing. *Nature* **233,** 125–126.

Whittingham, D. G. (1977). Fertilization in vitro and development to term of unfertilized mouse oocytes previously stored at −196 degrees C. *J. Reprod. Fertil.* **49,** 89–94.

Whittingham, D. G., Leibo, S. P., and Mazur, P. (1972). Survival of mouse embryos frozen to −196 degrees and −269 degrees C. *Science* **178,** 411–414.

Yoshida, N., and Perry, A. C. (2007). Piezo-actuated mouse intracytoplasmic sperm injection (ICSI). *Nat. Protoc.* **2,** 296–304.

CHAPTER FOUR

Archiving and Distributing Mouse Lines by Sperm Cryopreservation, IVF, and Embryo Transfer

Hideko Takahashi* *and* Chengyu Liu†

Contents

Abstract

The number of genetically modified mouse lines has been increasing exponentially in the past few decades. In order to safeguard them from accidental loss and genetic drifting, to reduce animal housing cost, and to efficiently distribute them around the world, it is important to cryopreserve these valuable genetic resources. Preimplantation-stage embryos from thousands of mouse lines have been cryopreserved during the past two to three decades. Although reliable, this

* Veterinary Research & Resource Section, and Genetics Engineering Facility, National Eye Institute, National Institutes of Health, Bethesda, Maryland, USA

† Transgenic Core Facility, National Heart, Lung, and Blood Institute, National Institutes of Health, Bethesda, Maryland, USA

Methods in Enzymology, Volume 476

ISSN 0076-6879, DOI: 10.1016/S0076-6879(10)76004-3

method requires several hundreds of embryos, which demands a sizable breeding colony, to safely preserve each line. This requirement imposes significant delay and financial burden for the archiving effort. Sperm cryopreservation is now emerging as the leading method for storing and distributing mouse lines, largely due to the recent finding that addition of a reducing agent, monothioglycerol, into the cryoprotectant can significantly increase the *in vitro* fertilization (IVF) rate in many mouse strains, including the most widely used C57BL/6 strain. This method is quick, inexpensive, and requires only two breeding age male mice, but it still remains tricky and strain-dependent. A small change in experimental conditions can lead to significant variations in the outcome. In this chapter, we describe in detail our sperm cryopreservation, IVF, and oviduct transfer procedures for storing and reviving genetically modified mouse lines.

1. Introduction

During the past three decades, the number of genetically modified mouse models for biomedical research has been growing at an exponential rate. Currently, the worldwide knockout mouse project (www.komp.org) is in the process of generating at least one null mouse line for essentially every mouse gene. Given the high cost of maintaining live animals and the fact that not all mouse lines are actively being used for research at a given time, it is important to cryopreserve some of these valuable animal models. The method for cryopreserving mouse embryos was first developed in the early 1970s by two independent laboratories (Whittingham *et al.*, 1972; Wilmut, 1972). Since then, thousands of mouse lines have been successfully cryopreserved as preimplantation-stage embryos in centralized animal resource centers, such as Jackson Laboratory (JAX), and in various core facilities, as well as individual research laboratories (Mazur *et al.*, 2008).

Generally speaking, cryopreservation of preimplantation-stage mouse embryos is efficient and reliable for banking mouse lines. Live mice can be recovered through the well-established embryo transfer procedures. However, one major drawback for embryo cryopreservation is that usually several hundreds of preimplantation-stage embryos are needed for safely cryopreserving a mouse line, and unfortunately many genetically modified mouse lines do not respond to superovulation well for yielding large number of embryos. Therefore, a relatively large breeding colony needs to be established first, which is not only time consuming, but also imposes significant financial burden up front (Mazur *et al.*, 2008). Because of the limitation on animal space and research budgets, embryo cryopreservation is not likely to be able to keep pace with the large number of mouse lines that are being generated each year. Cryopreservation of the male gametes, spermatozoa, can potentially overcome this problem because each suitable male can yield enough sperm to reconstitute a large number of offspring. The cryopreservation of human and live stock semen has been a common practice for several decades, but mouse sperm has remained difficult to freeze for quite some time. The introduction of 18% raffinose and 3% skim milk as a cryoprotective agent (CPA) is a significant breakthrough for mouse sperm cryopreservation (Kaneko *et al.*, 2006; Nakagata, 2000), although its success is still heavily dependent on the genetic background of the mice. Unfortunately, its efficiency is unacceptable for many important inbred strains, including the most popular mouse strain, C57BL/6 (Nakagata and Takeshima, 1993; Nishizono *et al.*, 2004; Songsasen and Leibo, 1997; Sztein *et al.*, 2000). Assisted reproductive techniques, such as partial zona dissection (PZD) (Anzai *et al.*, 2006; Kawase *et al.*, 2002; Nakagata *et al.*, 1997) and intracytoplasmic sperm injection (ICSI) (Kimura and Yanagimachi, 1995; Sakamoto *et al.*, 2005; Szczygiel *et al.*, 2002a), are often required for recovering the lines. Various attempts have been made to improve the IVF rate from cryopreserved inbred sperm (Bath, 2003; Liu *et al.*, 2009; Ostermeier *et al.*, 2008; Suzuki-Migishima *et al.*, 2009; Szczygiel *et al.*, 2002b; Taguma *et al.*, 2009; Takeo *et al.*, 2008), but the method that attracted the most attention is the discovery that inclusion of a reducing agent, more specifically monothioglycerol (MTG), in the freezing medium can substantially increase the postthaw IVF rate in many inbred strains including the C57BL/6 strain (Ostermeier *et al.*, 2008). This methodological advancement together with the Jackson Laboratory's efforts to educate (JAX Cryopreservation Workshop) and assist (JAX Sperm Cryopreservation Kit) has facilitated the rapid adoption of sperm cryopreservation as the preferred method for archiving and distributing genetically modified mice.

The mouse sperm cryopreservation method is simple, fast, and does not require sophisticated instrumentation, such as a controlled rate freezer, but this does not mean it is trivial. Small variations in experimental conditions, such as timing, temperature, and concentration, can dramatically change the outcome. The successfulness of sperm freezing experiments is commonly measured by the efficiency of thawed sperm in reconstituting the mouse lines through IVF and subsequent embryo transfer procedures. Therefore, in this chapter, we describe the methods for sperm cryopreservation, IVF, and embryo transfer used in our laboratories, which are largely based on the protocols developed by the Jackson Laboratory.

2. Sperm Cryopreservation

2.1. Preparing cryoprotective agent

1. Add 70 ml ultra pure water into a flask and heat to 37–50 °C on a stirrer/hot plate. Slowly add 18 g raffinose pentahydrate (Sigma Cat#: R-7630). Stir with a magnetic stir bar till the solution becomes clear.

2. Add 3 g of Difco dry skim milk (BD Diagnostics, Cat#: 232100) and stir till the dry milk is dissolved. Add more ultra pure water to bring the final volume to 100 ml.
3. Centrifuge at 13,000 rpm in a Beckman Coulter JA25.50 rotor for 60 min at 4 °C. Filter the supernatant through a sterile 0.22 μm filter unit (Millipore, Cat#: SCGPUO2RE).
4. Divide the filtered CPA into ten 15 ml disposable conical tubes, each containing 10 ml. Aliquoted CPA can be stored in a −80 °C freezer for 6 months. Once thawed, the CPA can be stored at 4 °C for 1 week.
5. Frozen CPA can be thawed by submerging the tubes in a 37 °C water bath. Before use, α-monothioglycerol (MTG, Sigma, Cat#: M6145) should be added at a final concentration of 477 μM.

2.2. Preparing cryopreservation straws

1. The 133-mm-long plastic straws from IMV Technologies, Inc. (Maple Grove, MN, Cat#: AAA201) are used for sperm cryopreservation. Before use, the cotton/PVA plug near one end (Fig. 4.1A) of the straw should be pushed inward till it is ~4.5 cm from the end, using a thin plastic rod (Piston Plastique from IMV, Ref. No. 7296). This 4.5 cm of

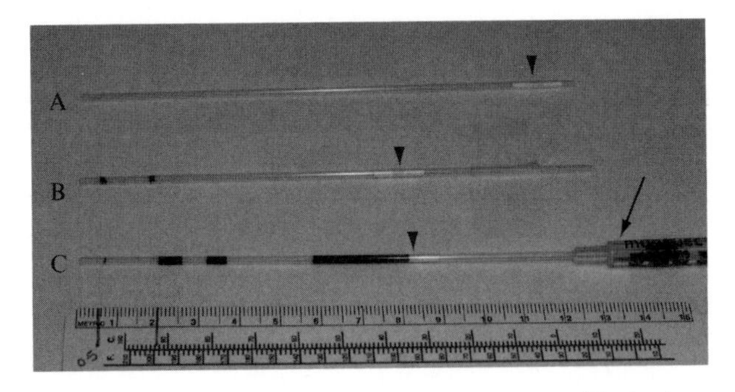

Figure 4.1 Preparing and loading cryopreservation straws. (A) A picture of a cryopreservation straw manufactured by IMV Technologies, Inc. (B) The cotton/PVA plug (arrow head) is pushed inward till it is ~4.5 cm from the end. Then, two lines are drawn using a marker on the straw, which are ~0.5 and ~2.0 cm from the open end, respectively. (C) Loading of CPA and sperm samples. A dark-colored solution is used in this example for easy visualization. Using the attached plastic syringe (arrow), a 2.0-cm-long segment of CPA is first drawn, and then a 2.0-cm-long segment of air is drawn into the straw. Next, two 0.5-cm-long segments of sperm samples, separated by a 0.5 cm segment of air, are drawn. Finally, more air is drawn into the straw till the first segment of CPA wets the cotton/PVA plug. The straw is now ready for sealing. Each 0.5 cm segment of sperm sample corresponds to 10 μl of sample, which is enough for one IVF dish.

space is necessary for labeling and sealing the straw. It is important to use labels and pens that can withstand long-term storage in liquid nitrogen (LN_2), such as the Brady labels (Brady Serve, Cat#: LAT-PTL 19-427).

2. At approximately 0.5 and 2.0 cm from the open end of the straw, draw two lines with a marker (Fig. 4.1B). Insert the plugged end of the straw into a 1 cc disposable syringe (Kendall, Ref. No. 8881501400), so that liquid can be drawn in or expelled out of the straw by pulling or pushing the plunger of the syringe, respectively (Fig. 4.1C).
3. Draw CPA/MTG solution into the straw until the solution reaches the 2.0 cm line. Then, draw a segment of air into the straw till the air reaches the 2.0 cm line. Place these straws on a clean surface until ready for loading sperm sample.

2.3. Collecting and loading sperm samples

1. Euthanize two male mice with proven breeding track records, preferably 2–6-month-old and that have been housed alone for 1 week. Quickly open the abdominal skin and body wall with a pair of scissors. As shown in Fig. 4.2, locate and expose the internal reproductive organs. Dissect out the cauda epididymis and vas deferens using a pair of watchmaker's No. 5 forceps (Roboz Surgical Instrument Co., Cat#: RS-4905) and a pair of dissecting scissors (Roboz, Cat#: RS-5882). Try to eliminate as much associated adipose, vascular, and other tissues as possible.
2. Place all four dissected vas deferentia and epididymides (two from each mouse) into a 35 mm culture dish containing 0.7 ml of 37 °C CPA/MTG. Under a stereomicroscope with transmitted light source, use a pair of watchmaker's forceps and a 26-gauge hypodermic needle to tear open the cauda epididymis by piercing the tissue several times. Semen inside each vas deferens can be expelled by gently pressing the tubule with the needle to slowly "walk" the semen out. For maximum sperm quality, the above two steps should be completed within 10 min.
3. Gently tap and swirl the dish for approximately 30 s and then leave the dish tilted as shown in Fig. 4.2C for 10 min at room temperature to allow sperm to swim out of the tissues. The dish should remain covered with a lid to minimize medium evaporation.
4. Assess the concentration and motility of the sperm under a high-magnification stereomicroscope. Discard the larger pieces of tissues and gently swirl the dish to mix the sperm suspension evenly.
5. Dip the open end of the straws prepared in Section 2.2 into the sperm suspension, and slowly draw in the sample until reaching the 0.5 cm mark. Then lift the straw and draw in air until reaching the 0.5 cm mark. A second aliquot, or even more aliquots, of sperm

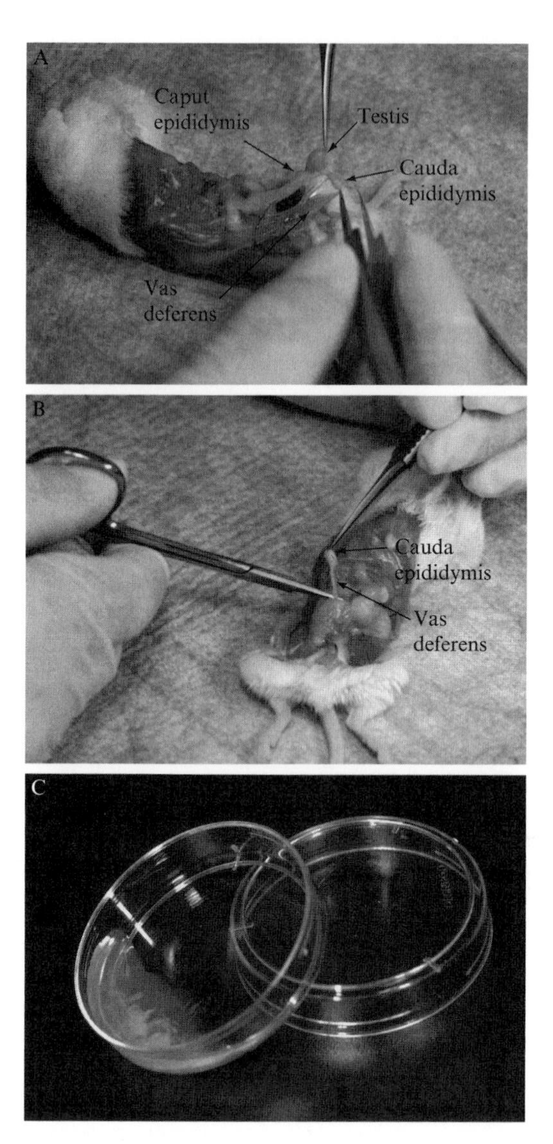

Figure 4.2 Collection of mouse sperm. (A) Exposed male internal reproductive organs, showing the testis, cauda epididymis, caput epididymis, and vas deferens. (B) Removing of the cauda epididymis and a portion of the vas deferens using scissors and forceps. (C) A sperm collection dish with torn epididymis and squeezed vas deferens in CPA.

sample can be drawn into the straw using the same method. Each 0.5-cm-long segment of sample corresponds to $\sim 10\ \mu l$, which is enough for one fertilization dish (see Section 4.2).

6. After finishing loading the sample, air is drawn into the straw until the first segment of CPA wets the cotton/PVA plug (Fig. 4.1C). Dry the outside of the straw with Kimwipes. Seal the open end of the straw using a heat sealer (American International Electric Inc., Model AIE—310 HD). Gently detach the straw from the syringe, and then seal the plugged end.

2.4. Freezing sperm samples

1. Dedicated LN_2 freezing box, ThermoSafe (Thermo Fisher, Cat#: 11-676-13) is available, but we still use a medium sized Styrofoam box. Before each use, we fill the box with about 9-cm-deep LN_2, and then place a 3.5-cm-thick Styrofoam board into the box, which should float on the LN_2 surface (Fig. 4.3A).
2. Lay the straws loaded with sperm on top of the Styrofoam raft (Fig. 4.3A). Cover the box with the matching Styrofoam lid and leave the box undisturbed for 10 min. Any straws that accidentally contacted LN_2 prior to the 10 min of gradual cooling should be discarded.
3. Open the lid of the Styrofoam box and quickly plunge the straws into the LN_2. Dip the metal cane with attached plastic goblet (Visoture Rond 13 mm Blanc, IMV, Ref. No. 006420) into the LN_2, and quickly transfer the frozen straws into the goblet using a pair of long forceps. To prevent the straws from falling out, a second goblet is inserted in the up-side-down orientation into the cane. As indicated in Fig. 4.3B, holes should be made in the closed end of the goblet to allow air to come out and LN_2 enter the goblets during storage. It is noteworthy that cassettes (Zanders Medical Supplies, Cat#: 16980/0601) specifically made for holding straws are available. Sealed straws can be loaded into these cassettes before placing on the Styrofoam raft for cooling.

3. Storing and Shipping Frozen Sperm Samples

The straws can be maintained in a LN_2 freezer for long-term storage. For each strain, the straws should be divided into two different canes and stored in two different LN_2 freezers preferably at two separate locations. Ideally, one set of samples is stored in your facility for easy access, while the other set is stored some distance away to reduce the possibility of losing all samples due to unpredictable disasters. It is highly recommended to store the second set of samples in a public or commercial repository, such as KamTech, Inc. (Gaithersburg, MD), JAX (Bar Harbor, ME), or MMRRC. Such repositories have reliable monitoring system, emergency power and supplies, and catastrophe-prevention measures to safeguard your valuable samples.

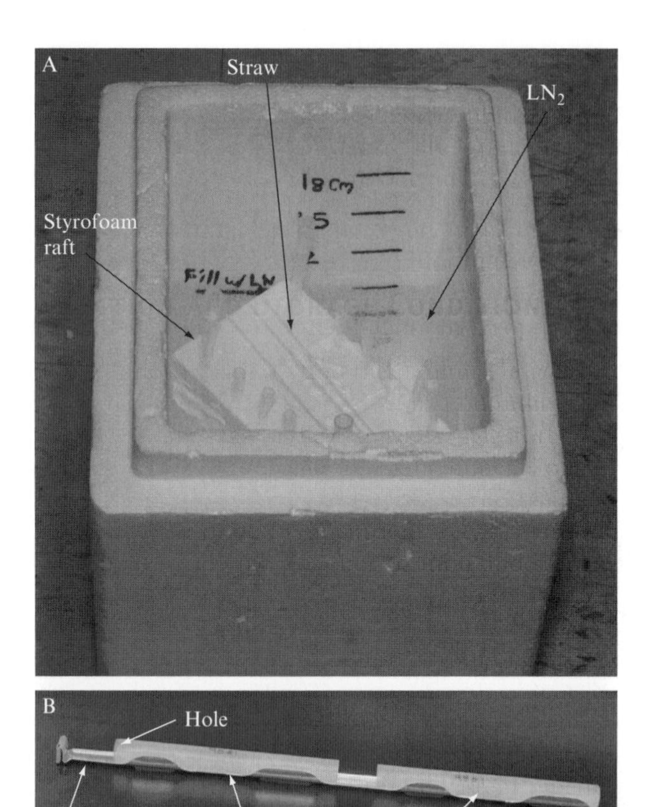

Figure 4.3 (A) Cooling of loaded straws in a Styrofoam box. The Styrofoam box must be covered with a matching lid (not shown) during the cooling period. (B) A metal cane with attached plastic goblets. After freezing, the straws are placed inside the bottom goblet, and then the top goblet is sliding down to completely enclose the straws, to prevent the straws from falling out. Holes are made on the closed end of the top goblet to let air out and LN_2 into the goblet.

Frozen sperm can be shipped around the world for reconstitution of mouse lines through IVF or artificial insemination (AI). It is strongly advised to keep the frozen sperm at LN_2 temperature at all times, even during transportation. Because LN_2 is a regulated hazardous material for transportation, an LN_2 dry shipper (Fig. 4.4) is often used for shipping cryopreserved samples. Dry shippers will not spill LN_2 if tipped over during transportation, but they can maintain the LN_2 temperature for up to 2 weeks if charged correctly. It is important to follow the manufacturer's instruction when filling the dry shipper with LN_2. Protective gloves and a face shield should be worn to prevent accidental injuries. The absorbent material in the dry

Figure 4.4 A photograph of an LN_2 dry shipper and its matching container for shipping.

shippers should be fully saturated with LN_2, but free flowing LN_2 should be poured out of the dry shipper before loading and shipping samples. When loading and unloading the dry shipper, it is critical to minimize the exposure of straws to non-LN_2 temperature. The lid/stopper of the dry shipper should be securely closed and fastened with a zip tie. Avoid wraping and boxing the shipper. Ideally it should be shipped in its matching shipping container (Fig. 4.4).

4. *In Vitro* Fertilization

4.1. Preparing egg donor mice

1. Three days prior to the scheduled IVF procedure, inject (i.p.) the egg donor mice with pregnant mare serum gonadotropin (PMSG, Sigma # G-4877). The number of females needed is dependent on the genetic background, age, as well as how many offspring are needed. For some strains, newly weaned young mice (19–24 days) are best, but for some other strains, mature females (6–8 weeks) are more consistent. The optimal dose of PMSG also varies depending on strain, age, and body weights, but in most cases 5 IU dissolved in 0.1 ml physiological saline (0.9% NaCl) is a good starting concentration. Avoid repeated thaw and freeze of the PMSG solution. Typically 5–15 usable eggs can

be expected from each superovulated female. We routinely inject PMSG late in the afternoon (6–8 p.m.) to avoid coming to work too early on the day of IVF.

2. On the day before the scheduled IVF procedure, 46–48 h after the PMSG injection, inject (i.p.) the same females with 5 IU of human chorionic gonadotropin (hCG, Sigma # C-1063) dissolved in 0.1 ml physiological saline.

4.2. Preparing culture and IVF dishes

Because IVF procedures usually begin early in the morning (13 h after hCG injection) and it takes time for the media, under mineral oil, to fully equilibrate with the gas contents in the incubator, we normally set up the culture and IVF dishes on the day before the scheduled IVF. Each IVF dish (Fig. 4.5) is made by adding one 250 μl and three 100 μl drops of Cook's IVF medium (Cooks Cat#: K-RVFE-50) into a 60 mm dish, and then quickly covering the media drops with embryo-tested mineral oil (Sigma Cat#: M8410). Each egg collection dish is made by adding 400 μl of IVF medium into a 35 mm tissue culture dish and then quickly covering the medium drop with mineral oil. All dishes should be placed in a 37 °C incubator with 5% CO_2, 5% O_2, and 90% N_2 (Robert Oxygen premixed gas, R0133502001) until use.

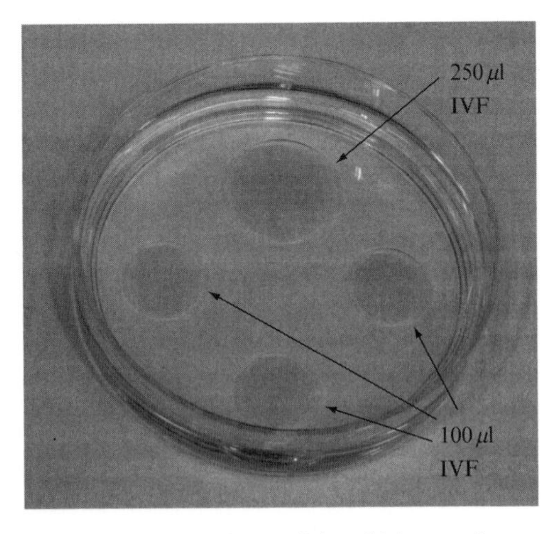

Figure 4.5 A photograph of a typical IVF dish, which contains one 250 μl and three 100 μl drops of IVF media cover under mineral oil.

4.3. Thawing cryopreserved sperm

1. Take the sperm-containing straw out of the LN_2 freezer using a pair of forceps, and immediately submerge the straw into a 37 °C water bath until all ice inside the straw disappears (20–40 s). Dry the straw with Kimwipes.
2. Cut the straw through the middle of the cotton/PVA plug using a pair of sharp scissors. Then, cut the unplugged (open) end of the straw near the sealed tip. Disinfect the open end with an alcohol wipe and then leave in air for a couple of minutes to allow the alcohol to evaporate.
3. Dip the disinfected end into the 250 μl drop in the IVF dish (Fig. 4.5), and then using a thin plastic rod (Piston Plastique from IMV, Ref. No. 7296), slowly push the cotton/PVA plug inward to dispense one segment of sperm sample ($\sim$10 μl) into the IVF medium. The other segment(s) of sperm sample contained in the same straw can be dispensed into other IVF dish(s) for fertilizing different batch(es) of eggs.
4. Incubate the IVF dishes in a 37 °C incubator containing 5% O_2, 5% CO_2, and 90% N_2 for $\sim$60 min before adding eggs for fertilization.

4.4. Collecting eggs and IVF

1. Thirteen hours after hCG injection (see Section 4.1), euthanize the superovulated females and open the abdominal cavity using a pair of scissors. Dissect out both oviducts from each female using a pair of watchmaker's forceps and dissecting scissors. To avoid damaging the oviducts, the ovary as well as a small piece of uterine horn can also be dissected.
2. Place the dissected oviducts into the 35 mm dish containing 400 μl of IVF medium prepared in Section 4.2. Using the fine forceps to hold the oviduct in place, and then use a 29-gauge needle to tear open the ampulla, which is the swollen and translucent segment of the oviducts. The cumulus mass, which usually contains several eggs surrounded by follicular cells, will extrude spontaneously into the media. Cumulus masses from up to five mice can be collected into the same dish, and each oviduct should be discarded immediate after processing.
3. Using a P1000 pipetman setting at 40–50 μl, carefully aspirate the cumulus masses with minimum volume of media and transfer them into the IVF medium drop that already contains the capacitated sperm (see Section 4.3). Alternatively, the eggs can be released directly from the ampulla into the IVF medium drop containing the incubated sperm.
4. Incubate the sperm and egg mix for 4–6 h in the incubator to allow fertilization to take place. During this incubation, the IVF dish can be taken out of the incubator briefly for monitoring the IVF process under a

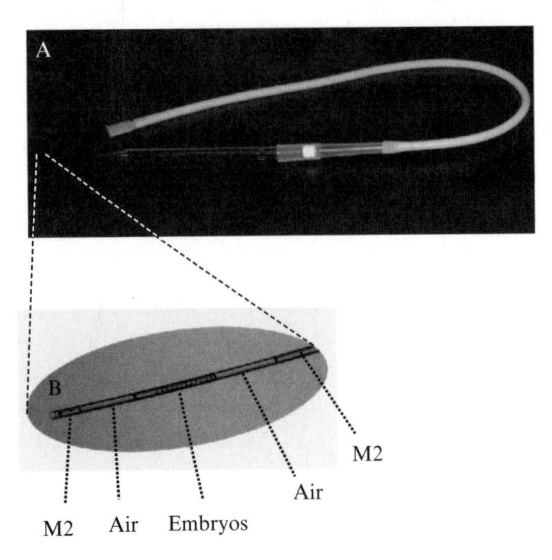

Figure 4.6 (A) A photograph of a mouth pipette for transferring embryos. It is assembled by connecting a pulled long glass Pasteur pipette through a filtered 1000 μl pipette tip and a rubber tubing to a plastic mouthpiece. (B) Microphotograph of the tip of the embryo transfer pipette with loaded embryos and air bubbles.

microscope, but avoid leaving the dishes outside the incubator for more than a few minutes. The eggs should be moving or spinning around because of the swimming of the bound sperm.

5. After the incubation, the cumulus masses usually dissemble and the follicle cells are no longer attached to the eggs. Therefore, individual eggs can be picked up and transferred using a mouth-controlled pipette (Fig. 4.6A), which is assembled by connecting the mouthpiece and rubber aspirator tubing (Sigma, Cat#: A5177) through a filtered 1000 μl pipette tip to a pulled long-tipped glass Pasteur pipette. Wash the eggs through two of the 100 μl drops of IVF media to eliminate debris and extra sperm, and leave them in the last (third) drop of medium.
6. Return the IVF dish into a 37 °C incubator with 5% CO_2 for culturing overnight.

5. Oviduct Transfer

1. On the day of IVF, select female Swiss Webster mice (Taconic Farm) with swollen and reddish genital areas, which indicates they are in estrus, and pair them up with vasectomized male mice for mating.

2. Check the mice for vaginal plugs the next morning. Mice with visible plugs can be used in Step 4 as surrogate mothers.
3. Examine the IVF dish using a stereomicroscope. Count the number of embryos that have reached the two-cell stage of development, and transfer them into an M2 drop covered under mineral oil.
4. Inject (i.p.) 2.5% Avertin solution at a dose of 0.017 ml/g of body weight to anesthetize the surrogate mothers. The Avertin solution is made by first dissolving 5.0 g of 2,2,2-tribromoethanol (Aldrich, Milwaukee, WI) in 5 ml *tert*-amyl alcohol (Aldrich), and then adding 195 ml of isotonic saline (0.9% NaCl solution). Aliquoted Avertin solution can be stored at −20 °C in the dark.
5. Load 10 two-cell stage embryos into the mouth operated transfer pipette. As shown in Fig. 4.6B, two air bubbles are sucked into the transfer pipette to mark the boundaries of the suspended embryos.
6. Check the depth of anesthesia at ~5 min after Avertin injection by gently pinching one of the hind paws. A supplemental dose of Avertin, usually a third of the initial dose, should be injected if the animal can still respond to this gentle pinch.
7. Remove the hair from a generous area on the dorsal lumbar back using a small hair clipper. Disinfect the clipped area by alternating applications of Betadine (Medline Industries, Mundelein, IL) and 70% alcohol.
8. Cut a dorsal midline incision (~1 cm long) in the cleaned skin area using a pair of sterilized scissors (Fig. 4.7). Move the skin incision to either side of the para-lumbar using a pair of iris forceps (Roboz Surgical Instrument Co., Cat#: RS-5130) to find the ovary and associated fat pad, which are slightly paler than the surrounding internal organs.
9. Make a small incision in the muscle body wall and then extending the cut to 5–10 cm long using the back of the scissors' blades. Using a pair of iris forceps grasp the fat pad and carefully pull out the fat pad, the ovary, the oviduct, and a segment of the uterine horn through the incision. Clamp the fat pad with a Dieffenbach clip (Roboz, Cat#: RS-7422) to hold the organs in place (Fig. 4.7).
10. Under a dissecting microscope, find the ampulla, which is the swollen and translucent segment of the oviducts, and usually can be found a few millimeters from the infundebulum. Use a 26-gauge needle to punch a hole in the oviduct wall between the ampulla and the ovary.
11. Insert the tip of the transfer pipette into the hole with the pipette opening pointing toward the ampulla. Gently blow into the mouthpiece to expel the two air bubbles and the embryos sandwiched between them. The presence of two air bubbles in the ampulla is a good indication that all embryos have been successfully implanted.
12. Carefully push the reproductive organs back into the abdominal cavity, and sew the abdominal wall incision with a cruciate suture utilizing

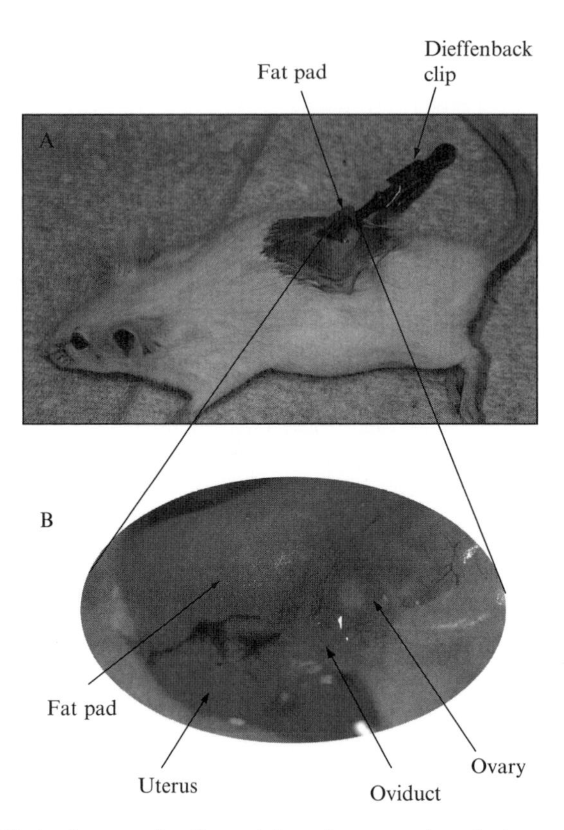

Figure 4.7 (A) A photograph of a recipient foster mother during embryo transfer procedures. The reproductive organs were pulled out through an incision on the back. A Dieffenbach clip is clamped on the fat pad to hold the reproductive organs in place. (B) Magnified view of the reproductive organs.

absorbable 5-0 Vicryl suture (Ethicon, Sommerville, NJ). Using a 1 cc disposable syringe with an attached 26-gauge needle, drop one or two drops of 0.25% Bupivacaine solution, which is a long-lasting local anesthetic, on the muscle at the surgical site. Finally, close the skin incision with two stainless steel surgical wound clips (Roboz, Cat#: RS-9260).

13. To aid the recovery, place the mice in a cage warmed by a circulating water blanket (GayMar Industries, Orchard Park, NY) or a carefully adjusted infra-red heating lamp (Bel-art Products, Pequannock, NJ).
14. After the mice wake up and start to move around, usually 30–60 min post-Avertin injection, return them to the animal room.

15. Remove the wound clip after 10–14 days, and pups will be born about 19 days post embryo transfer surgery. The offspring can be weaned, ear-tagged, and tail-biopsied when they are 18–21 days old.

6. Conclusions

Mouse line cryopreservation is not only important for saving animal space and research budget but also for safeguarding important strains against accidental loss due to diseases, contamination, human errors in screening and breeding colonies, and catastrophes such as fire, flood, and earthquake. Frozen embryos retain their genomes at a specific time point and hence prevent genetic drifting caused by long-term breeding and propagation. Cryopreservation of preimplantation embryos has been used during the past few decades for successfully archiving thousands of valuable mouse lines. Now, sperm cryopreservation is emerging as the leading method due to recent improvements in the efficiency and reliability of freezing and IVF procedures. The main advantage of sperm cryopreservation is that it requires much less "front end" investment in labor and animal space (Mazur *et al.*, 2008). Usually, two breeding age males can yield enough sperm to reconstitute a large number of offspring. The procedure itself is also simpler, faster, and does not require expensive instruments, such as a controlled rate freezer. However, sperm are haploid cells and therefore only half of the genetic material from the mouse is preserved. The thawed sperm must fertilize an egg, which brings in the other half of the genetic material, in order to reconstitute the mouse line. This could be problematic if the phenotypes of interest are dependent on multiple chromosomal loci or on an extensive breeding scheme. Currently, the efficiencies for sperm cryopreservation and IVF still varies dependent on the mouse strains and on the hands of the researchers. However, these disadvantages do not seem to prevent the rapid adoption of sperm cryopreservation for maintaining and distributing mouse lines. This can be partially attributed to the fact that a number of assisted reproduction techniques, such as PZD and ICSI, can help to recover mouse lines using imperfectly cryopreserved sperm. If fact, even freeze dried dead sperm have been used to recover mouse lines (Wakayama and Yanagimachi, 1998). Lastly, the method for mouse sperm cryopreservation is continually being improved (Liu *et al.*, 2009; Suzuki-Migishima *et al.*, 2009), which will undoubtedly make it more robust, reliable, and playing an even greater role in future archiving and distribution of genetically modified mouse lines.

ACKNOWLEDGMENT

This work was supported by the Intramural Research Program of NIH, NHLBI and NEI.

REFERENCES

Anzai, M., Nishiwaki, M., Yanagi, M., Nakashima, T., Kaneko, T., Taguchi, Y., Tokoro, M., Shin, S. W., Mitani, T., Kato, H., Matsumoto, K., Nakagata, N., *et al.* (2006). Application of laser-assisted zona drilling to in vitro fertilization of cryopreserved mouse oocytes with spermatozoa from a subfertile transgenic mouse. *J. Reprod. Dev.* **52**(5), 601–606.

Bath, M. L. (2003). Simple and efficient in vitro fertilization with cryopreserved C57BL/6J mouse sperm. *Biol. Reprod.* **68**(1), 19–23.

Kaneko, T., Yamamura, A., Ide, Y., Ogi, M., Yanagita, T., and Nakagata, N. (2006). Long-term cryopreservation of mouse sperm. *Theriogenology* **66**(5), 1098–1101.

Kawase, Y., Iwata, T., Ueda, O., Kamada, N., Tachibe, T., Aoki, Y., Jishage, K., and Suzuki, H. (2002). Effect of partial incision of the zona pellucida by piezo-micromanipulator for in vitro fertilization using frozen-thawed mouse spermatozoa on the developmental rate of embryos transferred at the 2-cell stage. *Biol. Reprod.* **66**(2), 381–385.

Kimura, Y., and Yanagimachi, R. (1995). Intracytoplasmic sperm injection in the mouse. *Biol. Reprod.* **52**(4), 709–720.

Liu, L., Nutter, L. M., Law, N., and McKerlie, C. (2009). Sperm freezing and in vitro fertilization in three substrains of C57BL/6 mice. *J. Am. Assoc. Lab. Anim. Sci.* **48**(1), 39–43.

Mazur, P., Leibo, S. P., and Seidel, G. E., Jr. (2008). Cryopreservation of the germplasm of animals used in biological and medical research: Importance, impact, status, and future directions. *Biol. Reprod.* **78**(1), 2–12.

Nakagata, N. (2000). Cryopreservation of mouse spermatozoa. *Mamm. Genome* **11**(7), 572–576.

Nakagata, N., and Takeshima, T. (1993). Cryopreservation of mouse spermatozoa from inbred and F1 hybrid strains. *Jikken Dobutsu* **42**(3), 317–320.

Nakagata, N., Okamoto, M., Ueda, O., and Suzuki, H. (1997). Positive effect of partial zona-pellucida dissection on the in vitro fertilizing capacity of cryopreserved C57BL/6J transgenic mouse spermatozoa of low motility. *Biol. Reprod.* **57**(5), 1050–1055.

Nishizono, H., Shioda, M., Takeo, T., Irie, T., and Nakagata, N. (2004). Decrease of fertilizing ability of mouse spermatozoa after freezing and thawing is related to cellular injury. *Biol. Reprod.* **71**(3), 973–978.

Ostermeier, G. C., Wiles, M. V., Farley, J. S., and Taft, R. A. (2008). Conserving, distributing and managing genetically modified mouse lines by sperm cryopreservation. *PLoS ONE* **3**(7), e2792.

Sakamoto, W., Kaneko, T., and Nakagata, N. (2005). Use of frozen-thawed oocytes for efficient production of normal offspring from cryopreserved mouse spermatozoa showing low fertility. *Comp. Med.* **55**(2), 136–139.

Songsasen, N., and Leibo, S. P. (1997). Cryopreservation of mouse spermatozoa. II. Relationship between survival after cryopreservation and osmotic tolerance of spermatozoa from three strains of mice. *Cryobiology* **35**(3), 255–269.

Suzuki-Migishima, R., Hino, T., Takabe, M., Oda, K., Migishima, F., Morimoto, Y., and Yokoyama, M. (2009). Marked improvement of fertility of cryopreserved C57BL/6J mouse sperm by depletion of Ca^{2+} in medium. *J. Reprod. Dev.* **55**(4), 386–392.

Szczygiel, M. A., Kusakabe, H., Yanagimachi, R., and Whittingham, D. G. (2002a). Intracytoplasmic sperm injection is more efficient than in vitro fertilization for generating mouse embryos from cryopreserved spermatozoa. *Biol. Reprod.* **67**(4), 1278–1284.

Szczygiel, M. A., Kusakabe, H., Yanagimachi, R., and Whittingham, D. G. (2002b). Separation of motile populations of spermatozoa prior to freezing is beneficial for subsequent fertilization in vitro: A study with various mouse strains. *Biol. Reprod.* **67**(1), 287–292.

Sztein, J. M., Farley, J. S., and Mobraaten, L. E. (2000). In vitro fertilization with cryopreserved inbred mouse sperm. *Biol. Reprod.* **63**(6), 1774–1780.

Taguma, K., Nakamura, C., Ozaki, A., Suzuki, C., Hachisu, A., Kobayashi, K., Mochida, K., Ogura, A., Kaneda, H., and Wakana, S. (2009). A practical novel method for ensuring stable capacitation of spermatozoa from cryopreserved C57BL/6J sperm suspension. *Exp. Anim.* **58**(4), 395–401.

Takeo, T., Hoshii, T., Kondo, Y., Toyodome, H., Arima, H., Yamamura, K., Irie, T., and Nakagata, N. (2008). Methyl-beta-cyclodextrin improves fertilizing ability of C57BL/6 mouse sperm after freezing and thawing by facilitating cholesterol efflux from the cells. *Biol. Reprod.* **78**(3), 546–551.

Wakayama, T., and Yanagimachi, R. (1998). Development of normal mice from oocytes injected with freeze-dried spermatozoa. *Nat. Biotechnol.* **16**(7), 639–641.

Whittingham, D. G., Leibo, S. P., and Mazur, P. (1972). Survival of mouse embryos frozen to −196 degrees and −269 degrees C. *Science* **178**(59), 411–414.

Wilmut, I. (1972). The effect of cooling rate, warming rate, cryoprotective agent and stage of development on survival of mouse embryos during freezing and thawing. *Life Sci. II* **11**(22), 1071–1079.

SECTION THREE

GAMETES AND EMBRYOS

CHAPTER FIVE

Isolation and Manipulation of Mouse Gametes and Embryos

Eveline S. Litscher *and* Paul M. Wassarman

Contents

Abstract

Many experimental approaches to answer questions about mammalian development begin with the isolation of mouse gametes and/or embryos. Here, methods used to isolate and store growing and fully-grown oocytes, ovulated (unfertilized) eggs, cleavage-stage embryos, and sperm from mice are described. Procedures used to carry out capacitation of sperm, binding of sperm to oocytes and ovulated eggs, and induction and assessment of the acrosome reaction *in vitro* are also described.

Department of Developmental and Regenerative Biology, Mount Sinai School of Medicine, New York, New York, USA

Methods in Enzymology, Volume 476
ISSN 0076-6879, DOI: 10.1016/S0076-6879(10)76005-5

1. Introduction

In mammals, oogenesis and spermatogenesis begin during early fetal development with formation of primordial germ cells from a small number of stem cells at an extragonadal site (Austin and Short, 1982). Primordial germ cells, in turn, become either oogonia in females or spermatogonia in males; cells that proliferate mitotically. Oogonia and spermatogonia then enter meiosis and become oocytes and spermatocytes, respectively. During meiosis crossing-over and recombination occur, and at the end of meiosis haploid gametes, eggs and sperm are produced.

Here, methods currently used to obtain and store growing and fully-grown oocytes, ovulated (unfertilized) eggs, cleavage-stage embryos, and sperm from mice are described.[1] This methodology should be useful to investigators interested in studying various aspects of oogenesis, spermatogenesis, fertilization, and embryogenesis in mice.

2. Materials, Media, and Solutions

The following alphabetical list includes materials, media, and solutions used to obtain, store, and manipulate mouse gametes and cleavage-stage embryos:

A23187 (Ca^{2+} ionophore): 40 m*M* stock solution in dimethylsulfoxide (DMSO), diluted 1:40 with PBS, pH 7.2, and mixed with M199-M to a concentration of 20 μM. Capacitated sperm are pipetted into this solution at a 1:1 dilution (10 μM final concentration). Used to induce sperm to undergo the acrosome reaction.

Ammonium acetate: 0.1 *M*, pH 9.0. Used to assess the status of the sperm's acrosome.

Antibody–gold conjugate: Antibody of choice conjugated to 5 nm gold particles, colloidal suspension (Sigma). Used to label the surface of sperm.

Blocking buffer: PBS/PVP with 0.05% Tween 20 and 5% goat serum. Used in gold-labeling of sperm.

CD-1 mice: Randomly bred Swiss mice (Charles River).

EGTA: 0.5 *M* ethyleneglycoltetraacetic acid (EGTA), pH 7.4 (stock solution), diluted 125× to 4 m*M* EGTA in M199-M. Used to prevent sperm from undergoing the acrosome reaction.

[1] Three particularly relevant sources for the methodology described here are Rafferty (1970), Nagy *et al.* (2003), and Wassarman and DePamphilis (2003).

Coomassie stain: 0.04% Coomassie brilliant-blue G-250 (BioRad), dissolved in 3.5% perchloric acid. Used to stain the sperm's acrosome.

Gelatin-coated glass slides: 0.3% gelatin (autoclaved), 0.01% chromium potassium sulfate, sterile filtered. Slides are dipped for several seconds and air-dried. Used to assess the status of the sperm's acrosome.

Human chorionic gonadotropin (hCG; Sigma): 5 International Units (IU)/ 100 μl/female mouse (see manufacturer's instructions for reconstitution). Used to obtain ovulated eggs.

Hyaluronidase: 1 mg/ml, dissolved in M199-M and prewarmed at 37 °C (from bovine testes, type VI-S; Sigma). Used to remove cumulus cells from ovulated eggs.

Light mineral oil: 225 ml light mineral oil is shaken vigorously with 25 ml M199, autoclaved, allowed to separate into aqueous and oil phases at room temperature (RT), and stored in the incubator at 37 °C at least 2 days prior to use. Used to culture mouse gametes and embryos.

Medium 199 (M199): Containing Earle's salts, L-glutamine, 2.2 g/l sodium bicarbonate, 25 m*M* HEPES buffer.

Medium 199-M (M199-M): M199 supplemented with 4 mg/ml bovine serum albumin (BSA; Sigma, fraction V, A-4503) and 30 μg/ml sodium pyruvate. Used to culture mouse gametes and embryos.

Mounting medium: PBS–30% glycerol (for Coomassie stain), SlowFade Antifade Kit (for PNA–FITC; Molecular Probes). Used to examine the status of the sperm's acrosome.

Pasteur pipettes: Borosilicate glass pipettes, 9 in. length, drawn to an internal diameter of ±150 μm (medium-bore), 150–200 μm (wide-bore), or not drawn (large-bore).

Phosphate buffered saline (PBS): 30 m*M* sodium phosphate, pH 7.2, 150 m*M* NaCl.

PBS–formaldehyde: PBS with 5% paraformaldehyde. Used as a fixative for sperm.

PBS/PVP: PBS containing polyvinylpyrrolidone-40 (PVP-40), 4 mg/ml, 0.02% sodium azide. Used for storage of oocytes, eggs, and embryos.

PBS/PVP–formaldehyde: PBS/PVP with 1% (a fixative for oocytes, eggs, and embryos) or 2% (a fixative for sperm bound to oocytes, eggs, and embryos) paraformaldehyde.

PBS/PVP–glutaraldehyde: PBS/PVP with 2.5% glutaraldehyde. A fixative used for gold-labeling of sperm.

Peanut agglutinin (PNA)–FITC (Arachis hypogaea) (Sigma): 15 μg/ml in PBS. Used to stain the sperm's acrosome.

Pregnant mare's serum (PMS; Sigma): 5 IU/100 μl/female mouse (see manufacturer's instructions for reconstitution). Used to obtain ovulated eggs.

Silver Enhancer Kit: Enlarges colloidal gold particles for signals visible by light microscopy. Contains silver enhancement solutions A, B, and sodium thiosulfate pentahydrate (SE-100, Sigma). Used to visualize gold-labeled antibodies.

Sterile steel needles: PrecisionGlide Needles 18G × 1½ in. and 30G × ½ in. (Becton Dickinson) connected to 1 ml syringes.
Sterile syringe filter: 0.2 μm, connected by flexible tubing and a mouthpiece.
Storage buffer: M199 containing 50 m*M* Tris–HCl, pH 7.5, 4 mg/ml PVP-40, 0.02% sodium azide. Used to store oocytes, eggs, and embryos.

3. Collection and Storage Conditions

Collection of mouse gametes and their culture *in vitro* are carried out in sterile-filtered medium M199-M at 37 °C in a humidified atmosphere of 5% CO_2 in air, unless otherwise indicated. M199-M, stored at 4 °C, should be used within 2 weeks of preparation. Light mineral oil saturated with M199 is used to cover drops of M199-M in sterile tissue culture dishes (60 mm diameter) and the dishes are incubated at least 30 min at 37 °C prior to addition of gametes and embryos. For fixation and storage of oocytes, eggs, and embryos, drops of PBS/PVP (with or without formaldehyde) and storage buffer are covered with light mineral oil at RT.

Oocytes, eggs, and embryos are transferred from one droplet to another under a dissecting microscope using a mouth-operated, medium-bore Pasteur pipette (drawn to an internal diameter of ~150 μm). The pipette is connected by flexible tubing via a sterile syringe filter (0.2 μm) to a mouthpiece. It is best to prefill a medium-bore Pasteur pipette with a small amount (a few microliters) of oil and an equal small amount of M199-M (as a visible oil-M199-M border line) before taking up small batches of cells (e.g., 5–10), and carefully monitoring the uptake as well as the downloading of cells. Capacitated sperm are transferred into drops of medium by Pipetman and microtips (e.g., 10 μl tips). Samples are always transferred through the oil layer covering the drops of M199-M.

3.1. Isolation of growing and fully-grown oocytes

To obtain growing oocytes, ovaries are excised from juvenile (<3 weeks old) female mice (Qi *et al.*, 2002; Sorensen and Wassarman, 1976). Juvenile mice possess an unusually large number of growing oocytes arrested in the Dictyate stage of first meiotic prophase (Wassarman, 1996). Fully-grown oocytes are excised from adult (>3 weeks old) females. The ovaries are placed in a tissue culture dish containing 5 ml of prewarmed (37 °C) M199-M and any fat and connective tissue are removed manually by using scissors and tweezers. Ovaries are poked extensively with fine steel needles (30G × ½ in.) under a dissecting microscope. Released, healthy oocytes are visible as shiny and round (sometimes oval) cells amid cellular debris and torn ovaries. Oocytes are washed by carefully transferring them through three 100 μl drops of M199-M. On average, 10–15 growing or fully-grown oocytes can be recovered from a single ovary (CD-1 females).

Typically, ovaries excised from 13- to 16-day-old juvenile female mice harbor a relatively large number of growing oocytes that are 60 ± 10 μm in diameter. For fully-grown oocytes (diameter ~80 μm), it is best to use ovaries from young adult females (3–4 weeks old) as the yield is similar to that of growing oocytes from juvenile mice and better than that from adult mice 5 weeks of age and older (Table 5.1; Fig. 5.1A).

Table 5.1 Relationship between age of juvenile female mice and diameter of their oocytes

Age of female (days)	Diameter of oocytes (~μm)
9	50
12	55
15	60
18	70
21	80

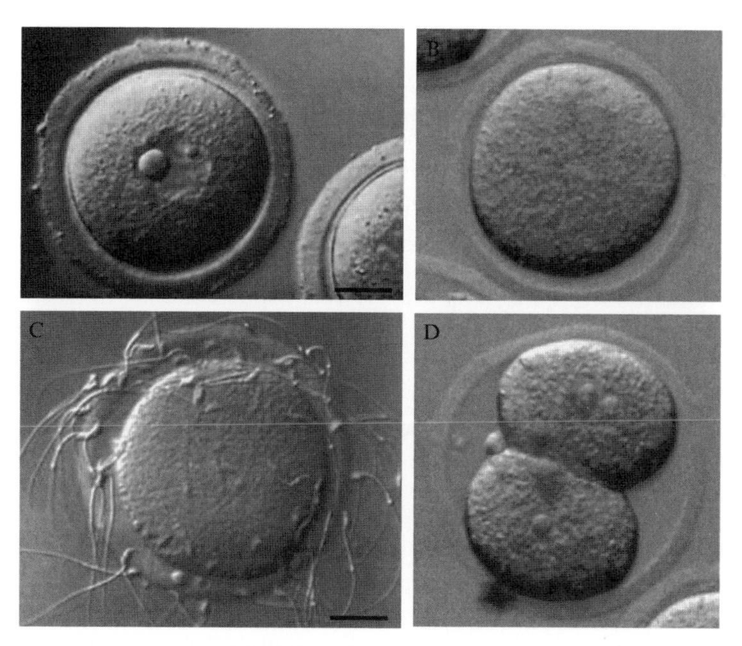

Figure 5.1 Light micrograph of (A) a mouse oocyte (bar ~20 μm), (B) an ovulated egg, (C) mouse sperm bound by their heads to the ZP of an ovulated mouse egg (bar ~20 μm) and (D) a two-cell embryo. Samples are fixed in 1% formaldehyde.

3.2. Isolation of ovulated eggs

Fully-grown oocytes undergo meiotic maturation, with emission of the first polar body, to become ovulated eggs (Wassarman, 1996). To obtain ovulated eggs arrested in metaphase II from superovulated adult female mice, 5 IU of PMS is injected intraperitoneally, preferably in the evening of day-1. This is followed 46–48 h later by the injection of 5 IU of hCG (day-3). In the morning of day-4 (16–18 h after injection of hCG), oviducts are excised and placed in a tissue culture dish containing 5 ml of prewarmed (37 °C) M199-M. The upper portion of the excised oviduct (i.e., the ampulla where ovulated eggs are located) is gently torn apart by using two pairs of tweezers (watchmaker's forceps, # 5) under a dissecting microscope. This results in the release of ovulated eggs surrounded by cumulus cells; so-called cumulus masses. Cumulus masses are collected by using a mouth-operated, wide-bore (not drawn) glass Pasteur pipette and are treated with hyaluronidase (1 mg/ml) in a 200 μl drop for 3–5 min at RT. Generally, such treatment results in the cumulus cells detaching and falling off the ovulated eggs, but sometimes gentle up-and-down pipetting may be required to free eggs from surrounding cumulus cells. Ovulated eggs are then transferred and washed through three 100 μl drops of M199-M. On average, 200–300 ovulated eggs can be recovered from 10 superovulated, 6–8-week-old female CD-1 mice (~20–30 eggs/mouse) (Fig. 5.1B).

3.3. Isolation of two-cell embryos

To obtain two-cell embryos from superovulated adult female mice, 5 IU of PMS is injected intraperitoneally, preferably in the afternoon of day-1, followed 46–48 h later by the injection of 5 IU of hCG (day-3). At the time of hCG injection, the female is caged overnight with a proven male breeder and checked the following morning (day-4) for a copulation plug (indicative of a successful mating, but not always visible or present). On the morning of day-5, oviducts are excised and placed in M199-M, as described above.

Two methods can be used to remove embryos from excised oviducts. In one procedure, a 1-ml syringe equipped with a sterile steel needle (30G × ½ in.) with a blunt (cutoff) tip is inserted into the oviduct opening and embryos are gently flushed out with M199-M. In the other procedure, oviducts are torn apart at several points along their length with watchmaker's forceps and embryos are released into the medium. Both procedures are done under a dissecting microscope. Isolated two-cell embryos are washed through three 100 μl drops of M199-M. Fifteen to 30 two-cell embryos can be obtained from one superovulated, 6–8-week-old female CD-1 mouse (Fig. 5.1D).

3.4. Isolation and capacitation of sperm

The cauda (tail) epididymis and the *vas deferens* are excised from a sexually mature (≥3 months old) male mouse and placed into a tissue culture dish containing 5 ml prewarmed M199-M/EGTA (EGTA is present to prevent sperm from undergoing the acrosome reaction). Any fat tissue is carefully removed with scissors and tweezers. Sperm from the cauda epididymis are released into the medium by puncturing and gently squeezing the organ with a sterile steel needle (18G × ½ in.). Sperm from the *vas deferens* are squeezed into the medium by using the same type of needle, but bent at a 90° angle. Sperm are collected from the culture dish with 1 ml pipette tips (Pipetman) into a 10 ml conical polypropylene tube and pelleted by low-speed centrifugation (~2000 rpm) for 5 min at RT. The supernatant (<5 ml) is discarded and 5 ml of fresh prewarmed M199-M, without EGTA, is added carefully on top of the sperm pellet without stirring it up. Inspecting a small sample (e.g., 50 μl) of the throw-away supernatant under a dissecting microscope helps determine the quality of sperm. In some cases, sperm density may still be high in the supernatant (indicative of high-motility sperm) and a second centrifugation cycle in a new 10 ml polypropylene tube will give another sperm pellet (often of better quality; i.e., without cellular debris, red blood cells, etc.).

Only capacitated sperm can fertilize ovulated eggs (Florman and Ducibella, 2006; Yanagimachi, 1994). Sperm are capacitated in M199-M for 1 h in the incubator at 37 °C. During this time, live sperm will swim up from the pellet into the medium ("swim-up" sperm), whereas dead sperm (no motility) will remain at the bottom of the tube. Sperm remain motile for 3–4 h after which time they become less lively and will eventually sink to the bottom of the tube. The final sperm concentration will vary depending upon the particular male mouse, but on average it is about 10^5–10^6 sperm/ml (based on 5 ml of "swim-up" sperm from one CD-1 male). Sperm prepared as described above cannot be stored and should be prepared fresh each time.

3.5. Storage of oocytes, ovulated eggs, and two-cell embryos

Isolated oocytes, eggs, and embryos, collected in M199-M and kept at 37 °C in a humidified atmosphere of 5% CO_2 in air, as described above, can be maintained in viable conditions for a day or two. For storage for up to 2 weeks, oocytes, eggs, and embryos in M199-M are transferred and washed through three 100 μl drops of PBS/PVP at RT prior to fixation for 1 h in a 100 μl drop of PBS/PVP–1% formaldehyde at RT. Following fixation, cells are washed through three 100 μl drops of PBS/PVP at RT, transferred to a 100 μl drop of storage buffer, and kept at 4 °C.

4. Sperm-Egg Interactions

4.1. Incubation of sperm with fully-grown oocytes, ovulated eggs, and two-cell embryos

Ovulated eggs or fully-grown oocytes and two-cell embryos, either stored in M199-M and kept at 37 °C for no more than 2 days, or kept in storage buffer at 4 °C, are washed through three 100 μl drops of M199-M at RT and transferred to the incubator for at least 30 min before being added to capacitated sperm (Williams *et al.*, 2006). Ten microliter drops of prewarmed M199-M are pipetted into a tissue culture dish and covered with oil. Alternatively, if a test substance is being evaluated, the sample is dissolved in a 10-μ l drop of M199-M and pipetted into a tissue culture dish, as above. Then 10 μl of capacitated "swim-up" sperm are added to the drop (through the oil layer; 1:1 dilution) and incubated for 15 min at 37 °C. Following this period, oocytes or eggs, and two-cell embryos are added to the 20 μl drop with as little medium as possible and incubated at 37 °C for an additional 35–45 min. Sperm motility is monitored under a dissecting microscope during this period and samples with sperm motility of less than 70% (i.e., less than 7 motile sperm out of 10 sperm) should be discarded. At the end of the incubation time, oocytes or eggs, and embryos with associated sperm are removed from the drops with a mouth-operated, wide-bore Pasteur pipette and washed through three 100 μl drops of M199-M. This step involves careful up-and-down pipetting (3–4 times) of groups of 5–10 cells in each drop until no more than one to two sperm remain attached to the two-cell embryos (negative control). Under these conditions, sperm bound to the oocyte or egg zona pellucida (ZP) remain bound, whereas sperm nonspecifically attached to oocyte or egg and embryo ZP are removed. Oocytes and eggs with bound sperm, as well as embryos, are then transferred to 10 μl drops of M199-M containing PBS/PVP–2% formaldehyde (1:1 by volume) covered with oil.

4.2. Analysis of sperm binding to fully-grown oocytes and ovulated eggs

The number of sperm bound per oocyte, egg, or embryo is determined by counting sperm tails in the largest diameter focal plane, using dark-field microscopy and 200× magnification. Typically, there are 12 oocytes or ovulated eggs and two two-cell embryos present per sample. The presence of embryos is essential in this assay since they serve as a negative control; sperm do not bind to the ZP of fertilized eggs or embryos. After determining the number of sperm bound to each oocyte or egg in one plane of focus, the highest and lowest values are discarded, and the remaining 10 values are used to calculate the average number of sperm bound per oocyte or egg (±S.D.). For example, when sperm are exposed to M199-M alone (no test substance), at sperm concentrations of about 5×10^5 sperm/ml, and incubated with oocytes or eggs and embryos, there are 30–50 sperm bound per oocyte or egg in the largest diameter focal plane (Fig. 5.1C).

5. Analysis of the Status of the Sperm's Acrosome

The acrosome is a large lysosome-like vesicle overlying the sperm's nucleus (Florman and Ducibella, 2006; Yanagimachi, 1994). Capacitated mouse sperm undergo the acrosome reaction (AR) shortly after binding to the egg ZP. The AR is a specialized form of cellular exocytosis. It involves multiple fusions between plasma membrane overlying the anterior region of the sperm head and the outer acrosomal membrane lying just beneath the plasma membrane. Small hybrid membrane vesicles form and the inner acrosomal membrane, overlying the sperm's nucleus, is exposed to the egg ZP (Cohen and Wassarman, 2001). Only acrosome-reacted sperm can penetrate the ZP, reach the plasma membrane, and fuse with eggs (i.e., fertilize eggs).

To assess the status of the sperm's acrosome (i.e., whether it is acrosome-intact or acrosome-reacted), capacitated "swim-up" sperm (see above) are incubated at 37 °C for 1 h in M199-M, either in 20–50 μl drops under oil in a tissue culture dish or in 0.5-ml Eppendorf tubes, in the presence or absence of a test substance. Sperm exposed to Ca^{2+} ionophore A23187 are used as a positive control since it induces sperm to undergo the AR *in vitro*. Capacitated sperm are pipetted into a solution of 20 μM A23187 at a 1:1 dilution to a final concentration of 10 μM (it should be noted that sperm incubated in A23187 become completely immotile within a few minutes of exposure). After incubation, sperm samples are pelleted in 0.5-ml Eppendorf tubes by centrifugation at ~10,000 rpm for 5 min in a microfuge at RT and fixed with PBS–5% formaldehyde (1:1 dilution with sperm sample) overnight at 4 °C. The following day, sperm are washed by centrifugation (as above) in 0.1 *M* ammonium acetate, pH 9.0, resuspended in 50 μl of 0.1 *M* ammonium acetate by gentle vortexing, pipetted onto gelatin-coated glass slides, and air-dried.

Several methods have been described to assess the status of the sperm's acrosome. Two of these methods, Coomassie staining and FITC–lectin staining of sperm, are described in the following sections.

5.1. Coomassie-staining of sperm

Slides are washed with distilled H_2O, methanol, H_2O, for 5 min each, Coomassie-stained for 5 min, and washed extensively with H_2O (Larson and Miller, 1999). Sperm can be observed directly on the glass slide or under a coverslip (sealed with clear nail polish for long-term storage) using PBS–30% glycerol as mounting medium, and are scanned in the presence or absence of an acrosome by bright-field microscopy at 400× magnification.

5.2. Lectin–FITC staining of sperm

Slides are washed twice with PBS for 5 min, incubated with PNA–FITC (*A. hypogaea*) for 30 min at RT (humid chamber), and rinsed with PBS for 15 min (Lybaert *et al.*, 2009). Samples are mounted with antifade mounting reagent and a coverslip (sealed with clear nail polish) and examined for the presence or absence of an acrosome by fluorescent microscopy at 400× magnification.

Sperm retaining an intact acrosome (Fig. 5.2A) will display a continuous blue (Coomassie)- or fluorescent green (PNA–FITC)-stained ridge overlying the sperm head. Sperm that have undergone the AR (Fig. 5.2B) will display either a patchy blue or patchy fluorescent green pattern, indicative of a partial AR, or no staining at all, indicative of a completed AR. Typically, A23187-treated sperm samples will be ~60–80% acrosome-reacted, whereas sperm exposed to M199-M alone will be ~15–25% acrosome-reacted (due to a so-called "spontaneous AR").

6. Visualization of Gold-Labeled Antibody on Sperm

Capacitated sperm are incubated with a sperm-binding probe (i.e., antigen of choice), fixed, washed and air-dried on gelatin-coated slides as described above (Williams *et al.*, 2006). Samples on slides are blocked with a

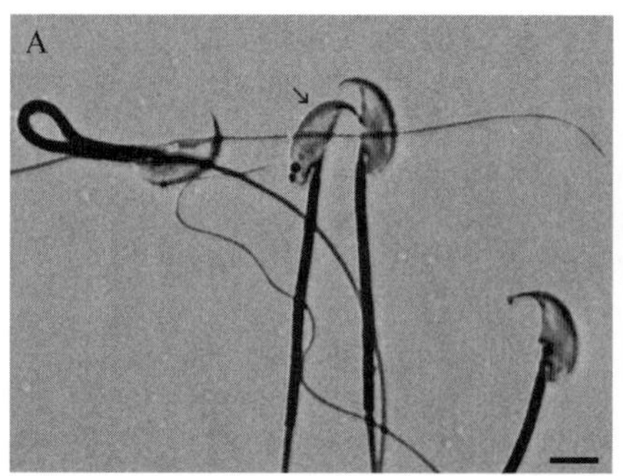

Figure 5.2 Light micrographs of (A) acrosome-intact and (B) acrosome-reacted mouse sperm, fixed in 2.5% formaldehyde and stained with Coomassie brilliant-blue G-250. Arrow: acrosomal cap (bar ~5.5 μm).

50 μl drop of PBS/PVP–0.05% Tween 20 containing 5% goat serum (blocking buffer) for 30 min at RT in a humid chamber, then rinsed in H_2O and air-dried for 15 min at RT. A diluted gold suspension (1:10) of antibody–gold conjugate (antibody of choice, coupled to 5 nm colloidal gold particles) in blocking buffer is pipetted onto the sperm sample (50 μl drop). The slide is incubated for 2 h at RT in a humid chamber on a slow moving shaker and washed afterward with PBS by submersion in a Petri dish for 5 min. The sample is fixed again with PBS/PVP–2.5% glutaraldehyde (50 μl drop) for 1 h at RT (humid chamber) and rinsed thoroughly with H_2O in a Petri dish. To enlarge colloidal gold labels for high-contrast signals that are visible by light microscopy, a 50 μl drop of silver stain solution (Silver Enhancer Kit) is applied onto the gold-labeled sperm samples for 15 min at RT in the dark (humid chamber). The slide is dipped in H_2O, fixed with 2.5% sodium thiosulfate pentahydrate (50 μl drop) for 3 min at RT, rinsed with H_2O, mounted with PBS/PVP–30% glycerol and a coverslip, and sealed with clear nail polish. Individual sperm are scanned by light microscopy for silver-enhanced colloidal gold–protein complexes and the number of particles (silver precipitate) associated with each sperm is determined.

ACKNOWLEDGMENTS

Our research was supported in part by the National Institutes of Health (NICHD), most recently by grant HD-35105.

REFERENCES

Austin, C. R., and Short, R. V. (eds.), (1982). Reproduction in Mammals. 1. Germ Cells and Fertilization, Cambridge University Press, Cambridge.

Cohen, N., and Wassarman, P. M. (2001). Association of egg zona pellucida glycoprotein mZP3 with sperm protein sp56 during fertilization in mice. *Int. J. Dev. Biol.* **45,** 569–576.

Florman, H. M., and Ducibella, T. (2006). Fertilization in mammals. *In* "The Physiology of Reproduction," (J. D. Neill, ed.), 3rd edn. pp. 55–112. Elsevier/Academic Press.

Larson, J. L., and Miller, D. J. (1999). Simple histochemical stain for acrosomes on sperm from several species. *Mol. Reprod. Dev.* **52,** 445–449.

Lybaert, P., Danguy, A., Leleux, F., Meuris, S., and Lebrun, P. (2009). Improved methodology for the detection and quantification of the acrosome reaction in mouse spermatozoa. *Histol. Histopathol.* **24,** 999–1007.

Nagy, A., Gertsenstein, M., Vintersten, K., and Behringer, R. (2003). Manipulating the Mouse Embryo: A Laboratory Manual. 3rd edn. Cold Spring Harbor Laboratory Press, Cold Spring Harbor, NY.

Qi, H., Williams, Z., and Wassarman, P. M. (2002). Secretion and assembly of zona pellucida glycoproteins by growing mouse oocytes microinjected with epitope-tagged cDNAs for mZP2 and mZP3. *Mol. Biol. Cell* **13,** 530–541.

Rafferty, K. A., Jr. (1970). Methods in Experimental Embryology of the Mouse. Johns Hopkins Press.
Sorensen, R. A., and Wassarman, P. M. (1976). Relationship between growth and meiotic maturation of the mouse oocyte. *Dev. Biol.* **50,** 531–536.
Wassarman, P. M. (1996). Oogenesis. *In* "Reproductive Endrocrinology, Surgery, and Technology," (E. Y. Adashi, J. A. Rock, and Z. Rosenwaks, eds.), Vol. 1, pp. 341–357. Lippincott-Raven Press.
Wassarman, P. M., and DePamphilis, M. L. (eds.), (2003). *In* Guide to Techniques in Mouse Development, Methods in Enzymology Vol. 225. Academic Press.
Williams, Z., Litscher, E. S., Jovine, L., and Wassarman, P. M. (2006). Polypeptide encoded by mouse ZP3 exon-7 is necessary and sufficient for binding of mouse sperm *in vitro*. *J. Cell. Physiol.* **207,** 30–39.
Yanagimachi, R. (1994). Mammalian fertilization. *In* "The Physiology of Reproduction 1," (E. Knobil and J. D. Neill, eds.), pp. 189–317. Raven Press.

CHAPTER SIX

Cryopreservation of Mouse Gametes and Embryos

Carlisle P. Landel*,†

Contents

Abstract

As the use of genetically engineered mouse models continues to expand, the need to cryopreserve strains of mice increases in parallel in order to preserve these unique research resources and provide a low-cost alternative to maintaining the large inventory of strains. This chapter discusses methods for the cryopreservation of mouse embryos, sperm, and oocytes, and briefly discusses other requirements for implementing a successful cryopreservation program.

* Department of Microbiology and Immunology, Thomas Jefferson University, Philadelphia, Pennsylvania, USA
† Kimmel Cancer Center, Thomas Jefferson University, Philadelphia, Pennsylvania, USA

Methods in Enzymology, Volume 476
ISSN 0076-6879, DOI: 10.1016/S0076-6879(10)76006-7

1. Introduction

Cryopreservation of mouse strains has a number of uses. First, it can provide security against the loss of a valuable research tool due to natural or man-made disaster, disease, breeding failure, genetic contamination, etc. Second, it can conserve resources, since mice in the freezer do not require food, bedding, labor, or animal housing space. Third, shipment of cryopreserved material can be more convenient than shipping live animals, since frozen samples in a liquid nitrogen dry shipper can travel in the face of environmental extremes of heat and cold that would prevent shipment of live animals; furthermore, recovery of live animals serves to rederive the strain, eliminating the time delay engendered by the need to quarantine incoming animals (Critser and Mobraaten, 2000; Glenister and Thornton, 2000; Landel, 2005; Marschall and Hrabe de Angelis, 1999; Mobraaten, 1986, 1999; Shaw and Nakagata, 2002). Fourth, foundation stocks can be stored and replenished periodically from the freezer, thus limiting genetic drift (Landel, 2005). Finally, cryopreservation can allow a transgenic facility to stockpile embryos for later microinjection (Keskintepe *et al.*, 2001; Landa and Slezinger, 1992; Leibo *et al.*, 1991).

This chapter will provide a number of methods in the cryopreservation of mouse lines. It will presume knowledge on the part of the reader with respect to the methodologies for embryo handling (i.e., collection, manipulation, and surgical implantation into pseudopregnant foster mothers).

2. Cryopreservation of Mouse Embryos

Successful cryopreservation of preimplantation mouse embryos, and indeed of cells of any kind, depends on preventing the formation of intracellular ice crystals. The methods involved almost always employ some form of cryoprotective agent, or CPA, usually small molecules that can permeate the cell and serve primarily to depress the freezing point of the solution but also to ameliorate the effects of high salt concentrations that occur as cells dehydrate (see below). A large number of protocols have been developed, but they can be broadly divided into two categories: equilibrium or "slow freeze" methods, and nonequilibrium vitrification or "fast freeze" methods. In the former, the increasing extracellular ice formation during a slow decrease in temperature increases the osmolarity of the solution, drawing water from the cell so that solidification of the cells occurs without any intracellular water to form ice. In the latter, the cells are first equilibrated with a very high concentration of CPA then quickly cooled by plunging them into liquid nitrogen. At this rapid rate of cooling, the CPA and the intracellular contents form a glass (vitrify) instead of crystallizing, and cellular viability is maintained.

The methods of embryo cryopreservation outlined below work at all of the preimplantation stages from pronuclear-stage fertilized eggs through the blastocyst stage. Embryos to be frozen can be collected from mated females (superovulation is usually employed to increase embryo yields) or can be produced by *in vitro* fertilization (IVF).

2.1. Equilibrium or "slow freeze" method

Since the publication of the first successful mouse embryo cryopreservation protocol by Whittingham *et al.* (1972), a large number of equilibrium methods have been published. All of them, however, share the following steps:

1. *Equilibration of the embryos in CPA.* In this step, the embryos are transferred to the CPA solution for a period of time that allows the CPA concentration to equilibrate between the intra- and extracellular space. If one observes this under the microscope, one will see that the volume of the embryo will at first decrease as water exits the cell(s) to balance the higher external osmotic pressure due to the CPA, but then will return to the original volume as the CPA flows into the cell to balance the internal and external CPA concentrations and water follows the CPA flow.
2. *Cooling of the sample to a temperature below the freezing temperature.* Solutions can be cooled below their freezing point (supercooled) without solidifying.
3. *Seeding ice crystal formation in the solution.* Once a solution has been supercooled, the introduction of ice crystals into the solution nucleates ice formation. Ice, which is a solid of pure water, removes water from the solution as it grows, resulting in an increase the concentration of the solutes in the remaining solution. Since the solutes (primarily salts) cannot cross the cell membrane, water flows from the cell under the osmotic pressure of the increased salt concentration. Ice growth continues until an equilibrium is achieved due to the freezing point depression of the solutes, so that the embryos are slightly dehydrated and floating in a slurry ice and concentrated medium.
4. *Cooling the sample at a slow rate (around 1 °C/min) until it reaches a temperature below that at which the solution solidifies (the eutectic point).* As the temperature drops, ice crystals grow, withdrawing more water from the solution and increasing the salinity, which in turn draws more water from the cell. The temperature must be ramped down slowly enough to allow water to exit the cell. Eventually, no water is left in the cell to form ice crystals, and the solution solidifies into an equilibrium mixture of ice crystals and hydrated salt crystals.

5. *Storing the sample in liquid nitrogen until recovery*. Once the sample reaches a temperature below the eutectic point of the solution, it can be transferred into liquid nitrogen for storage. At liquid nitrogen temperatures, the samples will remain viable, for all practical purposes, forever.

Recovering the frozen embryos then involves the following steps:

6. Removing the sample from liquid nitrogen and holding it at room temperature for a few moments to allow any liquid nitrogen in the sample to vent slowly to avoid explosive vaporization.
7. Rapidly warming the sample to room temperature, often by warming it in a water bath.
8. *Removing the CPA from the embryos*. If embryos are transferred directly into physiologically isotonic medium, water will quickly enter into the cells due to the osmotic forces generated by high intracellular CPA concentrations, causing the embryos to swell and lyse. In order to prevent this, an opportunity must be provided to allow the CPA to exit the cell. This is accomplished in one of several ways. First, the embryos can be transferred stepwise though solutions of decreasing CPA concentration. Second, isotonic medium can be added dropwise to the embryos in CPA, effectively lowering the external CPA concentration slowly enough to allow CPA to exit the cell. Finally, one can employ an "osmotic buffer," that is, a solution of high osmolarity that contains a solute that cannot permeate the cells, for example, sucrose. In this case, water and CPA will exit the cell, and embryos placed into the osmotic buffer and observed under a microscope will be seen to shrink drastically in volume. The embryos can then be transferred back to isotonic medium and rinsed free of the osmotic buffer, where they will return to their normal volume. At this point, the embryos can be transferred to pseudopregnant recipient females to recover live animals, or used for any other purpose.

The method outlined below was first described by Renard and Babinet (1984), and is utilized by several large repositories of cryopreserved mouse embryos including The Jackson Laboratory, the Mutant Mouse Resource Centers, and the European Mutant Mouse Archive at the MRC in Harwell, UK. It utilizes propylene glycol as the CPA, a seeding temperature of −7 °C, and a cooling rate of −0.5 °C/min to a final temperature of −35 °C. Other published methods also work equally well (for the most part), using different CPAs, equilibration times, seeding temperatures, cooling rates, and final temperatures. However, if one wishes to employ one of these other methods, one must adhere strictly to the protocol with regards to these parameters and not "mix and match." The method is suitable for all preimplantation mouse stages and for most mouse strains, though BALB/C mice do not recover well using this method for reasons that are not well understood.

2.1.1. Media, supplies, and equipment for equilibrium cryopreservation of embryos

2.1.1.1. Media and solutions

1. Buffered mouse embryo medium, for example, FHM, M2, or PB1. Media can be purchased from suppliers such as Millipore and Sigma or produced from published recipes.
2. *CPA*. 1.5 *M* propylene glycol (PROH) in embryo medium. Mix 4.4 ml medium and 0.6 ml PROH and filter-sterilize. Stable for 1 week at 4 °C.
3. *Osmotic buffer*. 1 *M* sucrose in embryo medium. Dissolve 1.7 g sucrose in 5 ml medium. Filter sterilize. Stable for 1 week at 4 °C.

2.1.1.2. Equipment and supplies

1. *Controlled-rate freezer*. Examples include: FTS Biocool IV, Cryologic Freeze Control system, Planar Kryo models, Thermo CryoMed freezer, etc.
2. Dissecting microscope.
3. Dissection tools for embryo removal.
4. Plastic semen straws, 0.25 ml. Manufactured by IMV.
5. Self-laminating labels, for example, Brady part LAT-17-361-2.5.
6. *Straw sealer*. There are several options: an electric heat sealer, PVA sealing powder, or Critaseal. The first two are available from agricultural supply houses such as AgTech; the latter is available from supply companies such as Fisher.
7. Aspirator to fill straws. Straws will fit into the hub of 1 ml Monoject syringes (Atlantic Healthcare, Portland); alternatively, one can use tubing to construct a fitting between a straw and a syringe.
8. Embryo transfer pipettes and pipettor.
9. *Metal plunger rod*. This is a metal rod, conveniently made from 19 or 20 gauge wire, about 150 mm long, that can fit inside the straw and be used to push the plug to expel the contents.
10. 35 mm Petri dishes.
11. Marking pen.
12. Scissors.
13. Kimwipes or equivalent.
14. Liquid nitrogen.
15. Small insulated container (Dewar bucket, foam ice bucket, or Styrofoam shipping box) for holding small volumes of liquid nitrogen.
16. Liquid nitrogen storage freezer and storage system for holding straws. Straw storage is most inexpensively done using a "goblet and cane" system, although there are other systems such as straw cassettes (produced by Minitube) or the CryoBioSystem produced by IMV.

2.1.2. Freezing method, equilibrium cryopreservation

1. Prepare the cooling apparatus: set to −7 °C.
2. Collect the embryos, screen carefully for abnormalities, and hold at room temperature in medium until ready to freeze.
3. Fill one 35 mm Petri dish with cryoprotectant and one with osmotic buffer; label the dishes.
4. Prepare the straws for loading with embryos (see Fig. 6.1).
   a. Push the plug into the tube until it is 75 mm from the end (Fig. 6.1A).
   b. Label the straw.
   c. Mark the straw with three marks at points 1, 2, and 3 as shown in Fig. 6.1B. This will help fill the straws with the appropriate volumes of osmotic buffer and cryoprotectant.
5. Fill the straws with media (Fig. 6.1C).
   a. Attach a Monoject syringe to the labeled end of the straw.
   b. Aspirate osmotic buffer into the straw until the meniscus reaches mark 1.
   c. Aspirate air until the meniscus reaches mark 2.
   d. Aspirate cryoprotectant until it reaches mark 3.

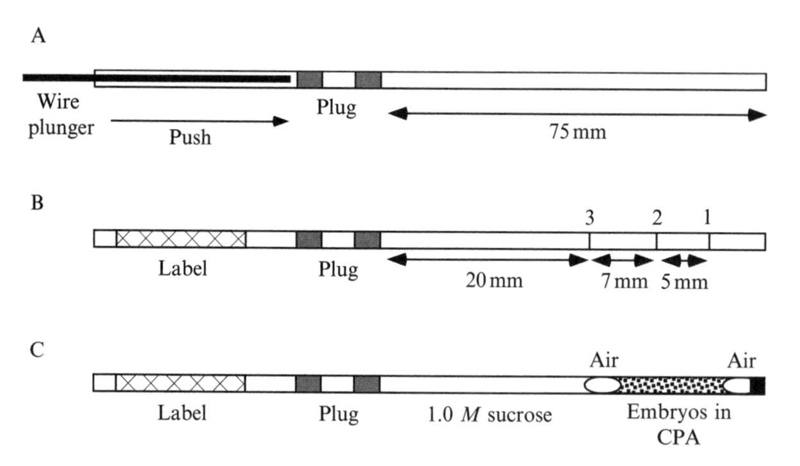

Figure 6.1 Preparation and loading of straws for equilibrium cryopreservation of embryos. (A) Using a metal push rod that fits inside the straw, push the plug into the straw so that it is 75 mm from the end. (B) Label the straw and mark the straw at 1, 2, and 3 as shown. (C) When filling the straw, a syringe or similar aspiration device is attached to the labeled end of the straw, and sucrose is drawn up until the meniscus reaches line 1. Air is then aspirated until the sucrose meniscus reaches line 2, followed by CPA until the meniscus reaches line 3. The whole solution is then drawn into the straw until it reaches the plug. Using an embryo transfer capillary, embryos are placed into the CPA fraction and the straw is then sealed.

e. Aspirate air again until the sucrose reaches the plug. This seals the straw.
f. Set filled straw aside and repeat until all straws are full. It is important to handle the straws gently so as not to inadvertently mix the sucrose and CPA fractions.

6. Load the embryos into the straws.
   a. Pipette the embryos into a dish containing CPA and let them equilibrate for 15 min at room temperature.
   b. Move the embryos into groups to be loaded into each straw.
   c. Pick up a group of embryos and place it into the cryoprotectant fraction of a straw. This can be visualized under a dissecting microscope. Take care to not blow lots of bubbles into the straw.
   d. Seal straw with a heat sealer or by dipping into sealing powder or Critaseal.
   e. Repeat until all straws are loaded.
7. Place loaded straws into the controlled-rate freezer and allow to equilibrate at −7 °C for 5 min.
8. Seed the straws by touching the sucrose fraction with a cotton swab dipped in liquid nitrogen or a pair of forceps cooled in liquid nitrogen. Ice will visibly form and begin to spread down the length of the straw.
9. After the ice has migrated into the CPA fraction, wait another 3 min and then cool the samples at 0.3 °C/min until they reach −30 °C.
10. Quickly plunge the samples into liquid nitrogen.
11. Transfer samples to the storage freezer.

2.1.3. Recovery method, equilibrium cryopreservation

1. Transfer the straw from the liquid nitrogen storage freezer to a smaller container of liquid nitrogen.
2. Using forceps, grasp the straw near the label and hold in air for 40 s then hold in water at room temperature until the ice disappears.
3. Wipe the straw dry.
4. Cut off the seal and then cut through the PVA plug, leaving about half the cotton plug in place to act as a plunger.
5. Using a metal rod, expel the entire liquid contents of the straw into a 35 mm Falcon dish. Do not let the plug drop into the dish.
6. Wait for 5 min. The embryos will shrink considerably.
7. Transfer the embryo to a drop of embryo medium in another dish. The embryos will regain their normal size.
8. Wash the embryos through a fresh drop of medium and then either transfer to a pseudopregnant recipient or culture to later stages in an appropriate medium.

2.2. Vitrification or "fast freeze" method

If a solution is cooled rapidly enough, it will form a glass, that is, an amorphous, noncrystalline solid, in a process known as vitrification. Since it is the formation of intracellular ice crystals that damage embryos during freezing, vitrification provides another route to cryopreserving cells (and embryos). In practice, a suspension of embryos must be placed in a very high concentration of CPA (on the order of 5 *M*) to suppress the freezing point and in a very small volume in order for heat transfer rates to be rapid enough for vitrification to occur.

The advantage of vitrification over equilibrium cryopreservation is that it does not require the use of a controlled-rate freezer and the actual freezing protocol is less complex. However, the samples themselves are much more delicate and must be stored and handled much more carefully because an aqueous glass can spontaneously crystallize at temperatures higher than about -130 °C (formally, this is known as the glass transition temperature). If such crystallization occurs, the embryos will be damaged and be lysed upon thawing. Thus, care must be taken to prevent any warming of the samples (in practice, they must be submerged in liquid nitrogen almost continuously) and upon thawing must be warmed very rapidly to minimize the time they reside between the glass transition temperature and the melting temperature. It is for this reason that most of the large archival repositories utilize equilibrium methods for embryo cryopreservation, although the practice in Japan is to use vitrification.

Vitrification of mouse embryos was first reported by Rall and Fahy (1985) and, as with equilibrium methods, a number of variations on this theme have been published. The common steps in all vitrification protocols are the following:

1. Equilibration of the embryos in CPA. Again, watching the embryos under a microscope during this step will reveal a volume excursion as CPA and water concentrations equilibrate across the cell membrane.
2. Very rapid cooling of the samples, usually by plunging directly into liquid nitrogen. Because the solution vitrifies, it does not change appearance.
3. Storing the samples in liquid nitrogen until recovery.

Recovering the embryos then involves:

4. Removing the sample from liquid nitrogen and holding it at room temperature for a few moments to allow any nitrogen in the sample to vent.
5. Rapidly warming the samples. This can be done by placing the sample in a water bath or by adding prewarmed osmotic buffer directly to the frozen sample.
6. Removal of the CPA. As is the case with equilibrium methods, care must be taken at this step to prevent osmotic lysis of embryos transferred into physiologically isotonic media; an osmotic buffer of sucrose is again conveniently employed here. Embryos in the osmotic buffer will again shrink and when subsequently placed into isotonic medium will reexpand.

The method outlined below is a modification of the original Rall and Fahy protocol that was published by Nakao *et al.* (1997).

2.2.1. Media, solutions, supplies, and equipment for embryo vitrification

2.2.1.1. Media and solutions

1. Buffered mouse embryo medium, for example, FHM, M2, or PB1. Media can be purchased from suppliers such as Millipore and Sigma or produced from published recipes.
2. DAP213: 2 *M* DMSO, 1 *M* acetamide, 3 *M* propylene glycol in mouse medium. The solution is made by mixing equal volumes of solutions A and B below. Aliquot at 0.5 ml and store at −80 °C.
   a. *Solution A*: Mix 2.3 ml mouse medium, 3.1 ml DMSO in 4.5 ml propylene glycol.
   b. *Solution B*: Dissolve 1.18 mg acetamide in 10 ml mouse medium.
3. 1 *M* DMSO solution. Mix 2 ml DMSO and 23 ml embryo medium. Aliquot at 1.0 ml and store at −80 °C.
4. Osmotic buffer, 0.25 *M* sucrose in mouse medium. Dissolve 1.7 g sucrose in 20 ml mouse medium. Aliquot at 1.0 ml and store at −80 °C.

2.2.2. Equipment and supplies for embryo vitrification

1. 60 mm Petri dishes.
2. V-bottom cryo tubes. NUNC 366656.
3. 20, 200 and 1000 μl micropipettors and tips.
4. Gel-loading micropipette tips, for example, MBP Gell 200, Molecular BioProducts #3621.
5. Ice bucket with ice or Nalgene Labtop Cooler (Catalog 5115-0012) precooled at −20 °C.
6. Cryo canes for vial storage.
7. Dewar or other liquid nitrogen container deep enough to submerge cryo vials on a cane.
8. Liquid nitrogen storage freezer with vial storage system.

2.2.3. Method for embryo vitrification

1. Prepare a bucket of ice or a block cooler. Place aliquot of DAP213 and labeled cryo vials on ice.
2. Place a series of 100 μl drops of 1 *M* DMSO solution into a Petri dish. One drop is used to wash the embryos from the collection medium. The others will hold the washed embryos.
3. Place a group of embryos into one of the drops. This rinses them free of the collection medium. After rinsing, divide the embryos between the

other drops so that each holds the number of embryos that one desires to aliquot a single storage vial. For example, if one wished store embryos in aliquots of 40 each and were to collect 120 embryos, they would first be placed together in the rinse drop and then divided among three of the remaining drops. Group the distributed embryos closely together in the drops to facilitate picking them up in the next step.

4. Using a 20 μl pipettor and a gel-loading tip, transfer the embryos in 5 μl of 1 *M* DMSO solution into a cryo vial and then put the vial on ice or into the block cooler at 0 °C and wait for at least 5 min.
5. Add 45 μl DAP213 at 0 °C into the freezing tube and equilibrate for 5 min on ice or in the block cooler.
6. Quickly fix the freezing tubes to a cane and plunge the samples directly into liquid nitrogen. After they equilibrate (stop boiling), transfer samples to a liquid nitrogen storage freezer.

2.2.4. Method for recovery of vitrified embryos

1. After removing the sample from liquid nitrogen, open the cap of the vial, quickly dump out any liquid inside, and let it stand at room temperature for 30 s.
2. Put 0.9 ml of PBS containing 0.25 *M* sucrose (prewarmed to 37 °C) into the cryo vial and thaw the sample quickly by pipetting the solution up and down, and then transfer the contents of the vial into a Petri dish.
3. Rinse the cryo vial with another 100 μl of sucrose solution and add this to the Petri dish.
4. Let the embryos stand in the sucrose solution for 10 min.
5. Recover the embryos, and transfer them into a drop of medium and wait for 5 min. This allows residual CPA to leave the embryos.
6. Wash the embryos through two more changes of fresh medium and then either transfer to a pseudopregnant recipient or culture to later stages in an appropriate medium.

3. Cryopreservation of Spermatozoa

The cryopreservation of mouse spermatozoa as a strategy for mouse strain cryopreservation is an attractive one, given the amount of material that can be obtained ($\sim 3 \times 10^7$ sperm/male) and the ease the freezing protocol itself. However, it must always be remembered that cryopreservation of sperm only preserves a haploid genome, and recovery can only be accomplished by using this material to fertilize oocytes to produce embryos. Thus, as opposed the recovery of cryopreserved embryos, where "what you freeze is what you get," recovered animals are at best one generation removed from that of the frozen material. Thus, for strain maintenance, it is appropriate only in cases where the strain background is a common inbred strain that will likely be extant at the time of recovery, and is inappropriate if the strain harbors more than two unlinked mutant loci. That is, if one imagines that the starting material were derived from a strain homozygous for multiple unlinked loci, then recovered animals would be compound heterozygotes that would require inbreeding to regenerate animals of the original genotype; the frequency of the desired genotype goes down exponentially as the number of loci is increased.

The other difficulty with sperm cryopreservation as a strategy lies with the more complex recovery of animals, since some sort of IVF step is involved. Until recently, the strategy was further complicated by strain variation in the efficiency of IVF, with the particular difficulty of recovery of viable sperm from C57BL/6 substrains. However, recent advances in handling the material (Bath, 2003; Suzuki-Migishima *et al.*, 2009) and adding antioxidants to the CPA (Ostermeier *et al.*, 2008) have largely ameliorated this strain problem. Nonetheless, the complexities of doing IVF for recovery remain.

The development of viable sperm cryopreservation protocols was complicated by a peculiarity of mouse spermatozoa cells: they are highly sensitive to high osmolarity and will not survive immersion in solutions with an osmolarity in excess of about 600 mOsm (Koshimoto and Mazur, 2002; Koshimoto *et al.*, 2000). In terms of developing viable protocols, then, vitrification in high concentrations of CPA and equilibrium cryopreservation in permeating cryoprotectants will not work, since the osmotic limit of the cells is exceeded in these methods. Instead, the methods rely on the use of the raffinose, a trisaccharide that is impermeable to the cell, as a CPA. Cooling rates on the order of 30 °C/min are achieved by placing the samples in the vapor phase over liquid nitrogen.

The method outlined below is based on Ostermeier *et al.*'s (2008) modification of Nakagata's (2000) protocol.

Sperm can be collected from individual males and frozen, or sperm from several males can be pooled. Sperm can be recovered throughout the reproductive lifetime of the males, but it is good practice to limit the age of the donor to between 10 and 20 weeks to maximize sperm count, and donors should be single-housed and not breeding for at least a week before collection.

3.1. Media, solutions, supplies, and equipment for sperm cryopreservation

3.1.1. Media and solutions

1. *CPA*. 18% raffinose, 3% nonfat dried milk, and 477 μM monothioglycerol (MTG) in water. Dissolve 18 g raffinose and 3 g nonfat dried milk powder in water. The final volume should be 100 ml. The solution will

require heating to 50 °C to dissolve the raffinose. Centrifuge the solution for 10 min at 10,000 × *g* to remove milk solids. Add 4 μl MTG, filter-sterilize, aliquot to 1 ml aliquots, and store at −80 °C. Thawed aliquots will require brief warming at 37 °C and mixing to redissolve any raffinose crystals that have formed.

2. 18% raffinose.

3.1.1.1. Supplies and equipment

1. Dissecting microscope.
2. Dissection tools for embryo removal.
3. Plastic semen straws, 0.25 ml. Manufactured by IMV.
4. Self-laminating labels, for example, Brady part LAT-17-361-2.5.
5. *Straw sealer.* There are several options: an electric heat sealer, PVA sealing powder, or Critaseal. The first two are available from agricultural supply houses such as AgTech; the latter is available from supply companies such as Fisher.
6. Aspirator to fill straws. Straws will fit into the hub of 1 ml Monoject syringes (Atlantic Healthcare); alternatively, one can use tubing to construct a fitting between a straw and a syringe.
7. *Metal plunger rod.* This is a metal rod, conveniently made from 19 or 20 gauge wire, about 150 mm long, that can fit inside the straw and be used to push the plug to expel the contents.
8. 35 mm Petri dishes.
9. Marking pen.
10. Scissors.
11. Kimwipes or equivalent.
12. Liquid nitrogen.
13. Small insulated container (Dewar bucket, foam ice bucket, or Styrofoam shipping box) for holding small volumes of liquid nitrogen.
14. Liquid nitrogen storage freezer and storage system for holding straws. Straw storage is most cheaply done using a "goblet and cane" system, although there are other systems such as straw cassettes (produced by Minitube) or the CryoBioSystem produced by IMV.

3.2. Method for sperm cryopreservation

1. Label straws at the plug end and mark them 7, 8, and 8.5 cm from the end with a pen as shown in Fig. 6.2A.
2. Place 1 ml of CPA per male in a small Petri dish and place the dish warming tray or in an incubator at 37 °C.
3. Sacrifice the male mouse. Remove both epididymides and *vas deferentia* and place in the dish containing CPA.

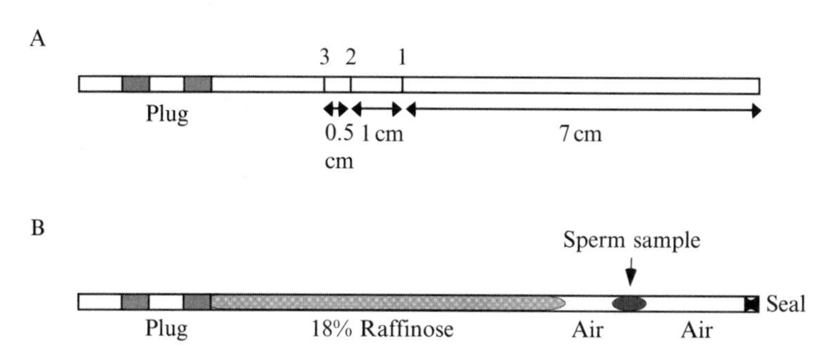

Figure 6.2 Sperm cryopreservation straws. (A) Preparation of the straws: straws are marked as shown. (B) Filling the straws: 18% raffinose is drawn into the straw until the meniscus reaches the mark 1, followed by air until the meniscus reaches mark 2, followed by sperm until mark 3 is reached. Air is then drawn in until the fluid column reaches the plug. The bottom is then sealed.

4. Under the dissecting scope, mince the epididymides to release the sperm. This can be accomplished with forceps and tuberculin needle, a pair of needles, or small scissors.
5. Next, squeeze the sperm out of the *vas deferens* either by *gently* running a pair of forceps down the vas or by "walking" a pair of forceps or 30 gauge needles along the vas to push out the sperm.
6. Incubate the sperm for 10 min at 37 °C. During this time, they will begin to distribute themselves through the solution.
7. Remove the epididymides from the plate and *gently* stir the dish with a pipette tip in order to mix the sperm and distribute them equally, and then load the straws as follows (see Fig. 6.2B):
   a. Attach a 1 cc Monoject syringe to the plug end of the tube.
   b. Draw 18% raffinose into the straw up to the 1st line at 7 cm. The raffinose simply acts as a weight to prevent the straw from floating in liquid nitrogen.
   c. Draw 1 cm of air into the straw (until the raffinose reaches the 2nd line).
   d. Wipe the straw with a tissue.
   e. Slowly draw 0.5 cm of sperm sample ($\approx$10 μl) into the straw.
   f. Slowly draw air into the straw until the raffinose contacts the plug. This seals the top of the straw.
   g. Seal the bottom of the straw with a heat sealer, Critoseal or sealing powder.
   h. Place the straw in a goblet or cassette.
   i. Repeat until all the desired number of samples is collected. Twenty samples should be more than enough for most purposes.
8. Place the goblet or cassette into the liquid nitrogen vapor phase above liquid nitrogen. This can be accomplished in a number of ways.

  a. The straws can be placed in the vapor phase at the top of a liquid nitrogen freezer.
  b. The samples can be placed into a charged liquid nitrogen dry shipper.
  c. The samples can be floated upon a raft of Styrofoam floating upon liquid nitrogen in the bottom of a Dewar bucket, Styrofoam box, ice bucket, etc.
  d. It is probable that the straws can also be placed into a rack in a −80 °C freezer and achieve an adequate cooling rate.
9. After 10 min at low temperature, the straws can then be placed into liquid nitrogen for storage.

3.3. Recovery of embryos from cryopreserved sperm by IVF

The recovery of mice from cryopreserved sperm requires utilizing this sperm to fertilize oocytes to produce embryos that can then be transferred to pseudopregnant foster mothers. The IVF technique itself, while straightforward in execution, seems to be difficult to learn to do consistently, and those attempting to produce embryos from frozen sperm should be confident in their ability to successfully perform IVF on a consistent basis with freshly collected sperm.

The method described below is essentially that first described by Sztein *et al.* (2000) as modified by Ostermeier *et al.* (2008).

3.3.1. Media, solutions, supplies, and equipment for IVF

1. IVF medium. HTF medium (Quinn *et al.*, 1985) or Mouse Vitro Fert medium (Cook Medical, K-RVFE 50).
2. Mineral oil (Sigma).
3. Dissecting microscope and dissection tools for oocyte and sperm collection.
4. 35 and 60 mm bacterial culture dishes.
5. Incubator in which IVFs and culture plates can be incubated in 5% O_2, 5% CO_2, 90% N_2.
6. Wide bore pipet tips (Rainin HR-250W).
7. 20 μl, 200 (or 100) μl, and 1 ml micropipettors.
8. 10 ml pipette and pipettor (for dispensing oil).
9. Insulin syringes (for use as dissecting needles).
10. Embryo transfer pipettes and pipettor.

3.3.2. IVF method

1. Prepare fertilization and sperm collection dishes and place these in the incubator at 37 °C under an atmosphere of 5% O_2, 5% CO_2, 90% N_2 (tri-gas) as shown in Fig. 6.3.

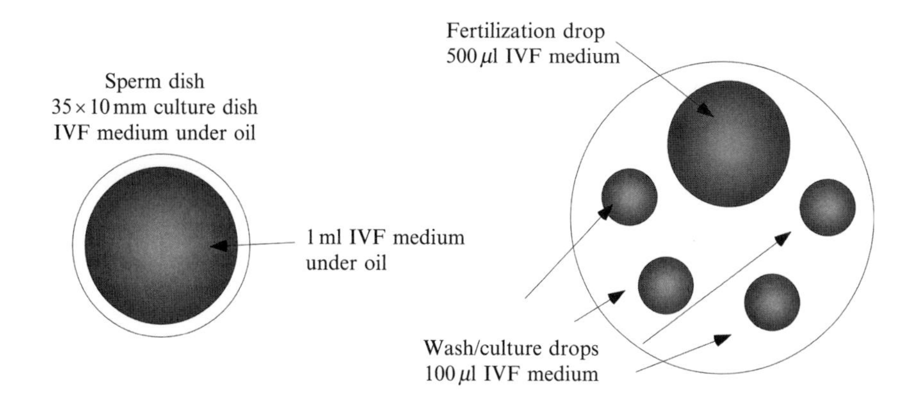

Figure 6.3 Diagram of fertilization and sperm collection dishes for IVF. Drops of IVF medium are placed under oil as indicated.

2. Thaw frozen sperm by removing a straw from liquid nitrogen, allow to warm for 10–15 s in room air, and then place in a 37–40 °C water bath until thawed.
3. Cut the seal from the bottom of the straw, and then using the metal push rod, push the plug to expel the 10 μl of frozen sperm from the straw into the fertilization drop in the dish.
4. Incubate 1 h in the incubator. Sperm motility should recover in this time.
5. Collect sperm from a male mouse for use as a control. Dissect the epididymides and *vas deferentia* into the sperm collection dish, mince the tissue to release the sperm, and incubate for 10–20 min to allow to the sperm to disperse. Remove the tissue and discard. After sperm is ready, aliquot 10 μl/fertilization drop using a wide-bore pipette tip for each control dish.
6. Sacrifice superovulated females 13 h post-hCG. Quickly dissect females and place oviducts under the oil in the fertilization dish, taking care to submerge the oviducts so that they do not float into the fertilization drops. Avoid excessive blood on the oviducts (rinse in IVF medium or blot blood away from tissue if present), as blood will poison the IVF.
   a. If you are using frozen sperm, use oocytes from 5 or fewer donors per dish.
   b. If you are using fresh sperm, you can use up to 10 donors per dish.
7. Grasp an oviduct with a pair of Dumont forceps, and using a tuberculin needle, tear the ampulla to release the "clutch" of oocytes and cumulus cells. Drag the clutch into the fertilization drop with the needle. Discard the oviduct. Repeat for all oviducts.

8. Incubate the fertilization dishes at 37 °C for about 4 h under a tri-gas atmosphere.
9. Using a glass transfer pipette, wash the oocytes through a wash drop in the fertilization dish and then place in the culture drop.
10. Culture overnight at 37 °C.
11. Count the number of two-cell and unfertilized eggs and calculate the fertilization rate (percentage of eggs that are fertilized).
12. Transfer, freeze, or culture the two-cell embryos.

4. Cryopreservation of Oocytes and Ovaries

The cryopreservation of female gametes has the same caveats as the cryopreservation of male gametes with respect to the preservation of only a haploid genome and of the necessity of having a source of appropriate male gametes for recovery. There are two strategies for accomplishing this: cryopreservation of oocytes, from which embryos are recovered by IVF, and cryopreservation of ovaries, in which recovery is accomplished via the surgical transfer of the thawed ovary to a histocompatible or immunocompromised host that is then bred to produce offspring. The former again requires proficiency at IVF, while the latter requires surgical skill and a source of host animals.

Oocyte cryopreservation requires the use of vitrification. It is believed that the more complex structure of the oocyte with its cortical granules interferes with the movement of water from the cell. The method for vitrification of oocytes is identical to the cryopreservation of embryos outlined in Section 2.2 above, with the following simple modifications: oocytes are removed from the cumulus mass with hyaluronidase and incubated for 10 min in 10% FBS in embryo medium before proceeding with the vitrification protocol.

This method can be extended to whole ovaries as described by Liu *et al.* (2008). In this case, the modification is to simply collect ovaries from 10-day-old mice and rinse them three times with medium before beginning the vitrification process, using twice the volume of DMSO and DAP213. Thawed ovaries are held at 37 °C in isotonic medium until surgical transfer to a 4-week-old recipient.

5. Troubleshooting

5.1. Troubleshooting embryo cryopreservation

Failure to recover intact embryos after freezing and thawing is almost always the result of either intracellular ice formation during the freezing process, improper storage, or osmotic shock upon thawing. Problems during the freezing process, in turn, usually revolve around a failure in the seeding process. If one observes embryos immediately upon thawing, one should see embryos of approximately normal size, with >90% intact. If most of the embryos are dead at this point, it is likely that seeding was not done properly. If all the embryos are dead, it is quite possible that the sample was improperly stored, and was allowed to thaw and refreeze in an uncontrolled manner or, in the case of a vitrified sample, warmed enough to allow crystallization. If embryos are initially intact, but are then lysed after transfer to culture medium, it is likely that they did not reside for a long enough time in the osmotic buffer. If the embryos are intact but no live animals are recovered, it is possible that the CPA has begun to degrade, although it is also possible that the particular strain does not survive cryopreservation. These possibilities can be differentiated by comparing the results to the recovery of control embryos frozen at the same time as the archived strain (see Section 5.2). The chemicals used for CPA should be replaced yearly.

5.2. Troubleshooting sperm cryopreservation

The sperm cryopreservation process is straightforward, and the failure to recover motile sperm after thawing is most likely due to uncontrolled thawing and refreezing of the sample. However, individual strains may not be successfully cryopreserved.

The IVF process for recovering embryos from the sperm is, on the other hand, a complex and delicate process fraught with possibilities for failure. Practitioners are urged to become proficient with the technique using freshly isolated spermatozoa before attempting to utilize frozen sperm, since the cryopreservation process will reduce viability of sperm from most inbred strains. It is important to perform a control IVF with fresh sperm to differentiate problems in the IVF process itself from problems originating with the cryopreserved sample.

If it develops that cryopreserved samples of one particular strain are not capable of generating embryos by IVF, all is not lost. Since the technique of intracytoplasmic sperm injection (ICSI) can produce viable embryos using nonmotile and even lysed sperm (Kuretake *et al.*, 1996), it may be possible to contract with a commercial provider of ISCI services in order to recover live animals from otherwise nonviable sperm.

6. Beyond the Methods: Other Considerations

In addition to the technical expertise required for cryopreservation, a number of other factors must be considered in cryopreserving mouse strains. These include storage and security, quality control, and record management.

6.1. Storage and security

Samples stored at liquid nitrogen temperatures will retain their viability for millennia (Dufrain, 1976; Mazur, 1976). However, uncontrolled thawing and refreezing of samples will result in their destruction, and it is therefore extremely important to insure that sample remains at cryogenic temperatures. Although mechanical cryogenic storage freezers exist, they are dependent upon an uninterrupted power supply, and storage in a liquid nitrogen Dewar system is therefore preferable, since a short-term fluctuation in delivery schedules is not as immediately devastating as a loss of power.

Liquid nitrogen storage systems are available for either liquid-phase or vapor-phase storage. The latter is extolled by some in order to avoid the potential cross-contamination of samples that might result from pathogens dispersing through liquid nitrogen in liquid storage. However, Pomeroy *et al.* (2009) argue convincingly that this risk is essentially nonexistent for cryopreserved germplasm. Given that vapor-phase storage units are to some degree dependent upon mechanical systems to mix the vapor phase and prevent thermal layering within the vapor, and because vapor phase has a lower heat capacity for rapid recooling of samples that are warmed to a small degree whenever storage racks are pulled from the freezer during sample introduction and removal, liquid-phase storage is probably preferable.

It is a good practice to divide all frozen stocks between at least two separate freezers, so that failure of one freezer does not result in the loss of all the stored material. Additional security can be obtained if some material is stored off-site, thus guarding against a catastrophic disaster such as fire.

Naturally, it is imperative that storage freezers are charged with liquid nitrogen at all times. Although automatic filling systems are available, it is preferable that these be used simply as backup and instead depend upon a schedule of manual fills that are recorded in a log book, since mechanical filling systems can fail. Alarm systems for low liquid levels add another layer of security.

In summary, as long as the liquid nitrogen is replenished, samples stored therein are, for all practical purposes, viable forever. All reasonable measure should be taken to insure that this occurs.

6.2. Quality control

The definitive demonstration of successful cryopreservation is the recovery of live animals from frozen material. Other measures of successful cryopreservation can be utilized such as recovery of intact embryos or motile sperm, viable *in vitro* culture of embryos, or successful IVF using recovered oocytes or sperm, and some practitioners may be confident enough in their abilities to substitute these less stringent criteria. Nonetheless, thawing an aliquot of frozen material for an assessment of quality control is good practice. However, this approach only provides data for a single aliquot frozen on a single day. In cases where the requisite inventory is collected in many sessions occurring over a period of many weeks or even months, as might happen if embryos are being collected as they become available from a small breeding colony, continued quality control for the duration of the project can be monitored by assessing the viability of aliquots of control embryos collected and frozen at the same time as the archival samples. This will reveal problems that may arise in the freeze process itself such as bad media or a faulty run of the controlled-rate freezer.

6.3. Data management

At the very least, a cryorepository requires an inventory system so that samples can be identified and located for retrieval. This inventory management system can be a complex electronic database or a simple paper ledger book, but however these records are kept, the system must be robust, since samples in the repository will retain viability for periods far longer than the careers—indeed, the lives—of the repository personnel. Thus, in addition to simple information about the contents and location of a particular sample, it may be useful to include information that will be useful for recovery of a particular strain such as unusual husbandry requirements or breeding characteristics in order to guide future generations of repository caretakers.

7. Summary

Cryopreservation is useful process for minimizing the resources devoted to maintaining and distributing mouse strains. Given the large number of different mouse strains that have been and continue to be developed, cryopreservation is the only viable method of maintaining this scientific resource without overwhelming the available capacity for mouse husbandry. This has led to the development of large national and international public and commercial cryorepositories of mouse strains, supplemented by any number of more local repositories based in individual labs and institutional core facilities. Taken together, these repositories will serve to safeguard this invaluable scientific resource.

ACKNOWLEDGMENTS

The work of the author is supported by funds from the National Cancer Institute and Thomas Jefferson University. I wish to thank Drs. Stanley Leibo and Larry Mobraaten and my former colleagues at the Jackson Laboratory Cryopreservation Program for continuing collaborative interaction and Jennifer Dunlap and Erin McDermott for assistance in preparing this manuscript.

REFERENCES

Bath, M. (2003). Simple and efficient in vitro fertilization with cryopreserved C57BL/6J mouse sperm. *Biol. Reprod.* **68,** 19.

Critser, J. K., and Mobraaten, L. E. (2000). Cryopreservation of murine spermatozoa. *ILAR J.* **41,** 197–206.

Dufrain, R. J. (1976). *In* "Basic Aspects of Freeze-Preservation of Mouse Strains," (O. Mühlbock, ed.), pp. 73–84. Gustav Fischer, Stuttgart, Germany.

Glenister, P. H., and Thornton, C. E. (2000). Cryoconservation—Archiving for the future. *Mamm. Genome* **11,** 565–571.

Keskintepe, L., Agca, Y., Pacholczyk, G. A., Machnicka, A., and Critser, J. K. (2001). Use of cryopreserved pronuclear embryos for the production of transgenic mice. *Biol. Reprod.* **65,** 407–411.

Koshimoto, C., and Mazur, P. (2002). The effect of the osmolality of sugar-containing media, the type of sugar, and the mass and molar concentration of sugar on the survival of frozen-thawed mouse sperm. *Cryobiology* **45,** 80–90.

Koshimoto, C., Gamliel, E., and Mazur, P. (2000). Effect of osmolality and oxygen tension on the survival of mouse sperm frozen to various temperatures in various concentrations of glycerol and raffinose. *Cryobiology* **41,** 204–231.

Kuretake, S., Kimura, Y., Hoshi, K., and Yanagimachi, R. (1996). Fertilization and development of mouse oocytes injected with isolated sperm heads. *Biol. Reprod.* **55,** 789–795.

Landa, V., and Slezinger, M. S. (1992). Production of transgenic mice from DNA-injected embryos cryopreserved by vitrification in microdrops. *Folia Biol. (Praha)* **38,** 10–15.

Landel, C. P. (2005). Archiving mouse strains by cryopreservation. *Lab. Anim.* **34,** 50–57.

Leibo, S. P., DeMayo, F. J., and O'Malley, B. (1991). Production of transgenic mice from cryopreserved fertilized ova. *Mol. Reprod. Dev.* **30,** 313–319.

Liu, L.-J., Xie, X.-Y., Zhang, R.-Z., Xu, P., Bujard, H., and Jun, M. (2008). Reproduction and fertility in wild-type and transgenic mice after orthotopic transplantation of cryopreserved ovaries from 10-d-old mice. *Lab. Anim.* **37,** 353–357.

Marschall, S., and Hrabe de Angelis, M. (1999). Cryopreservation of mouse spermatozoa: Double your mouse space. *Trends Genet.* **15,** 128–131.

Mazur, P. (1976). *In* "Basic Aspects of Freeze-Preservation of Mouse Strains," (O. Mühlbock, ed.), pp. 1–12. Gustav Fischer, Stuttgart, Germany.

Mobraaten, L. E. (1986). Mouse embryo cryobanking. *J. In Vitro Fert. Embryo Transf.* **3,** 28–32.

Mobraaten, L. E. (1999). Cryopreservation in a transgenic program. *Lab. Anim.* **28,** 15–18.

Nakagata, N. (2000). Cryopreservation of mouse spermatozoa. *Mamm. Genome* **11,** 572–576.

Nakao, K., Nakagata, N., and Katsuki, M. (1997). Simple and efficient vitrification procedure for cryopreservation of mouse embryos. *Exp. Anim.* **46,** 231–234.

Ostermeier, G. C., Wiles, M. V., Farley, J. S., Taft, R. A., and El-Shemy, H. A. (2008). Conserving, distributing and managing genetically modified mouse lines by sperm cryopreservation. *PLoS ONE* **3,** e2792.

Pomeroy, K. O., Harris, S., Conaghan, J., Papadakis, M., Centola, G., Basuray, R., and Battaglia, D. (2010). Storage of cryopreserved reproductive tissues: Evidence that cross-contamination of infectious agents is a negligible risk. *Fertil. Steril.* (in press).

Quinn, P., Kerin, J. F., and Warnes, G. M. (1985). Improved pregnancy rate in human in vitro fertilization with the use of a medium based on the composition of human tubal fluid. *Fertil. Steril.* **44,** 493–498.

Rall, W. F., and Fahy, G. M. (1985). Ice-free cryopreservation of mouse embryos at −196 degrees C by vitrification. *Nature* **313,** 573–575.

Renard, J. P., and Babinet, C. (1984). High survival of mouse embryos after rapid freezing and thawing inside plastic straws with 1-2 propanediol as cryoprotectant. *J. Exp. Zool.* **230,** 443–448.

Shaw, J. M., and Nakagata, N. (2002). Cryopreservation of transgenic mouse lines. *Methods Mol. Biol.* **180,** 207–228.

Suzuki-Migishima, R., Hino, T., Takabe, M., Oda, K., Migishima, F., Morimoto, Y., and Yokoyama, M. (2009). Marked improvement of fertility of cryopreserved C57BL/6J mouse sperm by depletion of Ca^{2+} in medium. *J. Reprod. Dev.* **55,** 386–392.

Sztein, J. M., Farley, J. S., and Mobraaten, L. E. (2000). In vitro fertilization with cryopreserved inbred mouse sperm. *Biol. Reprod.* **63,** 1774–1780.

Whittingham, D. G., Leibo, S. P., and Mazur, P. (1972). Survival of mouse embryos frozen to −196 degrees and −269 degrees C. *Science* **178,** 411–414.

CHAPTER SEVEN

Ovarian Follicle Culture Systems for Mammals

David F. Albertini* *and* Gokhan Akkoyunlu[†]

Contents

Abstract

Current approaches for the propagation and maintenance of mammalian ovarian follicles are reviewed in the context of the two dominant functional attributes of the follicle: sustaining oogenesis and regulating steroidogensis. Evidence is summarized indicating that the major utility for follicle cultures until recently has been to understand the steroidogenic properties of the follicle especially with regard to estrogen biosynthesis. Less stringent regulation of exposure to gonadotropins is tolerated in most systems when cell survival and differentiation are monitored for both the theca and granulosa components. In sharp contrast, establishing primary cultures of follicles for the maintenance of oogenesis is far more sensitive to a variety of experimental conditions and can in many cases compromise both the growth and maturative phases of oogenesis, a more recent focus for the applications of follicle culture in assisted reproductive technologies. Specific parameters that are discussed include media supplements, oocyte–granulosa interactions, culture atmosphere requirements, and maintenance of three-dimensional architecture. Limitations

* KU Cancer Center, Kansas University School of Medicine, Kansas City, Kansas, USA
† Department of Histology and Embryology, Faculty of Medicine, Akdeniz University, Antalya, Turkey

Methods in Enzymology, Volume 476
ISSN 0076-6879, DOI: 10.1016/S0076-6879(10)76007-9

based upon species variations and the need for microfluidic devices are discussed in the context of clinically relevant translational goals especially as they pertain to the emerging field of fertility preservation.

1. Background

1.1. Introduction

Removing cells and tissues from their *in vivo* context has been a widely popularized and productive experimental paradigm in contemporary biomedical research. The challenges confronted by complex tissue assemblies during *ex vivo* culture are formidable and often are a consequence of hypoxia due to a loss of vascularity and the removal of the mechanical and cellular parameters conferred within three-dimensional (3D) environments (Gomes *et al.*, 1999; Heise *et al.*, 2009; Torrance *et al.*, 1989; Yeung and Ng, 2000). The development of methods that would adequately support the growth and development of mammalian ovarian follicles required attention to each of these components and continues to engage the interests of laboratories probing the physiology of the follicle at a molecular and cellular level (Murray and Spears, 2000; Nayudu *et al.*, 2001; Picton *et al.*, 2003; Salustri and Martinozzi, 1980; Yoshimura *et al.*, 1992b). It is the purpose of this chapter to review the most recent advances in this field. To do so, a brief review of what is presently known about the physiology of the ovarian follicle in mammals will be provided followed by a detailed discussion of current 2D and 3D culture systems with focus on limitations most pertinent to what has become a clinically relevant application for this area of research. Among the recent clinical imperatives for follicle culture has been the necessity to derive developmentally competent human oocytes from cryopreserved ovarian cortex (Navarro-Costa *et al.*, 2005). The prospect of generating oocytes capable of undergoing *in vitro* maturation, fertilization, and early embryonic development is gaining support from advances in the field of follicle culture research and is shifting the emphasis from sustaining the endocrine aspects follicular differentiation to those properties of the follicle that nurture oogenesis through the growth and maturative phases of oocyte development (Hutt and Albertini, 2007). While evidence to support this idea has only been demonstrated in the mouse thus far, progress in this arena in human follicles and those from appropriate animal models represents a new and exciting area of assisted reproductive technologies (ARTs).

1.2. Physiology of the mammalian ovarian follicle

Like most functional tissue units within organs, the ovarian follicle is composed of several distinct cell types whose temporal and spatial integration define its physiological role in reproduction (Rodrigues *et al.*, 2008).

Two main functions are ascribed to the follicle. From a historical perspective, the role of the follicle as a secretory unit is best recognized and over the years, the diversity and complexity of the secreted substances and the regulation of hormone biosynthesis and release has been thoroughly investigated (Albertini *et al.*, 2001). Among these factors, steroid hormones of the family of estrogens and androgens were the first to be identified. Over the past 20 years, a new realization has been made with the discovery of many peptide and protein factors whose expression is highly regulated in a developmental context and whose secretion plays both a local and systemic function in the physiology of female reproduction. Interestingly, many of these protein factors are known to be members of the TGF-β super family of growth and differentiating ligands (Cortvrindt and Smitz, 2001; Eppig, 1977; Fukui *et al.*, 1987; Greenwald and Moor, 1989). These signaling molecules take origin from the distinct cell populations that constitute the ovarian follicle implicate and these factors have been widely implicated as paracrine and autocrine regulators of follicle development, viability, and differentiation (Kreeger *et al.*, 2006; Kwon and Lee, 1991; Meinecke and Meinecke-Tillmann, 1979). Figure 7.1 provides a simplistic scheme for understanding this aspect of follicle physiology.

The second major function of the ovarian follicle is the role it plays in the storage, differentiation, survival, and maturation of the oocyte (Rodrigues *et al.*, 2008; Smitz and Cortvrindt, 1998, 1999). Only in the last 10 years, it has become more broadly recognized that the fate of the follicle is determined by the oocyte as a result of the production of growth factors whose expression in most cases is a limiting and defining characteristic of the female germ line (Rodrigues *et al.*, 2008; Smitz and Cortvrindt, 2002; Sun *et al.*, 2004). GDF-9, BMP-15, and FGF-8 have emerged as central players in the regulation of follicle growth, in many cases acting as regulators of growth suppression during the slow preantral stages of follicle development or as differentiation repressors during the gonadotropin-dependent phases of antral follicle growth and differentiation (Albertini *et al.*, 2001; Hutt and Albertini, 2007). While the details of most of the reciprocal relationships between the major cell types of the follicle have yet to be fully uncovered, certain guiding principles have emerged that are of immediate relevance to the subject of this chapter. In general, these interactions fit the criteria ascribed to negative feedback or feed forward loops between the three major cell types found in the follicle: oocytes, granulosa cells, and thecal cells. Crucial to the design and implementation of follicle culture studies, it is important to note that the mass ratio of these cell types varies widely over the course of follicle development being dominated by the oocyte and granulosa in the earlier stages of folliculogenesis and being dominated by the theca and granulosa in the fully grown steroidogenically active mature dominant follicles (Jacquet *et al.*, 2005; Johnson *et al.*, 1995; Kreeger *et al.*, 2005). This developmental shift in cellular mass relationships

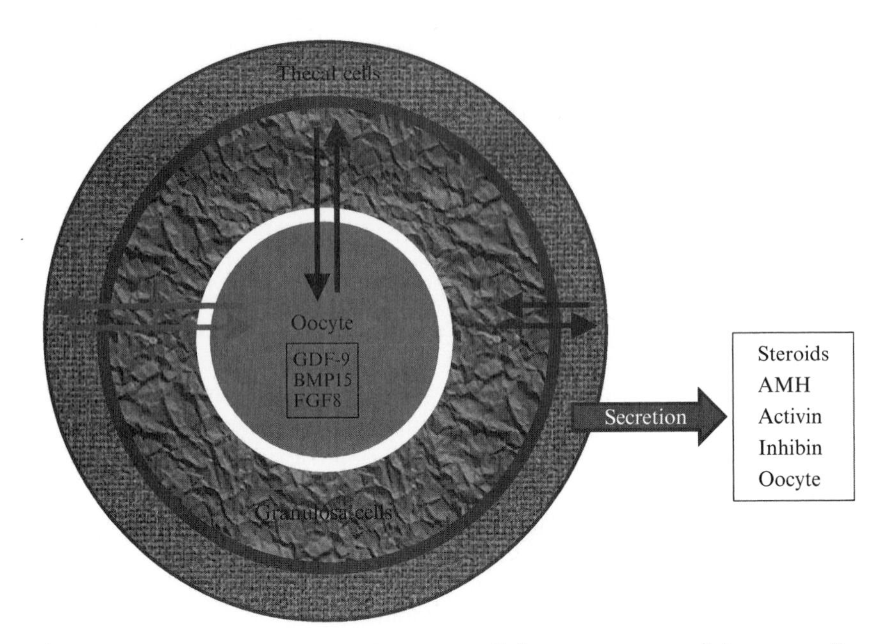

Figure 7.1 Diagram illustrating the major cellular components of the mammalian ovarian follicle and putative regulatory pathways for cell communication that underly follicular development. Note that oocyte specific factors that include GDF-9, BMP-15, and FGF-8 are believed to create a morphogen gradient from the oocyte outwards that likely inhibits granulosa cell differentiation, and perhaps proliferation, until receptors for FSH or LH become activated. The mechanisms that are used to transport these factors across the zona-pellucida to surrounding somatic cells are poorly understood but probably involve transzonal projections (*vide infra*).

reflects the two general aspects of the follicular physiology that are initially focused on the growth and development of the oocyte in preantral follicles and later take on the characteristics of an endocrine organ responsible for production of androgens by the theca that are then converted into estrogens by the aromatase activity of the granulosa (Hu *et al.*, 2002). From a historical perspective, it is the steroidogenic function of the follicle that has been most completely studied using 2D culture systems (Bishonga *et al.*, 2001; Carroll *et al.*, 1991; Cortvrindt *et al.*, 1998a). As we will see, this field has shifted focus to the preantral stages of development owing to the growing need to produce viable and developmentally oocytes in the area of ARTs.

1.3. Isolation of follicles for preparation of cultures

The objective of follicle culture experiments will vary according to what aspect of follicle functionality is subject to study. Variables that must be factored into follicle culture experiments include animal age and endocrine/

reproductive status (prepubertal, adult cycling, aged with erratic cycling) (Adam *et al.*, 2004; Adriaens *et al.*, 2004), follicle size and cellular composition (Carrell *et al.*, 2005; Cortvrindt *et al.*, 1996), desired outcome measures on either oocyte or somatic cell functions (Cortvrindt *et al.*, 1998b; Eppig and O'Brien, 1996), method of follicle isolation (Eppig, 1977; Eppig *et al.*, 1992; Fouladi-Nashta *et al.*, 1998), and media composition and supplements (Murray *et al.*, 2008; Nayudu *et al.*, 1994; Smitz *et al.*, 1996, 1998). Finally, the species of choice will measure into applying the best culture protocol as mammalian ovaries are striking variable with respect to their basic histological composition. For example, the ovaries of rodents are somewhat minimal in the relative content of connective tissue stroma, especially in young animals, whereas the ovaries of human and nonhuman primates and those of many of the domesticated species (bovine, ovine, caprine, canine) are stroma-rich and contain dense connective tissue in the cortex where most of the earliest stages of folliculogenesis are found. Accordingly, standard tissue dissociation methods are indicated in species with an abundance of ovarian stroma while physical methods for follicle isolation, drawing on microdissection strategies, suffice and work well when using rodent ovaries (Johnson *et al.*, 1995; Spears *et al.*, 1994; Sun *et al.*, 2005). Figure 7.2 provides an example of the differences in follicles isolated by these methods using the Rhesus monkey as a source of follicles.

Given the abundance of stromal collagen in the primate ovary, the use of a combination of collagenase, trypsin, and EDTA under neutral pH conditions is generally required to achieve dissociation of follicles from the ovarian extracellular matrix. Ovaries should first be trimmed of any adherent tissue and all antral follicles should be suctioned to remove enclosed oocytes and granulosa cells. The presence of the corpus luteum is not desirable as a potential source of cellular contamination and thus it is recommended that this structure be removed before mincing ovaries. Tissues are typically minced with a scalpel blade into pieces after the underlying medullary and blood vessel rich has been scraped away (also with a scalpel blade). One- to two-millimeter-thick strips measuring

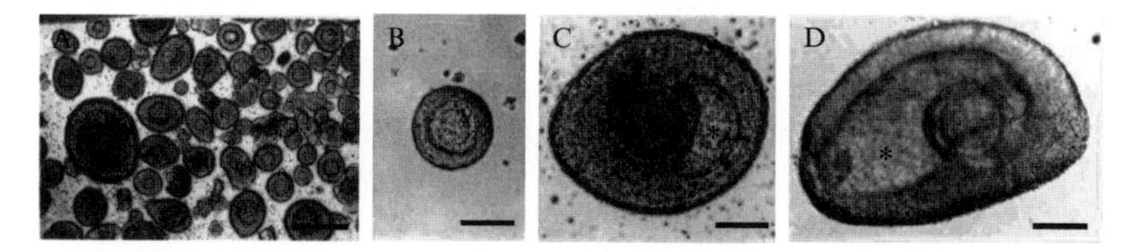

Figure 7.2 Rhesus monkey follicles released following collagenase-mediated tissue dissociation. Pooled follicles of different sizes (A) can be sorted into primary (B) and early antral (C, D) and, if enzyme exposure is prolonged, a clear loss of stromal attachment to the follicle basement membrane is observed as indicated by the smooth external contour of isolated follicles.

approximately 1–2 cm in length are incubated in tissue dissociation media with agitation for anywhere from 10 to 60 min. Follicles will be released at various times depending on their size, location, and amount of investing connective tissue they display. As shown in Fig. 7.2, preantral follicles of various sizes are retrieved using a suction pipette whose inner diameter can be adjusted to aid in sorting follicles into different size classes. Alternatively, nylon mesh of varying pore sizes can be used to sequentially filter the released follicles according to size and hence stage of development. Primary and even smaller antral follicles can be obtained in this way (Fig. 7.2B and C). It is critical to limit the time of enzyme exposure and also quench enzymatic activity as soon as follicles are released. This is typically accomplished with excess (10%) whole serum or bovine serum albumin exposure and transferring follicles out of dissociation solutions. Two adverse consequences of prolonged enzyme exposure have been identified: (1) complete removal of the theca-stroma attached to the basement membrane and (2) damage to the zona-pellucida and enclosed structures that connect the oocyte to the granulosa cells. For these reasons, it is best to select follicles that contain adherent stroma as this will facilitate plating and tend to support retention of the spherical shape of the follicle (Adriaenssens *et al.*, 2009; Johnson *et al.*, 1995; Nayudu *et al.*, 2003; Nogueira *et al.*, 2005). There is at present no facile way to evaluate the integrity of cell interactions within the follicle under noninvasive conditions. However, our preliminary studies do indicate that when oocytes are isolated from enzyme released follicles, they begin to exhibit alterations in both chromatin structure and phosphoprotein distribution that are consistent with a loss of cell contact at the oocyte–granulosa interface. Figure 7.3 illustrates the architecture of germinal vesicles from a freshly isolated follicle and one that had been subjected to enzyme dissociation for 30 min. A tendency to exhibit progressive chromatin condensation is associated with prolonged enzyme exposure and this is often accompanied by a change in the distribution of MPM-2 nuclear foci that are indicative of active transcriptional zones within the nucleus.

2. 2D Culture Systems

2D culture systems for ovarian follicles take their historical roots from the identification of theca stromal cells as cells of mesenchymal origin that demonstrate a high plating efficiency when follicles containing an investing stroma are placed on charged surfaces in the presence of serum (Rodrigues *et al.*, 2008; Smitz and Cortvrindt, 1999, 2002). This approach, generally, is most successful when physical dissection of follicles from the surrounding stroma is undertaken as enzymatic dissociation of mammalian ovary results is removal of the thecal investment of primary and early secondary follicles

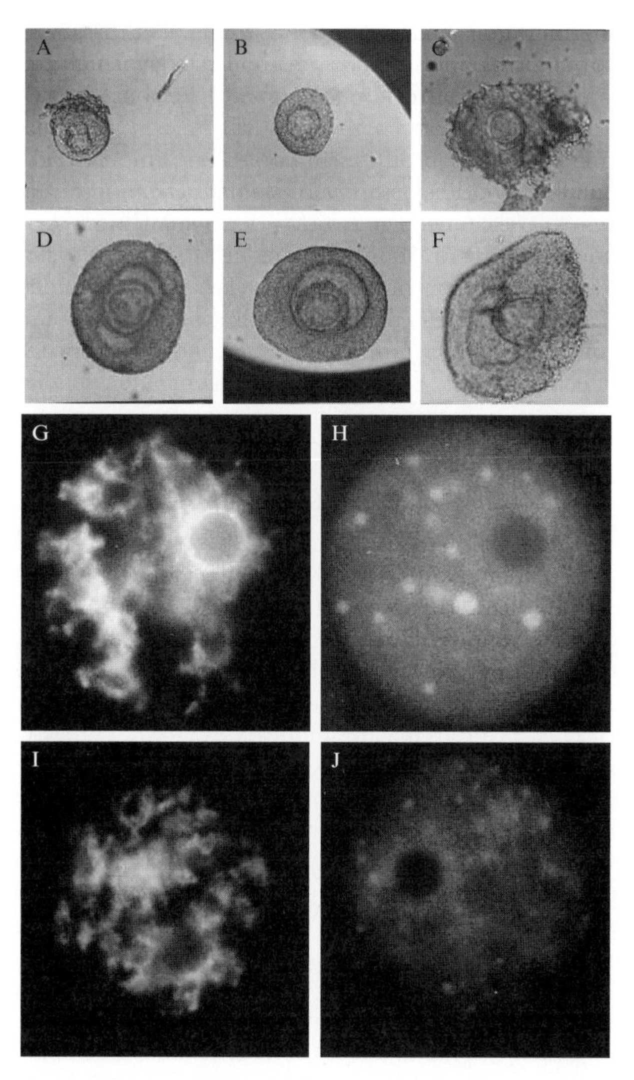

Figure 7.3 (A–F) Various stages of follicles released after enzyme exposure and different degrees of stroma associated at the follicle surface are indicated suggesting loss of thecal-stromal investments. Panel (F) shows that when stromal digestion is excessive, the perimeter of the follicle will become irregular in shape causing collapse of the antral cavity and dissociation of the cumulus mass from the follicle wall. Lower panels demonstrate status of germinal vesicle chromatin and nuclear phosphoprotein distribution (MPM-2 immunolabeling) in freshly isolated and enzyme-exposed oocytes of dissociated rhesus monkey follicles.

precluding efficient plating (Trounson *et al.*, 2001; Yoshimura *et al.*, 1992a). Typically, in studies on rodents, primary and secondary stage follicles are used and the relative abundance of these stages in prepubertal ovaries make this developmental age optimal for retrieving follicles that are both plentiful in numbers and relatively homogeneous in developmental status (Tsafriri *et al.*, 1989). Whether culture vessels are flat surfaced or conical, such as with various multiwell formats, the progression of cultures usually leads to the adherence and spreading of thecal stromal cells over the first 48 h to the substrate causing the transformation of follicles from a spherical ovoid configuration (Nayudu *et al.*, 2003; Nogueira *et al.*, 2005). Follicles cultured in this way are readily sustained for up to 10–12 days in humidified 5% CO_2/air with medium changes recommended every 48 h. Media composition plays an important part in the functional outcomes that can be assayed from this type of culture.

For example, as follicles acquire responsiveness to follicle-stimulating hormone (FSH), the principle growth and differentiating promoting factor in mammalian follicles, inclusion of FSH preparations that range from high purity (recombinant human FSH) to those with varying ratios of FSH and LH (luteinizing hormone) activity (human menopausal gonadotropin, equine chorionic gonadotropin) will yield strikingly different results (Adriaenssens *et al.*, 2009; Eppig *et al.*, 1992; Fouladi-Nashta and Campbell, 2006; Gomes *et al.*, 1999). Media formulations with FSH will stimulate granulosa cell proliferation and induction of various differentiation markers such as aromatase, LH receptors, and hyaluronic acid (Fouladi-Nashta and Campbell, 2006; Hu *et al.*, 2002; Kwon and Lee, 1991; Meinecke and Meinecke-Tillmann, 1979). While this response mimics that observed *in vivo*, and results in significant estradiol synthesis and release that is readily measured in conditioned medium samples, reports of premature luteinization evidenced by the production of progesterone have been made and raise concerns regarding the viability of enclosed oocytes (see below). Early stage rodent follicles have also been reported to exhibit strong growth and differentiation properties when media are supplemented with a variety of other growth factors such as epidermal growth factor (EGF), GDF-9, and BMP-15 (Cortvrindt *et al.*, 1997, 1998a; Picton *et al.*, 2003; Smitz *et al.*, 1998). Interestingly, and unlike FSH, these latter factors have been shown to encourage proliferation of granulose cells but appear to impair the tendency of granulosa cells to undergo premature luteinization. It is likely then that future formulations will involve using a sequential media paradigm that allows for reduplication of the slow growth and limited differentiation that would be expected of follicles at the primary and secondary stages in order to maximize functions related to supporting oogenesis followed by media that would bring follicle development to its physiological conclusion—the process of ovulation.

In the mouse system, achieving ovulation has become a routine and reproducible outcome that can be elicited by exposing follicles that have undergone prior exposure to FSH to a pulse of LH or hCG (human chorionic gonadotropin), two commonly employed *in vivo* ovulation inducing agents (Xu *et al.*, 2006; Yeung and Ng, 2000; Yoshimura *et al.*,

1992a). In such cases, the literature suggests that an efficiency for induced ovulation can range from 20% to 40% and release of the ovum within a mucified cumulus mass occurs some 12–14 h following administration of LH (or hCG) recapitulating the timing of follicle rupture that is seen *in vivo* for mice (Adam *et al.*, 2004; Bishonga *et al.*, 2001; Spears *et al.*, 1994; Trounson *et al.*, 2001). Significantly, ova collected from such mucified masses of cumulus cells can be successfully fertilized *in vitro*, cultured through preimplantation stages up to the blastocyst, and when transferred such embryos give rise to live young. Again, while this proof of principle has yet to be demonstrated for any other mammalian species, there are encouraging results from several laboratories and indications that the failure to achieve ovulation and embryo development in other animal models may be due to deficiencies in 2D cultures systems or the current protocols being used.

For example, studies by Johnson *et al.* (1995) showed quite clearly that one of the consequences of regular exposure to FSH is a precocious repression of oocyte transcription and resumption of meiosis that could be prevented by removing FSH between days 2 and 6 of culture. These and other studies reinforce the notion that maintenance of cell communication between granulosa cells and oocytes may be interrupted by FSH (Cortvrindt *et al.*, 1997, 1998b) which has been shown to reduce the integrity of cell–cell contacts within the follicle (Combelles *et al.*, 2004). Consistent with this is the finding that mouse oocytes grown in various culture systems usually do not achieve full size (Fig. 7.4) as would be indicated by a loss of somatic cell support and/or failure to sustain the transcriptional of oogenesis through to the end of the growth phase of oogenesis. These and other deficiencies intrinsic to existing 2D culture systems have prompted exploration of 3D systems in order to maintain biomechanical and architectural properties most consistent with folliculogenesis *in vivo* (Hu *et al.*, 2001; Murray *et al.*, 2008).

3. 3D Culture Systems

Efforts to design 3D culture systems for ovarian follicles have gained momentum in recent years (Carrell *et al.*, 2005; Gomes *et al.*, 1999; Hu *et al.*, 2001). While collagen overlays have been attempted in the past (Gomes *et al.*, 1999; Torrance *et al.*, 1989), most recent applications have drawn upon the use of calcium alginate gels due to their chemical inertness, ease of preparation, and stability in various media formulations (Heise *et al.*, 2005; Kreeger *et al.*, 2006; Xu *et al.*, 2006). Once again, in most cases where comparisons have been made directly between 2D and 3D cultures, this work has been confined to rodent models and with rare exception,

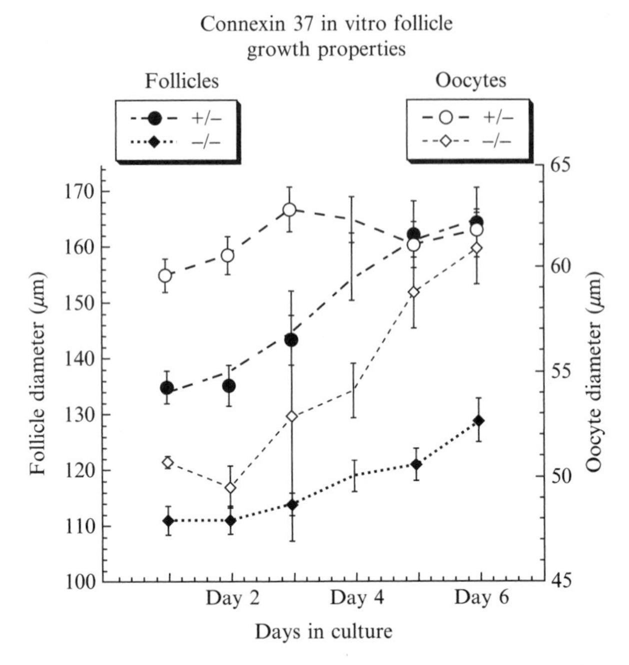

Figure 7.4 Graph depicting patterns of follicle and oocyte growth for mice heterozygous or homozygous null (−/−) for the CX 37 gene. Note that in lower two curves, heterozygous follicles show steady increments in both follicle and oocyte growth over a 6-day culture period. In contrast, loss of CX 37, a gap junction component needed for metabolic cooperation between oocytes and granulose cells, oocytes are typically arrested at about 50 μm diameters *in vivo* but when cultured grow considerably while follicle diameter remains about the same over 6 days.

endocrine and growth properties of the follicle, and not the enclosed oocyte, have been the focus of these investigations (Gomes *et al.*, 1999; Harris *et al.*, 2007). One of the more complete studies is that of Heise *et al.* (2005). In this work, juvenile rats were used to obtain 150–160 μm follicles that were cultured in either flat plates, suspension culture or after embedding follicles within calcium alginate. Another important parameter that was examined was the effect of FSH on the expression of connexin 43 (CX 43), a critical gap junctional integrator of development. In this regard, inclusion of FSH within alginate beads elicited a more penetrant and thorough expression pattern of CX 43 throughout the follicle when compared to exogenous exposure or exposure within suspension cultures. These studies raise the much neglected point of growth factor access to targets within the follicle and identify alginate beads as possibly deficient in allowing for the penetration of proteins known to be required for follicular growth *in vivo* (Albertini *et al.*, 2001).

An additional compounding factor in the long-term stability of follicle cultures is the optimization of gas and metabolite exchange over what can be protracted periods of time (days to weeks). Metabolite exchange in 3D cultures is likely to be limited as maintenance of cultures usually involves removal of minimal volumes of the culture vessel and thus would be expected to result in the accumulation of toxic wastes and metabolites (Murray and Spears, 2000; Murray *et al.*, 2008). Atmospheric oxygen is also used traditionally in these cultures and has only recently been addressed in a systematic way (Picton *et al.*, 2003; Smitz *et al.*, 1996).

Heise *et al.* (2009), again using the rat model, determined that given the changing environment that the follicle encounters as it proceeds through folliculogenesis, and gradually acquires a robust vasculature, oxygen tension would be expected to go from low to high levels of exposure. To test this, they compared the effects of static cultures with 20% oxygen to a dynamic range from 4% to 11% oxygen. Oocytes were evaluated for their ability to mature *in vitro* and their developmental potential was assessed by subjecting mature oocytes to parthenogenic activation. Both these measures of follicle integrity were improved after dynamic oxygen exposure when compared to the static atmospheric conditions that prevail in this field. Together with other studies focused on the ability of FSH to limit apoptosis and promote granulosa cell proliferation (Carrell *et al.*, 2005; Sun *et al.*, 2005; Tsafriri *et al.*, 1989), the advantages of 3D culture have become widely recognized and appreciated and will likely dominate this area of research for years to come.

4. Considerations Based on Species Variations

While the temptation to extrapolate between mammalian model systems is great owing to perceived similarities in tissue design principles and physiological control mechanisms, this has not been straightforward in the case of ovarian follicles (Nayudu *et al.*, 2003; Sun *et al.*, 2004). Few investigators have taken into serious consideration the vastly different parameters of size, tensegrity, cell–cell and cell–matrix interactions that are evident even between the more commonly used rodent models such as the rat and mouse. For example, at the time of ovulation, a mouse follicle approximates 400 μm in diameter whereas that of a rat approaches 800 μm, which on a volume basis exceeds two log orders despite the overall similar appearance of the follicle (Heise *et al.*, 2005). An additional complication with rodent models that exhibit polyovular estrous cycles is that depending on the age of the animals, the majority of follicles that would be isolated would be destined to undergo atresia based on the principles of follicle selection evidenced in these and other mammals (Rodrigues *et al.*, 2008).

Even in the case of the smaller classes of primary and secondary follicles that have begun to develop an antrum, significant variations exist in the pattern of cell adhesions, the integrity and cellular composition of the thecal investment, and the number of granulosa cell layers that are established prior to cavitation of the follicle (Rodrigues *et al.*, 2008). Although very little information currently exists as to satisfactory ways to achieve follicle growth between different mammalian species, the emerging principles of matrix bioengineering and microfluidics should guide future research in this area especially as it relates to the forces that are needed to sustain development of the oocyte.

5. Conclusions and Future Considerations

This chapter has sought to provide a set of guiding principles for the experimental use of ovarian follicle cultures. The major conclusions include: (1) considerations for tissue sourcing and isolation that will impact initial and prolonged activities of the follicle with respect to endocrine or germ line supporting functions, (2) comparisons of 2D and 3D culture systems which demonstrate clear differences in outcome measures depending in part on the species chosen for study, and (3) the need to further optimize current approaches using sequential media and culture strategies to accommodate the changing needs of the oocyte at successive stages of folliculogenesis. Continuing efforts in this direction are warranted and much needed given recent clinical imperatives.

For example, the follicle is an integrated unit and in a culture environment retains much of its functionality but so far has not been reliable for purposes of generating viable and developing oocytes. The emerging field of fertility preservation has recognized and called attention to the need to develop methods that would allow for the long-term storage of oocytes for couples seeking assistance with their reproduction. Women who undergo treatments for cancer, autoimmune diseases, or have been subject to environmental or occupational exposures that compromise ovarian function will as a result suffer from some degree of premature ovarian failure accelerating menopause onset and infertility. In such cases, there is widespread hope that such patients may be offered a new range of ARTs that include oocyte and ovarian tissue cryopreservation. Current efforts in many laboratories are aimed at optimizing culture systems that would support both the growth and maturative phases of oogenesis so that oocytes would be available for embryo production using traditional forms of ARTs. Achieving such a goal will require continued development of follicle culture models in mammalian organisms that are appropriate for translation to the human condition.

ACKNOWLEDGMENTS

We thank current and past members of the Albertini laboratory who have contributed to the work reported in this chapter and especially Alp Can, Susie Messinger, MaryJo Carabatsos, Lynda McGinnis, John Biggers, and the late Lorna Johnson. This work has been supported by the Hall Family Foundation, The ESHE Fund, and The Scientific and Technological Research Council of Turkey—TUBITAK (G. A.).

REFERENCES

Adam, A. A., *et al.* (2004). In vitro culture of mouse preantral follicles using membrane inserts and developmental competence of in vitro ovulated oocytes. *J. Reprod. Dev.* **50,** 579–586.

Adriaens, I., *et al.* (2004). Differential FSH exposure in preantral follicle culture has marked effects on folliculogenesis and oocyte developmental competence. *Hum. Reprod.* **19,** 398–408.

Adriaenssens, T., *et al.* (2009). Differences in collagen expression in cumulus cells after exposure to highly purified menotropin or recombinant follicle-stimulating hormone in a mouse follicle culture model. *Biol. Reprod.* **80,** 1015–1025.

Albertini, D. F., *et al.* (2001). Cellular basis for paracrine regulation of ovarian follicle development. *Reproduction* **121,** 647–653.

Bishonga, C., *et al.* (2001). In vitro growth of mouse ovarian preantral follicles and the capacity of their oocytes to develop to the blastocyst stage. *J. Vet. Med. Sci.* **63,** 619–624.

Carrell, D. T., *et al.* (2005). Comparison of maturation, meiotic competence, and chromosome aneuploidy of oocytes derived from two protocols for in vitro culture of mouse secondary follicles. *J. Assist. Reprod. Genet.* **22,** 347–354.

Carroll, J., *et al.* (1991). Effect of dibutyryl cyclic adenosine monophosphate on granulosa cell proliferation, oocyte growth and meiotic maturation in isolated mouse primary ovarian follicles cultured in collagen gels. *J. Reprod. Fertil.* **92,** 197–207.

Combelles, C. M., *et al.* (2004). Hormonal control of somatic cell oocyte interactions during ovarian follicle development. *Mol. Reprod. Dev.* **69,** 347–355.

Cortvrindt, R., and Smitz, J. (2001). In vitro follicle growth: Achievements in mammalian species. *Reprod. Domest. Anim.* **36,** 3–9.

Cortvrindt, R., *et al.* (1996). In-vitro maturation, fertilization and embryo development of immature oocytes from early preantral follicles from prepuberal mice in a simplified culture system. *Hum. Reprod.* **11,** 2656–2666.

Cortvrindt, R., *et al.* (1997). Assessment of the need for follicle stimulating hormone in early preantral mouse follicle culture in vitro. *Hum. Reprod.* **12,** 759–768.

Cortvrindt, R., *et al.* (1998a). Recombinant luteinizing hormone as a survival and differentiation factor increases oocyte maturation in recombinant follicle stimulating hormone-supplemented mouse preantral follicle culture. *Hum. Reprod.* **13,** 1292–1302.

Cortvrindt, R. G., *et al.* (1998b). Timed analysis of the nuclear maturation of oocytes in early preantral mouse follicle culture supplemented with recombinant gonadotropin. *Fertil. Steril.* **70,** 1114–1125.

Eppig, J. J. (1977). Mouse oocyte development in vitro with various culture systems. *Dev. Biol.* **60,** 371.

Eppig, J. J., and O'Brien, M. J. (1996). Development in vitro of mouse oocytes from primordial follicles. *Biol. Reprod.* **54,** 197–207.

Eppig, J. J., *et al.* (1992). Developmental capacity of mouse oocytes matured in vitro: Effects of gonadotropic stimulation, follicular origin and oocyte size. *J. Reprod. Fertil.* **95,** 119–127.

Fouladi-Nashta, A. A., and Campbell, K. H. (2006). Dissociation of oocyte nuclear and cytoplasmic maturation by the addition of insulin in cultured bovine antral follicles. *Reproduction* **131,** 449–460.

Fouladi-Nashta, A. A., *et al.* (1998). Maintenance of bovine oocytes in meiotic arrest and subsequent development In vitro: A comparative evaluation of antral follicle culture with other methods. *Biol. Reprod.* **59,** 255–262.

Fukui, Y., *et al.* (1987). Follicle culture enhances fertilizability and cleavage of bovine oocytes matured in vitro. *J. Anim. Sci.* **64,** 935–941.

Gomes, J. E., *et al.* (1999). Three-dimensional environments preserve extracellular matrix compartments of ovarian follicles and increase FSH-dependent growth. *Mol. Reprod. Dev.* **54,** 163–172.

Greenwald, G. S., and Moor, R. M. (1989). Isolation and preliminary characterization of pig primordial follicles. *J. Reprod. Fertil.* **87,** 561–571.

Harris, S. E., *et al.* (2007). Carbohydrate metabolism by murine ovarian follicles and oocytes grown in vitro. *Reproduction* **134,** 415–424.

Heise, M., *et al.* (2005). Calcium alginate microencapsulation of ovarian follicles impacts FSH delivery and follicle morphology. *Reprod. Biol. Endocrinol.* **3,** 47.

Heise, M. K., *et al.* (2009). Dynamic oxygen enhances oocyte maturation in long-term follicle culture. *Tissue Eng. Part C Methods* **15,** 323–332.

Hu, Y., *et al.* (2001). Effects of low O_2 and ageing on spindles and chromosomes in mouse oocytes from pre-antral follicle culture. *Hum. Reprod.* **16,** 737–748.

Hu, Y., *et al.* (2002). Effects of aromatase inhibition on in vitro follicle and oocyte development analyzed by early preantral mouse follicle culture. *Mol. Reprod. Dev.* **61,** 549–559.

Hutt, K. J., and Albertini, D. F. (2007). An oocentric view of folliculogenesis and embryogenesis. *Reprod. Biomed. Online* **14,** 758–764.

Jacquet, P., *et al.* (2005). Cytogenetic studies in mouse oocytes irradiated in vitro at different stages of maturation, by use of an early preantral follicle culture system. *Mutat. Res.* **583,** 168–177.

Johnson, L. D., *et al.* (1995). Chromatin organization, meiotic status and meiotic competence acquisition in mouse oocytes from cultured ovarian follicles. *J. Reprod. Fertil.* **104,** 277–284.

Kreeger, P. K., *et al.* (2005). Regulation of mouse follicle development by follicle-stimulating hormone in a three-dimensional in vitro culture system is dependent on follicle stage and dose. *Biol. Reprod.* **73,** 942–950.

Kreeger, P. K., *et al.* (2006). The in vitro regulation of ovarian follicle development using alginate-extracellular matrix gels. *Biomaterials* **27,** 714–723.

Kwon, H. B., and Lee, W. K. (1991). Involvement of protein kinase C in the regulation of oocyte maturation in amphibians (*Rana dybowskii*). *J. Exp. Zool.* **257,** 115–123.

Meinecke, B., and Meinecke-Tillmann, S. (1979). Effects of gonadotropins on oocyte maturation and progesterone production by porcine ovarian follicles cultured in vitro. *Theriogenology* **11,** 351–365.

Murray, A., and Spears, N. (2000). Follicular development in vitro. *Semin. Reprod. Med.* **18,** 109–122.

Murray, A. A., *et al.* (2008). Follicular growth and oocyte competence in the in vitro cultured mouse follicle: Effects of gonadotrophins and steroids. *Mol. Hum. Reprod.* **14,** 75–83.

Navarro-Costa, P., *et al.* (2005). Effects of mouse ovarian tissue cryopreservation on granulosa cell–oocyte interaction. *Hum. Reprod.* **20,** 1607–1614.

Nayudu, P. L., *et al.* (1994). Abnormal in vitro development of ovarian follicles explanted from mice exposed to tetrachlorvinphos. *Reprod. Toxicol.* **8,** 261–268.

Nayudu, P. L., *et al.* (2001). Progress toward understanding follicle development in vitro: Appearances are not deceiving. *Arch. Med. Res.* **32,** 587.

Nayudu, P. L., *et al.* (2003). In vitro development of marmoset monkey oocytes by pre-antral follicle culture. *Reprod. Domest. Anim.* **38,** 90–96.

Nogueira, D., *et al.* (2005). Effects of long-term in vitro exposure to phosphodiesterase type-3 inhibitors on follicle and oocyte development. *Reproduction* **130,** 177–186.

Picton, H. M., *et al.* (2003). Growth and maturation of oocytes in vitro. *Reprod. Suppl.* **61,** 445–462.

Rodrigues, P., *et al.* (2008). Oogenesis: Prospects and challenges for the future. *J. Cell. Physiol.* **216,** 355–365.

Salustri, A., and Martinozzi, M. (1980). In vitro culture of oocytes from growing mice: Behavior in defined medium and in the presence of follicular cells. *Boll. Soc. Ital. Biol. Sper.* **56,** 826–832.

Smitz, J., and Cortvrindt, R. (1998). Follicle culture after ovarian cryostorage. *Maturitas* **30,** 171–179.

Smitz, J., and Cortvrindt, R. (1999). Oocyte in-vitro maturation and follicle culture: Current clinical achievements and future directions. *Hum. Reprod.* **14**(Suppl 1), 145–161.

Smitz, J. E., and Cortvrindt, R. G. (2002). The earliest stages of folliculogenesis in vitro. *Reproduction* **123,** 185–202.

Smitz, J., *et al.* (1996). Normal oxygen atmosphere is essential for the solitary long-term culture of early preantral mouse follicles. *Mol. Reprod. Dev.* **45,** 466–475.

Smitz, J., *et al.* (1998). Epidermal growth factor combined with recombinant human chorionic gonadotrophin improves meiotic progression in mouse follicle-enclosed oocyte culture. *Hum. Reprod.* **13,** 664–669.

Spears, N., *et al.* (1994). Mouse oocytes derived from in vitro grown primary ovarian follicles are fertile. *Hum. Reprod.* **9,** 527–532.

Sun, F., *et al.* (2004). Preantral follicle culture as a novel in vitro assay in reproductive toxicology testing in mammalian oocytes. *Mutagenesis* **19,** 13–25.

Sun, F., *et al.* (2005). Aneuploidy in mouse metaphase II oocytes exposed in vivo and in vitro in preantral follicle culture to nocodazole. *Mutagenesis* **20,** 65–75.

Torrance, C., *et al.* (1989). Quantitative study of the development of isolated mouse pre-antral follicles in collagen gel culture. *J. Reprod. Fertil.* **87,** 367–374.

Trounson, A., *et al.* (2001). Maturation of human oocytes in vitro and their developmental competence. *Reproduction* **121,** 51–75.

Tsafriri, A., *et al.* (1989). Effects of transforming growth factors and inhibin-related proteins on rat preovulatory graafian follicles in vitro. *Endocrinology* **125,** 1857–1862.

Xu, M., *et al.* (2006). Identification of a stage-specific permissive in vitro culture environment for follicle growth and oocyte development. *Biol. Reprod.* **75,** 916–923.

Yeung, W. S., and Ng, E. H. (2000). Laboratory aspects of assisted reproduction. *Hong Kong Med. J.* **6,** 163–168.

Yoshimura, Y., *et al.* (1992a). Protein kinase C mediates gonadotropin-releasing hormone agonist-induced meiotic maturation of follicle-enclosed rabbit oocytes. *Biol. Reprod.* **47,** 118–125.

Yoshimura, Y., *et al.* (1992b). Direct effect of gonadotropin-releasing hormone agonists on the rabbit ovarian follicle. *Fertil. Steril.* **57,** 1091–1097.

CHAPTER EIGHT

Production of Mouse Chimeras by Aggregating Pluripotent Stem Cells with Embryos

Andras Nagy, Kristina Nagy, *and* Marina Gertsenstein

Contents

Abstract

Experimental mouse chimeras have served as immensely important research tools for studying many aspects of mammalian development ever since they first were produced over 50 years ago. Chimera studies have served as crucial assays in the era of modern mouse genetics that was triggered by the advent of mouse embryonic stem cells. Lately, chimeras are also used as proof of pluripotency and normality of induced pluripotent stem cells.

Mount Sinai Hospital, Samuel Lunenfeld Research Institute,Toronto, Ontario, Canada

Methods in Enzymology, Volume 476
ISSN 0076-6879, DOI: 10.1016/S0076-6879(10)76008-0

With this long history in mind, it may seem surprising that chimeras now have an ever-increasing role to play. The high-throughput mouse gene targeting projects are in the process of producing ES cell lines with a mutation in each of the close to 20,000 known protein coding genes. These will all be waiting for germline transmission through chimeras. Such a large-scale approach calls for simplified methods for generating germline transmitting chimeras. In this chapter, we will describe the currently most cost efficient and simple method; the aggregation of pluripotent stem cells with diploid or tetraploid mouse embryos. Since most of the large knockout projects are using the C57BL/6 background, we will pay special attention to cell lines derived from this inbred strain.

1. Introduction

Cultured mouse pluripotent stem cells can contribute to embryonic development and change into fully functional cells in the resulting chimeras (Evans and Kaufman, 1981; Okita *et al.*, 2007). The most common use of these chimeras is associated with mouse embryonic stem (ES) cell-mediated generation of mutant animals, where the mutation is generated in ES cells that are introduced into the mouse by chimeras containing ES cell-derived germ cells (Gossler *et al.*, 1986). The parallel development of powerful mutagenesis technology and conditions for generation of ES cells with exceptionally high developmental potential (Nagy *et al.*, 1993) not only revolutionized the genetics of the mouse during the past two decades, but is still accelerating our understanding of gene functions underlying normal development and pathological conditions.

One of the most powerful approaches to address gene function in the mouse is by gene targeting (Mansour *et al.*, 1988); the generation of null (or Cre recombinase excision-conditional null; Gu *et al.*, 1994) alleles in any gene of interest in ES cells and then in mice after germline transmission. Figure 8.1 shows the general protocol of introducing a targeted mutation into the mouse by the use of pluripotent cell lines.

By 2006, approximately one-fifth—or 4000—of the known mouse genes (Collins *et al.*, 2007b) had been knocked out and characterized for their respective mutant phenotypes. This effort by the scientific community demonstrated the power of gene targeting for understanding our life and diseases. Recognized by the upper echelons of scientific and political authority, a more orchestrated international coordination of targeting the rest of the genes has been initiated all across the globe. Somewhat similar to the human genome sequencing between 1995 and 2000, consortia have been formed in Europe (EUCOMM) and North America (KOMP and NorCOMM) (Collins *et al.*, 2007a,b) to generate this resource of mutant ES cell lines and provide them to the research community for phenotype analysis. This massive effort demands the optimization and standardization

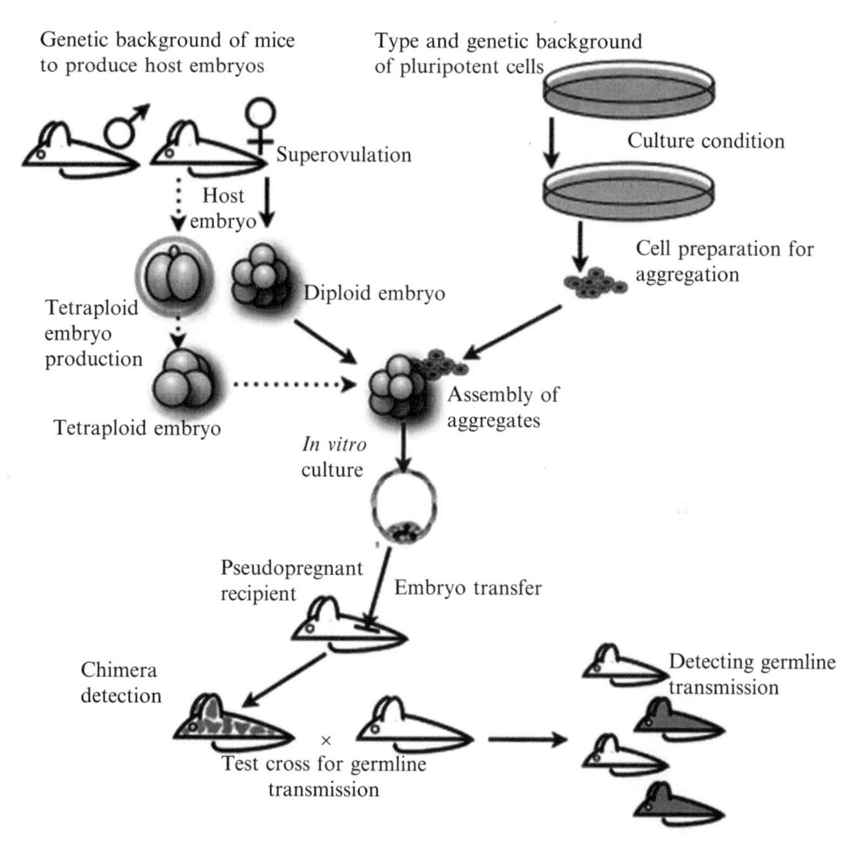

Figure 8.1 General protocol for mouse aggregation chimera production and germline transmission from pluripotent cell lines. The text shows the topics that will be discussed in this chapter.

of all parts of the targeted mutagenesis technology. Decisions have been made to move most of this effort to the C57BL/6 (B6) genetic background (Collins *et al.*, 2007b)—the gold standard for mouse studies. Highly germline competent and robust B6 ES cell lines had to be developed and associated procedures custom fitted, including the chimera production. The high-throughput knockout project pipelines are up and running and the plan is to by 2011, all the known protein coding mouse genes will be mutated and the corresponding quality-controlled ES cell lines made available to the scientific community through banks and distributing centers. There is no doubt that such a resource will create a high demand for germline transmitting chimera production to generate the corresponding mutant mouse lines.

The laboratory of Shinya Yamanaka recently showed that with a finite number of transcription factors, it is possible to reprogram somatic cells to a pluripotent state (Okita *et al.*, 2007; Takahashi and Yamanaka, 2006).

Since these iPS cells are developmentally equivalent to ES cells, the chimera assay has become the "acid test" for pluripotency in the mouse system. Chimeric embryos or animals can also be used for direct phenotype analysis (George *et al.*, 2007; Nagy and Rossant, 2001; Tam and Rossant, 2003) when created through the tetraploid complementation assay. Both iPS cells and tetraploid complementation have further added to the widening demand for chimera production.

2. Methods for Producing Chimeras: General Issues

The pioneering studies of Beatrice Mintz and Andrzej Tarkowski in the 1960s (Mintz, 1962; Tarkowski, 1961) showed that chimeric animals can be produced by aggregating two cleavage stage embryos. Aggregation chimeras between embryonic carcinoma cells and embryos were first generated by Colin Stewart (Stewart, 1982). This method was modified and further expanded to include ES cells and tetraploid embryos by Andras Nagy (Nagy *et al.*, 1990). Jacek Kubiak and Andrzej Tarkowski developed two-cell stage embryos blastomeres' electrofusion to generate tetraploid embryos (Kubiak and Tarkowski, 1985). All these early efforts have been vital for the development of the sophisticated chimera generation methods we have available today.

Most tetraploid embryos die shortly after implantation, but when complemented with diploid embryos, their contribution is primarily restricted to the extraembryonic tissues: the primitive endoderm of the yolk sac and the trophoblast layer of the placenta, and excluded from the primitive ectoderm lineage (Tarkowski *et al.*, 1977). On the other hand, ES cells have limited ability to contribute to the trophoblast lineage (Beddington and Robertson, 1989). When ES cells are complemented by tetraploid embryos, they colonize the embryo proper, the amnion, the allantois as well as the mesoderm layer of the yolk sac, while tetraploid cells are excluded from these lineages and restricted to the extraembryonic tissues; resulting in completely (Nagy *et al.*, 1993) or nearly completely (Eakin *et al.*, 2005) ES cell-derived embryos depending on the genetic background combination. Aggregation of tetraploid and diploid embryos can be used to segregate embryonic and extraembryonic phenotypes and rescue or bypass extraembryonic defects of specific mutations (Guillemot *et al.*, 1994; Rossant *et al.*, 1998). This technique also provides a rapid test for the developmental potential of ES or iPS cells and allows for the generation of mutant embryos directly from pluripotent cells (Carmeliet *et al.*, 1996). It is possible to derive viable and fertile animals carrying mutations and so speed up traditional breeding using characterized F1-hybrid ES cells (Eggan *et al.*, 2001; George *et al.*, 2007).

Currently, there are three distinct methods for generating mouse chimeras: (1) injection of pluripotent cells into the blastocoel cavity of 3.5 dpc (days post *coitum*) blastocyst stage embryos (Fig. 8.2A), (2) injection of cells under the zona pellucida of eight-cell stage (2.5 dpc) embryos (Fig. 8.2B), and (3) aggregation of cells with eight-cell stage (2.5 dpc) embryos after removal of the zona (Fig. 8.2C). All three techniques are similarly efficient when optimized and executed properly. Blastocyst injection and injection

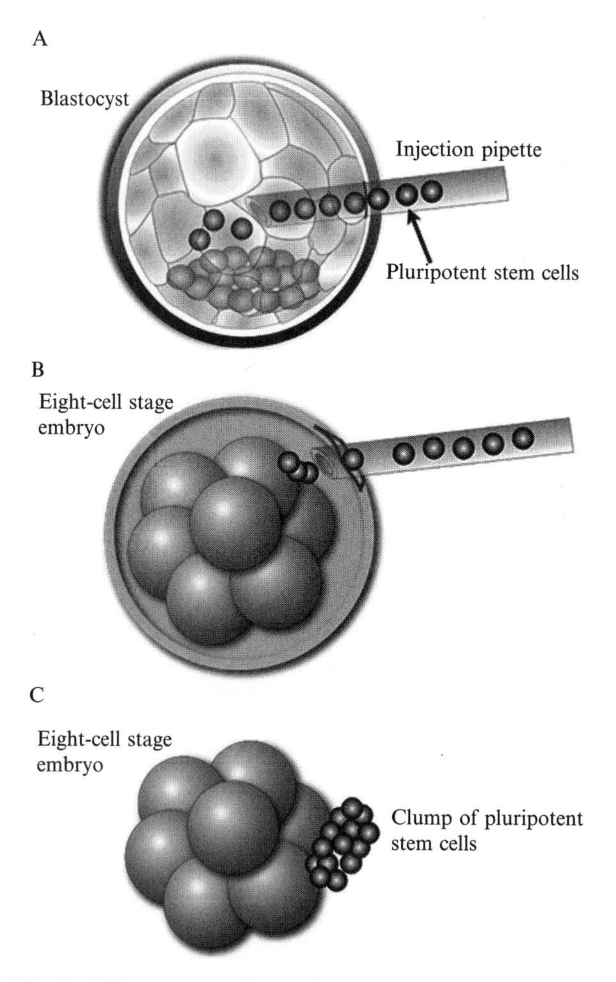

Figure 8.2 Three different ways of introducing pluripotent stem cells into mouse embryos for generating chimeras: blastocyst injection (A), eight-cell stage injection (B), and aggregation with zon-free eight-cell stage embryo (C).

under the zona requires high technical skills with micromanipulators, expensive equipment, and often inbred donor embryos but allows for careful selection of cells to be injected and immediate transfer of chimeric embryo. Aggregation with zona-removed embryos on the other hand does not require expensive equipment, is technically easier to perform, and allows for the use of robust and cheap outbred embryos. The drawback of this technique, however, is the difficulty to select for the "best looking" cells and the overnight culture of the aggregates requires optimization of the *in vitro* culture system. Here we detail the zona-free aggregation method and discuss the pros and cons of all the different means of generating chimeras.

3. Preparation of Donor Embryos

3.1. Genetic background

The combination of genetic background of the two chimera components and the means of chimera production (blastocyst injection vs. aggregation) both highly influence the efficiency of somatic and germline contribution of the pluripotent cell line. In case of ES cell injection into blastocysts, high germline transmission rates can be achieved by using C57BL/6 host blastocyst. If cells from "nonblack" ES cell lines, such as 129 inbred derived, are injected into C57BL/6 host embryo, the chimeras and their germline transmission can be readily recognized on the basis of simple coat color markers.

Obviously, the easy coat color readout is not an option when a B6 ES cell line is combined with B6 host embryos. In the search for suitable hosts for the increasing number of B6-targeted ES cell clones, several alternatives have been tested. BALB/c has been successfully used (Kontgen *et al.*, 1993; Ledermann and Burki, 1991; Lemckert *et al.*, 1997) but this background has never gained popularity due to its inferior ability to produce good quality embryos for injection.

Another alternative is the use of albino C57BL/6 host embryos. A spontaneous mutation occurred in the C albino locus in two independent breeding colonies; one (C57BL/6-Tyr$^{c\text{-}2J}$) in the Jackson Laboratory and the other (C57BL/6-Tyr$^{c\text{-}Brd}$) in the C57BL/6 colony of Dr. Alan Bradley (Liu *et al.*, 1998). Injection into albino C57BL/6-Tyr$^{c\text{-}Brd}$ blastocysts has been shown to be efficient for production of germline chimeras from large-scale knockout project (Pettitt *et al.*, 2009) using the JM8 B6 ES cell line and its variants. When VGB6.2 B6 ES cells were injected into eight-cell stage C57BL/6-Tyr$^{c\text{-}2J}$ embryos using the laser pulse for zona penetration, the chimeras resulted in completely ES cell-derived animals (Poueymirou *et al.*, 2007) at a reasonably high frequency. However, C57BL/6-Tyr$^{c\text{-}Brd}$ animals are not easily available, C57BL/6-Tyr$^{c\text{-}2J}$ purchased from the Jackson laboratory are relatively expensive; both coisogenic albino host options often require the maintenance of in-house breeding colonies that may or may not be feasible. A recent—and far less tested—option is the agouti B6 strain (C57BL/6J-A^{W-J}) (Pettitt *et al.*, 2009). The future use of these agouti B6 mice as donors for host embryos holds promise to make the detection of chimerism and germline transmission easy with blastocyst injected B6 ES cell lines.

Interestingly, the change to B6 ES cells did not affect the aggregation method. It allows the use of inexpensive and robust albino outbred eight-cell stage embryos, such as CD-1(ICR). Donor animals are available from several commercial suppliers (e.g., Harlan Sprague Dawley, Charles River Laboratories, or Taconic) and sufficient numbers of embryos for manipulations can easily be generated from these animals. Eight-cell embryos can be injected (Sato *et al.*, 2009) or aggregated (Gertsenstein *et al.*, 2010) to efficiently generate chimeras from B6 ES cells as shown earlier for 129 derived lines.

3.2. Superovulation

The number of embryos that can be recovered from naturally mated female donor mice depends largely on the genetic background, and is generally the same as can be expected from the first litter would the animal be let to give birth. A much larger number of embryos can be recovered if the female is superovulated by the administration of a hormone regimen outlined below.

Efficient superovulation depends on several variables: the strain background, age and weight of the females, the dose and the time of administration in relation to the light cycle in the animal room. The number of superovulated oocytes that actually become fertilized depends on the reproductive performance of the stud males. All these parameters should be considered together to achieve an efficient response to superovulation with all conditions optimized empirically in each actual setting.

As a general rule, prepubertal or older mice are given 5 IU (International Unit) of pregnant mare's serum gonadotropin (PMSG) on day 1 and a second injection of human chorionic gonadotropin (hCG) on day 3 after which they are immediately placed in the cage with the male. In most cases, mating will take place later the same afternoon or night, and a copulation plug can be detected the day after. Some genetic backgrounds respond poorly to superovulation, and for these, it is worthwhile trying if a slightly younger or older age of the female would be more successful. If this is not the case, the hormone dose can be varied from 2.5 to 10 IU.

The timing of the injections will influence the success as well, and most importantly, it will determine the stage at which the resulting embryos will have reached at the time of recovery. If the embryos are found to be at a too early or too late stage, the hormones can be given earlier or later correspondingly.

3.2.1. Materials

- PMSG: for example, NIH National Hormone & Peptide Program (http://www.humc.edu/hormones) or Folligon from Intervet or Sigma G4527 or Calbiochem 367222
- hCG: for example, Chorulon from Intervet or from Sigma C8554
- Hypodermic needle, 26- or 30-gauge, 1/2 in. and syringe, 1–3 ml
- CD-1 or ICR female mice (e.g., 6 weeks old)
- CD-1 or ICR male breeders individually caged

3.2.2. Method

1. Resuspend lyophilized powder at 500 or 1000 IU/ml in sterile PBS or provided diluent, aliquot and store at −20 or −80 °C.
2. Dilute an aliquot with sterile PBS to final concentration of 50 IU/ml.
3. Inject 5 IU PMSG (0.1 ml) intraperiotoneally between 1 and 2 pm on day 1.
4. Inject 5 IU hCG (0.1 ml) intraperiotoneally at 46–48 later (on day 3).
5. Place each female into a cage with one stud male after the administration of hCG.
6. Check for a copulation plug the next morning.

3.3. Recovering two- and eight-cell stage embryos from the oviduct

Mouse embryos travel down the female reproductive tract, entering the uterine horns 3–4 days after fertilization. Two-cell stage embryos collected at 1.5 dpc are used for generation of tetraploid embryos by electrofusion, while eight-cell stage embryos are routinely collected in the morning of 2.5 dpc, for aggregation the same day. Alternatively, to obtain eight-cell stage embryos, it is possible to inject hormones later in a day and/or collect two-cell stage embryos at 1.5 dpc and culture them overnight. Overnight culture slows the embryo development and a larger proportion remain at eight-cell stage early the next morning; this is beneficial for the efficient contribution of stem cells into chimeras including generation of completely ES cell-derived embryos (Cox *et al.*, 2010).

Embryos at both stages can be located in the oviduct coils. To recover two-cell stage embryos, the pregnant female mice are sacrificed at 1.5 dpc.

3.3.1. Materials

- Humidified incubator(s) at 37 °C, 6% CO_2 for embryo culture
- Stereomicroscope(s) with transmitted light, sterile Petri and tissue culture dishes (35, 60, 100 mm)
- Sterile 1 cc syringes, 26-gauge 1/2 in. needles, 30-gauge 1/2 in. needles

- Sterile PBS
- 70% EtOH
- *Sterile dissecting instruments*: forceps, fine watchmaker forceps, small scissors, very fine sharp-pointed scissors
- *Flushing needle*: the sharp tip of 30-gauge 1/2 in. needle is cut off and/or polished on a sharpening stone or sand paper. Flush with 70% ethanol before and after use.
- Bunsen or alcohol burner for pulling embryo-manipulating pipettes.
- *Embryo-manipulating pipettes*: Pasteur pipettes or glass microcapillaries (e.g., Drummond 1-000-0400 or 1-000-0500) are drawn by hand over the flame and broken flat with an inner diameter slightly larger than an embryo ($\sim$100 μm). It is very important to flame polish the tip to prevent damaging the embryos. Embryo-manipulating pipettes are connected through elastic rubber tubing 1/8 $\times$ 3/16 (e.g., latex VWR 62996-350 or tygon Fisher 14-169-1D) to an aspirator mouthpiece (round: Drummond 2-000-000 or Sigma A5177 Aspirator Tube Assembly; flat: Biotech Inc. MP-001-Y). Pasteur pipettes fit into standard 1000 μl pipettor tips and microcapillaries are inserted into silicone tips (Drummond 1-000-9003 or Sigma A5177).
- M2 (e.g., Millipore MR-015-D) is a HEPES-buffered medium used during embryo collection and other manipulations in room atmosphere. Aliquots are stored at 4 °C and brought to room temperature prior to use.
- KSOM-AA (e.g., Millipore MR-121-D) is a bicarbonate-buffered medium with nonessential amino acids used for embryo culture (Ho *et al.*, 1995; Lawitts and Biggers, 1992, 1993). Embryo culture medium should be gas-equilibrated by placing the tube with loose cap or prepared microdrop culture dishes in the incubator at least a few hours or ideally 16–20 h before use.
- Embryo-tested light mineral oil (e.g., Millipore ES-005-C or Sigma M8410) is used to overlay microdrops of embryo culture medium. We prefer to aseptically aliquot oil, store in the fridge and incubate with loose cap overnight in CO_2 incubator or prepare microdrop culture dishes the day before the experiment.
- Plugged 1.5 or 2.5 dpc CD-1/ICR embryo donor females.

3.3.2. Method

1. One day before embryo collection, prepare culture plates by placing 15 μl drops of KSOM-AA medium in a 35-mm tissue culture dish and overlay these with embryo-tested mineral oil.
2. On the day of the embryo collection, bring M2 medium to room temperature and pull embryo-manipulating pipettes.
3. Humanely sacrifice the embryo donor females following local animal welfare protocols.

4. Lay the animal on its back and wipe the abdomen with 70% EtOH.
5. Open the abdominal cavity and locate the ovaries.
6. Carefully pinch the fat pad above the ovaries with fine watchmaker forceps.
7. Make one cut across the top of the uterine horn and another right between the ovary and oviduct coils.
8. Lift out the oviduct and place it in a 35-mm tissue culture dish with sterile PBS.
9. Continue collecting the other oviducts.
10. Rinse the oviducts in the PBS by swirli the dish and place them in a new dish with M2 embryo culture medium.
11. Transfer one oviduct at a time into a small drop of M2 under stereomicroscope; insert the flushing needle attached to a 1-ml syringe filled with M2 into the infundibulum, and gently press the tip of the flushing needle against the bottom of the dish to hold it in place. Use the forceps to hold the needle in the right position. Flush M2 medium through the oviduct; observe its swelling. Proceed with the remaining oviducts keeping the time of manipulations to a minimum.
12. Collect the embryos and wash them through several fresh M2 drops to get rid of all debris and then several drops of equilibrated KSOM-AA medium. Transfer the embryos into the previously prepared KSOM-AA embryo culture dish and place it back in the incubator until ready for the next steps.
13. The time between euthanizing embryo donors and placing the embryos in the incubator should be kept to a minimum (ideally no more than 30 min), the number of donor females that can be handled during that time is determined by skills.

3.4. Production of tetraploid embryos by electrofusion

Each blastomere of the two-cell stage embryo harbors a full diploid set of chromosomes. Tetraploid embryos are produced by simply fusing the blastomeres by applying an electric current perpendicular to the cell membrane separating the two. The current must be adjusted so that it is just enough to lyse the membrane but not so high that one or both of the blastomeres are destroyed. The parameters may vary depending on the instrument and the mouse strain and need to be determined in a pilot experiment. We routinely apply one or two pulses of 30 V and 40 μs for the fusion of ICR embryos in a 0.3 *M* mannitol solution using a CF-150B cell-fusion BLS instrument with a 250-μm electrode chamber. The use of a nonelectrolyte solution together with an adjustable 1 MHz AC field set up at 0.5–1 V before the electric pulse is given greatly speeds up the process since it allows for an "automatic" line-up of the embryos in the exactly right position, allowing for the simultaneous fusion of a group of embryos. However, mannitol is toxic to the embryos if the exposure is long. The time the embryos spend in a nonelectrolyte solution should be kept to a minimum.

The fusion process does not take place immediately but develops slowly over a period of about 30 min after the electric current has been applied. Fusion is most efficient at the time when the embryos are close to the second division. This means that with the right timing, the embryos generally divide shortly after the fusion has taken place. For this reason, it is very important to follow the process carefully, examine each individual embryo with great care, and separate out those that have successfully fused. If this is not done in time, it will be impossible to tell whether an embryo is tetraploid and has gone through the second division, or if it has remained diploid at the two-cell stage. Incubation of 1.5 dpc embryos for at least 15 min before applying the electric pulse improves the fusion rate when generating tetraploid embryos.

3.4.1. Materials

- Stereomicroscope(s) with transmitted light. The use of two microscopes is convenient for the fusion of two-cell stage embryos but a single microscope is also sufficient.
- Cell-fusion instrument (e.g., CF-150B pulse generator) with 250 μm electrode chamber (BLS Ltd, Hungary, www.bls-ltd.com).
- 0.3 *M* mannitol (Sigma M4125) prepared in ultrapure embryo-tested water (e.g., Sigma W1503) containing 0.3% BSA (e.g., Sigma A3311), filtered and stored at −20 °C. Aliquots are thawed prior to use and not refrozen.
- Two-cell stage embryos
- M2 embryo culture media (Millipore MR-015D)
- KSOM embryo culture media with amino acids and glucose (Millipore MR-121D)
- Embryo-tested light mineral oil (Millipore ES-005-C or Sigma #M8410)
- 100 mm tissue culture dishes
- Mouth-pipetting device
- Humidified incubator, 37 °C, 5% CO_2

3.4.2. Method

1. Turn on the fusion machine and set up the parameters. Place the electrode chamber in the lid of a 100-mm dish.
2. Place one large drop of mannitol directly on the electrodes so that they are completely covered but the drop is not allowed to spill over on the dish from the electrodes (see Fig. 8.3).
3. Place another equally large mannitol drop, and three large M2 media drops on the dish close to the electrodes.

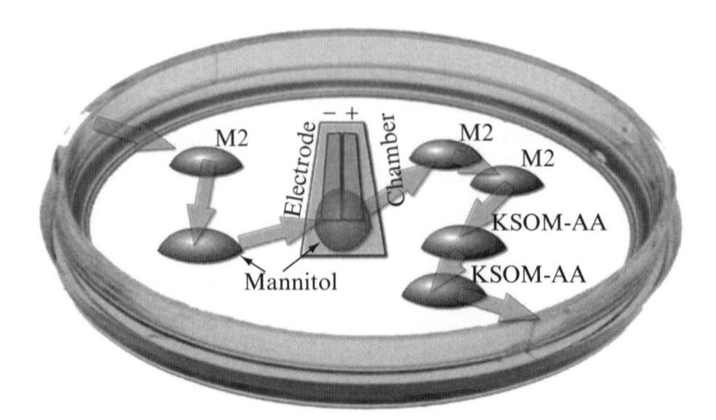

Figure 8.3 Arrangement of the fusion electrode chamber and different media drops on the 10 cm tissue culture dish lid.

4. Place the embryos in the first M2 drop. From here, work with a small group (e.g., 10 embryos) at a time through the fusion process. As experience is gained, the group size can be increased, but it is important that it never is larger than can be handled very swiftly.
5. Move a small group of embryos to the mannitol drop.
6. Completely empty your transfer pipette of culture media.
7. The embryos will first float up, then sink down in the mannitol.
8. Once the embryos have sunk down, pick them up and place them between the electrodes. The embryos should be well equilibrated in mannitol or they will float to the surface of the mannitol instead of resting between the electrodes. It is also important not to transfer M2 medium into the electrode chamber, otherwise, embryo orientation by the AC field will not work. However, embryos should be kept in mannitol for a minimum amount of time.
9. Apply a low AC current until the embryos are oriented—this should take no more than a few seconds.
10. Apply the fusion pulse as described above.
11. Immediately pick up the embryos and place them in the second M2 drop. Empty the transfer pipette of the media.
12. Pick up the embryos again and place in the last M2 media drop.
13. Repeat the process until all embryos have been subjected to the electric pulse.
14. Wash the embryos carefully in at least two drops KSOM-AA before placing them back in the incubator.
15. 30 min later, take out the dish from the incubator and carefully sort the embryos into separate drops depending on the success of fusion.
16. It is helpful to draw lines on the underside of the bottom of the microdrop dish to distinguish two groups of embryos; those subjected to the electric pulse but not yet fused, and successfully fused ones.

Embryos that do not fuse in 30–45 min should be placed back in the incubator for another 15–20 min of incubation. If they still show no sign of fusion, they can be subjected to another pulse exactly as described above.

During the fusion process, embryos with the following characteristics can be found and should all be sorted into separate drops:

- Clearly fused embryos: one large round blastomere.
- Embryos undergoing fusion: the two blastomeres are partially fused and give the embryo a scuba diving mask appearance.
- Embryos in which one blastomere has lysed: one small blastomere, debris, and a clear space between the zone and the cell.
- Embryos in which both blastomeres have lysed: empty zona with debris.

Only the first group should be regarded as tetraploid. Embryos in the second group should be incubated for an additional 10 min and examined again. In most cases, they will have fused completely by then. A second pulse can be applied to the embryos that did not fuse after 1- 1 ½ hours. After overnight incubation, the 'four cell' stage embryos are used for aggregation in the afternoon as described below. The development of fused embryos to the 'four-cell' stage should be at least 80% in optimal culture conditions. Embryos arrested at the 'one-cell' stage are not used for aggregations. 'two-cell' stage embryos are delayed and some may be aggregated later in the day after they have developed to the 'four-cell' stage. Aggregation of 'two-cell' stage tetraploid embryos is not recommended

4. Preparation of Pluripotent Cells

4.1. Pluripotent stem cell culture prior to aggregation

The proper culture of cells to be aggregated is of utmost importance at all times, but particularly before their introduction into mice to generate chimeras. Pluripotent cells should be kept in a subconfluent state and split at densities that are neither too high nor too low, that is, no more than 2 days without passaging at a ratio of 1:5–1:7, and the medium should be changed every day. Pluripotent cells are usually maintained on mitotically inactivated mouse embryonic fibroblasts (MEF) using FBS supplemented media and methods described in details in multiple publications (e.g., Nagy *et al.*, 2003) and briefly here. However, recent development of defined serum-free and feeder-free culture conditions with inhibitors of the WNT, FGF, and ERK signal transduction pathways (Chen *et al.*, 2006; Ying *et al.*, 2008) is changing the way of routine culture of pluripotent stem cells and should be kept in mind. Medium supplemented with the FGF receptor inhibitor SU5402, the MEK inhibitor PD184352, and the GSK3 inhibitor CHIR99021 commercially available from iSTEM Mouse ES cell Media, Stem Cell Sciences, Ltd. has been used for culture of B6 ES cells injected into eight-cell stage ICR embryos and successful generation of germline transmitting chimeras (Kiyonari *et al.*, 2010). Two inhibitors (MEK inhibitor PD0325901 and GSK3 inhibitor CHIR99021) have been shown to allow the derivation of previously elusive rat ES cells (Buehr *et al.*, 2008; Li *et al.*, 2008) and to produce germline competent NOD mouse ES cells (Nichols *et al.*, 2009). Our own work with B6 ES cells shows that chemically defined medium supplemented with inhibitors instead of classical FBS supplemented media provides a superior alternative to the maintenance of their undifferentiated state and consequent efficient generation of chimeras by the aggregation method (Gertsenstein *et al.*, 2010).

4.1.1. Materials

- Inverted microscope with phase contrast for ES cell culture observation
- Biosafety Cabinet for routine cell culture
- Humidified incubator(s) at 37 °C, 5% CO_2 for ES cell culture
- Sterile tissue culture dishes of various sizes
- Sterile Pasteur pipettes and plastic pipettes for tissue culture (1, 5, 10 ml)
- ES/iPS cell medium
  - Dulbecco's modified Eagle's medium (DMEM; Invitrogen 11960)
  - 15% FBS (tested lot)
  - 2 m*M* GlutaMAX™ (Invitrogen 35050) or L-glutamine (Invitrogen 25030)
  - 0.1 m*M* 2-mercaptoethanol (Invitrogen 21985-023)
  - 0.1 m*M* MEM nonessential amino acids (Invitrogen 11140)
  - 1 m*M* Sodium pyruvate (Invitrogen 11360)
  - 1000 U/ml LIF (Millipore ESG1107)
  - 50 U/ml penicillin and 50 μg streptomycin (Invitrogen 15140)
- 0.05% trypsin/EDTA (Invitrogen 25300)
- 0.1% gelatin in sterile water (Millipore ES-006B)
- Ca^{2+}/Mg^{2+}-free D-PBS (Millipore ES-1006-B or Invitrogen 14190)

4.1.2. Method

Three days before aggregation: thaw a vial of cells. If the cells were frozen properly, they should be subconfluent and ready for passage in 2 days. A vial of unknown viability should always be thawed earlier, for example 5 days before aggregation, to ensure their timely recovery and provision of the quantity of cells required for the experiment. Change the medium the day after thawing.

One day prior to the aggregation: subconfluent (70–80%) cultures are passaged on gelatinized plates as described below. This is sparser than usual passage produces the colonies of 8–15 cells that will be lifted by gentle trypsinization immediately before aggregation.

1. Add 0.1% gelatin to cover the surface of two to three 35- or 60-mm dishes.
2. Aspirate the medium from the subconfluent plate, add PBS, and remove right away.
3. Add trypsin to cover the cells (e.g., 1 ml per 60-mm dish) and place the dish in the incubator for 5 min.
4. Neutralize the trypsin with FBS-containing medium, resuspend by gentle pipetting, and ensure that a single cell suspension has been obtained.
5. Transfer the cell suspension to a tube and centrifuge at 200–300×*g* for 5 min.
6. Remove the supernatant and gently resuspend the pellet in fresh medium. The volume depends on cell density and the surface, for example, a subconfluent 60-mm dish is resuspended in 5 ml of medium.
7. Leave the tube undisturbed for 3–5 min to allow for the majority of large clumps and feeders to settle. Alternatively, place the cell suspension back into the original dish and put it back in the incubator for 10 min to allow the MEF to reattach (preplating).
8. Aspirate the gelatin solution from the plates prepared earlier.
9. Seed the cells from the top portion of the cell suspension in the tube or from carefully tilted original dish on new gelatinized plates using different dilutions (e.g., 1:10–1:50). For example, 0.2, 0.4, and 0.6 ml of 5 ml cell suspension from a subconfluent 60-mm dish are platted on three 60-mm plates.
10. Check the cell density under the microscope and adjust the volume.
11. Seed the rest of the cells on one or more dishes to serve as a backup and incubate overnight

On the day of aggregation, small colonies of 8–15 cells are lifted by gentle trypsinization immediately before the aggregation of embryos that have their zonae removed.

1. Feed cells with fresh media in the morning (optional but highly recommended).
2. Aspirate the medium and wash the cells first with PBS, and then with trypsin (this helps to loosen up the cells and minimize the amount of trypsin in the next step).
3. Add a minimal amount of trypsin to just cover the cells, for example, 0.3–0.5 ml per 60 mm plate, place in the incubator for 1–2 min or leave at room temperature. Check under the microscope, gently swirl the plate to detach the colonies, and tap at the microscope stage until all colonies are lifted and start floating. *Do not overtrypsinize, as cells will become sticky and hard to manipulate.*

4. As soon as the colonies are detached, quickly add medium to the plates (e.g., 4 ml per 60-mm dish). Do not pipette. If cell clumps are larger than the required 8–15 cells, a very gentle pipetting can be used, but care should be taken not to break the clumps into a single cell suspension. Loosely connected cell clumps are now ready for aggregation within the next 1–2 h and can be kept in the original plate. If required, for ease of transportation, suspension of clumps can be gently transferred into 5 ml tubes (e.g., Falcon 352063 or 352058). Keep the cells at room temperature, as they will start attaching to the plate if placed in the incubator.

5. Making the Aggregates

5.1. Preparation of aggregation plates

The aggregation of embryos and cells can only take place efficiently if they are confined at precise numbers in a small space that force the components of the aggregate together without restricting the supply of nutrients. This can best be accomplished by making small depressions in the plastic surface of a tissue culture plate once drops of culture media has been placed on it and those in turn have been covered by light mineral oil. The schematic in Fig. 8.4 explains how the process takes place. It is important to prepare the

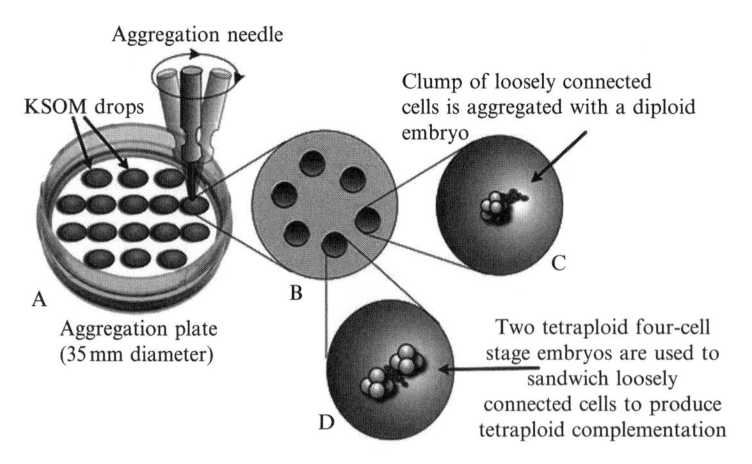

Figure 8.4 Preparation of aggregation plate and assembly of aggregation chimeras. (A) KSOM-AA drop arrangement in the aggregation plate and production of aggregation wells. (B) View of a single KSOM-AA drop with six aggregation well made by pressing the aggregation needle into the plastic. (C) View of one aggregation well containing a diploid embryo <-> stem cell aggregate. (D) View of one aggregation well containing two tetraploid embryos <-> stem cell aggregate.

aggregation plates well in advance of the experiment so that the media drops have ample time to get equilibrated in the 5% CO_2 in the incubator.

5.1.1. Materials

- 35 mm tissue culture dishes, 35 mm Easy Grip Falcon 35-3001 tissue culture plates are well suited for making depressions.
- M2 embryo culture media (Millipore MR-015D)
- KSOM embryo culture media with amino acids and glucose (Millipore MR-121D)
- Embryo-tested light mineral oil (Millipore ES-005-C Sigma #M8410)
- Humidified incubator, 37 °C, 5% CO_2
- Stereomicroscope(s) with transmitted light
- Aggregation needle DN-09 (BLS Ltd, www.bls-ltd.com)
- 70% EtOH

5.1.2. Method

1. The day before aggregation, place microdrops (~3 mm in diameter or 10–15 μl) onto 35-mm dish using 1 cc syringe filled with KSOM-AA medium or micropipettor. We recommend to place two rows of four drops in the middle of the dish, and two more rows of three drops each side at the side. Immediately cover the drops with embryo-tested mineral oil (Fig. 8.4A).
2. Wipe the aggregation needle with 70% ethanol. The needle can be autoclaved if required but ethanol washes are generally sufficient for keeping contaminations away. Press the needle into the plastic and make a slight circular movement (Fig. 8.4A). Do not press too hard or the plate will crack; however, not enough pressure will result in a too shallow depression. The goal is to create a small cavity with a smooth and transparent surface that is deep enough to hold the aggregate safely when moving the plate to the incubator.
3. Make 6–8 depressions per microdrop positioning them in a circle approximately halfway between the center and the edge (Fig. 8.4B). Do not make depressions too close to the edge—the embryos will be difficult to manipulate. Avoid the center, so that in the event that air bubbles are accidentally introduced to the drop, the embryos will remain visible.
4. Leave a few microdrops on the side without depressions to be used for the final selection of cell clumps. Forty to 60 depressions are usually made per plate to limit the time of embryo manipulations outside the incubator.
5. Place the aggregation plate in the incubator until needed.

5.2. Zona pellucida removal and placement of embryos in the aggregation plates

Preimplantation stage embryos are surrounded by a protective glycoprotein layer called the zone pellucida. This has to be removed prior to aggregation in order to give the components of the aggregate direct cellular contact with each other. Zona removal has historically been accomplished by pronase treatment, but others and we have found the use of acid Tyrode's superior.

5.2.1. Materials

- Acid Tyrode's solution (e.g., Sigma T1788) is aliquoted and stored at −20 °C. One aliquot is thawed and brought to room temperature prior to use (it can be kept at 4 °C for no longer than 2 weeks).
- Collected eight-cell or morula stage diploid embryos or fused tetraploid embryos
- Embryo-manipulating pipettes with fire-polished tips
- Embryo culture dishes prepared in advance
- M2 media
- Humidified incubator at 37 °C, 5% CO_2
- Stereomicroscope(s) with transmitted light. Frosted instead of transparent glass in the base of the microscope with transmitted light often provides better visualization of the zonae pellucidae, which is helpful for zona removal.

5.2.2. Method

1. Place three drops of acid Tyrode's in the lid of a 60-mm tissue culture dish. Alternatively, use bacteriological grade dishes to make sure that the "naked" embryos do not stick to the tissue culture-treated surface.
2. Place three drops of M2 culture media in the same dish.
3. Place the embryos in the first M2 drop. From here on, work with small groups of embryos until experience has been gained that allows for treating larger numbers at the same time. Move a few embryos at a time into the first acid drop taking care to transfer as little media as possible.
4. Immediately move the embryos to the second and then to the third acid drop.
5. Let the embryos settle and carefully "blow" acid on them with the pipette.
6. Carefully watch the embryos. Within a minute, the zona will dissolve (Fig. 8.5), and at this point, the embryos have to be immediately moved to the second and then the third media drop while taking care to transfer as little acid as possible.

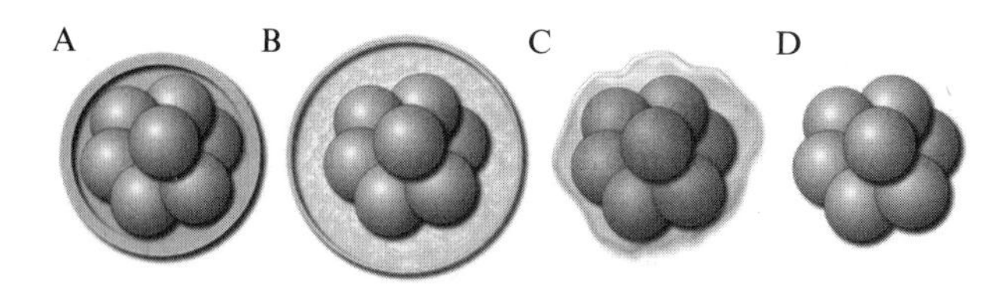

Figure 8.5 Phases of dissolving the zona pellucida of an eight-cell stage embryo in acid Tyrod's: (A) the embryo prior, (B) 25–40 s, (C) 40–60 s after transferred to acid, (D) the zona-free eight-cell stage embryo.

7. Proceed with the remaining embryos until they have all been treated.
8. Wash the embryos in three drops of KSOM media. They are now ready to be aggregated.

Once the zone has been removed from embryos, they become extremely fragile and sticky. It may take considerable practice to master the skill of handling, washing, and moving around zona-free embryos without losing them.

5.3. Aggregation

It is now time to place the zona-free embryos together with clumps of pluripotent stem cells. Assembling an aggregate in this manner requires some skills that can only be gained through patience and practice. It is important to dedicate this time to develop a good feel for how to handle the cells and embryos in order to be successful in the long run. Many problems that occur can be traced back to this crucial step (see Section 6).

Regarding the generation of completely ES cell-derived embryos/animals, two tetraploid embryos at the "four-cell stage" are aggregated with ES cells in a sandwich manner (Fig. 8.4D) as this method has proven to be more efficient in our case. However, if it is necessary for the experimental design to have tetraploid mutant embryos of different genotypes or the number of embryos is not sufficient, single tetraploid embryos may also be used.

5.3.1. Materials

- Cells and embryos prepared as described in the above sections
- Aggregation plate as described in Section 4.1
- Mouth-pipetting device
- Stereomicroscope(s) with transmitted light
- Humidified incubator, 37 °C, 5% CO_2

5.3.2. Method

1. Place one zona-free embryo in each aggregation well.
2. Pick up a large number of cell clumps of roughly the correct size and place them in a drop in the aggregation-dish that does not have any wells.
3. Make a second selection of clumps of the correct size and move one into each of the wells containing an embryo (Fig. 8.4C).
4. Place a second zona-free tetraploid embryo in each aggregation well that by now contains an embryo and a clump of cells (Fig. 8.4D).
5. Make sure that all embryos and cell clumps have good direct contact with each other. If this is not the case in an individual well, gently blow some KSOM media into the well to push the components together.
6. Place the dish in the incubator overnight.

The next day, the aggregates will have developed into blastocysts, and these are ready to be transferred into the reproductive tracts of a pseudopregnant female. In some cases, the development of the embryos is somewhat slower so that the blastocyst stage is only reached in the afternoon of the day after aggregation. This in itself is no reason for concern, but it is advisable not to transfer the embryos until the majority has reached at least the early blastocyst stage.

6. Embryo Transfer to Pseudopregnant Females

6.1. Production of pseudopregnant females

Chimeric embryos will only develop into fetuses and eventually full-term animals if they are placed in the natural environment of a receptive uterine tract. Mice ovulate and become receptive for pregnancy only upon physical mating, so prospective embryo recipient females are generally mated to sterile (vasectomized or genetically sterile) males. Typically outbred (ICR or CD-1) or hybrid (e.g., B6D2) females are used for this purpose. Vasectomized males can be purchased commercially or the surgery performed in the lab (Nagy *et al.*, 2003).

The female mouse estrous cycle is divided into four phases: proestrus (development of ovarian follicles), estrus (ovulation), metestrus (formation of corpora lutea), and diestrus (beginning of follicle development for the next ovulation and elimination of previous oocytes). Estrus can be induced by male pheromones. After being exposed to males, females enter estrus within 2 days, and most of them plug on the third night (the so-called "Whitten effect"). Females housed in groups and isolated from males tend to synchronize their estrous cycles. If these females are not mixed between stock cages, they can be used for mating on different days and so producing a relatively predictable number of plugs the next morning.

6.2. Embryo transfer

On the day following aggregation, the majority of embryos should have reached the blastocyst stage and ready to be transferred into the uteri of E2.5 pseudopregnant females.

The stage of pseudopregnancy is critical for successful embryo transfer. As a general rule to keep in mind, embryos should be placed in an *in vivo* environment that is equal to or less advanced than the embryos developmental stage. It is however entirely possible to place later stage embryos into an earlier pregnancy stage without ill effect as long as they are placed in the oviduct. The uterus is a hostile environment for embryos up until the time when the blastocyst stage embryos are supposed to enter at 2.5–3.5 dpc. Table 8.1 describes all ideal and possible combinations on which embryos stages can be placed.

In case of recipient shortage, aggregates can also be transferred into the oviducts of 0.5 dpc, the uteri of 3.5 dpc or cultured for an additional night and transferred the following day into 2.5 dpc uteri. All these options work, but others and we have found that the standard 2.5 dpc uterine transfer is the most reliably efficient option.

The implantation rate of zona-free embryos is lower than for those with an intact zona. If the plan is to dissect the embryos at midterm, we recommend to transfer 12–15 embryos per recipient. If, on the other hand, the embryos are planned to be left to term, this number should be increased to 18–22 to avoid extremely small litters that often cause problems at delivery. The number of pups and chimeras among them largely depends on the quality of ES cells. On average, ~30–50% of diploid embryo aggregates reach term and produce pups, and about 50% of those are

Table 8.1 Compatible stages of embryos and that of the pseudo-pregnant recipients for oviduct or uterus transfer.

	Pregnancy stage (dpc)			
	Oviduct transfer		Uterus transfer	
Embryo stage	Ideal	Possible	Ideal	Possible
One-cell	0.5	–	–	–
Two-cell	0.5	–	–	–
Four-cell	0.5	1.5	–	–
Eight-cell—morula	0.5	1.5	2.5	–
Blastocyst	0.5	1.5	2.5	3.5

chimeric to some degree, including some fully ES cell-derived animals based on coat color. Aggregation chimeras are most often either full transmitters or nontransmitters and rarely partial as it tends to be the case with blastocyst injection chimeras.

6.2.1. Materials

- Pseudopregnant females at 2.5 dpc
- Aggregated embryos after overnight incubation
- Embryo-manipulating pipettes with fire-polished tips, media, incubator, as described in Section 3.3.1
- Stereomicroscope(s) with transmitted and incident (e.g., fiber optics) light. If available, the use of two microscopes is convenient for embryo transfers.
- Anesthesia/analgesia according to local regulations
- Petri dishes, HEPES buffered media (e.g., M2)
- Sterile gauze tissues
- 30-gauge 1/2 in. needle, syringe
- Sterile PBS or tear-replacement drops/ointment
- *Surgical instruments* (e.g., Fine Scientific Tools—FST): sharp fine-pointed scissors, fine forceps (e.g., Dumont #5), straight or curved blunt forceps with serrated tips, baby Diffenbach serrefine clamp (e.g., FST18050-28); AUTOCLIP Wound Clip Applier (Becton Dickinson 427630) and AUTOCLIP Wound Clips, 9 mm (Becton Dickinson 427631).
- Heating pad or lamp

6.2.2. Method

Loading embryos in a pipette:

(a) Use the same type of pulled glass Pasteur pipettes for embryo transfer as for handling embryos in the previous protocols.
(b) Carefully take up a very small amount of M2 media in the pipette.
(c) Take up a very small air bubble, then a little more media.
(d) Take up the embryos in as little media as possible, and then load another small air bubble.

The bubbles will allow for a smooth and very well controlled movement of the embryos when they are placed in the uterus or oviduct.

1. Administer anesthesia/analgesia and prepare the mouse for aseptic surgery according to the local regulations (e.g., fur removal and preparation of skin).
2. Once the animal is fully asleep, place lubrication in the eyes.
3. Place the mouse in a ventral position on the lid of Petri dish covered with sterile gauze tissue.

4. Make a single middorsal skin incision ~0.5–1 cm along the vertebral column at the level of the last rib using sterile scissors and forceps. Slide the skin incision to expose the muscles of the left abdominal wall until the ovary or fat pad are visible. The fat pad can usually be located easily as a white area under the abdominal wall.
5. Make a 3–5 mm incision over the fat pad with fine dissection scissors or sharp forceps avoiding blood vessels approximately 4–5 mm caudal to the last rib. The incision should be similar in size to the ovary.
6. Grasp and exteriorize the ovarian fat pad with fine forceps. Hold the fat pad in position using a baby Diffenbach serrefine clip.
7. Check the ovary for signs of recent ovulation such as presence of corpora lutea (yellow bodies) forming of corpora hemorrhagica (bloody bodies) at the sites of oocytes extrusion from ovarian follicles after ovulation. The lack of corpora lutea and hemorrhagica means the recipient did not ovulate and should not be used for embryo transfer.
8. Using a 30-gauge 1/2 in. needle, holding it with the bevel upward, make a small hole in the uterine horn close to uterotubal junction. Make sure that the needle is inside the lumen before withdrawing it.
9. Without taking your eye off the site of needle insertion, pick up the transfer pipette and insert it into the hole. Slowly expel the embryos into the uterus, watching the air bubbles moving down.
10. Slowly withdraw the pipette and check in the media drop to make sure that all embryos have been transferred.
11. Remove the serrefine clip and use the blunt forceps to pick up fat pad and place the ovary, oviduct, and uterus back into the abdominal cavity.
12. Reposition the skin incision over the oviduct on the other side and transfer embryos to the other uterine horn the same way.
13. Close the skin using wound clips. Wrap the mouse in a tissue and place it back into a fresh cage on a heating pad or under a warming lamp. Monitor as required by local regulations.

Embryo transfer is a surgical procedure that requires the recipient animals to be adequately anesthetized. The method used will vary depending on the local regulations in the animal facility where the mice are housed. Commonly used are injectable Avertin, combinations of Ketamin and Xylazin, or inhalable vapors such as Isofluorane. Irrespective of the method, it is important to keep the following key points in mind when anesthetizing such small animals as mice:

(a) The body temperature can quickly drop, especially if ethanol is used to wash the incision area. The use of heating pads and warm light lamps are strongly recommended.
(b) Adequate level of anesthesia must be achieved before surgery is started. This can be gauged by pinching the animal's toes—it should not react.

On the other hand, the sleep should not be too deep as this could lead to impaired breathing and death.

(c) In most cases, the animal's eyes will remain open during anesthesia. It is highly recommended to apply lubrication to the cornea to prevent painful drying.

(d) The surgery should be performed with minimal invasiveness. That means, incisions should be kept as small as possible, and the procedure should be performed gently and swiftly.

(e) Care should be taken to keep the surgical area aseptic. Although mice are generally relatively resistant to post surgical infections, steps should be taken to avoid it.

7. Detection of Chimerism and Testing for Germline Transmission

Detection of chimerism in animals resulting from these experiments is most often—and easiest—done by examining the coat color. Chimeras between agouti 129 inbred cell lines and nonagouti B6 hosts can easily be judged around 8–10 days after birth, when the coat starts to grow. To detect germline transmission from the 129 cell line component, the mating partner for these chimeras should be B6. Appearance of any agouti F1 offspring form such mating proves germline transmission.

B6 ES cell lines, on the other hand, create a much more complex situation if the cells are injected into B6 blastocysts. As discussed earlier in Section 3.1, the switch to B6 ES cells did not affect the procedure for making chimeras by the aggregation method. In this case, the preferred host is the albino CD-1(ICR) outbred mouse line, which allows for easy detection of chimerism and germline transmission.

8. Discussion

The International Knockout Mouse Consortium (IKMC) decided to use C57BL/6N ES cell lines to introduce targeted mutations into all protein coding genes in the mouse (Austin *et al.*, 2004; Collins *et al.*, 2007b). These high-throughput gene targeting pipelines are expected to finish the targeting of close to 20,000 genes by 2011. Distribution centers (depositories) in Europe and North America have been established to make these mutant ES cell lines available to the scientific community. To set up such an operation has been a huge challenge; however, the result of this endeavor is going to create two even larger challenges: (1) the generation of mutant mice from this enormously large resource, and (2) the characterization of thousands of mutant phenotypes. An important component of meeting this need is to improve the method for generating germline transmitting chimeric mice from B6 ES cell lines. Our laboratory has been successfully using the simple and inexpensive morula aggregation method to generated germline transmitting chimeras with 129 and B6x129 F1-hybrid ES cells for almost 20 years. We have optimized this method for B6 ES cells as presented here, and we believe that this will make the IKMC and other B6 ES cell resources more accessible to the broader biomedical research community.

Just as chimeras became the ultimate pluripotency test to measure the developmental potential of mouse cells of different nature such as EC, EG, MAPC, and ES cells, it is now increasingly used to determine the developmental potential of iPS cells as well. The ever increasing activity of understanding and optimizing the process of iPS cell generation is expected to put additional demands on chimeric animal production in the future. Most importantly, nothing could be a better test for the safety of iPS-derived cells but the normality of fully iPS cell-derived animals obtained by tetraploid embryo complementation.

REFERENCES

Austin, C. P., Battey, J. F., Bradley, A., Bucan, M., Capecchi, M., Collins, F. S., Dove, W. F., Duyk, G., Dymecki, S., Eppig, J. T., *et al.* (2004). The knockout mouse project. *Nat. Genet.* **36,** 921–924.

Beddington, R. S., and Robertson, E. J. (1989). An assessment of the developmental potential of embryonic stem cells in the midgestation mouse embryo. *Development* **105,** 733–737.

Buehr, M., Meek, S., Blair, K., Yang, J., Ure, J., Silva, J., McLay, R., Hall, J., Ying, Q. L., and Smith, A. (2008). Capture of authentic embryonic stem cells from rat blastocysts. *Cell* **135,** 1287–1298.

Carmeliet, P., Ferreira, V., Breier, G., Pollefeyt, S., Kieckens, L., Gertsenstein, M., Fahrig, M., Vandenhoeck, A., Harpal, K., Eberhardt, C., *et al.* (1996). Abnormal blood vessel development and lethality in embryos lacking a single VEGF allele. *Nature* **380,** 435–439.

Chen, S., Hilcove, S., and Ding, S. (2006). Exploring stem cell biology with small molecules. *Mol. Biosyst.* **2,** 18–24.

Collins, F. S., Finnell, R. H., Rossant, J., and Wurst, W. (2007a). A new partner for the International Knockout Mouse Consortium. *Cell* **129,** 235.

Collins, F. S., Rossant, J., and Wurst, W. (2007b). A mouse for all reasons. *Cell* **128,** 9–13.

Cox, B. J., Vollmer, M., Tamplin, O., Lu, M., Biechele, S., Gertsenstein, M., Vancampenhout, C., Floss, T., Kühn, R., Wurst, W., *et al.* (2010). Phenotypic annotation of the mouse X chromosome. *Genome Res.* (in print).

Eakin, G. S., Hadjantonakis, A. K., Papaioannou, V. E., and Behringer, R. R. (2005). Developmental potential and behavior of tetraploid cells in the mouse embryo. *Dev. Biol.* **288,** 150–159.

Eggan, K., Akutsu, H., Loring, J., Jackson-Grusby, L., Klemm, M., Rideout, W. M., 3rd, Yanagimachi, R., and Jaenisch, R. (2001). Hybrid vigor, fetal overgrowth, and viability

of mice derived by nuclear cloning and tetraploid embryo complementation. *Proc. Natl. Acad. Sci. USA* **98,** 6209–6214.
Evans, M. J., and Kaufman, M. H. (1981). Establishment in culture of pluripotential cells from mouse embryos. *Nature* **292,** 154–156.
George, S. H., Gertsenstein, M., Vintersten, K., Korets-Smith, E., Murphy, J., Stevens, M. E., Haigh, J. J., and Nagy, A. (2007). Developmental and adult phenotyping directly from mutant embryonic stem cells. *Proc. Natl. Acad. Sci. USA* **104,** 4455–4460.
Gertsenstein, M., Nutter, L. M. J., Reid, T., Pereira, M., Stanford, W. L., Rossant, J., and Nagy, A. (2010). Efficient generation of germ line transmitting chimaeras from C57BL/6N ES cells by aggregation with outbred host embryos. *PLoS ONE* (in print).
Gossler, A., Doetschman, T., Korn, R., Serfling, E., and Kemler, R. (1986). Transgenesis by means of blastocyst-derived embryonic stem cell lines. *Proc. Natl. Acad. Sci. USA* **83,** 9065–9069.
Gu, H., Marth, J. D., Orban, P. C., Mossmann, H., and Rajewsky, K. (1994). Deletion of a DNA polymerase beta gene segment in T cells using cell type-specific gene targeting. *Science* **265,** 103–106.
Guillemot, F., Nagy, A., Auerbach, A., Rossant, J., and Joyner, A. L. (1994). Essential role of Mash-2 in extraembryonic development. *Nature* **371,** 333–336.
Ho, Y., Wigglesworth, K., Eppig, J. J., and Schultz, R. M. (1995). Preimplantation development of mouse embryos in KSOM: Augmentation by amino acids and analysis of gene expression. *Mol. Reprod. Dev.* **41,** 232–238.
Kiyonari, H., Kaneko, M., Abe, S., and Aizawa, S. (2010). Three inhibitors of FGF receptor, ERK and GSK3 establishes germline-competent embryonic stem cells of C57BL/6N mouse strain with high efficiency and stability. *Genesis* **48**(5), 317–327.
Kontgen, F., Suss, G., Stewart, C., Steinmetz, M., and Bluethmann, H. (1993). Targeted disruption of the MHC class II Aa gene in C57BL/6 mice. *Int. Immunol.* **5,** 957–964.
Kubiak, J. Z., and Tarkowski, A. K. (1985). Electrofusion of mouse blastomeres. *Exp. Cell Res.* **157,** 561–566.
Lawitts, J. A., and Biggers, J. D. (1992). Joint effects of sodium chloride, glutamine, and glucose in mouse preimplantation embryo culture media. *Mol. Reprod. Dev.* **31,** 189–194.
Lawitts, J. A., and Biggers, J. D. (1993). Culture of preimplantation embryos. *Methods Enzymol.* **225,** 153–164.
Ledermann, B., and Burki, K. (1991). Establishment of a germ-line competent C57BL/6 embryonic stem cell line. *Exp. Cell Res.* **197,** 254–258.
Lemckert, F. A., Sedgwick, J. D., and Korner, H. (1997). Gene targeting in C57BL/6 ES cells. Successful germ line transmission using recipient BALB/c blastocysts developmentally matured in vitro. *Nucleic Acids Res.* **25,** 917–918.
Li, P., Tong, C., Mehrian-Shai, R., Jia, L., Wu, N., Yan, Y., Maxson, R. E., Schulze, E. N., Song, H., Hsieh, C. L., *et al.* (2008). Germline competent embryonic stem cells derived from rat blastocysts. *Cell* **135,** 1299–1310.
Liu, P., Zhang, H., McLellan, A., Vogel, H., and Bradley, A. (1998). Embryonic lethality and tumorigenesis caused by segmental aneuploidy on mouse chromosome 11. *Genetics* **150,** 1155–1168.
Mansour, S. L., Thomas, K. R., and Capecchi, M. R. (1988). Disruption of the proto-oncogene int-2 in mouse embryo-derived stem cells: A general strategy for targeting mutations to non-selectable genes. *Nature* **336,** 348–352.
Mintz, B. (1962). Experimental study of the developing mammalian egg: Removal of the zona pellucida. *Science* **138,** 594–595.
Nagy, A., and Rossant, J. (2001). Chimaeras and mosaics for dissecting complex mutant phenotypes. *Int. J. Dev. Biol.* **45,** 577–582.
Nagy, A., Gocza, E., Diaz, E. M., Prideaux, V. R., Ivanyi, E., Markkula, M., and Rossant, J. (1990). Embryonic stem cells alone are able to support fetal development in the mouse. *Development* **110,** 815–821.

Nagy, A., Rossant, J., Nagy, R., Abramow-Newerly, W., and Roder, J. C. (1993). Derivation of completely cell culture-derived mice from early-passage embryonic stem cells. *Proc. Natl. Acad. Sci. USA* **90,** 8424–8428.

Nagy, A., Gertsenstein, M., Vintersten, K., and Behringer, R. (2003). Manipulating the Mouse Embryo: A Laboratory Manual. Cold Spring Harbor Laboratory Press, Cold Spring Harbor, NY.

Nichols, J., Jones, K., Phillips, J. M., Newland, S. A., Roode, M., Mansfield, W., Smith, A., and Cooke, A. (2009). Validated germline-competent embryonic stem cell lines from nonobese diabetic mice. *Nat. Med.* **15,** 814–818.

Okita, K., Ichisaka, T., and Yamanaka, S. (2007). Generation of germline-competent induced pluripotent stem cells. *Nature* **448,** 313–317.

Pettitt, S. J., Liang, Q., Rairdan, X. Y., Moran, J. L., Prosser, H. M., Beier, D. R., Lloyd, K. C., Bradley, A., and Skarnes, W. C. (2009). Agouti C57BL/6N embryonic stem cells for mouse genetic resources. *Nat. Methods* **6,** 493–495.

Poueymirou, W. T., Auerbach, W., Frendewey, D., Hickey, J. F., Escaravage, J. M., Esau, L., Dore, A. T., Stevens, S., Adams, N. C., Dominguez, M. G., *et al.* (2007). F0 generation mice fully derived from gene-targeted embryonic stem cells allowing immediate phenotypic analyses. *Nat. Biotechnol.* **25,** 91–99.

Rossant, J., Guillemot, F., Tanaka, M., Latham, K., Gertenstein, M., and Nagy, A. (1998). Mash2 is expressed in oogenesis and preimplantation development but is not required for blastocyst formation. *Mech. Dev.* **73,** 183–191.

Sato, H., Amagai, K., Shimizukawa, R., and Tamai, Y. (2009). Stable generation of serum- and feeder-free embryonic stem cell-derived mice with full germline-competency by using a GSK3 specific inhibitor. *Genesis* **47,** 414–422.

Stewart, C. L. (1982). Formation of viable chimaeras by aggregation between teratocarcinomas and preimplantation mouse embryos. *J. Embryol. Exp. Morphol.* **67,** 167–179.

Takahashi, K., and Yamanaka, S. (2006). Induction of pluripotent stem cells from mouse embryonic and adult fibroblast cultures by defined factors. *Cell* **126,** 663–676.

Tam, P. P., and Rossant, J. (2003). Mouse embryonic chimaeras: Tools for studying mammalian development. *Development* **130,** 6155–6163.

Tarkowski, A. K. (1961). Mouse chimaeras developed from fused eggs. *Nature* **190,** 857–860.

Tarkowski, A. K., Witkowska, A., and Opas, J. (1977). Development of cytochalasin in B-induced tetraploid and diploid/tetraploid mosaic mouse embryos. *J. Embryol. Exp. Morphol.* **41,** 47–64.

Ying, Q. L., Wray, J., Nichols, J., Batlle-Morera, L., Doble, B., Woodgett, J., Cohen, P., and Smith, A. (2008). The ground state of embryonic stem cell self-renewal. *Nature* **453,** 519–523.

CHAPTER NINE

Production of Cloned Mice from Somatic Cells, ES Cells, and Frozen Bodies

Sayaka Wakayama, Eiji Mizutani,[1] *and* Teruhiko Wakayama

Contents

Abstract

Somatic cell nuclear transfer (SCNT) has become a unique and powerful tool for epigenetic reprogramming research and gene manipulation in animals. Although the success rates of somatic cloning have been inefficient and the mechanism of

Center for Developmental Biology, RIKEN, Kobe, Japan
[1] Current address: Department of Animal Breeding and Reproduction, National Institute of Livestock and Grassland Science, Tsukuba, Japan

Methods in Enzymology, Volume 476
ISSN 0076-6879, DOI: 10.1016/S0076-6879(10)76009-2

reprogramming is still largely unknown, therefore, the nuclear transfer (NT) method has been thought of as a "black box approach" and inadequate to determine the detail of how genomic reprogramming occurs. However, only the NT approach can reveal dynamic and global modifications in the epigenome without using genetic modification, as well as can create live animals. At present, this is the only technique available for the preservation and propagation of valuable genetic resources from mutant mice that are infertile or too old, or recovered from carcasses, without the use of germ cells. This chapter describes a basic protocol for mouse cloning and embryonic stem (ES) cell establishment from cloned embryo using a piezo-actuated micromanipulator. This technique will greatly help not only in mouse cloning but also in other forms of micromanipulation such as intracytoplasmic sperm injection (ICSI) into oocytes or ES cell injection into blastocysts. In addition, we describe a new, more efficient mouse cloning protocol using histone deacetylase inhibitor (HDACi), which increases the success rates of cloned mice or establish rate of ES cells to fivefold.

1. Introduction

A new nuclear transfer (NT) method using nuclear injection with a piezo impact drive unit (hereafter termed "piezo unit") (Wakayama *et al.*, 1998) made mouse cloning possible. In this method, the donor nucleus were directly injected into oocyte cytoplasm; therefore, it does not matter whether donor cell are intact or broken, just need intact nucleus. Surprisingly, this method can resurrect normal mice from completely dead broken cells collected from frozen bodies kept in a freezer for 16 years at −20 °C. The weak point of this method is that it is technically very difficult and irritating, so it will probably take several months to attain sufficient technical skill to get useful data. Without hard practice, production of cloned mice is impossible. However, once the piezo unit is properly set up on the micromanipulator, it will greatly help not only in mouse cloning but also in other forms of micromanipulation such as intracytoplasmic sperm injection (ICSI) into oocytes (Kimura and Yanagimachi, 1995; Wakayama and Yanagimachi, 1998) or embryonic stem (ES) cell injection into blastocysts (Kawase *et al.*, 2001). Moreover, the piezo unit simplifies pipette preparation, as it allows one to use blunt tipped pipettes without any additional treatment.

Although, the success rate of animal cloning is still low, recently, we and other found that the efficiency of mouse cloning can be enhanced by up to sixfold through the addition of the histone deacetylation inhibitor (HDACi), such as Trichostatin A (TSA) or Scriptaid (SRC) into the oocyte activation medium (Kishigami *et al.*, 2006; Rybouchkin *et al.*, 2006; Van Thuan *et al.*, 2009), even though high concentrations of HDACi are toxic to embryonic development (Svensson *et al.*, 1998). This result suggested the possibility that nuclear reprogramming can be enhanced by chemical treatment, and this method will provide a new approach for practical improvements in mouse cloning techniques as well as new insights into the reprogramming of somatic cell nuclei.

On the other hand, NT technique has made it possible to create new types of ES cell lines from adult somatic cells via nuclear transfer (ntES cell lines) (Wakayama *et al.*, 2001, 2005c). We have shown that such ntES cell lines have the same differentiation potential as ES cells from fertilized blastocysts (Wakayama *et al.*, 2006). Moreover, cloned mice can be obtained from these ntES cell lines using a second NT procedure (Wakayama *et al.*, 2001, 2005b).

Interestingly, the ntES cell establishment rate is nearly 10 times higher than the success rate of cloned mice, even from so-called "unclonable" mouse strains (Wakayama *et al.*, 2001, 2005b). Therefore, we recommend the establishment of ntES cell lines at the same time as cloning to preserve the donor genome, because these lines can then be used as an unlimited source of donor nuclei for subsequent rounds of NT. In addition, this combination method allow us to preserve genetic resources even from infertile mutant mice or very aged mice without the use of germ cells (Mizutani *et al.*, 2008; Ono *et al.*, 2008; Wakayama *et al.*, 2005a, 2008).

Recently, we developed new NT techniques, which allowed us to resurrect normal mice from bodies kept frozen at −20 °C for up to 16 years without any cryoprotection (Wakayama *et al.*, 2008). Although we could not produce cloned offspring from the somatic cell, several ntES cell lines were established from the cell nuclei of most organs. Finally, healthy cloned mice were produced from these ntES cells by a second round of NT. Thus, this technique is applicable for the propagation of a variety of animals, regardless of age or fertile potential. If such an animal dies suddenly, just keep the body in a freezer, as it is clearly possible to resurrect animals from such conditions. Here we describe our improved approaches for the production of cloned mice and establishment of ntES cell lines from somatic cells.

2. Equipment

Inverted microscope with Hoffman optics from Olympus (Tokyo, Japan; model IX71)

Micromanipulator set from Narishige (Tokyo, Japan; model MMO-202ND, http://www.narishige.co.jp/)

Microforge (Narishige MF-900)

Warm plate (Tokay hit, http://www.ivf.net/ivf/tokai_hit_thermo_plate-o738.html)

Piezo impact drive system from Prime Tech Ltd. (Ibaraki, Japan; model MM-150FU, http://www.primetech-jp.com/en/01products/index.html)

Pipette puller (P-97) and glass pipette (B100-75-10) from Sutter Instrument Co. (Novato, CA, USA, http://www.sutter.com)

3. Mouse Strains

Oocyte: Usually B6D2F1 (C57BL/6 × DBA/2, about 2–3 months old) mice are used for oocyte collection, because B6D2F1 oocytes are very translucent and the metaphase spindle is easy to find; oocytes from outbred mice are stronger for *in vitro* manipulation and culture.

Donor cell: Usually hybrid strains, such as B6D2F1, B6C3F1, or B6129F1 can be used as donors, because hybrid mice are much better donor cells for the production of cloned mice. Inbred stains, such as C57BL/6 or C3H/He, are possible to use, but with a lower success rate for producing cloned mice than hybrid strains. The ICR (CD-1) strain is used for pseudopregnant surrogates, for lactating foster mothers and for vasectomized males.

4. Culture Media

4.1. Media for oocyte or embryo culture

Purchase distilled water (DW; Sigma-Aldrich, St. Louis, MO, USA; W1503)
Mineral oil (Sigma-Aldrich; M5310)
KSOM medium (Specialty Media, Millipore, Madison, WI, USA; MR-106-D, http://www.millipore.com/catalogue/item/mr-020p-5d)
M2 medium (Specialty Media; MR-015)
M2+hyaluronidase (Specialty Media; MR-051)

4.2. Media for donor cell culture and ntES cell establishment

Tail tip or embryonic fibroblast cells are cultured in Dulbecco's Modified Eagle's Medium (DMEM; termed EF medium).
CultiCell Medium (Dainippon Sumitomo Pharma, Tokyo, Japan; S2211101, http://www.ds-pharma.co.jp/english/index.html) is used for ntES cell establishment.
ES cell medium is used for maintaining ntES cells (Specialty Media; R-ES-101).
Phosphate-buffered saline (PBS) Ca^{2+}/Mg^{2+}-free (Sigma-Aldrich; BSS-1006-B). Purchase acidic Tyrode's solution (Specialty Media; MR-004-D) and trypsin solution (Specialty Media; SM-2003).

5. Chemicals

- *DMSO* (Sigma-Aldrich; D2650).
- *Cytochalasin B* (1 mg; Sigma-Aldrich; C6762). Add 2 ml DMSO into a vial with 1 mg cytochalasin B (500 μg; 100× CB stock solution). Divide into small tubes (10–20 μl) and store at −80 °C.
- *TSA* (1 mg; Sigma-Aldrich; T8552). Add 3.311 ml DMSO into a vial with 1 mg TSA (1 m*M*) and then take 2 μl of this stock and dilute with 198 μl of DMSO (10 μ*M*, 200× TSA stock solution). Divide into small tubes (10–20 μl) and store at −80 °C.
- *SCR* (1 mg; Sigma-Aldrich; S7817). Add 3.064 ml DMSO into a vial with 1 mg SCR (1 m*M*) and then take 5 μl of this stock and dilute with 95 μl of DMSO (50 μ*M*, 200× SCR stock solution). Divide into small tubes (10–20 μl) and store at −80 °C.
- $SrCl_2 \cdot 6H_2O$ (Sigma-Aldrich; S0390). Dissolved in DW at 100 m*M* and stored in aliquots at room temperature (10× stock solution).
- *EGTA* (Sigma-Aldrich; E8145). Dissolved in DW at 200 m*M* and stored in aliquots at 4 °C (100× stock solution).
- *Polyvinylpyrrolidone* (PVP; 360 kDa) (Irvine Scientific, Santa Ana, CA, USA; 99311, www.irvinesci.com).
- *Nuclear isolation medium* (NIM): 123.0 m*M* KCl, 2.6 m*M* NaCl, 7.8 m*M* NaH_2PO_4, 1.4 m*M* KH_2PO_4, 3 m*M* EDTA disodium salt, and 0.5 m*M* PMSF. The pH is adjusted to 7.2 by the addition of a small quantity of 1 *M* KOH.
- *Equine chorionic gonadotrophin* (eCG or PMSG; Sigma-Aldrich; G4527) and *human chorionic gonadotrophin* (hCG; Sigma-Aldrich; C8554). Dissolved in normal saline, stored in aliquots at −30 °C, and used at 5 IU/mouse.
- *Mercury* (Wako Pure Chemical Industries, Ltd., 133-01021, http://www.wako-chem.co.jp/english/).

6. Protocol

6.1. Preparation of enucleation and injection pipettes

Usually we make all pipettes by ourselves to save money and time (Fig. 9.1), but here we note supplying company for convenience.

Micropipettes can be ordered from several companies (e.g., Prime Tech Ltd., http://www.primetech-jp.com/en/01products/psk.html). For the holding pipette, the outside diameter (OD) should be smaller than that of the oocyte (e.g., OD 80 μm; inner diameter (ID) 10 μm). The ID of the enucleation pipette is 7–8 μm. The ID of the injection pipette depends on

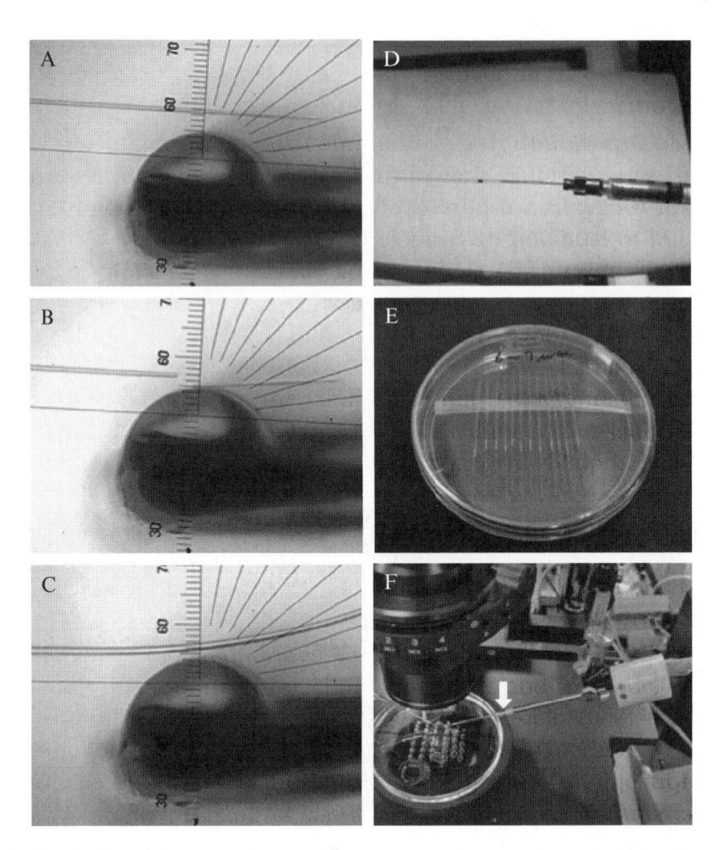

Figure 9.1 Making injection pipette. (A) Before cutting the tip of the injection pipette by microforge. One scale showed 10 μm. (B) After cut. (C) Bent pipette with bend approximately 300 μm from tip. (D) Inject a small amount of mercury into the pipette from the end using a 1 ml syringe with a 26-gauge needle. (E) Storage of pipette in the 10-cm dish. (F) Piezo device and pipette holder. The very important connection screw between the pipette and the pipette holder is shown by arrow.

cell type: 5–6 μm for cumulus cells and 6–7 μm for fibroblasts or ES cells. Ask the supplying company to bend all pipettes close to the tip (about 300 μm back) at 15–20 °C using a microforge (Fig. 9.1C).

When pipette were made or arrived from company, backload a small amount (about 3 m*M* long column) of mercury into the enucleation/injection pipette using a 1 ml syringe (Fig. 9.1D) and store in 10-cm dish at room temperature (Fig. 9.1E). *Note*: Mercury is toxic if absorbed by breathing or through the skin. Wear appropriate gloves and always use mercury in a working fume hood.

6.2. Preparation of media and dishes for manipulation

Place many 10–15 μl droplets of KSOM medium on the 6-cm dish and cover this with mineral oil. This medium can be used from oocyte collection to activation (Fig. 9.2A). This dish must be prepared before starting any other action.

Place about 15 μl droplets of three different media (M2, M2+CB, PVP) on the top of a 10-cm dish as shown in Fig. 9.2B and then cover this with mineral oil. For the enucleation medium, add 2 μl of CB stock solution to 198 μl of M2 medium (M2+CB medium: the final concentration of CB is 5 μg/ml). This chamber can be used for both enucleation and microinjection. Draw a line on the dish to distinguish these media.

For oocyte activation medium, add 10 μl of $SrCl_2$ stock solution (final concentration: 10 m*M*), 2 μl of EGTA stock solution (final concentration: 2 m*M*), 2 μl of CB stock solution (final concentration: 5 μg/ml) and 1 μl of TSA or SCR stock solution (final concentration: 50 or 250 n*M*, respectively) to 185 μl of KSOM medium (Kishigami and Wakayama, 2007). *Note*: If ES cells are used as sources of donor nuclei, TSA does not work well. If G2/M phase ES cells are used as donor nuclei, cytochalasin B must be omitted from the medium.

For reprogramming enhancement, cloned embryos should be cultured for an additional few hours with HDACi (TSA or SCR). Add 1 μl of TSA or SCR stock solution (final concentration: 50 or 250 n*M*, respectively) to 199 μl of KSOM medium.

Place about 15 μl droplets of activation medium (KSOM + EGTA + Sr + CB + HDACi) above the first line of 6-cm dish (Fig. 9.2C, above the line). Place about 15 μl droplets of HDACi medium (KSOM + TSA or SCR and KSOM) to enhance the reprogramming (Fig. 9.2C, between lines). Place about 15 μl droplets of KSOM medium for embryo washing (Fig. 9.2C, under the line). Cover those media with mineral oil.

Place about 15 μl droplets of KSOM medium for cloned embryo culture (Fig. 9.2D).

6.3. Collection of oocytes

Hormone injection. Inject eCG or PMSG (5 IU) into the abdominal cavity at 3 days before an experiment and then inject hCG (5 IU) 48 h later (1 day before an experiment). Usually we inject mice at 5–6 pm.

Collect oocyte-cumulus cell complexes from the oviduct ampullae at 14–15 h after hCG injection (usually we collect oocytes at 8–9 am) and move them into M2+hyaluronidase medium.

After 5 min, pick up the good oocytes, wash them in M2 medium three times and place in the KSOM medium prepared as above.

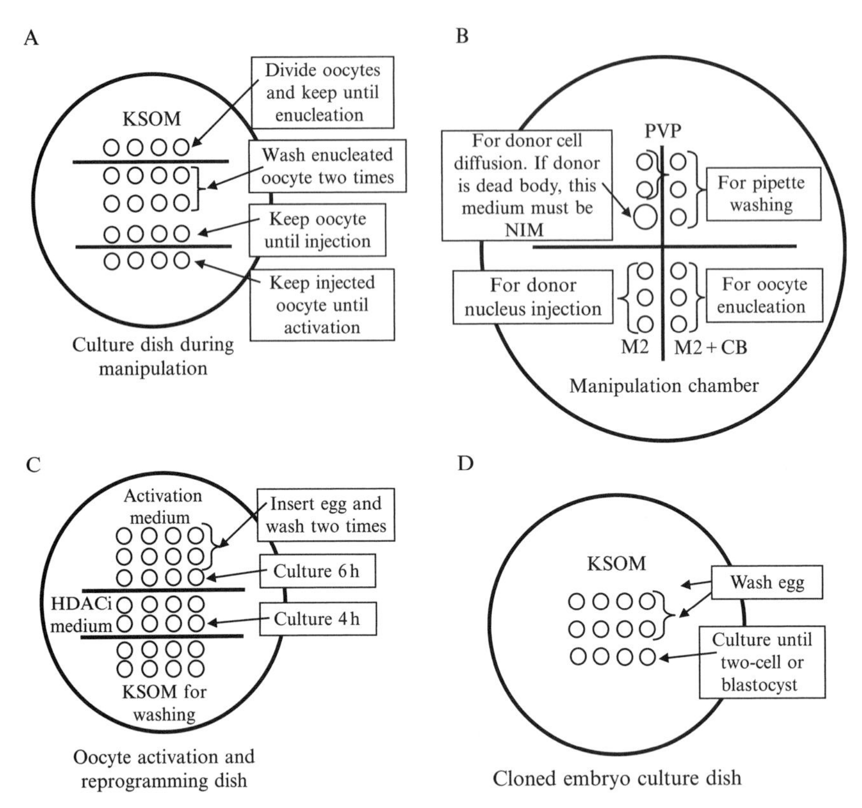

Figure 9.2 Manipulation, activation, and culture dish. (A) Oocytes are kept from the time of collection to just before activation in KSOM medium in a 6-cm dish. (B) The top of a 10-cm dish is used as a micromanipulation chamber. (C) Reconstructed oocytes are exposed to the activation medium for 6 h. Move embryos to the HDACi containing KSOM medium for additional 4 h to enhance the reprogramming. Wash embryos twice in KSOM medium after treatment. (D) Move embryos to the long-term culture dish until transfer into a recipient female.

Place oocytes into drops. The number of oocytes in a drop depends on each person's skill or type of experiment. Because we recommend that all oocytes in one drop must be manipulated within 15 min, from 10 (beginner) to 30 (skilled person) oocytes should be placed in each drop.

6.4. Preparation of donor cell

6.4.1. Tail tip fibroblasts

Tail tip cells must be prepared at least 2 weeks before NT. Cut the tail into sections at least 2 cm long and wash in 70% ethanol. In a sterile tissue culture hood, remove the skin and cut the tips into many small pieces on a 6 cm plastic dish. Culture the fragments in 10 ml DMEM under 5% CO_2 in air until used. There is no need to passage the cells (Wakayama and Yanagimachi, 1999).

6.4.2. ES cells

ES cell should be passaged 2 days before an experiment, or thawed at least 2 days before an experiment and cultured in 6-well dishes. We do not recommend cells used within 1 day of passage. ES cells are the most popular cell type for NT experiments because they have the best rate of producing full-term offspring (Rideout *et al.*, 2000; Wakayama *et al.*, 1999). However, each cell line shows different success rates, even from the same genetic background (Mizutani *et al.*, 2008). The number of passages of ES cell lines will also affect success rate. Note that these are pluripotent cells, not differentiated somatic cells, so they are not appropriate for genomic reprogramming experiments.

6.4.3. Cumulus cells

These cells are the easiest to prepare as nuclear donors because they can be used immediately after collection without washing, with no need to remove hyaluronidase from the medium (Wakayama *et al.*, 2000).

6.4.4. Sertoli cells

They are testicular sustentacular cells and the male counterparts of cumulus cells. Adult Sertoli cells are inappropriate donor cells because of their large size, but those collected from neonatal mouse testes (immature Sertoli cells) are small enough for injection and usually give better results than cumulus cells (Inoue *et al.*, 2003; Ogura *et al.*, 2000). Thus, the age of donors will affect the success rate. Use of newborn males younger than 6 days of age is recommended.

6.4.5. Frozen dead bodies

Collect the brain from the frozen body and break it into small pieces (1–2 mm^3) on the dry ice. Homogenize it under 500 μl of NIM in 1.5 ml tube by homogenizer pestle (Fig. 9.3), then filter it by cell strainer (BD Falcon 352235) and collect the supernatant. This medium contains many naked nuclei (Fig. 9.5E and F). Introduce a few microliters of suspension into NIM on the manipulation chamber instead of PVP medium. In addition to brain nuclei, 1–2 μl of blood cells from the frozen-thawed mouse tail could also be used for production of cloned mice. Collect blood and add to 500 ml of NIM to wash blood cells. Make a condensed cell suspension by centrifugation and introduce it into NIM in the manipulation chamber instead of PVP medium (Wakayama *et al.*, 2008).

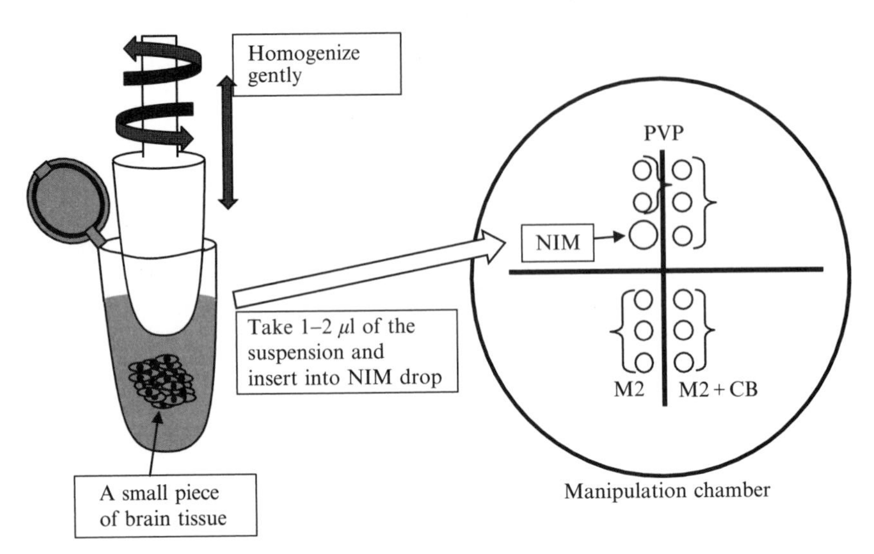

Figure 9.3 Collection of donor nuclei from frozen dead body. (A) Put small pieces (1–2 mm^3) of frozen brain tissue and homogenize it under 500 μl of NIM in 1.5 ml tube by homogenizer pestle, then filter it by cell strainer. Introduce a few microliters of suspension into NIM on the manipulation chamber instead of PVP medium.

6.5. Collection of single donor cell

For noncultured fresh cells, wash them in PBS to remove any enzyme by centrifugation (100 rpm or 3×*g* for 10 min) at least three times, except for cumulus cells, which can be used without washing.

For cultured cells, remove any culture medium from the dish and wash them in PBS (Ca^{2+}, Mg^{2+}-free). Remove the PBS, add trypsin medium then incubate 5–20 min. Add culture medium (including serum) and move the pipette in and out several times to select single cells. Wash cells with PBS by centrifugation at least three times, as above. *Note*: As trypsin is very toxic at the time of nuclear injection, the donor cells must be washed completely.

Make a very concentrated cell suspension in the medium. The final volume should be less than 10 μl. If the final concentration is too low, it is difficult to find an appropriate donor cell and the delay can be detrimental to the recipient oocyte awaiting NT.

Pick up 1–3 μl of condensed donor cell suspension and introduce them into a PVP droplet in the micromanipulation chamber. Mix the donor cells with PVP medium using sharp tweezers, gently but completely. We recommend to mix the donor cells with PVP medium for at least 30 s. If the donor cells are not mixed sufficiently in PVP medium, they will aggregate and it will be difficult to isolate single cells. Do not scratch the bottom of the chamber.

6.6. Setting up the micromanipulator

Attach the enucleation pipette to the pipette holder of the piezo unit. The top of the pipette holder must be screwed in tightly. Hang the piezo unit on the micromanipulator. If you cannot cut the zona pellucida by piezo pulses, usually not tightly connected between the pipette and pipette holder (Fig. 9.1F, arrow).

Expel any air and oil and a few drops of mercury from the enucleation pipette in the PVP medium. Wash both the inside and outside of the pipette using PVP medium. PVP will cover both the inside and outside of the pipette to keep the surface slick. Washing the pipette in PVP medium is very important and will affect not only oocyte survival rate but also embryo development after NT. Without this step, the pipette soils rapidly and needs to be changed.

Adapt and fit the pipette to the piezo unit. While expelling the air and mercury from the pipette in the PVP droplet, the piezo unit must be applied with high power (more than 10 units) and high speed (more than 10 units) for at least 1 min continuously.

Attach the holding pipette on the other side of the micromanipulator.

6.7. Enucleation of oocytes

Place one group of oocytes in a M2+CB droplet into the micromanipulation chamber and wait about 5 min before starting enucleation.

Locate the metaphase II (MII) spindle inside the oocyte. It can be recognized without any staining using Nomarski or Hoffman optics. Rotate the oocyte to place the spindle at between 2 and 4 o'clock or between 8 and 10 o'clock and then attach it firmly to the holding pipette (Fig. 9.4A). The room temperature is very important as the spindle microtubules will disperse and become unclear at room or lower temperature. However, the spindle will become visible again if the oocytes are cultured at 37 °C for 30 min before enucleation. Oocyte transparency also depends on the mouse strain: B6D2F1 is better than others are.

Cut through the zona pellucida using a few piezo pulses. There should be a slight negative pressure inside the pipette to enhance the piezo power. To avoid damaging the oocyte, ensure there is a large space between the zona pellucida and the oolemma: approximately as wide as the thickness of the zona pellucida.

Insert the enucleation pipette into the oocyte without breaking the oolemma (Fig. 9.4B) and remove the MII spindle by aspiration with a minimal volume of cytoplasm. The oocyte membrane and spindle must be pinched off slowly: do not apply piezo pulses to cut the membrane. The MII spindle is harder than the cytoplasm, so you can feel its consistency through the micromanipulator (Fig. 9.4C).

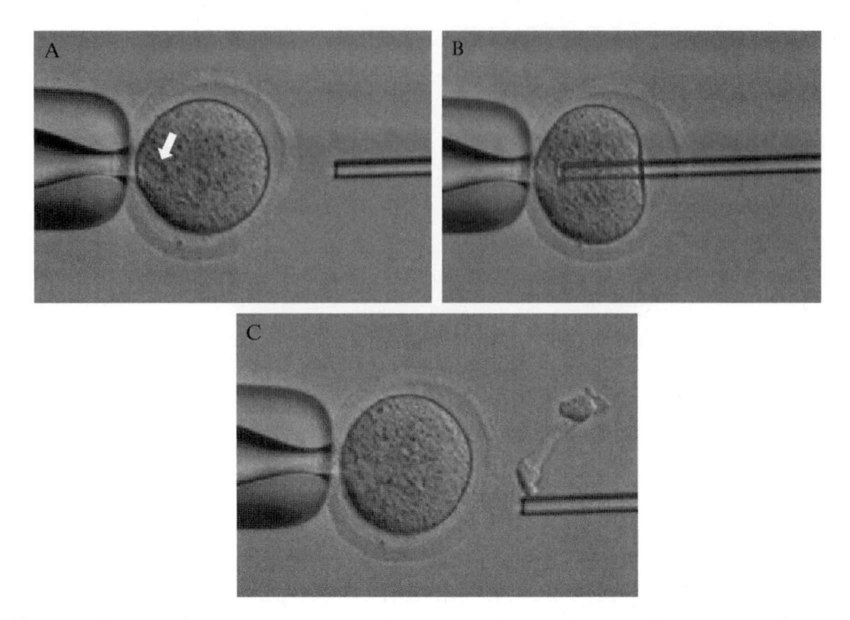

Figure 9.4 Enucleation. (A) Rotate the oocyte, locate the metaphase II spindle, and place it between the 8 and 10 o'clock position, or between the 2 and 4 o'clock position (see arrow). Then stabilize the oocyte on the holding pipette. (B) Insert enucleation pipette into the oocyte until near the metaphase II spindle. (C) Remove the spindle by suction without breaking the plasma membrane and gently pull the pipette away from the oocyte.

Wash the enucleated oocytes two times in KSOM to remove the cytochalasin B completely, and keep them in KSOM medium for at least 30 min in the incubator before starting donor cell injection.

If you feel tired at this point, take a short break before starting the next step. From the next step, there is no respite and it requires intense concentration.

6.8. Donor nucleus injection

Place about 10–20 enucleated oocytes into M2 medium. The number of oocytes per droplet depends on each individual's skill level. Each group should be finished within 15 min.

Remove the donor nuclei from the cells by gently aspirating them in and out of the injection pipette until each nucleus is clearly separate from any visible cytoplasmic material (Fig. 9.5A and B). Take up a few nuclei into the injection pipette. If the donor cells are not mixed sufficiently in PVP medium, they will aggregate and it will be difficult to isolate single cells.

ES cells are especially sensitive and fragile and it is better to make new ES cell suspension drops every 30 min.

Stabilize an enucleated oocyte using a holding pipette. Cut the zona pellucida using a few piezo pulses (Fig. 9.5C).

Reduce the power level of the piezo unit (power level 1–2 and speed 1). The oolemma is weaker than the zona pellucida and the survival rate of oocytes after injection will be better with this reduced power.

Push one nucleus forward until it is near the tip of the pipette and advance the pipette until it almost reaches the opposite side of the oocyte's cortex (Fig. 9.5D). Do not apply the piezo unit's power until the pipette reaches the opposite side. If the piezo power is applied with the tip of the pipette in the middle of the oocyte, the oocyte will die after injection.

Apply 1 weak piezo pulse to puncture the oolemma at the pipette tip. This is indicated by a rapid relaxation of the oocyte membrane. Expel the donor nucleus into the enucleated oocyte cytoplasm immediately with a minimal

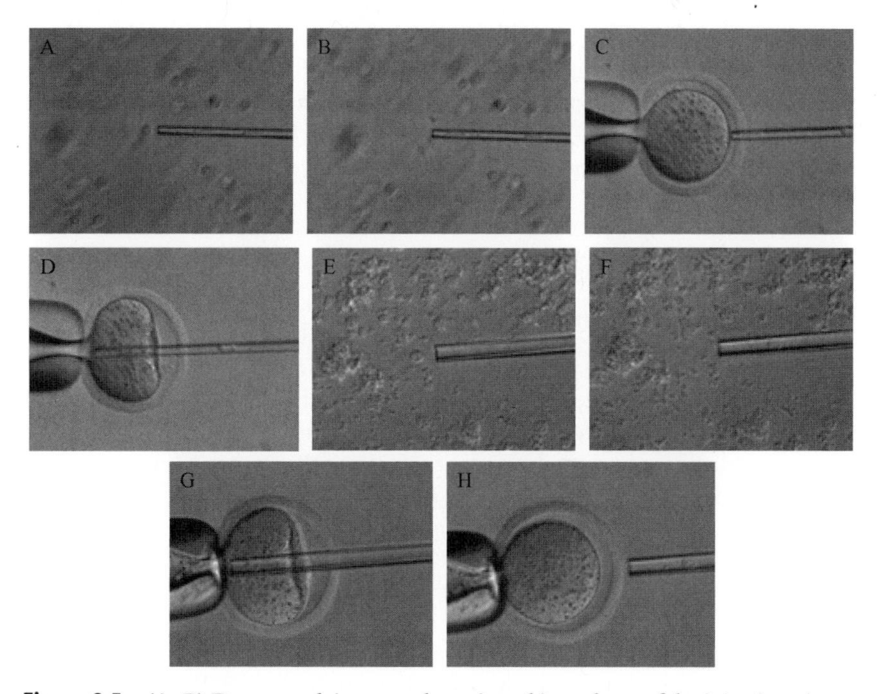

Figure 9.5 (A, B) Donor nuclei are gently aspirated in and out of the injection pipette until their nuclei are largely devoid of visible cytoplasmic material. (C) Hold the enucleated oocyte and cut the zona pellucida using piezo pulses. (D) Insert the injection pipette into the enucleated oocyte. Apply a single piezo pulse to break the membrane, and then inject the donor nucleus immediately (arrow). Gently withdraw the injection pipette from oocyte. (E, F) Collect naked and clear nuclei from NIM drop to large injection pipette (OD: 10–12 μm). (G, H) Inject naked nucleus into enucleated oocyte.

amount of PVP medium. Gently withdraw the injection pipette from the oocyte. If it is difficult to release donor nuclei from the pipette, probably the pipette is too dirty. It must be washed frequently using PVP medium by expelling some mercury and applying power from the piezo unit.

Each time, wash the injection pipette with PVP medium by expelling some mercury and applying power from the piezo unit. This washing step is essential to stop the pipette from getting sticky.

Keep the injected oocytes in this drop for at least 10 min, then transfer them into KSOM medium (Fig. 9.2A) and culture for at least 30 min in the incubator before activation. The process of NT should be perform under room temperature (25–26 °C), do not use warm plate for injection process.

If many of oocyte lysis after nuclear injection, probably you used too large a pipette, performed nuclear injection at higher room temperature or the pipette insertion was too shallow. A large pipette or warm temperatures increase the rate of oocyte lysis. The injection pipette must be inserted very deep into the oocyte before applying the piezo pulse. In addition, you need practice. If you are beginner, all oocytes will lyse immediately after injection. One month after starting practice, about half of your oocytes might survive. One year later, about 80% of oocytes will survive, if you continue to practice diligently. If oocytes are transferred to KSOM medium immediately after injection, 10–20% of them undergo lysis from the damage of injection. The oocyte membrane must be allowed to recover: this takes about 10 min.

6.9. Activation and embryo culture

Prepare the oocyte activation medium and drop (Fig. 9.2C and D) at least 30 min before use and equilibrate in a CO_2 incubator. We recommend to check the activation medium before starting the experiment, using intact oocytes. During strontium treatment, up to 10% of the oocytes will die (depend on mouse strain and age of oocyte) and the medium will become dirty. This is normal and the surviving oocytes are usually undamaged.

Transfer and culture each group of oocytes into drops of activation medium and wash twice, then culture for 6 h in a 5% CO_2 incubator at 37 °C (Fig. 9.2C).

Six hours after activation, all embryos must be washed at least once in HDACi medium to remove the cytochalasin B (Fig. 9.2C, between the line). Examine the rate of oocyte activation. If NT and activation are done properly, those oocytes should each possess two or three pseudopronuclei (Wakayama *et al.*, 1998). Then, culture embryos to enhance the reprogramming.

Three or four hours after additional culture, wash the cloned embryos in KSOM medium (Fig. 9.2C, under line) and move them to another dish (Fig. 9.2D), and culture to the two-cell stage (next day) or to the blastocyst stage (3 days later). *Note*: Because some of the chemicals used for activation can diffuse to other drops through the mineral oil and are embryotoxic, all embryos should be moved to different culture dishes for long-term culture. Some batches of mineral oil are toxic and need to be tested, such as control embryo culture before starting experiments.

If there is no pseudo-pronuclear formation after oocyte activation, probably because of the failure to break the donor cell membrane. The injection pipette must be smaller than the donor cell. If the donor cell has a tough cell membrane (e.g., tail tip fibroblasts), apply piezo power to break the donor cell membrane at the time of cell pickup.

6.10. Embryo transfer and caesarian section

Mate estrous ICR female mice with normal males on the same day or 1–2 days before the experiment: these will be used as foster mothers. Mate estrous ICR female mice with vasectomized males on the same day as the experiment: these will be used as pseudopregnant (surrogate) mothers.

Transfer the two-cell (24 h after NT) or four- to eight-cell (48 h) cloned embryos into oviducts of 0.5 days post copulation (dpc), or morulae/ blastocysts (72 h) or blastocysts (96 h) into the 2.5 dpc pseudopregnant female mice, respectively (Nagy *et al.*, 2003).

A cesarean section is required to recover the cloned mice fetuses securely. Euthanize the recipient female at 18.5 or 19.5 dpc. Remove the uterus from the abdomen and dissect out the cloned pups with their placentas. Wipe away the amniotic fluid from the skin, mouth, and nostrils and stimulate the pups to breathe by rubbing the back or pinching them gently with blunt forceps.

Transfer the cloned pups to the cage of a naturally delivered foster mother. Mix the cloned pups with bedding material from the foster mother's cage. Remove some pups from the foster mother's litter and then mix the cloned pups with the original pups.

All cloned mice to date have been born with abnormal and hypotrophic placenta and often die just after birth from respiratory failure. At this point, there is no way to avoid this lethal phenotype. If you have no success in getting full-term development, change the donor cell type from somatic to ES cells, or other hybrid mouse strains. Also, try making up new embryo culture medium. However, the most important solution is to keep practicing. Technical skill is essential. Do not give up!

6.11. Establishing ntES cell lines from cloned embryos

To make embryonic feeder cells, collect day 12.5–13.5 dpc fetuses from a pregnant mother as above and then cut off the head and internal organs in a 10 cm Petri dish containing PBS. Place the remaining bodies into a new 10-cm dish and mince them into very small pieces with sterile scissors. Add 25 ml DMEM and plate into large (175 cm^2) tissue culture flasks. One or two days later, split the cells 1:5 by trypsinization and allow them to grow to confluence. You can repeat this cell passage several times, but with increased passage number the support potential of ES cell culture will decrease.

Mitomycin C treatment. When the cells became confluent, treat with 10 μg/ml mitomycin C for 2 h in an incubator at 37 °C, 5% CO_2 in air. Wash the flask several times using PBS to remove mitomycin C and then collect the cells by trypsinization. Pellet the cells by centrifugation (1000 rpm or 3×*g* for 10 min). Aspirate the supernatant and gently resuspend the cell pellet in freezing medium (final concentration about 1×10^6 cells/ml) and divide into cryovials. Place the vials into a −80 °C freezer overnight; the next day the vials can be transferred to liquid nitrogen for long-term storage.

Thaw the vial quickly by warming and transfer the cells into a 15 ml tube filled with ES medium. Pellet the cells by centrifuging as before and then aspirate the supernatant and resuspend in fresh ES medium. Plate into a 96-well multidish and culture in the incubator until needed. These feeder cells should be prepared at least 1 day before starting ntES cell establishment.

Change the medium of the dish from DMEM to at least 200 μl of CultiCell medium before plating cloned embryos. The CultiCell medium does not contain fetal calf serum, which contains potential differentiation factors. Therefore, it is important to use this medium for establishment of new ntES or ES cell lines. Alternatively, you can use the new ES cell establishment "3i medium" which inhibits GSK3, MEK, and FGF receptor tyrosine kinases and enhances the ES cell establishment rate significantly (Ying *et al.*, 2008).

Remove the zona pellucida from cloned blastocysts (Day 4) using acidic Tyrode's solution. The zona pellucida will dissolve within 30 s and prolonged exposure to acid Tyrode's solution will decrease the quality and survival of embryos. Therefore, before dissolving the zona pellucida completely, you must pick up the embryos and wash them several times in M2 medium. The remaining thinned zonae are easily broken by repeated pipetting.

Plate each cloned blastocyst onto the feeder cell of 96-well multidishes one by one. Culture the multidish for 10–14 days in an incubator without changing the medium. During this period, the cloned blastocyst will attach to the surface of the feeder layer and the inner cell mass (ICM) can be seen to grow.

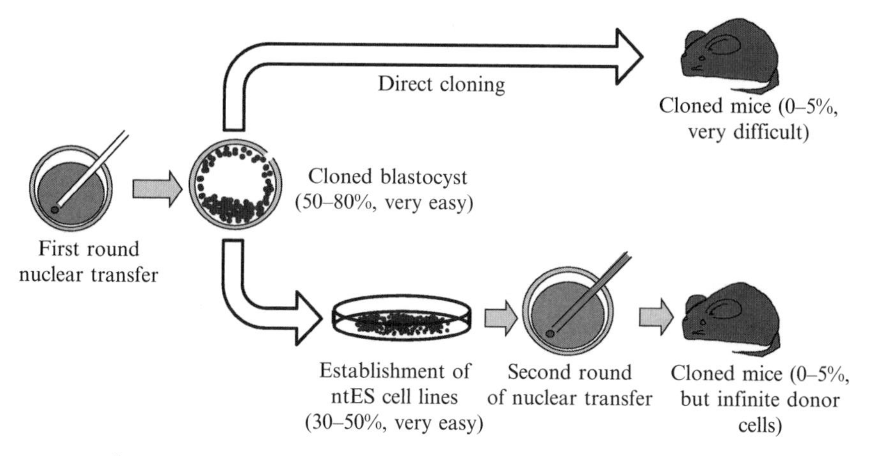

Figure 9.6 Two different approaches to generate cloned offspring.

Some of the wells should have clumps of large ICMs. When those clumps appear, treat them with trypsin and disaggregate the cells using a 200 μl pipette. Then replate the suspension into another well (preformed cell feeder) of the same multidish.

When ES-like cell colonies dominate the well, the cells should be expanded gradually to 48-well, 24-well, 12-well and then 12.5 cm^2 flask and 25 cm^2 flask by repeated passages several times. We do not use feeder cells from the 48-well dishes on. After the cell numbers have increased, the cells should be frozen-stored as usual for ES cells (Schatten *et al.*, 2005).

6.12. Production of cloned mice from ntES cell nuclei

Nuclear transfer is done using these ntES cells as donors, repeating steps 13.3. Unlike somatic cells, ntES cells (like ES cells in general) (Wakayama *et al.*, 2006) will divide indefinitely, so you can use them without limitation (Fig. 9.6). Moreover, the overall success rates of cloning from individuals are increased when ntES cell lines are used as intermediate nuclear donors (Mizutani *et al.*, 2008; Wakayama *et al.*, 2005b).

REFERENCES

Inoue, K., Ogonuki, N., Mochida, K., Yamamoto, Y., Takano, K., Kohda, T., Ishino, F., and Ogura, A. (2003). Effects of donor cell type and genotype on the efficiency of mouse somatic cell cloning. *Biol. Reprod.* **69,** 1394–1400.

Kawase, Y., Iwata, T., Watanabe, M., Kamada, N., Ueda, O., and Suzuki, H. (2001). Application of the piezo-micromanipulator for injection of embryonic stem cells into mouse blastocysts. *Contemp. Top Lab. Anim. Sci.* **40,** 31–34.

Kimura, Y., and Yanagimachi, R. (1995). Intracytoplasmic sperm injection in the mouse. *Biol. Reprod.* **52,** 709–720.

Kishigami, S., and Wakayama, T. (2007). Efficient strontium-induced activation of mouse oocytes in standard culture media by chelating calcium. *J. Reprod. Dev.* **53,** 1207–1215.

Kishigami, S., Mizutani, E., Ohta, H., Hikichi, T., Thuan, N. V., Wakayama, S., Bui, H. T., and Wakayama, T. (2006). Significant improvement of mouse cloning technique by treatment with trichostatin A after somatic nuclear transfer. *Biochem. Biophys. Res. Commun.* **340,** 183–189.

Mizutani, E., Ono, T., Li, C., Maki-Suetsugu, R., and Wakayama, T. (2008). Propagation of senescent mice using nuclear transfer embryonic stem cell lines. *Genesis* **46,** 478–483.

Nagy, A., Gertsenstein, M., Vintersten, K., and Behringer, R. (2003). Manipulating the Mouse Embryo. Cold Spring Harbor Laboratory Press, Cold Spring Harbor, NY.

Ogura, A., Inoue, K., Ogonuki, N., Noguchi, A., Takano, K., Nagano, R., Suzuki, O., Lee, J., Ishino, F., and Matsuda, J. (2000). Production of male cloned mice from fresh, cultured, and cryopreserved immature Sertoli cells. *Biol. Reprod.* **62,** 1579–1584.

Ono, T., Mizutani, E., Li, C., and Wakayama, T. (2008). Nuclear transfer preserves the nuclear genome of freeze-dried mouse cells. *J. Reprod. Dev.* **54,** 486–491.

Rideout, 3rd, W. M., Wakayama, T., Wutz, A., Eggan, K., Jackson-Grusby, L., Dausman, J., Yanagimachi, R., and Jaenisch, R. (2000). Generation of mice from wild-type and targeted ES cells by nuclear cloning. *Nat. Genet.* **24,** 109–110.

Rybouchkin, A., Kato, Y., and Tsunoda, Y. (2006). Role of histone acetylation in reprogramming of somatic nuclei following nuclear transfer. *Biol. Reprod.* **74,** 1083–1089.

Schatten, G., Smith, J., Navara, C., Park, J. H., and Pedersen, R. (2005). Culture of human embryonic stem cells. *Nat. Methods* **2,** 455–463.

Svensson, K., Mattsson, R., James, T. C., Wentzel, P., Pilartz, M., MacLaughlin, J., Miller, S. J., Olsson, T., Eriksson, U. J., and Ohlsson, R. (1998). The paternal allele of the H19 gene is progressively silenced during early mouse development: The acetylation status of histones may be involved in the generation of variegated expression patterns. *Development* **125,** 61–69.

Van Thuan, N., Bui, H. T., Kim, J. H., Hikichi, T., Wakayama, S., Kishigami, S., Mizutani, E., and Wakayama, T. (2009). The histone deacetylase inhibitor scriptaid enhances nascent mRNA production and rescues full-term development in cloned inbred mice. *Reproduction* **138,** 309–317.

Wakayama, T., and Yanagimachi, R. (1998). Development of normal mice from oocytes injected with freeze-dried spermatozoa. *Nat. Biotechnol.* **16,** 639–641.

Wakayama, T., and Yanagimachi, R. (1999). Cloning of male mice from adult tail-tip cells. *Nat. Genet.* **22,** 127–128.

Wakayama, T., Perry, A. C., Zuccotti, M., Johnson, K. R., and Yanagimachi, R. (1998). Full-term development of mice from enucleated oocytes injected with cumulus cell nuclei. *Nature* **394,** 369–374.

Wakayama, T., Rodriguez, I., Perry, A. C., Yanagimachi, R., and Mombaerts, P. (1999). Mice cloned from embryonic stem cells. *Proc. Natl. Acad. Sci. USA* **96,** 14984–14989.

Wakayama, T., Shinkai, Y., Tamashiro, K. L., Niida, H., Blanchard, D. C., Blanchard, R. J., Ogura, A., Tanemura, K., Tachibana, M., Perry, A. C., Colgan, D. F., Mombaerts, P., *et al.* (2000). Cloning of mice to six generations. *Nature* **407,** 318–319.

Wakayama, T., Tabar, V., Rodriguez, I., Perry, A. C., Studer, L., and Mombaerts, P. (2001). Differentiation of embryonic stem cell lines generated from adult somatic cells by nuclear transfer. *Science* **292,** 740–743.

Wakayama, S., Kishigami, S., Van Thuan, N., Ohta, H., Hikichi, T., Mizutani, E., Yanagimachi, R., and Wakayama, T. (2005a). Propagation of an infertile hermaphrodite mouse lacking germ cells by using nuclear transfer and embryonic stem cell technology. *Proc. Natl. Acad. Sci. USA* **102,** 29–33.

Wakayama, S., Mizutani, E., Kishigami, S., Thuan, N. V., Ohta, H., Hikichi, T., Bui, H. T., Miyake, M., and Wakayama, T. (2005b). Mice cloned by nuclear transfer from somatic and ntES cells derived from the same individuals. *J. Reprod. Dev.* **51,** 765–772.

Wakayama, S., Ohta, H., Kishigami, S., Thuan, N. V., Hikichi, T., Mizutani, E., Miyake, M., and Wakayama, T. (2005c). Establishment of male and female nuclear transfer embryonic stem cell lines from different mouse strains and tissues. *Biol. Reprod.* **72,** 932–936.

Wakayama, S., Jakt, M. L., Suzuki, M., Araki, R., Hikichi, T., Kishigami, S., Ohta, H., Van Thuan, N., Mizutani, E., Sakaide, Y., Senda, S., Tanaka, S., *et al.* (2006). Equivalency of nuclear transfer-derived embryonic stem cells to those derived from fertilized mouse blastocysts. *Stem Cells* **24,** 2023–2033.

Wakayama, S., Ohta, H., Hikichi, T., Mizutani, E., Iwaki, T., Kanagawa, O., and Wakayama, T. (2008). Production of healthy cloned mice from bodies frozen at −20 degrees C for 16 years. *Proc. Natl. Acad. Sci. USA* **105,** 17318–17322.

Ying, Q. L., Wray, J., Nichols, J., Batlle-Morera, L., Doble, B., Woodgett, J., Cohen, P., and Smith, A. (2008). The ground state of embryonic stem cell self-renewal. *Nature* **453,** 519–523.

CHAPTER TEN

Nuclear Transfer in Mouse Oocytes and Embryos

Zhiming Han,* Yong Cheng,* Cheng-Guang Liang,* *and* Keith E. Latham*,†

Contents

Abstract

Nuclear transfer methods provide an invaluable means of dissecting the genetic and epigenetic control of development, as well as the interactions between ooplasm and nucleus in the oocyte and early embryo. These procedures also provide novel means of manipulating animal genomes (e.g., through cloning with genetically engineered cells), and also have been applied for clinical purposes to treat infertility. This chapter reviews methods employed for a range of nuclear transfer techniques including germinal vesicle transfer, spindle transfer, intracytoplasmic sperm injection, pronuclear transfer, and somatic cell nuclear transfer.

* The Fels Institute for Cancer Research and Molecular Biology, Temple University School of Medicine, Philadelphia, Pennsylvania, USA

† Department of Biochemistry, Temple University School of Medicine, Philadelphia, Pennsylvania, USA

Methods in Enzymology, Volume 476

ISSN 0076-6879, DOI: 10.1016/S0076-6879(10)76010-9

1. Introduction

Nuclear transfer methods continue to gain ground as tools for dissecting basic developmental mechanisms in early embryos, as well as tools for applied and clinical purposes. The longest historical legacy for nuclear transfer has been in evaluating the developmental potential of nuclei from differentiated somatic cells. This approach was first taken in mammals using pronuclear transfer (PNT) and subsequently blastomere nuclear transfer (McGrath and Solter, 1983; Willadsen, 1986). With the advent of cloning by somatic cell nuclear transfer (review, Latham, 2004) and studies that demonstrated the ability of fully differentiated nuclei to support development of live progeny (Hochedlinger and Jaenisch, 2002; Lewitzky and Yamanaka, 2007), the question of developmental potency has been essentially put to rest. Cloning methods promise to support new approaches to endangered species preservation, livestock propagation, development of new genetic models, development of animals producing valuable biopharmaceutical, and novel stem cell-based methods for treating disease and injury, among other goals. In recent years, other microsurgical methods have continued to emerge and have been proposed as possible tools in the human for clinical purposes and in model organisms for research purposes. These include germinal vesicle (GV) transfer and spindle transfer, with these techniques wedded to methods for *in vitro* oocyte maturation (IVM) and intracytoplasmic sperm injection (ICSI) (Cheng *et al.*, 2009; Zhang *et al.*, 1999). This chapter summarizes these methods as applied in the mouse.

2. Equipment and Solutions

1. *Microscopes*: Stereomicroscope with a Gimbal-mounted mirror (Olympus) and an inverted microscope with Hoffman optics or relief phase contrast optics (Olympus IX71). We find the Olympus optics better for visualizing spindles in metaphase II (MII) oocytes. Available magnifications for inverted microscopes should include 4× and 20× objectives, 20× eyepiece.
2. *Micromanipulators and microinjectors*: For example, Micromanipulators, Narishige (MN-4); Microinjector, Narishige (IM-9B).
3. *Piezo pipet driver*: The Piezo-drill micromanipulation controller PMM-150 (Prime Tech Ltd, Ibaraki, Japan).
4. *Electrofusion device*: Electro cell manipulator (ECM 2001, BTX Inc., San Diego, CA) with dishes having 1 mm space between electrodes.

5. *Embryo culture incubator*: Humidified CO_2 incubator (Model 610, Fisher Scientific).
6. *Embryo culture chamber*: Modular plastic incubator (Billups-Rothenberg, Del Mar, CA).
7. *Micropipet puller*: Model P-87, Sutter Instruments Co., Novato, CA.
8. *Microforge*: De Fonbrune type, monocular or binocular, with 10× objective and 10× eyepiece.
9. *Pipet grinder*: Model A53100, Arenberg Sage Co.
10. *Glass pipets*: Pasteur pipet (Fisher, Cat. 13-678-20D).
11. *Glass needles*: Borosilicate tubing without filament (Sutter Instrument Co., Cat. B100-75-15).
12. *The preparation of pipets for the micromanipulation*: To make the pipets for the removal of the spindle–chromosome complex (SCC) (the inner diameter is about 10 μm) or injection (for injecting the cumulus cell nuclei the inner diameter is about 5 μm; for injecting the sperm head, the inner diameter is about 6–7 μm), the glass pipet is pulled by a pipet puller followed with tip breaking at 90° by the microforge. Bend the pipets 1–2 mm from the tip with about 20° angles by the microforge. The pipets for injection need to be washed with diluted dichlorodimethylsilane (Sigma, Cat. 440272) for one time and followed by five times washing with MilliQ water at least 1 day before use. To make the holding pipets, the glass pipet after tip breaking at 90° (the outside diameter is about 80–90 μm) is heated to smaller inner diameter (about 15–25 μm) by microforge. To make the pipets for the pronulear transfer, the bevel is made using the pipet grinder after tip breaking at 90°. The pipets need to be washed with 20% hydrofluoric acid quickly for three times (excessive time dissolves too much glass), washed with MilliQ water for five times and washed with 95% ethanol for three times in sequence before use. A sharp spike needs to be added to the tip of beveled pipet by microforge, and then broken off at the time of use. If the piezo pipet driver is employed, the spike is not required. To make the pipets for the removal of GV (the inner diameter is about 25 μm), the tip of the pipet should be blunt and heat polished slightly on the microforge.
13. *Culture dishes*: Falcon 6-cm (60 × 15 mm) Petri dishes (Becton Dickinson, Cat. 351007) dishes are used for oocyte and embryo culture. Make several drops in each dish and cover the drops with mineral oil. Culture dishes should be kept in the modular incubator and equilibrated with 5% CO_2/5% O_2/N_2 balance for normal embryo culture, or 5% CO_2/21% O_2/N_2 balance for cloned embryo culture.
14. *Manipulation chambers*: There are several kinds of manipulation chamber for different procedures, which are made with Falcon 10-cm (100 × 15 mm)

Petri dishes. For PNT, place one drop of 10% PVP for pipet washing and several drops of M2 medium containing 5 μg/ml cytochalasin B (CB) (Sigma, C-6762) and 0.2 μg/ml demecolcine (Sigma, D-7385) in the manipulation dish and cover the drops with mineral oil (Fisher, Cat. O121-1). The latter drops are in two rows staggered to allow access to each row from the side. For somatic cell nuclear transfer, two kinds of manipulation dishes need to be made. For the removal of SCC, place one drop of 10% PVP and several drops of HCZBG medium containing 2.5 μg/ml CB in the manipulation dish and cover the drops with mineral oil. For the injection, place one drop of 10% PVP, several drops of 3% PVP, and several drops of HCZBG in the manipulation dish and cover the drops with mineral oil. For spindle transfer, use the same setup as for PNT. For GV removal and transfer, make several drops of M2 medium containing 5 μg/ml CB and 0.2 m*M* 3-isobutyl-1-methylxanthine (IBMX). Cover the drops with mineral oil. For the ICSI manipulation chamber, make one drop of 10% PVP for injection pipet washing. Make several drops of 10% PVP for sperm head collection. Make several drops of HCZBG for oocyte injection. Align the sperm head collection drops and oocyte injection drops side by side. Cover the drops with mineral oil.

15. *Media*: CZB-HEPES (HCZBG) (Gao, 2006) is the general oocyte or embryo manipulation medium. When doing ICSI and donor nuclear injection for cloning, this medium can be used without any modification. When removing the SCC, 2.5 μg/ml CB should be added. When doing PNT, 5 μg/ml CB and 0.2 μg/ml demecolcine should be added. CZB (Chatot *et al.*, 1989) can be used for short-term oocyte culture. MEM-α (Gibco, 12561) + 20% (v/v) FCS is used for oocyte maturation. MEM-α (Gibco, 12561) + 20% (v/v) FCS and 0.2 m*M* IBMX is used to inhibit maturation and maintain the immature oocyte at the GV stage. For oocyte activation, Ca^{2+}-free CZB medium + 5 μg/ml CB, 1 m*M* glutamine (Gibco, 21051), and 10 m*M* $SrCl_2 \cdot 6H_2O$ (Sigma, 25521) is used. Either HCZBG or M2 medium can be used for oocyte manipulation during cloning. CZB medium is used for normal embryo culture (Chatot *et al.*, 1989) and can also be used for sperm suspension. Cell suspension medium for SCNT is 3% PVP in HCZBG (Gao, 2006). MEM-α medium (Sigma, M-4536) supplemented with 0.5% BSA and 1 m*M* glutamine is used for SCNT embryo culture (Gao and Latham, 2004; Gao *et al.*, 2004). Higher concentrations of PVP for injection can be harmful to the constructs. For electrofusion, the fusion medium is prepared with 275 m*M* mannitol, 0.05 m*M* $CaCl_2$, 0.1 m*M* $MgSO_4$, and 0.3% BSA (Han *et al.*, 2005).

3. Procedures

3.1. Germinal vesicle transfer

GV transfer achieves the largest substitution of ooplasm possible by microsurgery. This procedure (Fig. 10.1) transfers the donor genome with the smallest amount of accompanying ooplasm into new cytoplasm environment at the GV stage. GV transfer is useful in many circumstance, such as avoiding defective meiosis or evaluating genetic and epigenetic effects in development.

Mouse follicles are primed to grow with a large antrum by the injection of 5 IU equine chorionic gonadotropin (eCG) in abdominal cavity for 48 h. Cumulus cell-enclosed immature GV oocyte cumulus cell-oocyte complexes (COCs) are released with follicular fluid after puncturing ovarian follicles using a fine gauge needle or forceps. COCs are collected in M2 medium with 0.2 m*M* IBMX. IBMX prevents GV breakdown (GVBD) and maintains oocytes at an immature stage. COCs are cultured in MEM-α medium containing 20% fetal bovine serum (FBS) and 0.2 m*M* IBMX for 2 h. Culture in MEM-α + IBMX prompts GV oocytes to develop

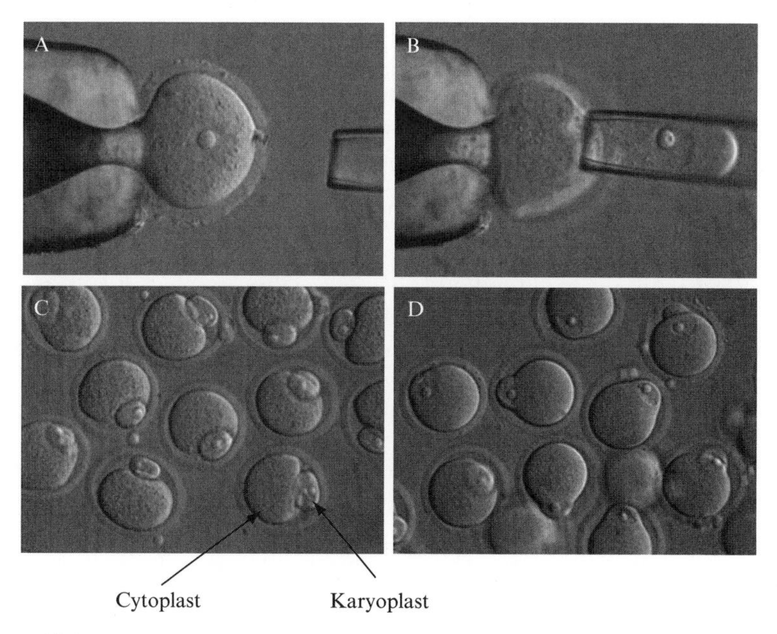

Figure 10.1 Steps in the GV transfer method: (A) GV oocytes before manipulation, (B) aspiration of GV into pipet. The volume is exaggerated here for clarity, (C) cytoplast–GV karyoplast couplets, (D) fused couplets.

perivitelline spaces, facilitating manipulation and enucleation, and providing room to receive the transferred GV. Prior to the GV removal, the surrounding cumulus cells are stripped off repeated aspiration with a smaller pipet at the diameter about 75 μm. Care should be taken to ensure the pipet is not so narrow that oocyte lysis occurs. Denuded GV oocytes are cultured in MEM-α supplemented with 0.2 m*M* IBMX, 5 μg/ml CB, and 0.2 μg/ml demecolcine for half an hour.

Enucleation and transplantation of GV are performed in manipulation drops of M2 medium supplemented with 0.2 m*M* IBMX, 5 μg/ml CB, and 0.2 μg/ml demecolcine. Immature oocytes are placed in manipulation drops for 10 min before the GV removal. A slit is made in the zona pellucida by pressing a sharp glass meedle (size and shape similar to an injection pipet) through the perivitelline space against the holding pipet. A blunt-tip micropipet with the inner diameter about 25 μm is inserted to remove the GV. The micropipet is rinsed in 10% PVP before GV removal to reduce sticking of the membranes within the pipet, then two or three droplets of mineral oil are aspirated to help control the fluid flow during the manipulation. A GV with a minimal amount of surrounding cytoplasm (GV karyoplast) is aspired into the micropipet and transferred to the perivitelline space under the zona pellucida of another GV cytoplast from which the GV was previously removed. Oocyte–karyoplast couplets are thoroughly rinsed to remove CB and demecolcine, and then cultured in MEM-α medium with 0.2 m*M* IBMX for 1 h prior to the electrofusion.

The couplets are aligned between the two microeletrodes for electrofusion with an electro cell manipulator (ECM 2001, BTX Inc.). Couplets can be aligned with brief application of a low-voltage AC field, and then a single 900 V/cm direct current (DC) fusion pulse is delivered for 10 μs. After the DC pulse, reconstituted oocytes are cultured in MEM-α with IBMX for 1 h for the fusion.

After fusion, reconstituted GV oocytes are thoroughly rinsed to remove IBMX, and then cultured in MEM-α medium containing 20% FBS up to 14 h for nuclear maturation as evidenced by the extrusion of the first polar body (*in vitro* maturation).

3.2. Spindle transfer

Spindle transfer is accomplished using procedures that combine the spindle removal procedure from the cloning method with the electrofusion method described in detail for PNT (below). The spindle transfer pipet is not beveled. Electrofusion settings for spindle transfer are the same as above.

3.3. Intracytoplasmic sperm injection

Since mouse oocyte plasma membranes are extremely sensitive, the oocytes are easily destroyed following conventional manual ICSI method without a piezo instrument to facilitate rapid membrane penetration. On the contrary, piezo-actuated ICSI can get nearly 100% survival rate for a skilled person. Some protocols have been published and the procedure for the ICSI has been stated (Kimura and Yanagimachi, 1995; Yoshida and Perry, 2007). But some details can be clarified for the purpose of applying this method quickly and efficiently, especially for the beginners.

The injection pipet for ICSI is similar to that for somatic cell nuclear transfer except that the external diameter is bigger, and this can be adjusted for different strains (e.g., 6.5 μm for BDF1 and C57BL/6 sperm, 7.0 μm for DBA/2 sperm). It is noted that the size of sperm is different for different mouse strains. The ideal pipet should have the inner diameter that is a little bit bigger than the sperm head. A pipet with a narrow diameter will make it difficult to collect the sperm head and it will be easy for the head to stick inside the pipet. The holding pipet is the same as described above.

Matured oocyte (MII oocyte) preparation is same as described in section 3.5 below. For the sperm preparation, we highly recommend using fresh sperm collected from cauda epididymides, which will give higher embryonic developmental rate compare to the frozen-thawed sperm. Most labs use "Epididymides Dicing" method to collect sperm. This method dissects cauda epididymides in sperm collection medium. The sperm suspension will flow out by itself. With this approach, however, the blood, debris, and other cells associated with the epididymides can be released into medium. This will contaminate the medium and affect sperm motility and quality. Here, we introduce a method that results in less contamination and greater ease of sperm collection.

First, prepare one piece of sterilized Whatman filter paper and a glass pipet with a polished end (the shape of the end should be a ball with 1 mm in diameter). Males are used at the age of 10–12 weeks. Isolate cauda epididymides and put them on the filter paper. Gently roll over the cauda epididymides to remove the visible blood and fat. Put the cauda epididymides on the edge of the filter paper and fold the paper. Cut the surface of the cauda epididymides with fine scissors. Wearing gloves, pinch the cauda epididymides with fingers, and gently squeeze it. At this time a small drop of sperm should be pushed out. Use the polished glass pipet to collect the sperm and transfer it immediately to 1 ml balanced CZB medium in an Eppendorf tube. Return the tube to the CO_2 incubator. The sperm suspension can be used 15 min later. Because most of the active sperm will swim up in 15 min, the sperm located at the upper part of the tube are ideal for use.

Load the sperm into the droplet of 10% PVP in HCZBG. The final concentration of PVP is 7–8% after sperm loading. Within a few minutes, the sperm will disperse in the PVP-HCZBG drops and be ready for use. Transfer MII oocytes to the HCZBG drops. Position the injection pipet in the PVP-HCZBG drop containing sperm (*Caution*: The holding pipet should not be inserted into the drop at this time. Contact of the holding pipet with PVP can block the pipet). Because sperm motility is inhibited by PVP, they can be easily aspirated. Select intact sperm with head and tail. Sperm with good quality should show progressive, regular movements with a sigmoidal beat. Draw the sperm head into the injection pipet until the head-midpiece boundary touches the pipet tip. Apply the piezo pulses with intensity of 3–6 and frequency of 2 to separate the head from tail. Try to use the lowest intensity and frequency settings possible to minimize the damage to the sperm DNA. Draw the sperm head further into the injection pipet. Then repeat this procedure for another sperm. For a skilled person, 5–10 sperm heads can be collected and positioned within the pipet at intervals of about 100 μm to increase the injection speed and efficiency. Aspirating more than 10 sperm heads at one time is not recommended, because this will increase the PVP volume inside injection pipet, which will decrease the efficacy of the piezo pulses. Because long exposure of sperm in PVP may adversely affect embryo development, it is recommended to load fresh sperm in new PVP-HCZB drops every half an hour.

After sperm aspiration, move the injection pipet to the HCZBG drop containing eggs. Focus the microscope on the oocytes until spindles can be observed very clearly. Lower the holding pipet to the same level of the oocytes. Orient the selected oocyte with holding and injection pipets so that the MII spindle is located at 12 or 6 o'clock. This position will let the injection pipet penetrate the oolemma without damaging the spindle. Hold the oocyte firmly with the holding pipet with gentle negative pressure. Use the injection pipet to touch the oocyte to make sure the oocyte is held firmly in place. Move the sperm head until it is about 50 μm from the tip. Move the injection pipet to make it touch with zona at the 3 o'clock position. Apply a little bit negative pressure within the injection pipet and apply a piezo pulse (intensity = 3–6 and frequency = 2, same as sperm head collection) at the same time to make a hole in the zona. Sometimes the debris that cut from the zona will be sucked into the injection pipet. In order to inject sperm head alone, the zona debris should be expelled before injection. Now the tip of injection pipet should be in the perivitelline space.

Switch the parameter of piezo instrument to a lower intensity (about 1–2) and set for single pulse. Push the injection pipet gently until it reaches near to the opposite side of the oocyte, almost to the holding pipet position. At the same time, apply positive pressure within injection pipet to push the sperm head to the tip. Give a very slight negative pressure within injection pipet, and apply the piezo pulse. Now the oocyte plasma membrane should return to the original position, which indicates the membrane has been penetrated. If the oolemma does not return, another piezo pulse needs to be applied. After penetrating the oolemma, gently apply a slight positive pressure within injection pipet to push out the sperm head. Once the sperm head has been pushed into the cytoplasm, the injection pipet should be withdrawn immediately and quickly to avoid excessive medium being injected into the cytoplasm. Release the oocyte from the holding pipet, and deposit injected oocytes in one region of the droplet until remaining oocytes are injected.

When 15–20 oocytes are injected, let them recover for 5–10 min in the manipulation medium at room temperature. Transfer all the finished oocytes back to the culture dish and wash thoroughly at least four times to remove the HEPES completely. Transfer the culture dish to the CO_2 incubator for further culture. The oocytes will be activated by the endogenous sperm factors.

3.4. Pronuclear transfer

The PNT procedure (Fig. 10.2) was originally described by McGrath and Solter (1983). A key innovative aspect of the method was the avoidance of any step to penetrate the oolemma, relying instead upon aspiration of a membrane bound karyoplast that was then fused with the recipient cell. Fusion was accomplished originally with heat-inactivated Sendai virus, but as other labs adopted the procedure to other species, electrofusion became more widely applied. The procedures required specially crafted microtools

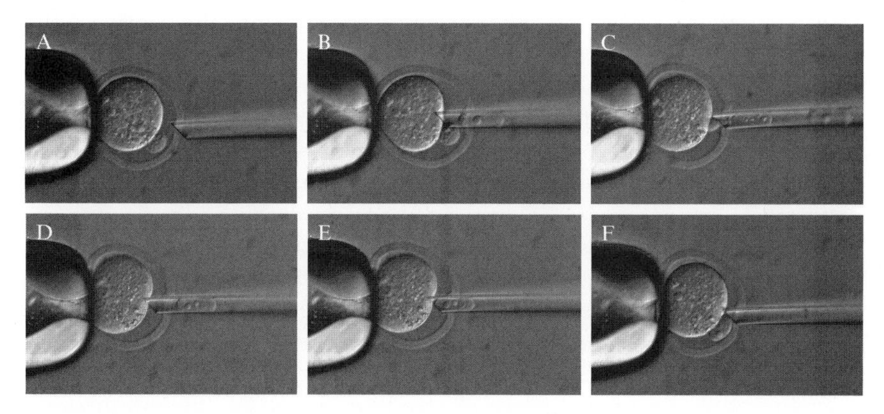

Figure 10.2 Steps in the PN transfer method: (A) embryo before procedures, showing paternal (left) and maternal pronuclei in same plane as polar body, (B) aspiration of maternal PN, (C) drawing karyoplast and 1st polar body into pipet, (D) insertion of pipet with donor karyoplast into previous opening in zona, (E) insertion of karyoplast, (F) PN transfer couplet prior to fusion.

(described above), particularly a holding pipet with correct size aperture and a PNT pipet that is beveled, sharpened if required, and treated to prevent membrane adhesion to the inner walls as described above. This basic method remains the core of the procedure today, although new equipment—chiefly the piezo pipet driver and the BTX electrofusion device—has permitted recent modifications to simplify and streamline the process (Han *et al.*, 2005).

Embryos to be used are selected initially on the basis that they have formed the two pronuclei (PN) and are clearly fertilized. Embryos of abnormal morphology or granularity should be avoided. Prior to PNT, zygotes are treated for at least 30 min with 5 μg/ml CB and 0.2 μg/ml demecolcine.

Embryos are placed in a series of drops (M2 or HCZB) under oil on the stage of an inverted microscope equipped with suitable optics such as relief phase contrast. The embryo is oriented with the second polar body at approximately 2 o'clock position and the two PN visible in the same or nearly the same plane so that their relative sizes can be discerned. The maternal and paternal PN are distinguished on the basis that the maternal PN is smaller and located closer to the polar body. If necessary, the PN can be "nudged" with the PNT pipet after penetrating the zona to confirm apposition of the tip near the pronucleus. Zona penetration requires either that the PNT pipet be sharpened as described above, or that it be employed using a piezo pipet driver. The piezo pipet driver allows very rapid zona penetration with no harm to the zygote. With the piezo, care must be taken to keep the intensity of pulses as low as possible and to avoid transmitting the pulse to the oolemma, which would lyse the cell.

The side of the oolemma is pressed inward and the PNT pipet tip bevel is positioned next to the PN of choice without penetrating the oolemma. Mild aspiration pressure is applied to draw the PN into the lumen of the pipet, and then the pressure is equalized. The pipet is withdrawn from the perivitelline space and the karyoplast is drawn slightly further into the pipet to avoid contact with the oil when moving the pipet between drops. Typically, at the start of an experiment the first karyoplast removed is discarded, so that there is a ready recipient for each additional karyoplast aspirated.

The donor karyoplast is inserted into the perivitelline space of the recipient through the original opening made during the previous enucleation step, and the couplets are then washed through several drops of culture medium without CB and demecolcine and allowed to recover for at least 30 min. Embryos are then transferred to the electrofusion chamber, being placed carefully between the two electrodes. Because of different medium densities embryos can be washed through a droplet of electrofusion medium or allowed to equilibrate within the pipet before being expelled between the electrodes, to avoid them floating. Electrofusion is accomplished most easily using a combination of a brief AC pulse to orient the couplets within the electric field, followed by a single DC pulse of 900 V/cm (10 μs), such as is allowed using the BTX system. Only a single pulse should be required. Additional pulses can be given, but this may diminish viability. Where viability is the endpoint being measured, this risk may not be acceptable for the needs of the experiment. After the pulse is delivered, embryos are washed carefully through several drops of medium and allowed to recover in the incubator. Fusion should be complete within about 1 h. The reconstructed PNT embryos are cultured in CZB medium for further studies.

3.5. Somatic cell nuclear transfer

Oocytes are obtained from superovulated females. We prefer B6xD2 F1 females at 8–12 weeks of age. Superovulation is induced by injection of 5 IU eCG (Calbiochem, San Diego, CA) followed 48 h later by 5 IU human chorionic gonadotropin (hCG; Sigma-Aldrich, St. Louis, MO). Excessive hormone dose can be detrimental to oocyte quality. Additionally, aging of oocytes should be avoided. For oocyte isolation, prepare culture drops (CZBG) in a Petri dish and cover with oil. Keep the culture dishes in the incubator for at least 30 min before collecting oocytes. Prepare the drops of M2 medium and M2 medium containing 120 U/ml of hyaluronidase (ICN Pharmaceuticals, Costa Mesa, CA) at room temperature. Collect the oviducts into an M2 medium drop. Release the COCs from oviducts in M2 medium and transfer COCs to M2 medium containing 120 U/ml of hyaluronidase at room temperature. Remove cumulus cells by aspiration with a pipet (inner diameter of 200–300 μm) after 5 min of enzyme treatment. Wash the oocytes with M2 medium extensively to remove both cumulus cells and enzyme, then wash them through several changes of CZBG medium and culture in CZBG at 37 °C incubator until use. Excessive hyaluronidase treatment can be detrimental to oocyte quality.

Many kinds of somatic cells can be used as the donor cells for SCNT. In many studies, cumulus cells are used as the donor cells. After removing the cumulus cells from COCs, collect the cumulus cells into a 1.5 ml Eppendorf tube. Add 1 ml of HCZBG and wash once by 1000×*g* centrifugation for 3 min. Resuspend the cell pellets in 30–50 μl of 3% PVP solution and keep the cells on ice until use.

The first step in the manipulation (Fig. 10.3) is the removal of the SCC. Place one 30–40 μl drop of 10% PVP and five to ten 20 μl drops of HCZBG medium containing 2.5 μg/ml CB in a Falcon 100 × 15 mm dish and cover the drops with mineral oil. Transfer 20–30 MII stage oocytes to HCZBG medium containing 2.5 μg/ml CB in the manipulation dish and pretreat the oocytes for at least 5 min at room temperature before removal of the SCC. Set up the holding pipet and the pipet used for the SCC removal. The holding pipet is connected to a 50 ml syringe filled with air, and the pipet

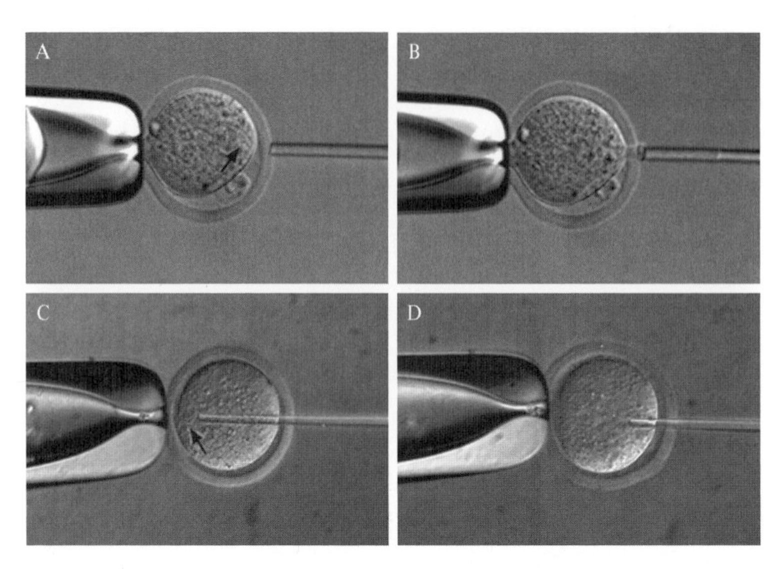

Figure 10.3 Steps in the SCNT method: (A) spindle positions at about 3 o'clock before removal (the arrow points the SCC), (B) SCC removal, (C) injection of donor nucleus (the arrow points the donor nucleus), (D) withdrawal of injection pipet.

used for SCC removal is connected with a microinjector filled with MilliQ water. The inner diameter of the pipet used for the removal of the SCC is about 10 μm. Add 1–2 mm (distance of fill starting at pipet shoulder) of mercury (Sigma, Cat. 83359) into the pipet before connecting to microinjector. Move the mercury forward to the tip of the pipet in the 10% PVP drop and aspirate a small volume of 10% PVP into the pipet. Move the pipet to the HCZBG drop containing CB where the pretreated oocytes are, and aspirate a small volume of HCZBG into the pipet. To locate the position of SCC, rotate the oocyte by the pipet used for the removal of the SCC. Hold the oocyte by the holding pipet once the position of SCC has been determined. Make a hole in the zona pellucida close to the SCC by applying the piezo pulses to the tip of the pipet used for the removal of the SCC. Gentle suction to the zona will help the pipet penetrate. Stop the piezo pulse immediately once a hole is made, even once most of the thickness of the zona has been traversed. Traversing the zona only partially may allow the pipet to be inserted easily without additional pulses, reducing risk of damage to the oocyte. Move the pipet forward to touch the oolemma and position the tip adjacent to the underlying SCC. Remove the SCC by aspiration without penetrating the plasma membrane. The removal process for one group of oocytes should be within 10–15 min. Return the SCC-depleted oocytes to the CZBG culture drops to recover in the incubator before injecting donor nuclei, about 1 h.

The next step is to inject the donor cell nucleus. Prepare the manipulation dish for the injection. Place one 30–40 μl drop of 10% PVP, six to ten 20 μl drops of 3% PVP, and six to ten 20 μl drops of HCZBG in a Falcon 100 × 15 mm dish and cover the drops with mineral oil. Add one group of the MII stage oocytes without SCC to the HCZBG drop and add the cumulus cells into the 3% PVP drop. Prepare and position the holding and injection pipets as described above. The inner diameter of the injection pipet should be about 5 μm with a flat opening. Add 1–2 mm of mercury to the pipet before connecting to the microinjector. Move the mercury forward to the tip of the pipet in the 10% PVP drop and aspirate a small volume of 10% PVP into the pipet. Wash the pipet thoroughly with 10% PVP before collecting the cumulus cell nuclei. If the injection pipet becomes sticky during the experiment, it can be washed again with the 10% PVP solution. Move the pipet to the 3% PVP drop containing the cumulus cells. Select round cumulus cells with a diameter of 10 μm and aspirate into the pipet with membrane breaking assisted by a piezo pulse to the tip of the pipet. After aspirating six to ten cumulus cell nuclei, move the injection pipet to the HCZBG drop containing the SCC-depleted oocytes. Inject a single nucleus into the single cytoplasm gently by breaking the membrane with piezo pulse and withdraw the injection pipet quickly, as described in detail above for ICSI. To improve survival rate, the piezo pulse for breaking the membrane should be as low intensity as possible, and the nucleus should be injected into the cytoplasm with as little medium or PVP as possible.

After the constructs are made, they need to be activated. Prepare the activation medium 1 h before moving the reconstructed oocytes to the activation medium. Add CB and glutamine to 9 ml Ca^{2+}-free CZBG medium and equilibrate them in the 37 °C incubator for 10 min. Add 1 ml 10× strontium stock to 9 ml of equilibrated medium to a final concentration of 5 μg/ml CB, 1 m*M* glutamine, and 10 m*M* of strontium. Prepare the 100 μl drops of activation medium in the culture dish and cover with mineral oil. Keep the culture dishes in the 37 °C incubator for at least 30 min before transferring the reconstructed oocytes to the drops. Wash the reconstructed oocytes through three to four drops and keep them in the final drop for 5–6 h. It is important to keep the strontium media equilibrated at the correct pH in order to avoid precipitation.

The cloned constructs after activation are cultured to observe development. Prepare the culture drops in the culture dishes at least 2 h before moving the embryos into the culture medium. For cumulus cell cloned embryos, use MEM-α supplemented with 0.5% BSA and 1 m*M* glutamine for culture. Transfer the activated oocytes with pseudo-pronuclei to culture medium and culture the cloned embryos in a Billups-Rothenberg modular incubator filled with 5% CO_2/21% O_2/N_2 balance at 37 °C. Other media can be used as well, such as KSOM medium. We find that development in these other media is often less than with MEM-α, unless constructs were exposed to dimethylsulfoxide during the activation period. Because DMSO can exert unexpected epigenetic effects and induce stress responses, we prefer to avoid using this reagent.

ACKNOWLEDGMENTS

Work in the authors' laboratory is supported by grants from the National Institutes of Health, NICHD HD43092, HD52788, and NCRR RR18907.

REFERENCES

Chatot, C. L., Ziomek, C. A., Bavister, B. D., Lewis, J. L., and Torres, I. (1989). An improved culture medium supports development of random-bred 1-cell mouse embryos in vitro. *J. Reprod. Fertil.* **86,** 679–688.

Cheng, Y., Wang, K., Kellam, L. D., Lee, Y. S., Liang, C. G., Han, Z., Mtango, N. R., and Latham, K. E. (2009). Effects of ooplasm manipulation on DNA methylation and growth of progeny in mice. *Biol. Reprod.* **80,** 464–472.

Gao, S. (2006). Protocols for nuclear transfer in mice. *Methods Mol. Biol.* **325,** 25–33.

Gao, S., and Latham, K. E. (2004). Maternal and environmental factors in early cloned embryo development. *Cytogenet. Genome Res.* **105,** 279–284.

Gao, S., Czirr, E., Chung, Y. G., Han, Z., and Latham, K. E. (2004). Genetic variation in oocyte phenotype revealed through parthenogenesis and cloning: Correlation with differences in pronuclear epigenetic modification. *Biol. Reprod.* **70,** 1162–1170.

Han, Z., Chung, Y. G., Gao, S., and Latham, K. E. (2005). Maternal factor controlling blastomere fragmentation in early mouse embryos. *Biol. Reprod.* **72,** 612–618.

Hochedlinger, K., and Jaenisch, R. (2002). Monoclonal mice generated by nuclear transfer from mature B and T donor cells. *Nature* **415,** 1035–1038.

Kimura, Y., and Yanagimachi, R. (1995). Intracytoplasmic sperm injection in the mouse. *Biol. Reprod.* **52,** 709–720.

Latham, K. E. (2004). Cloning: Questions answered and unsolved. *Differentiation* **72,** 11–22.

Lewitzky, M., and Yamanaka, S. (2007). Reprogramming somatic cells towards pluripotency by defined factors. *Curr. Opin. Biotechnol.* **18,** 467–473.

McGrath, J., and Solter, D. (1983). Nuclear transplantation in the mouse embryo by microsurgery and cell fusion. *Science* **220,** 1300–1302.

Willadsen, S. M. (1986). Nuclear transplantation in sheep embryos. *Nature* **320,** 63–65.

Yoshida, N., and Perry, A. C. (2007). Piezo-actuated mouse intracytoplasmic sperm injection (ICSI). *Nat. Protoc.* **2,** 296–304.

Zhang, J., Wang, C. W., Krey, L., Liu, H., Meng, L., Blaszczyk, A., Adler, A., and Grifo, J. (1999). In vitro maturation of human preovulatory oocytes reconstructed by germinal vesicle transfer. *Fertil. Steril.* **71,** 726–731.

CHAPTER ELEVEN

Culture of Whole Mouse Embryos at Early Postimplantation to Organogenesis Stages: Developmental Staging and Methods

Jaime A. Rivera-Pérez,* Vanessa Jones,† *and* Patrick P. L. Tam†

Contents

Abstract

In vitro culture of whole mouse embryos enables the maintenance of growth and morphogenesis of postimplantation embryos outside the uterine environment. This technological advent facilitates the observation of the development of embryos in real time whereby cell lineage and tissue morphogenesis can be

* Department of Cell Biology, University of Massachusetts Medical School, Worcester, Massachusetts, USA
† Embryology Unit, Children's Medical Research Institute, Sydney Medical School, University of Sydney, Westmead, New South Wales, Australia

Methods in Enzymology, Volume 476
ISSN 0076-6879, DOI: 10.1016/S0076-6879(10)76011-0

traced with appropriate vital cell lables and molecular markers. Embryos in culture are also amenable to direct experimental manipulations for elucidating developmental mechanisms of embryogenesis, germ layer formation, and embryonic patterning. This chapter outlines protocols for culturing mouse embryos at the immediate postimplantation period. We also present a system of developmental staging so that the outcome of different embryo culture studies may be assessed properly with reference to the precise developmental stage of the embryos used for the specific experiments.

1. Introduction

After implantation, mouse embryos establish intimate anatomical and functional association with maternal uterine tissues, an absolute requirement for full-term development. As the developing postimplantation embryo is not physically accessible, direct observation or experimentation poses a technical challenge. Noninvasive visualization of embryonic development can be achieved by whole animal imaging using ultrasound or magnetic resonance technologies. These imaging techniques can be adapted for experimental manipulations such as guided trans-uterine or trans-placental microsurgery or introduction of molecular and virus-based reagents to study specific aspects of mouse development. These *in vivo* manipulations, however, lack the necessary resolution to study embryos of the precise stages within a few days after implantation. The alternative venue is to observe the progression of embryonic development in real-time and to experiment directly on these embryos in a controlled culture system that mimics the intrauterine environment.

The techniques for culturing whole postimplantation mouse embryos are derived from protocols originally designed to culture rat embryos (Tam, 1998). The goal is to provide the embryo with essential nutrients and embryotrophic serum factors in conjunction with a physiological gaseous composition (principally oxygen levels that match the transition of anerobic to aerobic mode of respiration), pH and osmolarity to support optimal growth (cell proliferation and differentiation, tissue expansion), and morphogenesis. Typically, the culture medium is formulated as chemically defined bicarbonate buffered solution (e.g., Dulbecco's modified Eagles medium (DMEM), Hams F10/F12 medium) supplemented with high serum content (50–75%, v/v). However, undiluted serum can also be used and is preferable for embryos at the pre- and early gastrulation stages. Other supplements such as glucose, essential amino acids, growth factors, and specific embryotrophic factors or differentiation inducers may be included as required in specific circumstances. Other culture parameters are related to the choice between static and constantly agitating methods of culture. In addition, the type of culture vessels are dictated by the nature of the experimental analysis (e.g., whether time-lapse or real-time imaging and observation is required) and duration and developmental end-points (e.g., how advanced the embryos will have to develop) of the culture experiments.

The current methods of whole embryo culture can support the growth and morphogenesis of mouse embryos in a 6-day window (i.e., E5.5–E10.5: embryonic (E) age) of postimplantation development. However, the nominal age of a litter of embryos is at best an approximate index of embryonic development and littermates are often at distinctly different stages of development. In view of the rapid development of embryos during this period of postimplantation development, it is an essential requisite that the initial developmental stage of cultured embryos be as precisely defined as possible for a meaningful interpretation of the outcome of the experiment and for comparison among experiments performed in different laboratories. To this end, we present in this chapter a standardized reference for staging whole embryos for the culture experiment. This system of staging the development of the embryos covers those dissected from pregnant mice between E5.5 and E8.25, the gestational ages when most *in vitro* culture of postimplantation embryos commences.

Presently, whole embryo culture methods are able to support the *in vitro* development of E5.5–E8.0 embryos for a period of 2–3 days. During this time, the majority (>80%) of cultured embryos develop in a manner that matches the *in vivo* counterparts in regard to size and protein content, morphological features, and molecular marker expression. However, it is yet not feasible to sustain the development of embryos *in vitro* continuously for the whole duration of postimplantation development from pregastrulation (E5.5) to early organogenesis (E10.5). In practice, the methods outlined in this chapter are optimal for growing embryos in periods from pregastrulation to the early gastrulation (E5.5–E6.5), from early gastrulation (E6.5) to early neurulation and anterior body formation (E8.5) and from late gastrulation (E7.5) to early organogenesis (E10.0). Embryos grown for longer than these age ranges, that is, maintained for more than 2.5 days, become retarded in growth and develop abnormalities. The outcome of such longer term culture experiments should be evaluated with caution.

The ability to remove postimplantation embryos from the intrauterine confine and grow them in a normal manner *in vitro* has opened unlimited opportunities to study the development of live embryos directly under controlled experimental conditions. Cultured embryos, wild type or mutant, can be monitored for their developmental progression periodically in a time-course mode or continuously in real-time. These studies can yield useful information on the impact of normal and aberrant gene activity on embryogenesis and morphogenesis. Cultured embryos may also be subject to pharmacological treatments to alter the molecular or biochemical activity, for example, enhancing or suppressing signaling activity, metabolic activity, and cellular/organelle functions. In combination with experimental manipulations, such as transplantation of cells (tagged to express lacZ or a variety of fluorescent proteins) and cell labeling (painting/microinjecting with fluorescent dyes, iontophoretic injection of histochemical labels, or electroporation/lipofection of expression vectors), the tissue fates and morphogenetic behavior of specific individual cells or cell population can be traced in the live embryo (Bildsoe *et al.*, 2007; Fossat *et al.*, 2010; Franklin *et al.*, 2007; Khoo *et al.*, 2007). In addition, studies of cell differentiation and morphogenesis can be performed by activating reporters in specific cell populations or tissue domains in the embryo. This can be achieved by global or gene locus-specific recombinase activity and photoactivation or conversion of fluorescent protein (Nowotschin and Hadjantonakis, 2009). Similar studies can be performed on embryos with engineered genetic modification (knockout mutants and transgenics), if the use of appropriate cell/tissue tags could be integrated into the experimental strategy. Furthermore, embryos of any genotype may be manipulated molecularly to express gene constructs encoding for specific proteins and their normal isoforms, variants, and chimeric/aberrant versions, noncoding RNA and shRNA/RNAi (Loebel *et al.*, 2010). Similar to the strategy for introducing labeling constructs, these expression vectors can be introduced to the embryo by electroporation or lipofection, targeted to the whole embryo or specific tissue layer or focally to a defined cell population (Tanaka *et al.*, 2010). The effect of overexpressing or ectopically expressing normal, dominant negative, or mutant forms of the gene products, or suppressing the gene function by knockdown may reveal the functional attribute of the genes or molecular components in controlling embryonic development. In essence, whole embryo culture provides a versatile tool for embryological analysis of cellular and molecular function in mouse development.

2. Developmental Staging of Postimplantation Mouse Embryos Dissected Between E5.5 and E8.25

2.1. Rationale for a standard staging system

Because the mouse embryo undergoes major morphogenetic changes and cell differentiation rapidly during gastrulation and the formation of organ primordia, the reliability of the outcome of embryo culture experiments depends on using correctly staged embryos at the beginning of the experiment. It is common among researchers to use the gestational age, for example, E7.5 or 7.5 dpc (days post *coitum*) as a proxy for developmental stages. This approach, however, can be confounded by the variation in development among embryos in the same litter and from different litters of nominally the same gestational age. Dissection of a litter at E7.5, for example, will yield embryos that span from primitive streak to neural plate stages (Downs and Davies, 1993).

Here, we propose a system of staging criteria to enable the assessment of the developmental stage of pre-primitive streak (pre-PS) embryos. Combining this staging system with an established one for embryos at the primitive streak and the neural plate stages (Downs and Davies, 1993), it provides a reference to standardize the designation of the development stage of mouse embryos dissected between E5.5 and E8.25, which are the most commonly used materials for initiating whole embryo culture experiments. It is important to clarify that the developmental stages of the embryo based on recognizable milestones of embryogenesis are representative snapshots of a continuum of development.

2.2. Overview of morphology and developmental milestones

The early postimplantation conceptus is cylindrically shaped and it is conventionally known as the egg cylinder (Fig. 11.1). The egg cylinder is enclosed by the parietal yolk sac which is composed of trophoblasts cells adhered to the maternal decidual tissue and a layer of loosely organized parietal endoderm, which is separated from the trophoblast by a thick basement membrane known as Reichert's membrane (Reichert's membrane is frequently used as a synonym for parietal yolk sac). The parietal yolk sac has to be removed in order to access the egg cylinder and to allow the embryo to grow properly *in vitro*. The egg cylinder is composed of the epiblast, the tissue that forms the germ layers; the extraembryonic ectoderm which abuts the epiblast proximally and the visceral endoderm which covers both the epiblast and the extraembryonic ectoderm (Fig. 11.1). The extraembryonic ectoderm contributes to the chorion, which together with the ectoplacental cone forms the placenta.

The visceral endoderm layer is divided into two populations, the extraembryonic visceral endoderm which overlays the extraembryonic ectoderm and the embryonic visceral endoderm which covers the epiblast. The visceral endoderm was traditionally considered an extraembryonic tissue. However, recent evidence suggests that this population of cells can contribute to both embryonic (gut) and extraembryonic (yolk sac) endodermal tissues (Kwon *et al.*, 2008). Within the embryonic visceral endoderm epithelium, there is a group of tall columnar cells that display unique expression of *Hhex*, *Cerl*, *Lefty1*, and other molecular markers that distinguish them from the rest of the embryonic visceral endoderm cells (Pfister *et al.*, 2007; Rivera-Pérez *et al.*, 2003; Thomas *et al.*, 1998). In embryos dissected at E5.5, this group of

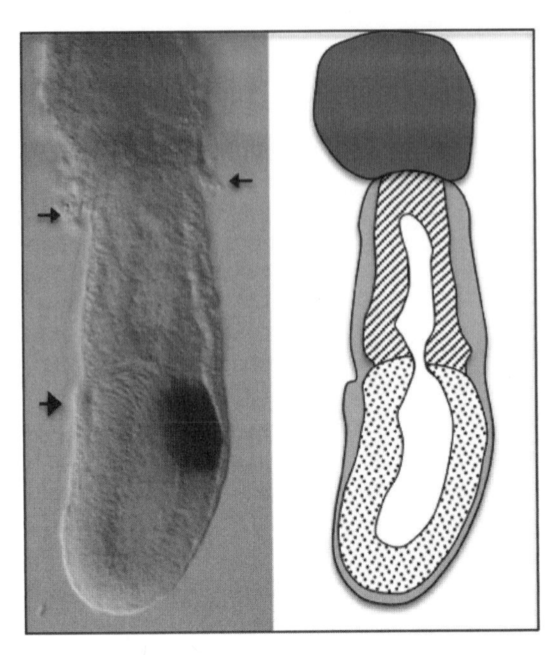

Figure 11.1 Mouse embryo dissected at E6.5 and a diagram showing its different components. The conceptus is composed of the epiblast (stippled), extraembryonic ectoderm (striped), ectoplacental cone (dark gray), and visceral endoderm (light gray). The visceral endoderm overlaying the extraembryonic ectoderm is known as extraembryonic visceral endoderm and the one overlaying the epiblast as embryonic visceral endoderm. The embryo has been hybridized with *Brachyury* to reveal the future location of the primitive streak. The anterior visceral endoderm is indicated by the big arrow. The parietal yolk sac, also known as Reichert's membrane has been removed but remnants are indicated by small arrows.

cells is located at the distal tip of the egg cylinder and it is usually referred to as distal visceral endoderm (DVE). At later stages, these cells occupy a position opposite to the primitive streak and are then called anterior visceral endoderm (AVE) (Rossant and Tam, 2009; Srinivas *et al.*, 2004; Thomas and Beddington, 1996). The DVE and AVE can be distinguished morphologically under the dissecting microscope by their thickness and serve as markers to determine the developmental stage of pre-PS embryos.

The next developmental milestone is the appearance of the primitive streak on one side of the epiblast opposite to the AVE. The primitive streak is a region of the epiblast where cells undergo an epithelial to mesenchymal transition in preparation for gastrulation. The primitive streak becomes evident in embryos dissected at E6.5 or later. Concurrent with the appearance of the primitive streak, the embryo undergoes gastrulation to form the three embryonic layers: ectoderm, mesoderm, and endoderm. Between

E7.5 and E8.5, the embryo develops the anterior neural (head) folds, the heart and early vasculature, the embryonic foregut and hindgut invaginations (the anterior and posterior intestinal portal), and the paraxial mesoderm which begins to segment into somites (Downs and Davies, 1993). These features mark the beginning of organogenesis.

2.3. Key morphological criteria for staging postimplantation embryos

2.3.1. Pregastrulation stages

The vast majority of embryos dissected between E5.0 and E6.5 lack a primitive streak. At these egg cylinder stages, differences in the morphology of the embryonic visceral endoderm layer, and changes in the shape and size of the egg cylinder can be used to determine the developmental stage of the embryos (Rivera-Pérez and Magnuson, 2005; Rivera-Pérez *et al.*, 2003; Tam and Loebel, 2007).

2.3.1.1. DVE stage (~E5.5 to ~E5.75): Fig. 11.2A A morphologically distinct DVE is visible at the distal tip of the egg cylinder. The tall columnar cells of the DVE can be twice as tall as the rest of the visceral endoderm covering other areas of the epiblast. In some embryos, the DVE may be tilted to one side of the distal region of the embryo. The proamniotic cavity is clearly evident and is confined to the epiblast compartment of the egg cylinder. In CD-1 embryos, the length of the egg cylinder averages 185 μm (Rivera-Pérez *et al.*, 2003).

2.3.1.2. AVE stage (~E5.75 to ~E6.0): Fig. 11.2B In these embryos, the AVE is found extending distal from the epiblast–extraembryonic ectoderm boundary and covers one half of the epiblast. The proamniotic cavity expands into the extraembryonic ectoderm region. When viewing the embryo with the AVE facing one side, the egg cylinder appears flattened in the anteroposterior dimension. In the larger embryos, the AVE forms a bulge at the anterior epiblast/extraembryonic ectoderm boundary (see Fig. 11.1) rather than a shield covering the anterior half of the epiblast as in the younger embryos. This is due to the formation of a crescent of AVE cells that extends from an anteroproximal to a mediolateral position of the egg cylinder.

2.3.1.3. Pre-PS stage (~E6.0 to ~E6.5): Fig. 11.2C and D By this stage, the anteroposterior axis, marked at the anterior pole by the AVE, has lengthened to become the longer of the two orthogonal axes in the transverse plane of the egg cylinder. This feature distinguishes the pre-PS embryos from the AVE embryos. Embryos at this stage display an intact

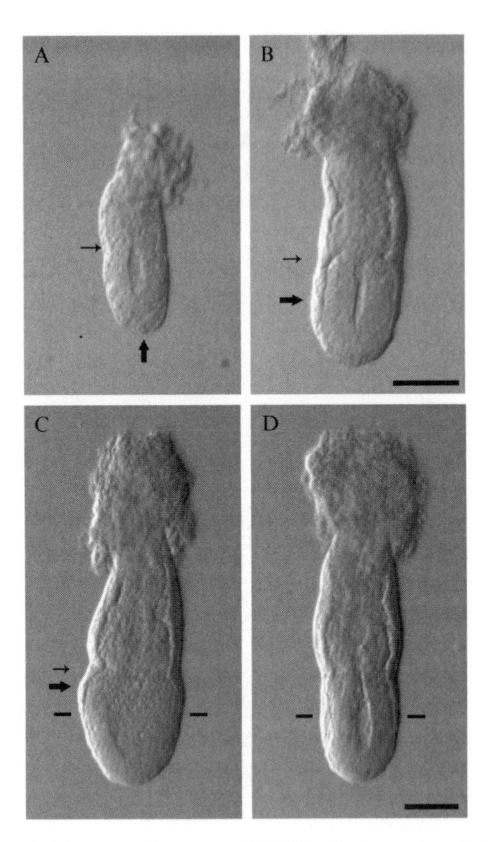

Figure 11.2 Pre-primitive streak stages. (A) Distal visceral endoderm (DVE) stage embryo. In these embryos, a thickening of the visceral endoderm is visible at the tip of the egg cylinder (thick arrow). The proamniotic cavity is restricted to the epiblast region. (B) Anterior visceral endoderm stage (AVE) embryo. A thickening of the visceral endoderm extends from the tip of the epiblast to the boundary between the epiblast and the egg cylinder (thick arrow). The proamniotic cavity is visible as a narrow slit when the embryo is viewed with the AVE facing one side. (C, D) Prestreak stage embryo. Side view (C) with the AVE facing left and posterior view (D) of the same embryo. The axis marked on one end by the AVE (thick arrow in A) and situated at the mid-level of the proamniotic cavity (marker by the two lines in C) is longer than its orthogonal axis (marked by two lines in D). The thin arrow marks the epiblast/extraembryonic ectoderm boundary in (A)–(C). Panels (A), (B) and (C), (D) are shown at the same magnification. The scale bars equal 100 μm.

epiblast which has not yet formed the primitive streak in the posterior epiblast. These embryos, however, express *Brachyury* in the posterior epiblast which marks the prospective site where the primitive streak will form (Rivera-Pérez and Magnuson, 2005).

2.3.2. Primitive streak and neural plate stages

These stages are based on the staging system proposed by Downs and Davies (1993). Primitive streak stages are based on the length of the primitive streak and the emergence of the head process. Neural plate stages are identified mainly by the shape of the anterior part of the ectoderm as it flattens and expands to form the primordium of the head folds and the development of the allantois.

2.3.2.1. Early streak (ES) stage (~E6.5 to ~E6.75): Fig. 11.3A An incipient primitive streak is located in the epiblast opposite to the AVE and at the boundary between the epiblast and extraembryonic ectoderm of the egg cylinder. The nascent primitive streak extends to a maximum of 50% of the proximodistal length of the epiblast. The primitive streak is difficult to discern at this stage and is visualized better by rotating the egg cylinder around its proximodistal axis.

2.3.2.2. Mid streak (MS) stage (~E6.75 to ~E7.25): Fig. 11.3B The primitive streak spans 51–100% of the length of the epiblast. The newly formed mesodermal layer can be seen extending about half way around the circumference of the egg cylinder. A population of extraembryonic mesoderm appears between the epiblast and the receding extraembryonic ectoderm.

2.3.2.3. Late streak (LS) stage (~E7.0 to ~E7.5): Fig. 11.3C Embryos at this stage have a distinct condensation that bulges from the distal tip of the primitive streak—the head process. The posterior amniotic fold is evident at the proximal end of the primitive streak. This fold produces a narrowing of the proamniotic cavity at the epiblast/extraembryonic ectoderm boundary forming the proamniotic canal.

2.3.2.4. Neural plate, no allantoic bud (OB) stage (~E7.25 to ~E7.75): Fig. 11.3D In these embryos the anterior and posterior amniotic folds have fused to form the amnion. This closure and the proximal displacement of the extraembryonic ectoderm by extraembryonic mesoderm lead to the creation of three cavities; the amniotic, exocoelomic, and ectoplacental cavities and three membranes; the amnion, visceral yolk sac, and chorion. There is no sign of the allantois.

2.3.2.5. Neural plate, early allantoic bud (EB) stage (~E7.5 to ~E8.0): Fig. 11.3E A distinctive characteristic of embryos at this stage is the flattening of the anterior region of the ectoderm and the formation of a small allantoic bud protruding from the proximal end of the primitive streak into the exocoelomic cavity. The node is evident at the distal tip of the egg

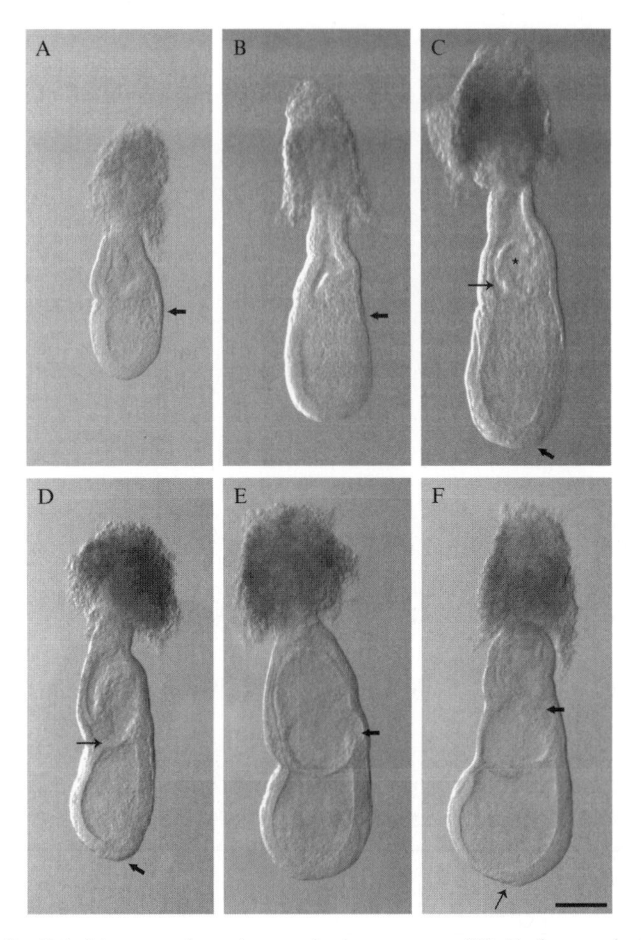

Figure 11.3 Primitive streak and neural plate stages. (A) Early streak (ES) stage embryo. The primitive streak appears as a small wedge at the epiblast/extraembryonic ectoderm boundary (arrow). (B) Mid streak (MS) stage embryo. The primitive streak extends from half to the full length of the epiblast and wraps around half the circumference of the egg cylinder (arrow). The extraembryonic mesoderm appears at the base of the primitive streak. (C) Late streak (LS) stage. The posterior amniotic fold produced by extraembryonic mesoderm is prominent at the base of the primitive streak (asterisk). The proamniotic canal has not closed (thin arrow). The head process is evident at the tip of the egg cylinder (thick arrow). (D) No allantoid bud (OB) stage. In these embryos the amniotic folds have fused to form the amnion (thin arrow) but the allantois is not yet evident. The head process is prominent at the tip of the epiblast (thick arrow). (E) Early allantois (EB) bud stage. A small allantois bud is present at the base of the primitive streak (arrow). (F) Late allantois (LS) bud stage. The allantois bud has increased in size and projects into the exocoelomic cavity (thick arrow). The node is readily discerned by a small indentation at the tip of the epiblast (thin arrow). Scale bar equals 200μm for panels (A)–(C) and 250μm for panels (D) and (E).

cylinder and is marked on the outside by an indentation (the nodal pit) and a condensed strip of tissue extends anteriorly from the node to form the head process in the midline of the embryo.

2.3.2.6. Neural plate, late allantoic bud (LB) stage (~E8.0 to ~E8.25): Fig. 11.3F The anterior ectoderm is beginning to form a broad plate with a midline crease, the future neural groove. The allantoic bud has increased in size and projected into the exocoelomic cavity. The node pit is visible as a small indentation at the distal tip of the embryo.

3. Materials

3.1. Solutions and reagents required

- DMEM (high glucose content) (12100-061, Gibco Labs, Invitrogen, USA). Before DMEM is used for culture, add 10 ml each of 200 m*M* stock glutamine (Gibco Labs, Invitrogen) and 5000 μg/ml penicillin/streptomycin stock solution (Gibco Labs, Invitrogen) to 1 l of DMEM.
- 70% ethanol.
- PB1 medium prepared according to Table 11.1. Add the chemical components in the correct order (as written). The pH should be between 7.3 and 7.4. The solution is sterilized by passing through a 0.22 μm

Table 11.1 The composition of PB1

Components	Concentration (g/l)
NaCl	8.0
KCl	0.2
$Na_2HPO_4 \cdot 12H_2O$	2.86
KH_2PO_4	0.2
$CaCl_2 \cdot 2H_2O$	0.13
$MgCl_2 \cdot 6H_2O$	0.1
Sodium pyruvate[a]	0.036
Phenol red[a]	0.01
Penicillin[a]	0.06
Glucose	1.0
Bovine serum albumin	4.0
pH	7.3–7.4
Osmolarity	286–292 mOsmol/l

[a] See reagents for method of preparation.

millipore filter and stored in 50 ml aliquots at 4 °C. The solution should be used within 2 weeks.
- Phosphate-buffered saline (PBS).
- Penicillin (Sigma), add 599 mg to 100 ml of 0.9% (w/v) NaCl. Store at −20 °C.
- Phenol red (Sigma), add 5.2 ml of 0.5% solution phenol red to 258 mg of $NaHCO_3$ (BDH Chemicals) dissolved in 14.8 ml of dH_2O. It can be kept for 2 weeks at 4 °C.
- 4% paraformaldehyde stored at 4 °C.
- 100% rat serum—refer to Section 3.3.1 for the procedure on collection of rat serum.
- Sodium pyruvate (Sigma), dissolve 170 mg of sodium pyruvate in 20 ml of 0.9% (w/v) NaCl. Dilute 1:50 in 0.9% (w/v) NaCl before use. It can be kept for 2 weeks at 4 °C.
- Isofluorane (Webster Veterinary, USA. Cat. No. 07-836-6544) used as anesthetic.

3.2. Equipment

- Blunt fine forces (Fine Sciences Tools Cat. No. 11064-07)
- Dissecting microscope (Leica Microsystems)
- Forceps (#5 forceps, Fine Sciences Tools Cat. No. 91150-20)
- Glass culture bottles, thin walled (BTC Engineering, Milton, Cambridge, UK)
- Iridectomy scissors (Aesculap Cat. No. OC461)
- 19-gauge butterfly catheter (Fisher Cat. No. B387359)
- Pasteur pipettes (Fisher Cat. No. 0344824)
- Plastic transfer pipettes (Fisher Cat. No. 1371121)
- Plastic snap-cap tubes (5 ml, Fisher Cat. No. 352063)
- Rotating bottle culture incubator (BTC Engineering)
- Scissors (Fine Sciences Tools Cat. No. 91460-11)
- Serrated scissors (Fine Sciences Tools Cat. No. 14110-15)
- Sharpening stone, fine-grained (Fine Sciences Tools Cat. No. 29008-22)
- 20 ml syringes
- Tissue culture wares: tissue culture dishes (60 mm, Cat. No. 25010; Corning Glassworks), 4-well chamber slides (Lab-Tek™ Chamber Slides™, Nunc), 4-well plates (Nunc 176740)
- Water-jacketed CO_2 incubator, 5% CO_2 in air at 37 °C (e.g., Forma Scientific)
- 56 °C water bath
- 5% CO_2, 5% O_2, 90% N_2 gas mixture and gas regulator
- 5% CO_2, 20% O_2, and 75% N_2 gas mixture and gas regulator
- 5% CO_2, 40% O_2, and 55% N_2 gas mixture and gas regulator

3.3. Preparation of handling and culture medium and serum supplement

3.3.1. Preparation of rat serum

Rat serum is a basic component of the mouse embryo culture media. To isolate serum, rats are anesthetized using a continuous flow of isofluorane emanating from a vaporizer. This ensures that the animals will be in deep anesthesia during the exsanguination procedure. Before the exsanguination procedure ensure that the rat is fully anesthetized by testing for reflexes with toe and tail pinching. If the rat is fully anesthetized it will not respond to the stimuli.

For exsanguination, lay the rat on its back and wet the abdomen generously with 70% ethanol. Make transverse incision on the skin and abdominal wall using the serrated scissors. Push the guts to one side to expose the dorsal aorta. Use blunt forceps to remove the fatty tissue over the vessel. The dorsal aorta is a silvery vessel located at the back of the abdominal cavity. Using a butterfly catheter connected to a 20-ml syringe, pierce the dorsal aorta with the bevel of the needle facing down. If the incision is done correctly, blood should appear on the catheter tube. Allow the heart of the rat to push the blood into the syringe as much as possible. Apply only minimal force on the plunger as you pull it out. This step is critical to obtain serum free of hemolysis. Remove as much blood as possible. At this time, cut through the diaphragm of the exsanguinated rat to ensure euthanasia. Once the maximal amount of blood is recovered, remove the needle and slowly deposit it in a 15-ml falcon tube, allowing it to slide down the wall of the tube. Spin the tube at 1200 × *g* for 5 min and place it on ice to allow it to coagulate. This will separate the cellular components of the blood from the serum. After centrifugation, squeeze the fibrin clot formed with sterile forceps and spin again. After the second spinning, collect as much serum as possible avoiding the cells packed at the bottom of the tube. The serum is divided into three categories depending on its level of hemolysis. The best quality serum should have a transparent amber color. The poor quality serum shows a rusty dark red color. Aliquots of the same quality may be pooled together and stored frozen at −80 °C where they can last for a period of 1 year. They are heat inactivated at 56 °C for half an hour prior to use.

Heat inactivation can also be done immediately after collection. After heat inactivation, the serum is spun at 1200 × *g* for 5 min and dispensed in 0.5, 1.0, and 5.0 ml aliquots. They can be stored frozen for similar duration as the unheated serum. Check that the rat serum does not contain any particulate matters. If the rat serum was not frozen properly it can easily grow fungus which has an adverse effect on the development of the embryo.

Culturing younger embryos usually require a couple of milliliters of serum while that for more advanced embryos can use 5 ml or more depending on the number of embryos being cultured.

3.3.2. Preparation of handling medium

This is prepared according to the formulation in Table 11.1.

3.3.3. Preparation of culture medium

1. Prior to dissection of mouse embryos, prepare the required culture media:
   a. 100% rat serum: for pregastrulation to ES embryos
   b. 75% rat serum:25% DMEM: for ES–LS embryos
   c. 50% rat serum:50% DMEM: for embryos more advanced than LS stage
2. The prepared media is equilibrated for 1–2 h at 37 °C and in the appropriate gaseous conditions:
   a. 5% CO_2, 5% O_2, 90% N_2 for pregastrulation to LS embryos culturing in rotating bottles
   b. 5% CO_2, 20% O_2, 75% N_2 for embryos more advanced than LS to early-somite stage culturing in rotating bottles
   c. 5% CO_2, 40% O_2, 55% N_2 for embryos more advanced than early-somite stage culturing in rotating bottles
   d. 5% CO_2 with the balance as air for all static cultures in hanging drop or in Nunc 4-well chamber slide
3. Keep some prepared media for replenishing/changing the culture during the course of the experiment.

4. Culture Protocols

4.1. Culture of embryos at DVE and AVE stages

4.1.1. Special aspects of culture protocol

Embryos at these stages are cultured for short periods of time in hanging drops of 100% rat serum. DVE embryos can be cultured for a period of 7 h. This is enough to monitor the shifting of the DVE cells from its distal to its lateral position to form the AVE. AVE embryos can be cultured for 1 day until an incipient primitive streak is observed. It is important to maintain the ectoplacental cone attached to the egg cylinder for proper development. Longer periods of culture are not recommended because the embryos tend to develop an expanded proamniotic cavity which interferes with closing of the amniotic folds and proper development.

4.1.2. Culture protocol

1. Prior to dissection of mouse embryos, prepare required culture medium (see Section 3.3.3). The medium is then placed in a Nunc 4-well chamber slide, which is put into a CO_2 incubator at 37 °C to equilibrate.
2. *Dissection*: Pregnant mice are killed by cervical dislocation and placed in a supine position. Wet the abdomen with 70% ethanol and make a transverse cut through the skin and underlying muscles. Pull the skin back exposing the thorax and use forceps and scissors to hold and cut through the abdominal wall. Using forceps grab one of the uterine horns and cut it at the vaginal end. The implantation sites appear as beads on a string along the length of the uterus. Lift the severed end, cut away the mesentery and fat, and then severe the ovarian end. Place the uterine horn in a Petri dish containing PBS and cut the implantation sites individually. Repeat this procedure with the other uterine horn. Using forces rip apart the uterus layer on each implantation sites to free the decidua. Once all decidua are isolated, transfer them to a fresh Petri dish containing dissection media: DMEM, 20 m*M* HEPES, 10% FCS and antibiotics.
3. *To dissect out the embryos*: Working under a dissecting microscope, orient the deciduum with the wider side facing left and locate the embryonic crypt which is parallel to the long axis of the deciduum. Locate the center of the deciduum and impale it using forceps at a point slightly off to the wider side. Pierce the deciduum with the closed forceps and split the upper part in two lobes by opening the forceps.
4. Grab one lobe of the opened deciduum with forceps in a position as close as possible to the ectoplacental cone (identified as a bloody spot located close to the center of the deciduum). Grasp the opposite lobe with another pair of forceps. Pull the forceps apart and split the deciduum into two halves. The egg cylinder should be attached to one deciduum half. The tip of the cylinder can be seen next to a darkened spot of necrotic tissue produced during the decidual reaction at implantation. This landmark is useful in case that the ectoplacental cone is not visible, especially for dissecting embryos at E5.5 as they appear translucent. Scoop out the egg cylinder using the tip of one side of the forceps.
5. Using the fine forceps, grab the parietal yolk sac in a position next to the ectoplacental cone. Care should be taken not to damage the embryo while doing so. With a second pair of forceps, grab the parietal yolk sac at a point adjacent to the first forceps and pull the parietal yolk sac apart and deflect it around the egg cylinder. Remove the excess yolk sac using forceps.
6. To culture the embryos:

   Option 1: Place a 90 μl drop of rat serum in the cavity of the inner surface of the cap of a 5 ml snap-cap plastic tube and 3 ml of PBS in the tube.

Transfer the dissected embryo into the drop of serum using a polished Pasteur pipette attached to a rubber bulb. Carefully, place the cap on the tube without tightening it. Place the tube in the incubator.

Option 2: Transfer the embryo to 0.5 ml of serum in the well of a 4-chamber slide.

4.2. Culture of embryos at pre-primitive streak, primitive streak, and neural plate stages

4.2.1. Special aspects of culture protocol

Embryos at these stages can be cultured optimally for 2 days. Pre-PS and ES embryos are usually cultured in static conditions in CO_2 incubators. Embryos at MS or more advanced stages are cultured in rotating bottles with continuous gassing.

4.2.2. Culture protocol

1. Prior to dissection of mouse embryos, prepare required culture media. Embryos aged E6.0–E7.0 are cultured in 100% rat serum, while embryos aged E7.5–E8.5 are cultured in 75% RS:25% DMEM, which is supplemented with penicillin, streptomycin, and glutamine.
2. The prepared media is then placed in a Nunc 4-well chamber slide (for static culture) or in bottles (for rotating culture), which is then equilibrate at 37 °C with 5% CO_2 gas mixture.
3. *Dissection*: Pregnant mice are killed by cervical dislocation and placed in a supine position. Cleanse the abdomen with 70% ethanol and make a horizontal cut through the skin and underlying muscles. Pull the skin back exposing the thorax and use forceps and scissors to hold and cut through the peritoneum. Using forceps, grasp the cervical segment of the uteri and sever it from the vagina using scissors.

   Option 1: The two uterine horns are still connected by the cervical stump, lift this clear of the peritoneal cavity. Pull away any fat or mesometrium using forceps. Holding the cervical stump with forceps, split one of the uterine walls using scissors along the anti-mesometrial side and extend down the uterine horn. Be very careful not to nick any of the decidua along the way. The uterine walls open to expose the decidua which can be freed by carefully sliding the point of a pair of forceps underneath the decidua and pinching it off the uterine wall. The decidua are then gently transferred (with fine forceps) to fresh dissection media (PB1 supplemented with 10% FCS) in a 60 mm Petri dish. Repeat the procedure for the other uterine horn until all decidua are removed.

   Option 2: Remove the uterus from the animal and place it in a 60 mm Petri dish containing PB1 supplemented with 10% FCS. Cut the uterine horn between the implantation sites. Dissect the uterine wall to expose and free the decidua. Using a plastic pipette, transfer the decidua to another 60 mm plate containing fresh PB1 supplemented with 10% FCS.

4. *To dissect out the embryos*: Working under a dissecting microscope, cut the pear-shaped decidua into halves along its long axis with iridectomy scissors. The cutting is positioned slightly off the mid-plane for E6.5 deciduas and slanted more to one side for E7.5 decidua. E8–E8.5 embryos are considerably more expanded so cut further away from the mid-plane of the decidua to ensure the visceral yolk sac remains intact. With the proper cut, the embryo will be left embedded in one half of the deciduum. The half-deciduum is then pinned down with forceps and the embryo is slowly scooped out by another pair of forceps. Repeat this procedure until all embryos are removed from the decidua. To remove the parietal yolk sac from the embryo, first grasp the membrane near the distal region of the embryo with a pair of forceps. Using another pair of forceps, grasp the membrane at a site adjacent to the first forceps. Peel the parietal yolk sac off the embryo by separating the two forceps while holding on the membrane. The parietal yolk sac can then be trimmed close to its attachment to the ectoplacental cone by further dissection. Transfer the dissected embryo to the equilibrated culture media in the 4-well Nunc chamber slides or culture bottles. The embryos are now ready for further experimental manipulations.
5. To culture the embryos:

   For static culture: The embryos are transferred to fresh medium in the 4-well chamber slide which is placed in the CO_2 incubator.

   For rotating culture: A hollow rotating drum with outlets to hold culture bottles is placed inside the culture chamber set to 37 °C. Put 2 ml of culture media in each small 15-ml culture bottle. Insert the culture bottles with hollow rubber corks into the slots on the rotating drum. An external supply of gas mixture is connected to the rotating drum so that the gas is fed continuously into the culture bottle.
6. Two milliliters of culture media is enough to culture up to three embryos for 24 h. If culturing for more than 24 h, replace the culture media with fresh media on the morning of the following day and again in the late afternoon for culturing overnight. Changing the rat serum more frequently encourages better development of the embryo.
7. The special mixture gas is used to culture embryos and it varies depending on the age of the embryos to be cultured:
   a. 6.0–7.0 dpc on 5% CO_2, 5% O_2, and 90% N_2
   b. 8.0–9.5 dpc on 5% CO_2, 20% O_2, and 75% N_2
   c. 9.5 dpc on 5% CO_2, 40% O_2, and 55% N_2

Embryos aged from E7.0 up are kept rotating at all times during the culture period. If the embryos appear to have developed poorly during the *in vitro* culture, check the quality of the rat serum or the gas supply.

8. Embryos are harvested at the end of the culture period by removing them from the culture vessel and rinsing through three washes of PBS to remove the serum. They are then processed for further analysis as required.

ACKNOWLEDGMENTS

J. A. R-P was supported by NIH grant GM087130 and P.P.L.T. by NHMRC grant 372102.

REFERENCES

Bildsoe, H., Franklin, V. J., and Tam, P. P. L. (2007). Fate-mapping technique: Using carbocyanine dyes for vital labeling of cells in gastrula-stage mouse embryos cultured in vitro. *Cold Spring Harb. Protoc.* 10.1101/pdb.prot4915.

Downs, K. M., and Davies, T. (1993). Staging of gastrulating mouse embryos by morphological landmarks in the dissecting microscope. *Development* **118,** 1255–1266.

Fossat, N., Loebel, D. A. F., Jones, V., Khoo, P.-L., Bildsoe, H., and Tam, P. P. L. (2010). Marking cells for imaging to analyse morphogenetic behaviour and cell fates in mouse embryos. *In* "Imaging in Developmental Biology: A Laboratory Manual," (R. Wong, J. Sharpe, and R. Yuste, eds.), Cold Spring Harbor Laboratory Press, Cold Spring Harbor, NY (in press).

Franklin, V. J., Bildsoe, H., and Tam, P. P. L. (2007). Fate-mapping technique: Grafting fluorescent cells into gastrula-stage mouse embryos at 7–7.5 days post coitum. *Cold Spring Harb. Protoc.* 10.1101/pdb.prot4892.

Khoo, P.-L., Franklin, V. J., and Tam, P. P. L. (2007). Fate-mapping technique: Targeted whole embryo electroporation of DNA constructs into the germ layers of 7–7.5 dpc mouse embryos. *Cold Spring Harb. Protoc.* 10.1101/pdb.prot4893.

Kwon, G. S., Viotti, M., and Hadjantonakis, A. K. (2008). The endoderm of the mouse embryo arises by dynamic widespread intercalation of embryonic and extraembryonic lineages. *Dev. Cell* **15,** 509–520.

Loebel, D. A. F., Radziewic, T., Pwer, M., Studdert, J. B., and Tam, P. P. L. (2010). Generation of mouse embryos with small hairpin RNA-mediated knockdown of gene expression. *Methods Enzymol.* (in press).

Nowotschin, S., and Hadjantonakis, A. K. (2009). Use of KikGR a photoconvertible green-to-red fluorescent protein for cell labeling and lineage analysis in ES cells and mouse embryos. *BMC Dev. Biol.* **9,** 49.

Pfister, S., Steiner, K. A., and Tam, P. P. (2007). Gene expression pattern and progression of embryogenesis in the immediate post-implantation period of mouse development. *Gene Expr. Patterns* **7,** 558–573.

Rivera-Pérez, J. A., and Magnuson, T. (2005). Primitive streak formation in mice is preceded by localized activation of *Brachyury* and *Wnt3*. *Dev. Biol.* **288,** 363–371.

Rivera-Pérez, J. A., Mager, J., and Magnuson, T. (2003). Dynamic morphogenetic events characterize the mouse visceral endoderm. *Dev. Biol.* **261,** 470–487.

Rossant, J., and Tam, P. P. L. (2009). Blastocyst lineage formation, early embryonic asymmetries and axis patterning in the mouse. *Development* **136,** 701–713.
Srinivas, S., Rodriguez, T., Clements, M., Smith, J. C., and Beddington, R. S. (2004). Active cell migration drives the unilateral movements of the anterior visceral endoderm. *Development* **131,** 1157–1164.
Tam, P. P. L. (1998). Post-implantation mouse development: Whole embryo culture and micro-manipulation. *Int. J. Dev. Biol.* **42,** 895–902.
Tam, P. P. L., and Loebel, D. A. F. (2007). Gene function in mouse embryogenesis: Get set for gastrulation. *Nat. Rev. Genet.* **8,** 368–381.
Tanaka, S., Yamaguchi, Y. L., Jones, V. J., and Tam, P. P. L. (2010). Analyzing gene function in whole mouse embryo and fetal organ in vitro. *Methods Enzymol.* (in press).
Thomas, P., and Beddington, R. (1996). Anterior primitive endoderm may be responsible for patterning the anterior neural plate in the mouse embryo. *Curr. Biol.* **6,** 1487–1496.
Thomas, P., Brown, A., and Beddington, R. S. P. (1998). Hex: A homeobox containing gene revealing peri-implantation asymmetry in the mouse embryo and an early transient marker of endothelial precursors. *Development* **125,** 85–94.

CHAPTER TWELVE

In Utero and *Exo Utero* Surgery on Rodent Embryos

Valérie Ngô-Muller* *and* Ken Muneoka†

Contents

Abstract

Mammalian development has been best characterized using the mouse model. Direct intervention of the postimplantation mouse embryo *in utero* represents one of many experimental approaches that can be used to probe mammalian embryogenesis. Experimental access to the mouse embryo is difficult, but techniques have been developed to circumvent some of the challenges of operating on the embryo *in vivo*. Experimental studies have been carried out on postimplantation stage embryos from E8.5 to term, so much of the gestational period is accessible for experimentation. One approach that has helped to enhance embryo accessibility was the development of surgical techniques based on the finding that embryonic development continued normally *exo utero*. *Exo utero* development refers to the surgically created condition in which the embryo develops outside of the uterine cavity, yet within the female abdominal cavity and attached, via the placenta, to the uterus. Using this approach it is feasible to carry out precise surgical manipulations of the mouse embryo without compromising embryo viability associated with

* CNRS EAC4413, Functional and Adaptative Biology, Physiology of the Gonadotrope Axis, Université Paris Diderot, Paris, France
† Division of Developmental Biology, Department of Cell and Molecular Biology, Tulane University, New Orleans, Louisiana, USA

Methods in Enzymology, Volume 476
ISSN 0076-6879, DOI: 10.1016/S0076-6879(10)76012-2

postsurgery uterine contractions. In this chapter we review technical aspects of both *in utero* and *exo utero* surgical approaches and how these surgeries are used in conjunction with other experimental applications.

1. Introduction

Mouse embryo surgery techniques were developed more than 20 years ago and initially applied to mammalian limb development and regeneration studies (Muneoka *et al.*, 1986, 1989; Wanek *et al.*, 1989). Since then, surgical access to the rodent embryo has become progressively more feasible with modification to existing techniques, with the integration of novel experimental approaches, and the development of new techniques. In addition to limb studies, surgeries on rodent embryos have been used to study a wide range of developing tissues and organs including the nervous system, musculoskeletal system, digestive system, hematopoietic system, reproductive system, and, as well, the wound healing response that transitions from scarless healing in the embryo to the scar forming response of mature skin. A variety of experimental approaches have been adapted for use in with rodent embryo surgery including targeted injections of various substances, *in vivo* transfection studies, cell or tissue transplantation, amputation or other injury models including targeted laser ablation, microcarrier bead delivery of growth factors, use of genetically modified strains, and the introduction of specific mechanical constraints on the embryo/fetus. There are a growing number of important publications that utilize direct embryonic manipulations in rodent models, and we apologize in advance if we have overlooked any specific study. Our goal in this chapter is to outline the technical challenges associated with embryo surgery in rodents and to provide examples of how this approach can impact selected fields of studies.

The two approaches generally used are *in utero* or *exo utero* surgery and both are demanding procedures that require some level of expertise. The postimplantation embryo is encased in its extraembryonic membranes (amnion and yolk sac) within the tubular uterus. The embryo can be accessed by injection passing through the layers of the uterine wall (perimetrium, myometrium, and endometrium) and the extraembryonic membranes. Intrauterine embryo injections can be successfully carried out on mouse embryo stages as early as E9.5 (see Shimogori *et al.*, 2004). For direct surgery on the embryo, *in utero* studies require opening and closing the uterus and extraembryonic membranes, and this approach is restricted to late embryonic/fetal stages (E14.5 and later) because early embryos are too fragile to survive the postsurgical forces resulting from the contracting uterus. *Exo utero* surgery is based on the finding that embryonic development is not perturbed after the uterine tube is opened but not sutured close (Muneoka *et al.*, 1986). The embryos remain attached to the open uterus via the placentae and develop suspended within the abdominal cavity of the female, much like an ectopic pregnancy in humans. When embryos are exposed in this manner, it is possible to perform various embryo surgeries at early embryonic stages. Injection experiments using *exo utero* surgery have been carried out on stages as early as E8.5 (Serbedzija *et al.*, 1994) and direct surgery on the embryo can be carried out on E11.5 embryos and older (Wanek *et al.*, 1989). While technically demanding, direct manipulation of the rodent embryo is possible and, in combination with other experimental approaches, provides another avenue of experimental studies of mammalian development. Here, we review procedures for mammalian embryo surgery, and highlight technical innovations that have been published using this approach.

2. Preparation for Abdominal Surgery

Experiments are carried out on timed-pregnant females (generally mouse) under general anesthesia and requires opening and closing the abdomen, thus it is categorized as a major surgery. Experimental manipulations are carried out under a dissection microscope that needs to be located in an isolated part of the lab, preferably away from ventilation and within an isolated area, such as a hood enclosure (Fig. 12.1). The animal preparation

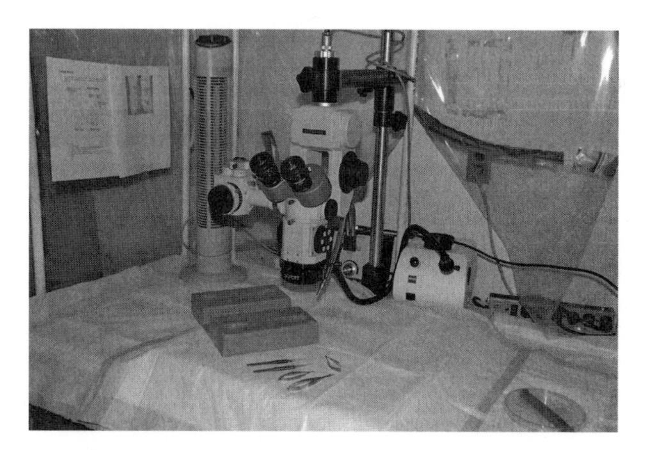

Figure 12.1 Surgery set up with accessories for *exo utero* surgery. A foot-focusing surgery scope is suspended by a boom arm over the surgery area. The surgery scope is equipped with a side arm for video imaging. The area is enclosed within a simple cage made of a PVC tubing frame and covered with clear vinyl. A bilevel stage used of *exo utero* surgery is shown beneath the surgery scope. The female is placed on the lower level of the stage and layers of paper towels are used to raise the mouse so that the abdomen is at the upper level. After opening the abdomen the body wall is retracted at the upper level. This bilevel stage allows the abdominal cavity of the female to be filled with saline so that the embryos remain submerged during the surgery.

area should be removed from the operating area to minimize contamination by hair or cage bedding. In the animal preparation area the animal is anesthetized, shaved, and scrubbed for surgery.

There are a number of different approaches to anesthesia that have been used for various studies on embryonic and fetal rodents that we review here. In all cases the anesthetic target is the pregnant female and not the embryo/fetus, although the embryo/fetus is exposed to maternal levels of the drug. The extent to which the circulating anesthetic itself is influencing normal development is largely unknown; however, isoflurane anesthesia has been linked to enhanced levels of apoptosis in the hippocampus and retrosplenial cortex of fetal brains when administered for 1 h to near term pregnant rats (Wang *et al.*, 2009). For the most part, it is assumed that general anesthesia of the pregnant female is not a significant factor influencing embryonic development because (1) treated postnatal pups display normal morphology and behavior, (2) normal embryo/fetal movements continue during an operation suggesting that the anesthetic is not influencing embryo/fetal physiology, and (3) development of the nervous system, the target of anesthetics, is late and extends into postnatal stages. Nevertheless, the potential influence of the circulating anesthetic is something that investigators should consider in the context of any specific experimental approach, especially if the experimental outcome is compared to similar experiments in other embryonic systems (e.g., chick or amphibian) carried out without anesthesia.

Anesthesia with ketamine/xylazine or pentobarbital induces prolonged anesthesia (30–45 min with ketamine/xylazine; >45 min with pentobarbital) and is administered by intraperitoneal (i.p.) injection. For mice, ketamine/xylazine is administered at a dose of 100 mg/kg of ketamine and 10 mg/kg xylazine (80 mg/kg ketamine and 8 mg/kg xylazine for rats), a working solution is made by combining 1.0 ml of a stock solution of ketamine (100 mg/ml) with 0.1 ml of a stock solution of xylazine (100 mg/ml) and diluted with sterile water to a final volume 10.0 ml. The pregnant female is weighted and 0.01 ml/g body weight of the working solution is administered. Reversal of ketamine/xylazine anesthesia can be obtained by injecting the antagonist yohimbine (1.0 mg/kg, s.c.) when surgery has been completed. Alternatively, the pregnant female mouse may be anesthetized with sodium pentobarbital (Nembutol) at a dose of 50 mg/kg body weight. A working solution is made by combining 1.0 ml of a stock solution of sodium pentobarbital (50 mg/ml) with 9 ml of sterile water. The pregnant female is weighted and 0.01 ml/g body weight of the working solution is administered.

Contraction of the myometrium during and after embryo/fetal manipulation can compromise the procedure and result in poor survival. Contraction of the myometrium is induced by stress so how the mouse is handled during the administration of any anesthetic can influence the outcome of the surgery. For i.p. injections, the pregnant mouse is held by the tail and allowed to grasp the cage lid with its forepaws while the injection is made in the lower abdominal area with a sharp needle. The angle of the injection should be close to parallel with the body axis to avoid hitting organs. Injections in the subcutaneous space will be obvious as a bubble will form at the injection site. The delivery of the anesthetic is much slower from s.c. injections. Injections that hit either the digestive tract or the uterus will also be obvious as the anesthetic solution will flow out of the anus or vagina, respectively. Injections that hit an organ within the abdomen will also be obvious by the presence of blood once the abdomen is open. Surgeries in which there is internal bleeding as a result of injecting the anesthetic should be abandoned.

Isoflurane vapor is another anesthetic approach for surgeries on pregnant female mice. Isoflurane is a gas inhalant produced by an isoflurane vaporizer system that generates a precise mixture of isoflurane with oxygen. Initial anesthesia is induced in a specialized chamber with 2.5% isoflurane and the mouse is anesthetized within 5 min. The mouse is then transferred to the operation site and fitted with a nose cone and anesthesia is maintained with 2% isoflurane for the duration of the surgery. Following the surgery the mouse recovered rapidly after removing the nose cone. Use of isoflurane may improve animal recovery from surgery, since waking up is almost immediate.

Once anesthetized the female is placed on her back and the abdomen is shaved along the midline extending from the sternum to roughly 5 mm anterior to the vaginal opening, using a rodent hair clipper. Care must be taken during shaving to avoid the ventrolateral rows of teats. The shaved area is scrubbed with povidone iodine solution (Betadine surgical scrub) followed by 70% ethanol and wiped dry. During anesthesia ophthalmic ointment (Ilium Chloroint) can be applied to the eyes to prevent drying of the corneas. The animal is transferred to the operating area for surgery. At this point the animal should be deeply anesthetized and unresponsive to toe or tail pinching with forceps.

3. *In Utero* Manipulation of Embryos

Most studies where *in utero* manipulation was applied were performed on mouse embryos, though a few studies have been applied to rat embryos (Ackman *et al.*, 2009; Bai *et al.*, 2003; de Nijs *et al.*, 2009; Wang *et al.*, 2006). *In utero* surgery requires that the abdomen is opened to access the uterus and after the surgical procedure the abdomen is closed and the animal is allowed to recover. The operating area should be cleaned with 70% ethanol and the operating scope should be wiped down ahead of time. The animal is placed on its back on a clean paper towel. To open the abdomen an initial large midline incision of the belly skin is made from the posterior to the anterior with microdissection scissors (4 in., straight, McPherson-Vanna, Roboz

Surgical Instruments). If the incision is made on the midline there will be no ruptured vessels and virtually no bleeding of the skin. Blunt forceps should be used to handle the skin. The skin retracts exposing the peritoneum, and a midline suture, the linea alba, is visible. A second incision is made along the linea alba to open the abdomen and if the incision does not stray from the midline, there will be no bleeding. Care should be taken not to damage organs within the abdominal cavity, and in particular, adipose tissues located in the caudal region directly beneath the peritoneum. These abdominal incisions are done without the aid of the dissection microscope. At this point there should be little to no bleeding caused by the two incisions.

With the abdomen open the uterine horns can be found in the lateral regions of the abdominal cavity and simply pulled out onto sterile damp gauze placed on the female's ventral surface (Fig. 12.2A). If the

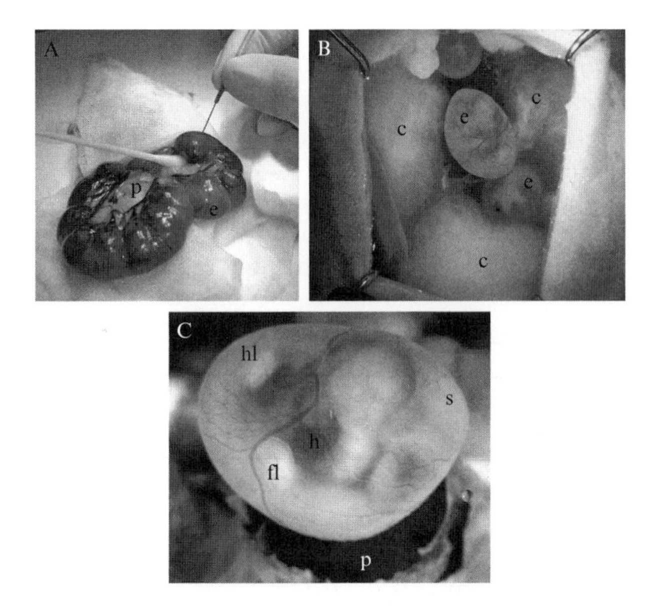

Figure 12.2 *In utero* and *exo utero* surgery of mouse embryos. (A) Following midline laparotomy, uterine horns are pulled out and placed on damp gauze on the abdominal surface of the female mouse. Embryos may be manipulated through the uterine wall by using an injection pipette. (B) For *exo utero* surgery the uterus is cut along the anti-placental side exposing embryos (e) for surgery. The abdomen is filled with saline to submerge the embryo and the embryo is positioned with sterile cotton balls (c). Two embryos are shown at E12.5; the right side of the top embryo is shown with its right hindlimb bud protruding from the extraembryonic membranes, and the caudal end of a neighboring embryo (bottom) shows both hindlimbs and the tail. (C) An E11.5 embryo is shown attached to the placenta (p). The developing skull (s), heart, and both forelimb (fl) and hindlimb (hl) are indicated. Yolk sac vessels can be seen converging on the umbilical vessels in the developing abdominal region. (A) From Chen *et al.* (2009), (B) from Muneoka *et al.* (1990), (C) from Muneoka *et al.* (1986).

myometrium is relaxed the uterine horns can be grasped with forceps and moved without damage to the embryos. If the uterus is contracted the embryos will appear as individual beads on a chain and the uterus should be moved by carefully grasping in the space between adjoining embryos. The embryos are enclosed within the amnion and the amnionic cavity provides a hydrostatic buffer to protect the embryo from physical damage caused by uterine contractions. Manipulating the embryo requires rupturing of the amnion and the loss of this hydrostatic buffer, thus the extent of uterine contraction is important for embryo survival following any manipulation. This is especially true for experiments on early stage embryos, for example, E9.5–E10.5. Myometrium relaxation can be achieved with i.p. injection of fentanyl/droperidol (0.05 ml of 0.04 mg/ml fentanyl and 2 mg/ml droperidol) or Rytodrine (ritodorin hydrochloride, Sigma-Aldrich, St. Louis, MO, USA) prior to surgery, or by direct application onto the uterus once the abdomen is open.

In utero manipulations generally involve injections into the embryo that must pass through the uterine wall and the extraembryonic membranes (yolk sac and amnion). The injection should avoid any blood vessels. There are no blood vessels in the amnion and, with the exception of the placental side of the uterus, there are few blood vessels in the uterine walls. The yolk sac has both large and small vessels cursing through the tissue that connect with the embryonic circulatory system via the umbilical arteries and veins. Large yolk sac vessels are visible through the uterine wall and can be avoided when selecting an injection route.

Since there are many embryos on each uterine horn, it may be necessary keep track of specific embryos and the quality of each procedure. Video recording of the procedure is useful for this purpose. Embryo manipulation is best performed using a stereo zoom surgical microscope, ideally controlled with a foot-focusing attachment to allow the investigator use of both hands during the procedure. The amount of total time the uterus is maintained outside the abdomen should be kept to a minimum and the uterus should be kept moist the entire time. Draping damp gauze over regions of the uterus not being operated on is one way to prevent desiccation of uterine tissue. Various types of embryo manipulations are outlined in the following section.

Once the surgery is completed the uterine horn is returned to the abdominal cavity and if the second horn is used the procedure is repeated. After the operation, the abdominal cavity is closed with suture. The peritoneum is closed using a running suture (6.0, Ethicon) from the sternum toward the posterior. The abdominal skin is closed with a similar running suture. After surgery the mouse is placed in a cage that has been previously warmed up on a heating pad. We typically place the cage so that the heating pad warms half of the cage so that during recovery the mouse is free to move to an optimal temperature. For collecting embryos/fetuses at prenatal stages the female is euthanized and the abdomen is opened to collect the specimens. For postnatal stages the female is allowed to give birth and the neonates are collected at the appropriate stage. If postnatal survival is compromised then fostering of newborns should be considered.

4. *Exo Utero* Manipulation of Embryos

Precise manipulations of the rodent embryo can be carried out using surgical techniques developed based on the *exo utero* survival of embryos within the confines of the abdominal cavity (Muneoka *et al.*, 1986). There are a number of previously published technical reviews on *exo utero* surgery that are of value to the new investigator (Hatta *et al.*, 2004; Muneoka *et al.*, 1990; Ngo-Muller and Muneoka, 2000b; Yamada *et al.*, 2008). The general operation involved making a longitudinal incision of the entire length of the uterus so that the embryos remain attached to the uterus but are not contained within the uterine cavity, and the exposed uterus is returned to the abdominal cavity where development continues *exo utero*. In the original study, embryos from E9.5 to E13.5 were found to develop normally to term (Muneoka *et al.*, 1986). In a subsequent study by Serbedzija *et al.* (1994) *exo utero* survival of embryos that have received injections into the amnionic cavity as early as E8.5 has been reported, thus it would appear that *exo utero* embryo manipulations are possible throughout mouse development. For our own studies on limb and digit development and regeneration, we have focused ours studies on E11.5 and later. Early stage embryos are surrounded by a layer of decidual tissue that obscures visualization of the embryo and removal of this layer compromises embryo survival. In general, our experience is that the survival rate of manipulated embryos increases with later stages and with less invasive manipulations. The various types of manipulations that have been accomplished using the *exo utero* approach are summarized below.

Preparation of pregnant mice for *exo utero* surgery is identical to that outlined above for *in utero* surgery. *Exo utero* surgery is performed within the abdominal cavity of the female mouse, so a specialized operating stage that has two functional levels has been developed (see Fig. 12.1). After the mouse is prepared for surgery and the abdomen is opened, it is placed in the lower level of the operating stage and retraction hooks are used to retract the abdominal skin and peritoneum at the upper level (Fig. 12.2B). The uterus is teased away from the digestive tract and an angled iridectomy scissors is used to cut the uterine wall on both horns along the antiplacental side where there are very few blood vessels. To minimize damage to the embryos the uterus must be relaxed so that it can be lifted away from the embryos with fine forceps. The angled iridectomy scissors is positioned parallel to the plane of the uterine horn with its tip pointed slightly away from the embryo so as not to inadvertently rupture the extraembryonic membranes. This incision is typically made from a central position along the uterine horn first toward the ovaries then toward the common uterus. The uterine wall will contract to the placental side of the uterus and the embryos will now be exposed for manipulation. This part of the procedure must be done carefully so as to not to damage extraembryonic membranes or the placental connections. The end result is an embryo encased within its extraembryonic membranes firmly attached by the placenta to the incised uterus (Fig. 12.2C).

Exo utero surgery is a lengthier procedure than *in utero* manipulation and all embryos are not manipulated in a single female. In some cases where embryo surgery is compromising embryo survival, removing all unoperated embryos can dramatically improved the survival of operated embryos (Shikanai *et al.*, 2009). We have routinely removed embryos to make room available for surgical considerations and because embryos closer to the ovaries tend to have a higher survival rate (Muneoka *et al.*, 1989). There are two different techniques that have been reported for removing embryos from the uterine horn during *exo utero* surgery. We routinely remove all embryos but four, leaving two embryos in each horn in positions toward the ovarian end of the uterus. Embryos and placentae are removed by placing a dry cotton-tipped applicator at the placental–uterine junction and gently rolling it across the placenta. This procedure separates the placenta from the uterus and causes a small amount of bleeding from the uterus. Bleeding is controlled by applying direct pressure with the cotton-tipped applicator at the former placental attachment site. The cotton absorbs the blood and clotting usually occurs in a few seconds. Yamada *et al.* (2008) report removing untreated embryos from the uterus by cutting embryo enclosed in its embryonic membranes and leaving the placenta attached to the uterus.

Once embryos are removed and any bleeding is controlled, the abdominal cavity is flushed with saline to remove any tissue debris that might induce a postsurgical fibrotic response. To facilitate flushing of the abdominal cavity we use an intravenous pack containing sterile saline to fill the abdominal cavity and a vacuum connected to an aspirator system to remove excess saline. The flow rate of the intravenous pack should be regulated with a variable tube clamp and the aspirator system should be gentle enough and regulated so as to not damage tissues. Once the abdominal cavity is flushed, it is filled with sterile saline and the embryos are maintained submerged in saline during the operation. For older stage embryos it may not be necessary to keep the embryos submerged; however, they should not be allowed to dry during the operation. For each embryonic manipulation the embryo needs to be positioned within the abdominal cavity. Embryos can be moved within the abdominal cavity by using forceps to grasp the retracted uterine wall away from the placenta and gently move the entire uterine horn. Once positioned, small balls of sterile cotton are used to maintain the position of the uterine horn and the embryo itself (see Fig. 12.2B). Note that maintaining the embryo position without securing the uterine horn will result in damaging the placental connection as the uterine horn will tend to return to its normal position within the abdomen.

There are two major types of surgical manipulations that can be made using the *exo utero* method. The first involves targeted injection and does not necessarily require opening of the extraembryonic membranes. The second involves direct surgical manipulation of the embryo itself and this requires opening and closing of extraembryonic membranes. For targeted injections the approach is similar to other vertebrate embryos with the exception that the needle passes through the yolk sac and the amnion prior to entering the embryo, and targeting the injection site is much easier and more precise with the uterine wall removed. A sharp needle is required for injections through the extraembryonic membranes. The amnion in particular is very resilient. Pinching the extraembryonic membranes with sharp forceps while injecting can stabilize the amnion making it easier to penetrate. The embryo is floating within the amnionic cavity with its ventral surface facing the placenta, so the embryo may need to be transiently repositioned for some targeted injections. Positioning of the embryo within the amnionic cavity can be accomplished by gently pushing down on the extraembryonic membranes with a blunt glass probe until it touches the embryo and turning it to the desired position. Use of both hands for some manipulations is sometimes necessary and important, so having an operating scope with a foot-focusing device can improve the precision of the operation.

Opening the extraembryonic membranes is required for direct manipulations of the embryo, and this approach can also be used for targeted injections. The most challenging aspect of this approach is suturing to close the membranes after the embryo surgery; however, this is simply a matter of practice and experience. The extraembryonic membranes are opened by grasping with fine forceps and making a small cut with straight iridectomy scissors, or by grasping with two pairs of fine forceps and gently tearing the tissue. Rupturing the yolk sac will be obvious but the amnion is very thin and translucent and it is not always obvious that it has been cut. It is important to keep track of the edges of the amnion, as the amnion will have to be sutured closed at the end of surgery. There are a number of major and minor vessels cursing throughout the yolk sac (see Fig. 12.2C) and it is important to avoid yolk sac vessels when selecting a site to open the extraembryonic membranes. It is optimal to have no bleeding from the yolk sac however, should it occur it can be controlled by gently pinching the ruptured vessel with fine forceps. As an aside, the major yolk sac vessels are easily accessible for systemic injections into the embryo. Once the extraembryonic membranes are ruptured, amnionic fluid will leak out and the amnion will collapse onto the embryo. The embryo can be positioned relative to the extraembryonic membrane incision by gentle pushing with blunt glass probes. For example, during surgery on the embryonic limb, the limb rudiment can be positioned to protrude through the incision and after the surgery the limb can be repositioned back within the membranes either by gently pushing with blunt glass probes or by pulling the membranes over the limb. After the embryo surgery is completed the extraembryonic membranes are sutured to close the incision. A single suture (10.0 monofilament suture, Ethicon) that passes through the yolk sac and amnion on both edges of the incision is sufficient. Closure of the amnion in particular is important since ruptures of the amnion heal spontaneously and can cause constriction abnormalities of protruding embryonic tissues (e.g., tail; K. Muneoka, unpublished). A similar response in humans is associated with constriction abnormalities observed in amniotic band syndrome (Kawamura and Chang, 2009).

After all embryo surgeries are completed, cotton balls that were used to stabilize and position the embryos should be removed and uterine horns returned to their initial position. The abdominal cavity is flushed extensively with saline to remove any blood clots or other tissue debris. Injection of sterile Hank's solution just before closing is reported to prevent abdominal adhesions postsurgery (Yamada *et al.*, 2008). The skin and peritoneum are released from the retractors and closed with running suture as described for *in utero* surgery.

5. Postsurgery

We typically group house similarly treated females after surgery because they seem to do better than in isolation. Abdominal surgery is a major surgery but operated mice generally appear well-groomed and are moving freely about the cage within 24 h. Mice that appear ungroomed or display restricted mobility should be watched carefully and euthanized if their condition does not improve. Poor recovery after surgery is generally associated with the formation of abdominal adhesions. For collecting experimental samples the female mouse is euthanized and embryos are identified based on their position along the uterine horn. For direct manipulations the presence of the extraembryonic suture will help to identify experimental embryos. If development to postnatal stages is required, it is necessary to have available a foster female, as the operated female will not be able to give birth since the uterus was cut open. For fostering it is important to collect the pups at or very close to term. Timing is important since collecting pups prematurely will make fostering difficult, but neonate survival can be compromised by waiting too long since separation of the placenta from the uterus will eventually occur. When the abdomen is opened, term neonates encased in their extraembryonic membranes and connected to the uterus are intermingled with the digestive tract and fills the abdominal cavity. The neonates need to be carefully separated from each other and from the digestive tract to establish their order on the uterine horns. On occasion we have observed term neonates completely outside the extraembryonic membranes and connected to the placenta by only the umbilical cord. To collect neonates, the placenta is separated from the uterus with a dry cotton-tipped applicator and moved to a clean Kimwipe in a Petri dish. Using forceps the extraembryonic membranes are removed and the umbilical cord is cut with scissors. Gloves should be worn and newborns should not be physically handled by the investigator. Under a dissection microscope the newborn should be gently cleaned with cotton-tipped applicators. Breathing can be stimulated by gently rubbing the base of the tail with a cotton-tipped applicator. The experimental neonates can be kept in a group and placed on a heating pad or under a lamp until they are all breathing regularly and moving about. At this point they can be presented to the foster female. The foster female's own litter is removed and she is kept in a dark room. Bedding from the foster female's cage is removed and mixed with the experimental newborns prior to introducing the foster pups with the foster female.

In our various studies we find that embryo death, when it is observed, occurs soon after the surgery and that by term dead embryos are highly regressed and appear as small white balls of tissue encased within the extraembryonic membranes. Embryo survival can be highly variable, but in general we find that survival tends to be an all or nothing phenomenon for any individual pregnancy. This suggests that postsurgical survival of embryos is linked to systemic events associated with the maternal surgery (e.g., abdominal adhesions) rather than the embryonic manipulation itself. For this reason, success using the *exo utero* approach is equally dependent on the development of good surgical skills as well as the microsurgical techniques required to manipulate the embryo.

6. Experimental Approaches

6.1. Injections

Injection studies have been described using either the *in utero* or *exo utero* method. Injections generally utilize glass needles made from micropipettes of varying size. Yamada *et al.* (2008) described how to make glass micropipettes with a beveled point using a microforge in order to perform successful microinjection with high embryo viability. The micropipette can be connected to a mouth controlled pressure system or to an automated hydrolic (mineral oil) microinjection system (e.g., UltraMicro Pump, WPI Inc.) fitted with a Hamilton-type syringe that allows precise control over injection volume. It is often useful to coinject a vital dye (e.g., 0.05% Nile blue sulfate or 1% Fast Green) in order to monitor the injection procedure. Injection studies include use of markers such as carbon particles for establishing fate maps (Muneoka *et al.*, 1989) or lipophillic tracers such as DiI (CellTracker; Molecular Probes) to characterize cell migration patterns (Ngo-Muller and Muneoka, 2000a; Serbedzija *et al.*, 1994; Ting *et al.*, 2009; Yoshida *et al.*, 2008). Injection of virus has been used to study cell lineage (Turner *et al.*, 1990) or to study the targeted effect of a specific virus on development (Naruse and Tsutsui, 1989; Ogawara *et al.*, 2002). Targeted injection of purified growth factors or signal transduction antagonist directly into the embryo has been used to study signaling during normal and abnormal development (Hatta *et al.*, 2002; Mathijssen *et al.*, 2000; Shinohara *et al.*, 2004).

Nucleic acid injections followed by electroporation have been used extensively in order to induce targeted gene expression in the injected and electroporated area. This approach allows the use of wild type, but also mutant and transgenic rodent models, to target the expression of a selected gene with a time-framed and site-specific surgery. Electroporation has been applied to injections of plasmids encoding marker genes for cell labeling studies (Fig. 12.3A–C) and/or functional studies (Garcia-Frigola *et al.*, 2007; Kawauchi *et al.*, 2006; Navarro-Quiroga *et al.*, 2007; Okada *et al.*, 2007; Saba *et al.*, 2003; Saito and Nakatsuji, 2001; Soma *et al.*, 2009; Takiguchi-Hayashi *et al.*, 2004), plasmids encoding short hairpin RNA to create RNA interference (Ackman *et al.*, 2009, Bai *et al.*, 2003; de Nijs *et al.*, 2009; Friocourt *et al.*, 2008; Wang *et al.*, 2006), and dual-fluorescence reporter/sensor plasmid for single-cell detection of microRNAs (De Pietri Tonelli *et al.*, 2006). The combined use of electroporation of a CRE recombinase construct into a conditional transgenic mouse strain has also been used to label specific cells and to follow their fate into adult stages (Nakahira *et al.*, 2006). Depending on the strength of the promoter, DNA concentrations of 1–10 $\mu g/\mu l$ is used with an injection volume of 1–2 μl. For any specific experimental design the electroporation parameters must be established empirically, but is associated with the electrode shape and the embryonic stage. Shimogori and Ogawa (2008) detailed how to build extrafine electrodes for successful targeted early embryonic stages (E9.5–E11.5) whereas tweezer-type electrodes are generally used for later stages (E12.5 onward). The use of GFP constructs with electroporation has been recently reviewed (Harvey *et al.*, 2009) and electroporation of retinal ganglion cells is detailed in a video article by Petros *et al.* (2009).

Cells have been introduced into the embryo by targeted injection for use as *in vivo* reporters or to characterize the behavior of stem cells in the embryonic and adult environment. Introduction of fibroblasts into the

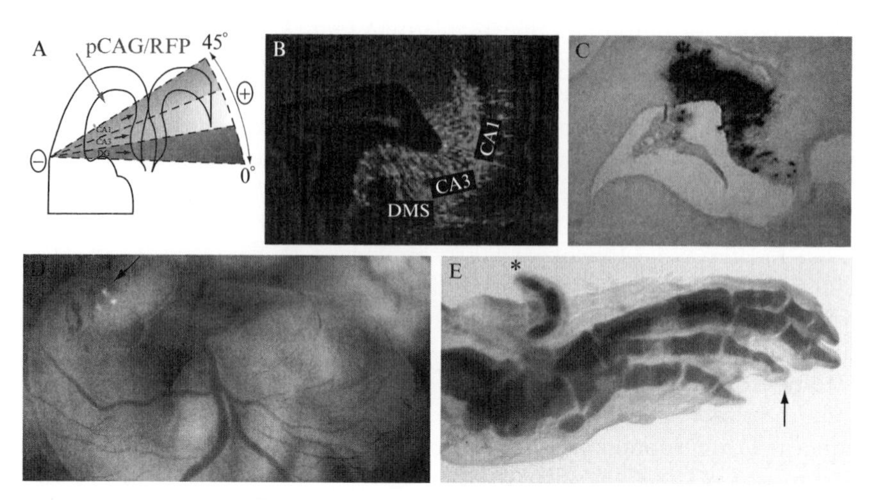

Figure 12.3 *In utero* and *exo utero* manipulation of the mouse embryo. (A–C) *In utero* electroporation of the E14.5 hippocampus with a plasmid encoding red fluorescent protein under the control of the chicken β-actin promoter (pCAG/RFP) as a lineage tracer. (A) Efficient transfection of specific areas of the hippocampal anlage (CA1, CA3, and dentate gyrus (DG)) is accomplished by manually changing the orientation of the electric field. (B, C) Two days after electroporation, widespread expression of RFP (B) and RFP mRNA (C) are present in the ventricular zone, the initial differentiating fields of the CA1–CA3 regions, and the dentate migratory stream (DMS). (D) Targeted injection of labeled 3T3 cells (arrow) into the zeugopodial region of the E12.5 hindlimb was used to explore regional growth dynamics during development. Cells were coinjected with fluorescent latex beads to identify the injection site. (E) Digit amputations at E13.5 combined with grafts of the amputated digit to the dorsal surface of the limb was used to characterize the extent of the regenerative response (arrow) in relation to the level of the amputation (*). (A–C) From Navarro-Quiroga *et al.* (2007), (D) from Trevino *et al.* (1992), (E) from Reginelli *et al.* (1995). (See Color Insert.)

embryonic mouse limb proliferate and differentiate in a position-dependent manner (Trevino *et al.*, 1992, 1993). On the other hand, the injection of cells that are known to secrete high levels of specific hormones have been used to experimentally perturb embryogenesis (Nimura *et al.*, 2008; Zhang *et al.*, 1998). Targeted injection of genetically labeled liver stem cells into the embryonic liver results in chimeric livers that persist to adult stages and can be used for both the investigation of liver development and regeneration (Shikanai *et al.*, 2009). Introduction of allogenic stem cell preparations into the embryo represents an animal model for the development of therapies for the early management of hereditary defects that are diagnosed prenatally. Efficient engraftment of adult hematopoietic cells that had been transduced with a retroviral vector expressing EGFP into E14.5 embryos has been reported (Rio *et al.*, 2005). Recent studies demonstrate that cell transplantations at progressively earlier embryos stages resulted in higher levels of chimerism (Chen *et al.*, 2009). Some clinically relevant studies include the rescue of a genetic mouse model for autosomal recessive osteopetrosis, a human disorder associated with defective osteoclasts, with allogenic fetal liver cell transplantation (Tondelli *et al.*, 2009), and the rescue of a mouse model of osteogenesis imperfecta with transplantation of adult bone marrow cells (Panaroni *et al.*, 2009).

6.2. Embryonic surgery

In many instances, experimental design calls for direct surgery on the embryo. For early stage embryos, such studies are best done using the *exo utero* approach since it eliminates having to incise and suture the uterus, and avoids postsurgery complications arising from uterine contractions. Embryo surgery in mice is similar to operating on the later stage chick embryo and we recommend using operations on the chick for training and to pilot technical aspects of any experimental design. Besides the obvious differences between embryogenesis of birds and mammals, an important value of the mouse model is the extensive genetics and the availability of mutant and transgenic lines for functional studies. The most invasive studies to date include amputations of the limb or digit to study regenerative responses (Reginelli *et al.*, 1995; Wanek *et al.*, 1989). Amputation studies have also been carried out on mice with targeted mutation to identify genes that are functionally required for a regenerative response (Han *et al.*, 2003). Similarly, embryo/fetal surgery has been used to investigate changes in the wound healing response during development and such studies have led to the discovery of specific signaling pathways and cellular responses critical for the scarring response (Ferguson and O'Kane, 2004; Hopkinson-Woolley *et al.*, 1994; Kishi *et al.*, 2006; Martin *et al.*, 2003; Whitby and Ferguson, 1991; Wilgus *et al.*, 2008).

While technically challenging, it is possible to transplant tissues between mouse embryos to study cell–cell interactions during development. Examples include studies of the interaction between anterior and posterior tissues during mouse limb development (Wanek *et al.*, 1989), grafts of digits in association with amputation studies (Reginelli *et al.*, 1995), and exploring the diastema region of the jaw as a permissive site for the development of a transplanted tooth germ (Song *et al.*, 2008). The transplantation of the mouse tooth germ represents a good example of the technical challenges. These operations involve the isolation of appropriately staged tooth germ from a genetically marked donor tissue (e.g., *Rosa26*) so that cellular contribution can be monitored. The engraftment site selected was the mandibular diastema, a toothless region of the oral cavity between the incisor and the molar. There is evidence from mutant analyses of tooth formation in this region suggesting that the diastema was capable of supporting the development of the tooth germ. The mouse tooth germ is at the bud stage at E13.5, a stage that is easily accessible for *exo utero* embryo surgery. After isolation from donor tissue, a small pin is placed in the graft to maintain the orientation of the tooth germ and also to facilitate handling of the tissue during transplantation. The graft tissue is stained with a vital dye for visualization during implantation and maintained in PBS at 4 °C. The graft site is prepared by making a small incision at the selected location of the mandibular diastema and the graft is immediately inserted into the site (Fig. 12.4A). The graft is oriented and fixed in place by pushing the pin into the host tissue, and inserting a second pin helps to stabilize the graft in place. The presence of the grafted tissue can be identified by EGFP expression in neonatal mice (Fig. 12.4B) and histological analyses verify the development of a well-developed ectopic tooth germ (Fig. 12.4C and D). Postnatally, the ectopic tooth germ erupts and forms a normal incisor in the ectopic position (Fig. 12.4E).

Amputation, wound healing and tissue grafting surgeries cause significant trauma to the embryo and can compromise embryo survival. In our experience, these types of embryo surgeries can have a high level of success from E13.5 and later, whereas similar manipulations at earlier stages is more

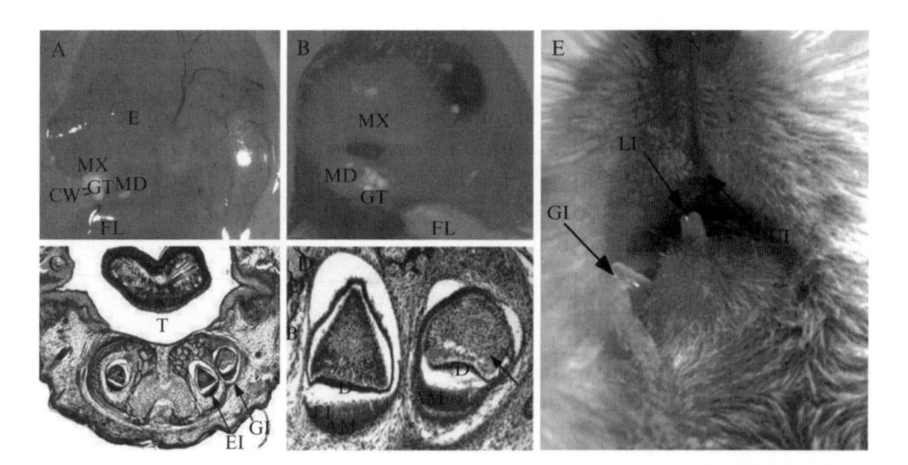

Figure 12.4 Ectopic tooth development in the mouse. (A) *Exo utero* surgery was used to implant an E13.5 incisor tooth germ from a *Rosa26-Egfp* donor (GT) into the mandibular region (MD) of an unlabeled host. The graft was implanted into a cut made in the diastema region and held in place using copper wire (CW). (B) Neonates with grafts were identified by GFP expression. (C, D) Histological analyses of neonatal mandibles indicated the presence of a well-developed grafted incisor germ (GI) positioned next to the endogenous incisor germ (EI). (E) The oral region of a mouse postweaning showing an ectopic incisor that formed from the graft (GI) next to the endogenous lower incisor (LI). AM, ameloblast; B, alveolar bone; D, dentin; E, eye; FL, forelimb; MX, maxillary; N, nasal; T, tongue; UL, upper incisor. From Song *et al.* (2008).

challenging yet feasible (see Wanek *et al.*, 1989). A less invasive experimental approach involving the implantation of microcarrier beads soaked in purified growth factors has been used to investigate specific responses of developing tissues. Such targeted application of FGF2 during the formation of the skull vault can modify the expression of a number of genes critical for maintaining the proliferation/differentiation balance important for normal development and implicated in a variety of craniosynostosis syndromes (Iseki *et al.*, 1997; Mikura *et al.*, 2009). We have used implantation of microcarrier beads containing FGF4 to modify cell migration patterns and induce digit bifurcation (Fig. 12.5A) during digit development (Ngo-Muller and Muneoka, 2000a). Stable positioning of agarose microcarrier beads can be difficult since the bead tends to exit the tissue via the wound created during implantation. To circumvent this problem we pin the bead onto the tip of an electrolytically sharpened tungsten needle and allow it to dehydrate just prior to implantation. The bead collapses onto the needle tip and creates a much smaller entry wound during implantation. After implantation the bead hydrates at the desired position and its increased size stabilizes it in the tissue and prevents any movement within the implantation wound. We have used microcarrier bead implantation in conjunction with targeted injection of the lipophillic cell marker, DiI (Fig. 12.5B), to demonstrate that a source of FGF4 has a very localized effect on the distribution of cell during induced digit bifurcation (Ngo-Muller and Muneoka, 2000a). This study demonstrates how multiple targeted manipulations can be successfully combined using an *exo utero* approach. Other surgical manipulations that have been carried out on mouse embryos

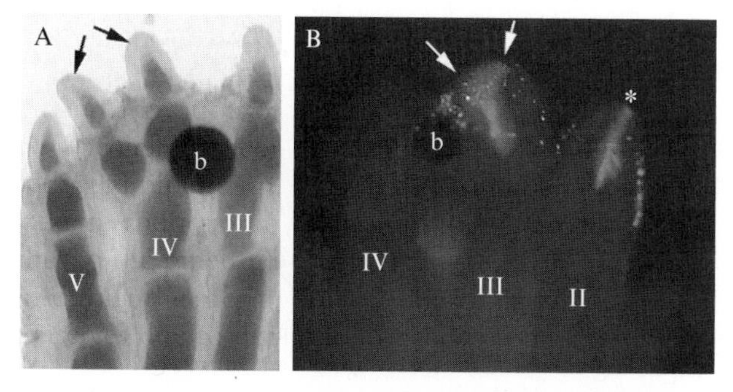

Figure 12.5 Targeted application of growth factors: (A) microcarrier beads (b) soaked in a high concentration of purified FGF4 implanted between digits III and IV of a E12.5 hindlimb induces the bifurcation of the distal region of digit IV (arrows). (B) Microcarrier bead implantation into the region between digits III and IV and in conjunction with targeted injections of DiI into digits II and III demonstrates a localized effect on the movements of digit III cells (arrows) with the cells of digit II (*) completely unaffected. From Ngo-Muller and Muneoka (2000a,b). (See Color Insert.)

include laser surgery as a way to correct a polydactylous defect or to induce arhinencephaly (Naruse and Kameyama, 1990; Naruse and Keino, 1995), using sutures to restrain movements of the upper and lower mandibles or the leg to investigate the role of fetal movements in have on joint formation (Habib *et al.*, 2005, 2007; Hashimoto *et al.*, 2002), and using a surgical approach to experimentally induce spina bifida aperta (Inagaki *et al.*, 1997).

7. Summary

Classical embryologists focused on lower vertebrate embryos, such as amphibians and birds, which are easily maintained and experimentally manipulated. On the other hand, molecular geneticists have used the mouse model to develop powerful tools for studying the function of specific genes during development. To explore development, especially as it pertains to human health issues, there is clearly a need to expand on, and develop, new strategies that enhance our ability to directly access the postimplantation mammalian embryo. The number of studies using *in utero* or *exo utero* approaches have increased over the past 25 years, and it is now clear that we can successfully probe the *in utero* environment of the mammalian embryo both classically (amputation, tissue transplantation, bead implantation) and genetically (electroporation, viral transduction). Advancing technologies in molecular genetics and *in vivo* imaging will soon add a new dimension to our ability to explore the postimplantation mammalian embryo.

References

Ackman, J. B., Anikszteјn, L., Crépel, V., Becq, H., Pellegrino, C., Cardoso, C., Ben-Ari, Y., and Represa, A. (2009). Abnormal network activity in a targeted genetic model of human double cortex. *J. Neurosci.* **29,** 313–327.

Bai, J., Ramos, R. L., Ackman, J. B., Thomas, A. M., Lee, R. V., and LoTurco, J. J. (2003). RNAi reveals double cortin is required for radial migration in rat neocortex. *Nat. Neurosci.* **6,** 1277–1283.

Chen, X., Gong, X. L., Katsumata, M., Zeng, Y. T., Huang, S. Z., and Zeng, F. (2009). Hematopoietic stem cell engraftment by early-stage in utero transplantation in a mouse model. *Exp. Mol. Pathol.* **87,** 173–177.

de Nijs, L., Léon, C., Nguyen, L., Loturco, J. J., Delgado-Escueta, A. V., Grisar, T., and Lakaye, B. (2009). EFHC1 interacts with microtubules to regulate cell division and cortical development. *Nat. Neurosci.* **12,** 1266–1274.

De Pietri Tonelli, D., Calegari, F., Fei, J. F., Nomura, T., Osumi, N., Heisenberg, C. P., and Huttner, W. B. (2006). Single-cell detection of microRNAs in developing vertebrate embryos after acute administration of a dual-fluorescence reporter/sensor plasmid. *Biotechniques* **41,** 27–32.

Ferguson, M. W. J., and O'Kane, S. (2004). Scar-free healing: From embryonic mechanisms to adult therapeutic intervention. *Phil. Trans. R. Soc. Lond. B* **359,** 839–850.

Friocourt, G., Kanatani, S., Tabata, H., Yozu, M., Takahashi, T., Antypa, M., Raguénès, O., Chelly, J., Férec, C., Nakajima, K., and Parnavelas, J. G. (2008). Cell-autonomous roles of ARX in cell proliferation and neuronal migration during corticogenesis. *J. Neurosci.* **28,** 5794–5805.

Garcia-Frigola, C., Carreres, M. I., Vegar, C., and Herrera, E. (2007). Gene delivery into mouse retinal ganglion cells by in utero electroporation. *BMC Dev. Biol.* **7,** 103.

Habib, H., Hatta, T., Udagawa, J., Zhang, L., Yoshimura, Y., and Otani, H. (2005). Fetal jaw movement affects condylar cartilage development. *J. Dent. Res.* **84,** 474–479.

Habib, H., Hatta, T., Rahman, O. I., Yoshimura, Y., and Otani, H. (2007). Fetal jaw movement affects development of articular disk in the temporomandibular joint. *Congenit. Anom. (Kyoto)* **47,** 53–57.

Han, M., Yang, X., Farrington, J. E., and Muneoka, K. (2003). Digit regeneration is regulated by *Msx1* and BMP4 in fetal mice. *Development* **130,** 5123–5132.

Harvey, A. R., Ehlert, E., de Wit, J., Drummond, E. S., Pollett, M. A., Ruitenberg, M., Plant, G. W., Verhaagen, J., and Levelt, C. N. (2009). Use of GFP to analyze morphology, connectivity, and function of cells in the central nervous system. *Methods Mol. Biol.* **515,** 63–95.

Hashimoto, R., Kihara, I., and Otani, H. (2002). Perinatal development of the rat hip joint with restrained fetal movement. *Congenit. Anom. (Kyoto)* **42,** 135–142.

Hatta, T., Moriyama, K., Nakashima, K., Taga, T., and Otani, H. (2002). The role of gp130 in cerebral cortical development: *In vivo* functional analysis in a mouse *exo utero* system. *J. Neurosci.* **22,** 5516–5524.

Hatta, T., Matsumoto, A., and Otani, H. (2004). Application of the mouse *exo utero* development system in the study of developmental biology and teratology. *Congenit. Anom. (Kyoto)* **44,** 2–8.

Hopkinson-Woolley, J., Hughes, D., Gordon, S., and Martin, P. (1994). Macrophage recruitment during limb development and wound healing in the embryonic and foetal mouse. *J. Cell Sci.* **107,** 1159–1167.

Inagaki, T., Schoenwolf, G. C., and Walker, M. L. (1997). Experimental model: Change in the posterior fossa with surgically induced spina bifida aperta in mouse. *Pediatr. Neurosurg.* **26,** 185–189.

Iseki, S., Wilkie, A. O. M., Heath, J. K., Ishimaru, T., Eto, K., and Morriss-Kay, G. M. (1997). Fgfr1 and osteopontin domains in the developing skull vault are mutually exclusive and can be altered by locally applied FGF2. *Development* **124,** 3375–3384.

Kawamura, K., and Chung, K. C. (2009). Constriction band syndrome. *Hand Clin.* **25,** 257–264.

Kawauchi, D., Taniguchi, H., Watanabe, H., Saito, T., and Murakami, F. (2006). Direct visualization of nucleogenesis by precerebellar neurons: Involvement of ventricle-directed, radial fibre-associated migration. *Development* **133,** 1113–1123.

Kishi, K., Ohyama, K., Satoh, H., Kubota, Y., Tanaka, T., Imanishi, N., Nakajima, H., Kawamura, K., and Nakajima, T. (2006). Mutual dependence of murine fetal cutaneous regeneration and peripheral nerve regeneration. *Wound Repair Regen.* **14,** 91–99.

Martin, P., D'Souza, D., Martin, J., Grose, R., Cooper, L., Maki, R., and McKercher, S. R. (2003). Wound healing in the PU.1 null mouse—Tissue repair is not dependent on inflammatory cells. *Curr. Biol.* **13,** 1122–1128.

Mathijssen, I. M., van Leeuwen, J. P., and Vermeij-Keers, C. (2000). Simultaneous induction of apoptosis, collagen type I expression and mineralization in the developing coronal suture following FGF4 and FGF2 application. *J. Craniofac. Genet. Dev. Biol.* **20,** 127–136.

Mikura, A., Okuhara, S., Saito, M., Ota, M., Ueda, K., and Iseki, S. (2009). Association of tenascin-W expression with mineralization in mouse calvarial development. *Congenit. Anom. (Kyoto)* **49,** 77–84.

Muneoka, K., Wanek, N., and Bryant, S. V. (1986). Mouse embryos develop normally *exo utero*. *J. Exp. Zool.* **239,** 289–293.
Muneoka, K., Wanek, N., and Bryant, S. V. (1989). Mammalian limb bud development: *In situ* fate maps of early hindlimb buds. *J. Exp. Zool.* **249,** 50–54.
Muneoka, K., Wanek, N., Trevino, C., and Bryant, S. V. (1990). *Exo utero* surgery. *In* "Post-Implantation Mammalian Embryo," (A. Copp and D. Cockroft, eds.), pp. 41–60. IRL Press, Oxford, UK.
Nakahira, E., Kagawa, T., Shimizu, T., Goulding, M. D., and Ikenaka, K. (2006). Direct evidence that ventral forebrain cells migrate to the cortex and contribute to the generation of cortical myelinating oligodendrocytes. *Dev. Biol.* **291,** 123–131.
Naruse, I., and Kameyama, Y. (1990). Fetal laser surgery in genetic polydactyl mice. *Teratology* **41,** 731–735.
Naruse, I., and Keino, H. (1995). Apoptosis in the developing CNS. *Prog. Neurobiol.* **47,** 135–155.
Naruse, I., and Tsutsui, Y. (1989). Brain abnormalities induced by murine cytomegalovirus injected into the cerebral ventricles of mouse embryos *exo utero*. *Teratology* **40,** 181–189.
Navarro-Quiroga, I., Chittajallu, R., Gallo, V., and Haydar, T. F. (2007). Long-term, selective gene expression in developing and adult hippocampal pyramidal neurons using focal *in utero* electroporation. *J. Neurosci.* **27,** 5007–5011.
Ngo-Muller, V., and Muneoka, K. (2000a). Influence of FGF4 on digit morphogenesis during limb development in the mouse. *Dev. Biol.* **219,** 224–236.
Ngo-Muller, V., and Muneoka, K. (2000b). *Exo utero* surgery. *Methods Mol. Biol.* **135,** 481–492.
Nimura, M., Udagawa, J., and Otani, H. (2008). Adrenocorticotropic hormone affects nonapoptotic cell death of undifferentiated germ cells in the fetal mouse testis: *In vivo* study by exo utero transplantation of corticotropic tumor cells into embryos. *Congenit. Anom. (Kyoto)* **48,** 81–86.
Ogawara, M., Takahashi, M., Shimizu, T., Nakajima, M., Setoguchi, Y., and Shirasawa, T. (2002). Adenoviral expression of protein-L-isoaspartyl methyltransferase (PIMT) partially attenuates the biochemical changes in PIMT-deficient mice. *J. Neurosci. Res.* **69,** 353–361.
Okada, T., Keino-Masu, K., and Masu, M. (2007). Migration and nucleogenesis of mouse precerebellar neurons visualized by in utero electroporation of a green fluorescent protein gene. *Neurosci. Res.* **57,** 40–49.
Panaroni, C., Gioia, R., Lupi, A., Besio, R., Goldstein, S. A., Kreider, J., Leikin, S., Vera, J. C., Mertz, E. L., Perilli, E., Baruffaldi, F., Villa, I., *et al.* (2009). *In utero* transplantation of adult bone marrow decreases perinatal lethality and rescues the bone phenotype in the knockin murine model for classical, dominant osteogenesis imperfecta. *Blood* **114,** 459–468.
Petros, T. J., Rebsam, A., and Mason, C. A. (2009). *In utero* and *ex vivo* electroporation for gene expression in mouse retinal ganglion cells. *J. Vis. Exp.* http://www.jove.com/index/Details.stp?ID=1333.
Reginelli, A. D., Wang, Y., Sassoon, D., and Muneoka, K. (1995). Digit tip regeneration correlates with regions of *msx1* (formerly *Hox7*) expression in fetal and newborn mice. *Development* **121,** 1065–1076.
Rio, P., Martinez-Palacio, J., Ramirez, A., Bueren, J. A., and Segovia, J. C. (2005). Efficient engraftment of in utero transplanted mice with retrovirally transduced hematopoietic stem cells. *Gene Ther.* **12,** 358–363.
Saba, R., Nakatsuji, N., and Saito, T. (2003). Mammalian BarH1 confers commissural neuron identity on dorsal cells in the spinal cord. *J. Neurosci.* **23,** 1987–1991.
Saito, T., and Nakatsuji, N. (2001). Efficient gene transfer into the embryonic mouse brain using in vivo electroporation. *Dev. Biol.* **240,** 237–246.

Serbedzija, G. N., Bronner-Fraser, M., and Fraser, S. E. (1994). Developmental potential of trunk neural crest cells in the mouse. *Development* **120,** 1709–1718.

Shikanai, M., Asahina, K., Iseki, S., Teramoto, K., Nishida, T., Shimizu-Saito, K., Ota, M., Eto, K., and Teraoka, H. (2009). A novel method of mouse *ex utero* transplantation of hepatic progenitor cells into the fetal liver. *Biochem. Biophys. Res. Commun.* **381,** 276–282.

Shimogori, T., and Ogawa, M. (2008). Gene application with *in utero* electroporation in mouse embryonic brain. *Dev. Growth Differ.* **50,** 499–506.

Shimogori, T., Banuchi, V., Ng, H. Y., Strauss, J. B., and Grove, E. A. (2004). Embryonic signaling centers expressing BMP, WNT and FGF proteins interact to pattern the cerebral cortex. *Development* **131,** 5639–5647.

Shinohara, H., Udagawa, J., Morishita, R., Ueda, H., Otani, H., Semba, R., Kato, K., and Asano, T. (2004). Gi2 signaling enhances proliferation of neural progenitor cells in the developing brain. *J. Biol. Chem.* **279,** 41141–41148.

Soma, M., Aizawa, H., Ito, Y., Maekawa, M., Osumi, N., Nakahira, E., Okamoto, H., Tanaka, K., and Yuasa, S. (2009). Development of the mouse amygdala as revealed by enhanced green fluorescent protein gene transfer by means of *in utero* electroporation. *J. Comp. Neurol.* **513,** 113–128.

Song, Y., Yan, M., Muneoka, K., and Chen, Y. (2008). The mouse embryonic diastema region is an ideal site for the development of ectopically transplanted tooth germ. *Dev. Dyn.* **237,** 411–416.

Takiguchi-Hayashi, K., Sekiguchi, M., Ashigaki, S., Takamatsu, M., Hasegawa, H., Suzuki-Migishima, R., Yokoyama, M., Nakanishi, S., and Tanabe, Y. (2004). Generation of reelin-positive marginal zone cells from the caudomedial wall of telencephalic vesicles. *J. Neurosci.* **24,** 2286–2295.

Ting, M. C., Wu, N. L., Roybal, P. G., Sun, J., Liu, L., Yen, Y., and Maxson, R. E., Jr. (2009). EphA4 as an effector of Twist1 in the guidance of osteogenic precursor cells during calvarial bone growth and craniosynostosis. *Development* **136,** 855–864.

Tondelli, B., Blair, H. C., Guerrini, M., Patrene, K. D., Cassani, B., Vezzoni, P., and Lucchini, F. (2009). Fetal liver cells transplanted *in utero* rescue the osteopetrotic phenotype in oc/oc mouse. *Am. J. Pathol.* **174,** 727–735.

Trevino, C., Calof, A., and Muneoka, K. (1992). Position specific growth regulation of 3T3 cells in the developing mouse limb. *Dev. Biol.* **150,** 72–81.

Trevino, C., Anderson, R., and Muneoka, K. (1993). 3T3 cell integration and differentiative potential during limb development in the mouse. *Dev. Biol.* **155,** 38–45.

Turner, D. L., Snyder, E. Y., and Cepko, C. L. (1990). Lineage-independent determination of cell type in the embryonic mouse retina. *Neuron* **4,** 833–845.

Wanek, N., Muneoka, K., and Bryant, S. V. (1989). Evidence for regulation following amputation and tissue grafting in the developing mouse limb. *J. Exp. Zool.* **249,** 55–61.

Wang, Y., Paramasivam, M., Thomas, A., Bai, J., Kaminen-Ahola, N., Kere, J., Voskuil, J., Rosen, G. D., Galaburda, A. M., and Loturco, J. J. (2006). DYX1C1 functions in neuronal migration in developing neocortex. *Neuroscience* **143,** 515–522.

Wang, S., Peretich, K., Zhao, Y., Liang, G., Meng, Q., and Wei, H. (2009). Anesthesia-induced neurodegeneration in fetal rat brains. *Pediatr. Res.* **66,** 435–440.

Whitby, D. J., and Ferguson, M. W. J. (1991). The extracellular matrix of lip wounds in fetal neonatal and adult mice. *Development* **112,** 651–668.

Wilgus, T. A., Ferreira, A. M., Oberyszyn, T. M., Bergdall, V. K., and Dipietro, L. A. (2008). Regulation of scar formation by vascular endothelial growth factor. *Lab. Invest.* **88,** 579–590.

Yamada, M., Hatta, T., and Otani, H. (2008). Mouse *exo utero* development system: Protocol and troubleshooting. *Congenit. Anom. (Kyoto)* **48,** 183–187.

Yoshida, T., Vivatbutsiri, P., Morriss-Kay, G., Saga, Y., and Iseki, S. (2008). Cell lineage in mammalian craniofacial mesenchyme. *Mech. Dev.* **125,** 797–808.

Zhang, H., Hatta, T., Udagawa, J., Moriyama, K., Hashimoto, R., and Otani, H. (1998). Induction of ectopic corticotropic tumor in mouse embryos by *exo utero* cell transplantation and its effects on the fetal adrenal gland. *Endocrinology* **139,** 3306–3315.

SECTION FOUR

FERTILIZATION

CHAPTER THIRTEEN

Enhancement of IVF in the Mouse by Zona-Drilling

Kevin A. Kelley

Contents

Abstract

Cryopreservation of mouse sperm has become an essential method for the long-term storage of novel, genetically modified mouse lines. Cryopreserved sperm from most hybrid lines can be effectively used for *in vitro* fertilization (IVF) of mouse oocytes. Unfortunately, IVF recovery with cryopreserved sperm from inbred lines is very inefficient. This is especially troublesome since many transgenic lines are created on the popular C57Bl/6 inbred strain. Cryopreserved sperm from C57Bl/6 inbred and genetically modified lines is generally very inefficient when used in standard IVF recovery experiments, with fertilization rates that can be lower than 10%. Assisted reproductive techniques have been developed to improve the IVF efficiencies of cryopreserved inbred sperm.

Department of Developmental and Regenerative Biology, Mouse Genetics Shared Resource Facility, Mount Sinai School of Medicine, New York, New York, USA

Methods in Enzymology, Volume 476
ISSN 0076-6879, DOI: 10.1016/S0076-6879(10)76013-4

These techniques include zona-drilling, which introduces a hole into the zona pellucida (ZP) surrounding mouse oocytes, using a chemical solution (acid Tyrode's), mechanical disruption (partial zona dissection or piezo-driven micropipette drilling), or laser photoablation. By allowing direct access of the sperm to the cytoplasmic membrane, zona-drilling can improve the efficiency of IVF fertilization rates with inbred sperm to greater than 90%, thus improving the chances of recovering mouse lines on inbred backgrounds that are maintained with cryopreserved sperm. The technique described in this chapter makes use of a piezo controller to mechanically disrupt the ZP, resulting in dramatic increases in the fertilization efficiency of cryopreserved sperm.

1. Introduction

Advances in techniques for the manipulation of the mouse genome have resulted in a tremendous explosion in the number of genetically modified mouse lines since the first transgenic mouse was described (Gordon *et al.*, 1980). Pronuclear injection, gene targeting in ES cells, ENU mutagenesis, and more recently, modification of the mouse genome with lentiviral vectors, have subsequently produced thousands of novel lines of mice. The rapid expansion of genetically modified mouse lines resulted in the need for development of methods of cryopreservation in order to be able to reliably bank embryos and/or gametes. From the researcher's point of view, the availability of cryopreservation techniques is an important tool for managing colonies and maintaining stocks of mice that can be used to recover valuable lines that may be lost due to pathogen outbreaks or environmental stress, such as sustained power failures during blackouts, HVAC failures, or natural disasters such as floods. An important consideration at many institutions is the demand that is put on available housing space by the sheer numbers of genetically modified mouse lines that are available for use in research. Cryopreservation programs are becoming increasingly important as they offer institutions the ability to open up extremely valuable housing space. In addition, cryopreserved embryos and/or gametes provide a convenient alternative to shipment of live animals for the transfer of novel mouse lines between institutions.

The cryopreservation of mouse embryos was first accomplished in the early 1970s, and provided a method for long-term storage of mouse lines (Whittingham, 1971; Whittingham *et al.*, 1972). While embryo cryopreservation is an important method for maintaining lines, there are some limitations that are inherent to the technique, including specialized equipment for most protocols (controlled rate freezers), as well as a requirement for a large cohort of immature females for superovulation to produce the embryos to be frozen. On the other hand, isolation of sperm does not require specialized equipment or the need for expanding a colony to generate enough immature females to serve as embryo donors. Unfortunately, cryopreservation techniques for mouse sperm proved to be very difficult to refine. It was not until 1990 that mouse sperm was successfully cryopreserved (Tada *et al.*, 1990), lagging significantly behind the cryopreservation of sperm from numerous other mammalian species, including human sperm that had first been successfully cryopreserved in the 1950s (Bunge and Sherman, 1953). The 18% raffinose, 3% skim milk cryoprotectant that is commonly used in most mouse sperm cryopreservation techniques was first introduced in 1991 (Takeshima *et al.*, 1991).

While the original raffinose/skim milk sperm cryopreservation technique has worked well for many hybrid strains of mice, there has been variable success with inbred and genetically modified strains. For example, cryopreserved sperm from inbred C57Bl/6J mice, which is one of the most commonly used mouse strains, proved to be very difficult to use for recovery through *in vitro* fertilization (IVF) (Nakagata and Takeshima, 1993). This difficulty with recovery of inbred strains has been lessened by recent refinements of the sperm cryopreservation method (Ostermeier *et al.*, 2008), but there are many thousands of lines that were cryopreserved during the preceding 17 years using the original method. As a result, there are numerous C57Bl/6-based mouse lines that have been cryopreserved using the older method that may prove to be difficult to recover by IVF.

Clearly, the newer refined cryopreservation protocol is a great advance; however, techniques that improve the IVF efficiencies of cryopreserved sperm are critical for recovery of lines frozen using the original protocol. To this end, there have been several methods developed that are designed to enhance the yields of fertilized eggs with sperm preparations from inbred lines. Intracytoplasmic sperm injection (ICSI) has been an important tool for assisted reproductive techniques in many species, including hamster, cow, human, and most recently, mouse (Goto *et al.*, 1990; Kimura and Yanagimachi, 1995; Lanzendorf *et al.*, 1988; Palermo *et al.*, 1992; Uehara and Yanagimachi, 1976). ICSI depends upon the injection of isolated sperm heads directly into the cytoplasm of meiotic stage oocytes, and was difficult to adapt to mouse oocytes until a new method was developed that used a piezo-driven micropipette (Kimura and Yanagimachi, 1995). ICSI has been successfully used in human IVF clinics, and with sperm from mouse lines that do not function well in standard IVF experiments, including transgenic lines with sperm defects as well as inbred strains. Unfortunately, ICSI requires a significant amount of skill, and is a labor-intensive technique, requiring the injection of isolated sperm heads into individual mature oocytes. Even with these considerations, ICSI is an important tool in the mouse embryologist's arsenal of assisted reproductive techniques.

Another method that was developed to improve fertilization efficiencies was the use of acid Tyrode's to make small holes in the zona pellucida (ZP).

Zona-drilling with acid Tyrode's was initially found to be useful for samples with low sperm counts (Conover and Gwatkin, 1988; Gordon and Talansky, 1986). Subsequently, zona-drilling mediated by acid Tyrode's has been used to successfully rescue lines with defective sperm (Ahmad *et al.*, 1989). This technique appears to circumvent problems with sperm binding to the ZP, and allows direct interaction of sperm with the cytoplasmic membrane.

While the administration of acid Tyrode's to the surface of the ZP to create a hole is effective, there are some technical difficulties associated with it. Controlling the flow of acid Tyrode's, and thus the size of the hole, can be difficult to master, and there are concerns about the effect that exposure to acid Tyrode's might have on further embryo development (Khalifa *et al.*, 1992). These issues have resulted in several more refined techniques for the introduction of holes in the zona, including laser photoablation, partial dissection, and mechanical disruption of the zona with a piezo-driven micropipette. Lasers have been successfully used to introduce holes in the ZP, and oocytes treated in this manner have been shown to have increased fertilization rates in IVF assays (Germond *et al.*, 1996; Liow *et al.*, 1996).

Partial zona dissection (PZD) is another technique that involves treating oocytes with a sucrose solution to shrink them, thus increasing the perivitelline space. The oocytes are then attached to culture dishes, and a partial dissection of the zona is accomplished using a 30-gauge needle. PZD has been shown to improve fertilization rates of cryopreserved C57Bl/6 sperm (Nakagata *et al.*, 1997).

Although PZD has been shown to be effective in improving IVF rates, there are some disadvantages with this technique. The oocytes need to be treated with a sucrose solution to shrink them sufficiently enough for the microdissection to be done, and the resulting fertilized embryos must be cultured to the morula or blastocyst stage prior to embryo transfer to pseudopregnant hosts due to the potential loss of blastomeres through the relatively large hole in the zona during the early stages of development (Kawase *et al.*, 2002; Nakagata *et al.*, 1997). In a more refined technique, partial ZP incision (the ZIP technique) creates a smaller hole in oocytes by mechanical disruption of the zona with a piezo-driven micropipette. The ZIP technique results in increased fertilization rates, especially with cryopreserved sperm (Kawase *et al.*, 2002). The use of a piezo drill to introduce holes in the ZP of mouse oocytes prior to IVF is a relatively simple procedure that can be adapted to most mouse embryology labs that are set up to perform ICSI. In fact, the manipulations required for zona-drilling are simpler, and less time consuming than the ICSI technique, and can produce a higher number of fertilized embryos. The following protocol details the preparation of zona-drilled oocytes, the IVF of these oocytes, and the surgical reimplantation of the resulting embryos.

2. Materials

2.1. Mice

C57Bl/6J (stock number 000664), B6C3 F1 hybrid (stock number 100010), B6D2 F1 hybrid (stock number 100006), FVB/NJ (stock number 001800), and C3H/HeJ (stock number 000659) mice used as embryo or sperm donors were purchased from Jackson Laboratories. C57Bl/6NTac mice used as embryo or sperm donors, and SW mice used as embryo hosts were purchased from Taconic Farms. All mice were maintained in micro-isolator cages (Allentown Caging) on a 14 h:10 h light:dark cycle with access to an automatic watering system, and *ad libitum* access to PicoLab Rodent Diet 20 (product code 5053, LabDiet). All mice were euthanized by CO_2 inhalation.

2.2. Hormones

Embryo donors are superovulated as described in Section 3 with pregnant mare's serum (PMS; National Hormone and Peptide Program, http://www.humc.edu/hormones/material.html) and human chorionic gonadotropin (HCG; Sigma catalogue number CG10). PMS and HCG are reconstituted to 1 International Unit (IU)/μl from lyophilized stocks with sterile saline (0.9% NaCl). Stocks containing 50 IU of each hormone are prepared and stored at −80 °C. Dilutions of each hormone to 5 IU/0.1 ml are prepared immediately before injection by addition of sterile saline to a final volume of 1 ml.

2.3. Piezo controller

Piezo-driven zona-drilling was performed with a Prime Tech PMM controller (model PMAS-CT150). The piezo controller uses a piezoelectric effect to generate vibrations in a micropipette holder. The resulting vibrations of an attached micropipette are performed at sufficient intensity to locally disrupt the ZP. Techniques using a piezo controller, such as ICSI, require mercury to be loaded in the piezo-driven pipette. Mercury is not necessary, however, for the zona-drilling technique described here. This is an important biosafety consideration. Very good results can be obtained with pipettes that are simply air-filled.

2.4. Microscopes, microforge, and pipette puller

All oocyte isolations and embryo washes are performed using a Nikon SMZ2 dissecting microscope to visualize the oocytes/embryos at low magnification (10–63×). Piezo-drilling of oocytes is performed using a

Nikon Diaphot microscope equipped with Nomarski differential interference contrast (DIC) optics at 600× (15× eyepieces, 40× DIC objective). The holding and piezo-drilling pipettes are controlled with Narashige coarse (MM-188) and fine (MO-188) manipulators. Holding pipettes are prepared from hand-drawn pipettes (1B100-6, 1 mm OD/0.58 mm ID, World Precision Instruments, Inc.) that are forged to the final dimensions on a microforge (MF-1, Technical Products International, Inc.). The piezo-drilling pipettes are prepared on a Sutter Instruments Model P-80/PC pipette puller. These pipettes are prepared from TWF100-6 pipettes (thin-wall 6 in. pipettes with 1 mm OD, 0.75 mm ID containing an internal filament, World Precision Instruments, Inc.) using the following program: heat = 715, pull = 95, velocity = 40, time = 15 (ramp value for the pipettes in this program is 696). Surgeries are performed in a laminar flow hood, using a Nikon SMZ10 stereomicroscope on a boom stand.

2.5. Culture media

FHM (Millipore/Specialty Media, catalogue number MR-024-D) is a specially formulated HEPES-buffered media that is used when oocytes or embryos are to be isolated or manipulated outside of a 5% CO_2 incubator. FHM with hyaluronidase (Millipore/Specialty Media, catalogue number MR-056-F) is used to dissociate oocytes from cumulus cells in isolated cumulus masses. KSOM + AA media (Millipore/Specialty Media, catalogue number MR-121-D) is bicarbonate-buffered, and used for the culture of embryos in a 5% CO_2 incubator at 37 °C. HTF (Millipore/Specialty Media, catalogue number MR-070-D) is a formulation based on human tubal fluid that is used for IVF of mouse oocytes (Quinn, 1995). Research VitroFert (K-RVFE-50, Cook Australia) is a media that has been developed to optimize mouse IVF. All incubations in a 5% CO_2 incubator at 37 °C are performed with the appropriate media covered with embryo-tested mineral oil (Sigma M8410) in either 6-well tissue culture dishes for IVF or 35 mm tissue culture dishes for *in vitro* embryo development.

2.6. Anesthetic

Avertin anesthesia is used for all surgical transfer of embryos to the reproductive tract of pseudopregnant SW female mice. Avertin is prepared from 2,2,2-tribromoethanol (Sigma-Aldrich, catalogue number T48042) and *tert*-amyl alcohol (Sigma-Aldrich, catalogue number 249486) as described by Nagy *et al.* (2003). Briefly, a stock solution (100%) is prepared by dissolving 10 g of 2,2,2-tribromoethanol in 10 ml of *tert*-amyl alcohol in a 50 °C water bath until it is fully dissolved. A working solution of 2.5% avertin is prepared by a 1:40 dilution of the 100% stock with warm (37 °C) water. The diluted solution is sterilized through a 0.2-μm filter and stored at

4 °C for 2–3 weeks before being replaced. The avertin is administered intraperitoneally at 0.016 ml/g of body weight.

2.7. Sperm cryopreservation buffer

Raffinose (Sigma R0250) and skim milk (DIFCO 0032-17-3) are used to prepare 18% raffinose, 3% skim milk sperm cryopreservation buffer as described by Nagy *et al.* (2003). A 6% skim milk solution is prepared in a volume of 300 ml and centrifuged for 1 h at 15,000×*g*. Two hundred milliliters of the supernatant is removed from the pellet, to which 72 g of raffinose is added. After the raffinose dissolves, the volume is adjusted to 400 ml, and the solution is sequentially passed through 0.45 μm and then 0.22 μm filters. Aliquots (1.2 ml) are frozen and stored at −80 °C.

3. Methods

3.1. Sperm isolation and freezing

For each male, one 1.2 ml vial of sperm cryopreservation buffer stored at −80 °C is thawed for 5 min at 37 °C. Both cauda epididymides with several millimeters of attached *vas deferens* are isolated into 1 ml of the 18% raffinose, 3% skim milk sperm cryopreservation buffer in a 35 mm tissue culture dish. Using two 26-gauge tuberculin syringe needles, several cuts are made across the epididymis, and any sperm in the attached *vas deferens* is pushed out the cut end with one needle. The dish is placed in a 37 °C oven for 10 min to allow the sperm to be released from the cut epididymis into the cryoprotectant. Using wide-bore pipette tips (1–200 μl, product number 1011-8810, USA Scientific) to prevent damage to the sperm, 0.1 ml of the sperm suspension is pipetted into each of 10 cryopreservation vials (1.8 ml NUNC, product code 377267) that are then placed in liquid nitrogen vapor for 10 min. This is conveniently done using a liquid nitrogen Dewar flask (D-3000, Cole-Parmer Instrument Company), with two cryostorage boxes stacked on the bottom. Liquid nitrogen is added until it reaches approximately three-quarters of the way up the bottom box. The vials to be frozen are placed in the top box, which has no lid, for 10 min. The vials are then transferred to cryostorage boxes in the liquid phase of a liquid nitrogen freezer for long-term storage.

3.2. Superovulation and oocyte isolation

Oocyte donors are maintained in an animal room with the lights on at 7 am and off at 9 pm. The females are superovulated by intraperitoneal injection of 5 IU of PMS/female at 5:30 pm (day 0), followed by 5 IU of

HCG/female 48 h later (day 2). At 8:00 am on the morning of day 3, the females are euthanized, and the oviducts are removed and collected into a 35-mm tissue culture dish containing FHM media at room temperature. Individual oviducts are sequentially transferred to a depression slide or 35 mm tissue culture dish containing FHM media with hyaluronidase (prewarmed to 37 °C), and observed on a dissecting microscope (Nikon SMZ2). Each oviduct is immobilized behind the ampulla (the swelling containing the cumulus mass of cells) with a #3 forcep (Fine Science Tools), and the outer wall of the ampulla is opened by tearing with a second #3 forcep to release the cumulus mass. After all cumulus masses have been released into the hyaluronidase media, the dissolution of the oocytes from the cumulus cells is carefully monitored. Oocytes are generally more difficult to dissociate from cumulus cells than fertilized eggs. Individual oocytes are recovered and washed through three changes of FHM media. At this point, the oocytes are either washed through several changes of Cook's Research VitroFert media and used immediately for an IVF assay with sperm that has been allowed to capacitate for 1 h (standard IVF assay with zona-intact oocytes) or they are transferred to FHM media in an injection chamber for piezo-drilling of the ZP.

3.3. Piezo-driven zona-drilling

Isolated oocytes are transferred into FHM media in a glass slide injection chamber that is then placed on the injection microscope. The use of glass or plastic injection chambers will depend upon the type of optics that are used on the injection microscope. A forged holding pipette is used to immobilize the oocytes while a piezo-driven pipette is used to introduce a hole into the ZP. As seen in Fig. 13.1, oocytes are oriented so that the portion of an oocyte that has the largest perivitelline space is nearest to the piezo-driven pipette. If this is not done, damage to the cytoplasmic membrane is likely to happen, resulting in lysis of the oocytes. To make a hole in the ZP, the piezo-driven pipette is passed tangentially to the surface of the ZP several times (usually three passes are sufficient) while the piezo controller is activated (Fig. 13.2A). With the profile of the pipette tip that is used (which is similar to an injection pipette used for pronuclear injection), the speed and intensity settings of the controller are 9 and 7, respectively. These high speed and intensity settings work well for this particular tip profile, with the ZP having the appearance of "melting" away as the pipette is passed near it. For the wider opening pipette profiles used in other protocols, such as the ZIP technique (Kawase *et al.*, 2002), however, these settings are too high, and would result in a high percentage of lysed oocytes. In fact, if the pipette tips are damaged during the zona-drilling (e.g., by accidentally touching them against the holding pipette), they should be replaced immediately.

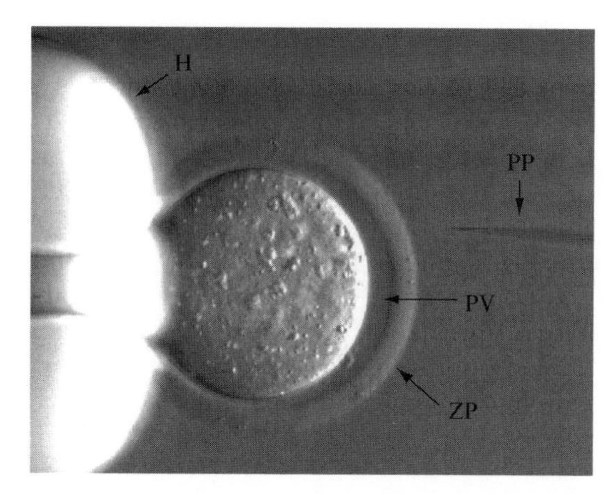

Figure 13.1 Proper orientation of unfertilized oocyte for piezo-mediated zona-drilling. The oocyte is immobilized on a holding pipette (H) in an orientation that places the region with the widest perivitelline space (PV) near the piezo-activated pipette (PP). The piezo controller can then be used to incise a hole in the zona pellucida (ZP).

Each tangential pass is made closer to the oocyte, with the removal of more of the ZP each time (Fig. 13.2B–D). The final pass results in a hole in the ZP that is approximately 40 μm in length (Fig. 13.2D). This calculation is based on the radius of an oocyte being approximately 50 μm, with the hole in the ZP being inscribed by a central angle of approximately 45°, which represents 1/8th of the circumference of the oocyte, or about 40 μm.

The introduction of holes in the ZP using a piezo controller is a relatively efficient procedure. There are, however, several considerations to keep in mind. First, the total removal of cumulus cells from all oocytes with hyaluronidase can be difficult to accomplish. Some of the isolated oocytes will have cumulus cells attached to the surface of the ZP (Fig. 13.3A), and activation of the piezo-pipette near these cells will result in their lysis. The cellular debris from the lysed cumulus cells will attach to the piezo-activated pipette tip (Fig. 13.3B), significantly reducing its effectiveness for the introduction of holes into the ZP. In most cases, if cellular debris attaches to the pipette, it will need to be replaced. Another problem that can arise during the preparation of zona-drilled oocytes is the lysis of the oocytes themselves due to disruption of the cytoplasmic membrane by the piezo-activated pipette. This generally occurs with oocytes that have a very small perivitelline space (Fig. 13.3C). Even with these considerations, we usually observe loss of only 1–2% of oocytes from disruption of the cytoplasmic membrane. After holes have been introduced into the ZP of all oocytes, they are washed through several changes of Cook's Research VitroFert media, and placed into IVF drops (VitroFert) with sperm that has capacitated for 1 h.

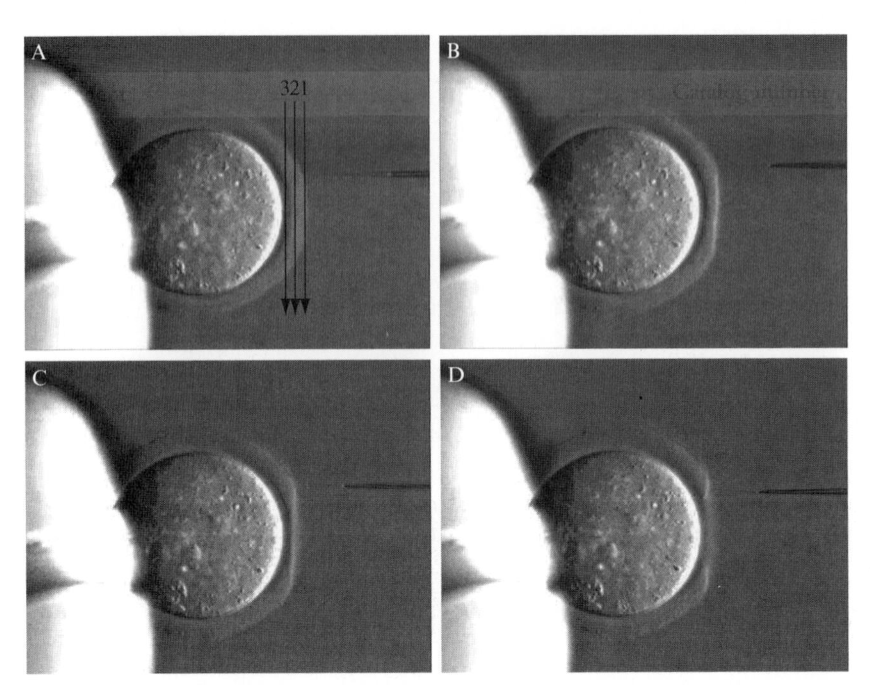

Figure 13.2 Zona-drilling of mouse oocytes. (A). Introducing a hole into the zona pellucida (ZP) is accomplished by successive tangential passes with a piezo-activated pipette. Generally, three passes are required to make a hole that completely penetrates the ZP, indicated by the three numbered arrows. The results of the first, second, and third passes of a piezo-activated pipette are shown in (B), (C), and (D), respectively. Each pass produces additional thinning of the ZP, and on the final pass, a hole has been introduced through the ZP of the oocyte.

3.4. *In vitro* fertilization and culture

IVF assays for either zona-intact (standard assay) or zona-drilled oocytes are set up in 250 μl drops of Cook's Research VitroFert that is overlaid with embryo-tested mineral oil. These drops can be conveniently prepared in 6-well tissue culture dishes to accommodate multiple IVF assays. The IVF assays can also be set up in 35 mm tissue culture dishes, but the lower profile of the dishes makes it more difficult to work with as the mineral oil will nearly fill the dish to cover the 250 μl drop of IVF media.

One hour prior to the anticipated addition of oocytes to the IVF assay, vials of frozen sperm are thawed rapidly in a 37 °C water bath for 2 min. Care should be taken to allow residual liquid nitrogen to evaporate from the tube prior to thawing so that the rapidly expanding gas does not forcefully eject the lid. A small aliquot ($\sim$5 μl) of the frozen sperm is placed in a Makler counting chamber (Sefi-Medical Instruments) to determine the quantity of sperm to add to each assay. The sperm is not highly motile in

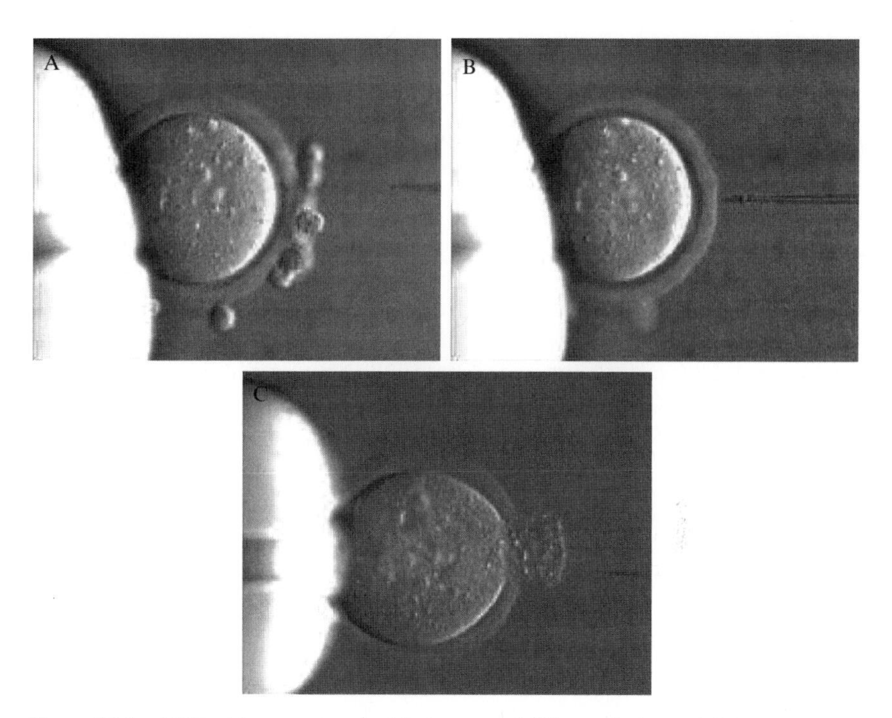

Figure 13.3 Difficulties associated with the zona-drilling technique. (A) Some oocytes will have cumulus cells that are still attached to the ZP. Attempts to introduce a hole into the ZP with the piezo-activated pipette will result in lysis of the cumulus cells, resulting in the attachment of cellular debris (B) to the pipette. (C) In some oocytes, the perivitelline space is too small, resulting in lysis of the cytoplasmic membrane.

the cryopreservation solution, and it is usually not necessary to inactivate the sperm prior to counting. This can be done easily, however, by incubating an aliquot of the sperm at 50 °C for 5 min, if necessary. The sperm concentration is calculated by counting sperm heads located within horizontal strips of 10 boxes in the counting grid. This is repeated in four to five horizontal strips, and the counts are then averaged. The result is the total sperm count in millions per milliliter. This count does not differentiate between motile and nonmotile sperm. Using wide-bore pipette tips, sperm from the thawed sample is then added to a concentration of 1000–2000 sperm/μl (2.5–5.0 $\times$ 10^5 sperm total) to each 250 μl IVF drop in the 6-well dish.

The sperm are allowed to capacitate for approximately 1 h before the addition of either zona-intact or zona-drilled oocytes. After a 4–6-h incubation of capacitated sperm and oocytes, the oocytes are collected and washed through several changes of KSOM + AA media that has been equilibrated in a 5% CO_2 incubator. They are then transferred to microdrops of KSOM + AA media under mineral oil (prepared in a 35-mm tissue culture dish), and incubated overnight for development of two-cell stage embryos, or longer for development to the morula/blastocyst stage. Fertilization efficiencies are calculated by determining the percentage of oocytes that develop to the two-cell stage after IVF. If the embryos are cultured past the two-cell stage, they should be transferred to fresh media drops each day.

3.5. Embryo transfers

Embryos at the two-cell stage are transferred to the oviduct of day 0.5 pseudopregnant SW female mice, under avertin anesthesia (oviduct surgeries are described in Nagy *et al.*, 2003). The mice are pseudopregnant as judged by the presence of a vaginal plug the day after mating (day 0.5) with vasectomized SW males. Zona-intact, zona-drilled, or zona-free (acid Tyrode's treated) embryos at the morula or blastocyst stage can also be transferred to the uterus of day 2.5 pseudopregnant SW females (uterine surgeries are described in Nagy *et al.*, 2003).

4. Results and Discussion

Two commercially available IVF media were tested for their effectiveness in supporting the fertilization of zona-intact and zona-drilled oocytes using frozen, thawed sperm. HTF (Specialty Media) and Research VitroFert (Cook, Australia) were compared side-by-side in IVF assays, using KSOM media (Specialty Media) as a control. Successful IVF was judged by development of the oocytes to the two-cell stage (Fig. 13.4). VitroFert supported the development of the highest percentage of fertilized embryos

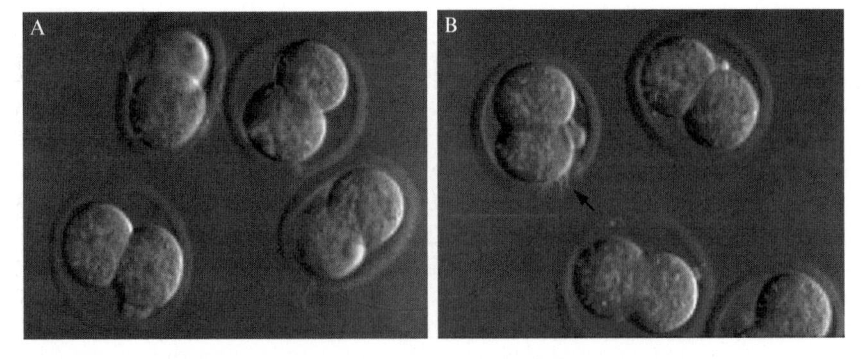

Figure 13.4 Fertilized embryos derived from IVF with zona-intact (A) or zona-drilled (B) oocytes. The oocytes were isolated from superovulated C57Bl/6J females. The arrow in (B) indicates an opening in the ZP that resulted from zona-drilling.

from either C57Bl/6J (33%) or B6C3 F1 (68%) zona-intact oocytes (Table 13.1). As expected, zona-intact B6C3 F1 hybrid oocytes yielded higher fertilization rates than zona-intact C57Bl/6J oocytes. C57Bl/6J oocytes that underwent zona-drilling also had the highest rates of fertilization (96%) when the IVF was performed in VitroFert (Table 13.1). The two-cell embryos that develop from the zona-drilled oocytes are indistinguishable from two-cell embryos derived from zona-intact oocytes (Fig. 13.4). The most notable observation from the tests of the different media was the dramatic increase in fertilization rates of zona-drilled oocytes in all media. This observation is consistent with previous reports that have shown that introducing a hole into the ZP increases fertilization rates of mouse oocytes (Conover and Gwatkin, 1988; Germond *et al.*, 1996; Gordon and Talansky, 1986; Kawase *et al.*, 2002; Liow *et al.*, 1996). Based on the tests of the effects of the IVF media on fertilization rates in these two strains, VitroFert was used in all subsequent IVF assays.

The effect of sperm concentration was evaluated for fertilization of zona-drilled C57Bl/6J oocytes. The introduction of a hole in the ZP has been found to be beneficial for enhancing the efficiency of IVF with samples that have low sperm counts (Gordon and Talansky, 1986). The piezo-activated zona-drilling technique described in this chapter requires between 5×10^2 and 1×10^3 sperm/μl for maximal fertilization efficiency (Table 13.2), which is consistent with the results determined by Liow *et al.* (1996), using laser photoablation. The sperm concentration that is used in all subsequent IVF assays is between 1 and 2×10^3 sperm/μl.

To further test the effectiveness of the zona-drilling technique, frozen sperm from a number of transgenic and knockout lines was used in IVF assays with both zona-intact as well as zona-drilled oocytes from various inbred and hybrid strains. These results indicated that zona-drilling was very effective at increasing the fertilization rate of oocytes from a variety of background strains (Table 13.3). For example, the combined results from six different lines on C57Bl/6J backgrounds revealed an increase from 36% of oocytes that were successfully fertilized with zona-intact oocytes to 83% that were fertilized with the same sperm samples when zona-drilled oocytes were used (Table 13.3). Two other inbred lines (C57Bl/6NTac and C3H/HeJ) and the hybrid B6C3 F1 line showed similar increases in fertilization rates after zona-drilling (Table 13.3). While FVB/NJ oocytes also yielded higher fertilization rates after zona-drilling, the effect was not as great as with the other lines tested (Table 13.3). The greatest effect was observed for one transgenic line on a C3H/HeJ background for which only 8% of zona-intact oocytes were fertilized. After zona-drilling, the fertilization rate for this sperm sample increased to 97% of treated oocytes (Table 13.3).

The viability of embryos that were derived from zona-drilled oocytes was further tested by transfer of two-cell embryos into the oviduct of pseudopregnant hosts. Eight distinct transgenic and knockout lines on

Table 13.1 Effect of IVF media on fertilization rates using zona-intact (ZI) and zona-drilled (ZD) oocytes

	C57Bl/6J[a]			B6C3 F1[b]		
	KSOM + AA	HTF	VitroFert	KSOM + AA	HTF	VitroFert
ZI IVF[c]	7/41[d] (17%)[e]	8/42 (19%)	14/42 (33%)	41/111 (37%)	59/111 (53%)	69/101 (68%)
ZD IVF[f]	30/44 (68%)	36/44 (82%)	43/45 (96%)	49/53 (92%)	54/55 (98%)	47/49 (96%)

[a] IVF with C57Bl/6J oocytes and cryopreserved sperm.
[b] IVF with B6C3 F1 hybrid oocytes and cryopreserved sperm.
[c] ZI IVF refers to zona-intact oocytes used in the IVF incubation.
[d] Number of two-cell embryos after 1 day *in vitro* (DIV) culture/number of oocytes used for IVF.
[e] Percentage of oocytes that developed to the two-cell stage after 1 DIV.
[f] ZD IVF refers to the use of zona-drilled oocytes in the IVF.

Table 13.2 Effect of sperm concentration on fertilization rates of zona-intact and zona-drilled C57Bl/6J oocytes

	Zona-drilled oocytes				Zona-intact oocytes
Sperm concentration[a] (μl^{-1})	1×10^2	5×10^2	1×10^3	2×10^3	2×10^3
Fertilization rate[b]	4/34[c] (12%)[d]	28/34 (82%)	34/35 (97%)	35/35 (100%)	5/30 (17%)

IVF assays were prepared with thawed, cryopreserved sperm using VitroFert media.

[a] Sperm concentration calculated from counts of thawed sperm samples in a Makler counting chamber.

[b] Fertilization rates determined by number of oocytes that develop to the two-cell stage after a 4-h incubation with sperm, and overnight culture in KSOM + AA media.

[c] Number of two-cell embryos after 1 day *in vitro* (DIV) culture/number of oocytes used for IVF.

[d] Percentage of oocytes that developed to the two-cell stage after 1 DIV.

Table 13.3 Fertilization rates of transgenic and knockout lines using zona-intact (ZI) and zona-drilled (ZD) oocytes[a]

	Oocyte donor strain				
	C57Bl/6J (6)[b]	C57Bl/6NTac (2)	C3H/HeJ (1)	FVB/NJ (2)	B6C3 (3)
ZI IVF	215/635[c] (36%)[d]	24/101 (24%)	13/485 (8%)	76/624 (12%)	56/376 (15%)
ZD IVF	365/441 (83%)	200/213 (94%)	149/153 (97%)	91/147 (62%)	209/218 (96%)

IVF assays were prepared with thawed, cryopreserved sperm using VitroFert media.

[a] Data presented in the table represents individual experiments in which ZI and ZD oocytes were used in side-by-side comparisons of fertilization efficiency with the same sperm sample.

[b] The number in parentheses after each strain name is the number of different transgenic or knockout lines tested.

[c] Number of two-cell embryos after 1 day *in vitro* (DIV) culture/number of oocytes used for IVF.

[d] Percentage of oocytes that developed to the two-cell stage after 1 DIV.

Table 13.4 Recovery of live mice from zona-drilled (ZD) oocytes

Background	ZD oocytes	Two-cell	Surgeries (# two-cell)	Births (% of transferred two-cell)
C3H/HeJ	153	149 (97%)[a]	4 (149)[b]	20 (13%)
C57Bl/6J	79	46 (58%)	2 (46)	2 (4%)
C57Bl/6J	101	99 (98%)	3 (99)	4 (4%)
C57Bl/6J	40	37 (92%)	1 (27)	5 (18%)
C57Bl/6J	104	77 (74%)	1 (35)	4 (11%)
FVB/NJ	103	70 (68%)	2 (70)	4 (6%)
B6C3 F1	72	68 (94%)	2 (68)	5 (7%)
B6D2 F1	98	93 (95%)	1 (25)	5 (20%)
Total	750	639 (85%)	16 (519)	49 (9%)

IVF assays were prepared with thawed, cryopreserved sperm using VitroFert media.
[a] Number in parentheses is the percentage of oocytes that developed to the two-cell stage after IVF with thawed, cryopreserved sperm from transgenic or knockout lines.
[b] Number in parentheses is the number of two-cell stage embryos transferred into the oviduct of pseudopregnant day 0.5 SW female mice.

several inbred (C3H/HeJ, C57Bl/6J and FVB/NJ) and hybrid (B6C3 and B6D2) backgrounds were recovered from cryopreserved sperm. The success rates for recovery ranged from 4% to 20% of transferred two-cell embryos that resulted in live births (Table 13.4). The overall efficiency that was seen across all lines was an average of 9% of transferred two-cell embryos that resulted in live births. This is in contrast to recovery rates of around 40–50% for transferred two-cell embryos that are isolated from natural matings (data not shown). The lower than expected recovery rates were examined in more detail by allowing fertilized oocytes to develop further *in vitro*. Zona-intact and zona-drilled oocytes from C57Bl/6J and B6C3 mice were cultured for 3 days *in vitro* (DIV) after IVF with cryopreserved sperm. As seen previously, there was a dramatic increase (to 96%) in the number of two-cell stage embryos that developed from the zona-drilled oocytes in both strains (Table 13.5). By 3 DIV, 46% and 47% of the two-cell C57Bl/6J and B6C3 embryos, respectively, developed to the morula stage (Table 13.5). This is in line with expectations for the recovery rate of the fertilized embryos, and thus defects in development to at least the morula stage do not seem to be responsible for the lower recovery rates observed.

It has been reported by Khalifa *et al.* (1992) that attempts to perform assisted hatching of blastocysts with acid Tyrode's solution can result in failure of the blastocysts to properly hatch. The expanded blastocysts only partially hatch, and are trapped in the opening introduced into the ZP by the acid Tyrode's solution. To investigate whether this same phenomenon is occurring with zona-drilled oocytes used for IVF recoveries, development of preimplantation embryos was followed for several more DIV.

Table 13.5 *In vitro* preimplantation development of zona-intact (ZI) versus zona-drilled (ZD) oocytes after IVF

	C57Bl/6J			B6C3 F1		
	Two-cell[a]	Four-cell[b]	Morula[b]	Two-cell[a]	Four-cell[b]	Morula[b]
ZI IVF	14/42 (33%)	5/14 (36%)	2/14 (14%)	69/101 (68%)	40/69 (58%)	33/69 (48%)
ZD IVF	43/45 (96%)	22/43 (51%)	20/43 (46%)	47/49 (96%)	22/47 (47%)	22/47 (47%)

IVF assays were prepared with thawed, cryopreserved sperm using VitroFert media.

[a] Fertilization rate was assessed after 1 DIV culture by determining the percentage of zona-intact or zona-drilled oocytes that developed to the two-cell stage.

[b] Development to the four-cell (2 DIV) or morula (3 DIV) stage was assessed relative to the number of two-cell embryos.

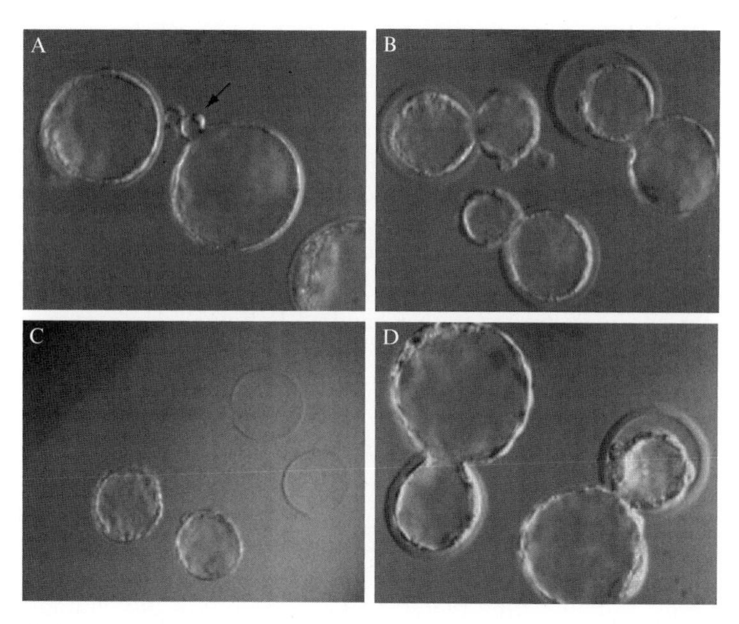

Figure 13.5 Development and hatching of zona-intact (A and C) and zona-drilled (B and D) B6C3 F1 hybrid blastocysts. (A) Blastocysts that develop from IVF with zona-intact oocytes are apparent by 4 days *in vitro* (DIV) culture, and are beginning to hatch through the ZP (arrow) (600×). In contrast, blastocysts derived from zona-drilled oocytes (B) are hatching to a greater extent by 4 DIV (600×). Complete hatching of zona-intact blastocysts (C) is observed by 5 DIV, as evidenced by the presence of empty zonas (300×). The zona-drilled blastocysts remain trapped in the ZP (D) at 5 DIV culture (600×).

Embryos that develop from IVF using zona-intact B6C3 oocytes (Fig. 13.5A and C) and C57Bl/6J oocytes (not shown) develop to fully expanded blastocysts by 4 DIV culture and are completely hatched by 5 DIV culture. In contrast, the majority of embryos that are derived from zona-drilled B6C3 oocytes (Fig. 13.5B and D) and C57Bl/6J oocytes (not shown) develop to hatching blastocysts by 4 DIV culture, but remain trapped in the ZP by 5 DIV culture. Indeed, many of the blastocysts have still failed to properly hatch by 6 DIV culture, although a low percentage (10–15%) do eventually hatch. This observation is consistent with the lower recovery rates from transferred two-cell embryos derived from zona-drilled oocytes, and it appears that the lower recovery rates are due to this failure of blastocysts to properly hatch. Khalifa *et al.* (1992) hypothesized that the holes introduced by zona-drilling affect the natural thinning of the ZP due to expansion of the blastocyst, since the blastocyst can expand out through the artificially introduced hole rather than pressing uniformly against the ZP. In addition, they suggest that a proteinase that is produced by the mural trophectoderm that can induce lysis of the ZP during hatching (Perona and

Wassarman, 1986) may be lost through the hole in the ZP, or diluted by the culture media, severely restricting the natural hatching process.

While the increased fertilization efficiency may seem to be offset by the potential problems with blastocyst hatching, it is important to keep in mind that the zona-drilling technique described here has been successfully used to recover transgenic and knockout mouse lines that could not be recovered using standard IVF with zona-intact oocytes. For example, a transgenic line on a C3H/HeJ background was successfully recovered with the birth of 20 pups using cryopreserved sperm and zona-drilled oocytes (Table 13.4). This line could not be recovered with a standard IVF procedure, since only 13 two-cell embryos developed from 485 zona-intact oocytes (Table 13.3). The zona-drilling technique increased the rate of fertilization with this sperm from only 8% with zona-intact oocytes to 97% with zona-drilled oocytes, generating sufficient embryos for recovery surgeries.

Clearly, improvements in the hatching efficiency of blastocysts derived from zona-drilled oocytes would be an important refinement of this technique. One possible approach would be to treat either morula or blastocysts derived from zona-drilled oocytes with acid Tyrode's to completely remove the ZP prior to transferring the embryos to the uterus of pseudopregnant females. This combination of two assisted reproductive techniques could dramatically enhance the recovery of embryos derived from zona-drilled oocytes. This has been shown to be the case in recovery of one knockout line from trapped blastocysts that developed from IVF of zona-drilled B6C3 oocytes. Two attempts to recover this line using zona-intact oocytes resulted in only 7–9% of the oocytes becoming fertilized and developing to the two-cell stage, with no births from one surgery that received 19 two-cell embryos (Table 13.6). Zona-drilled oocytes were used for a third attempt, and while the fertilization efficiency rose to 94%, there were no births after transfer of 75 two-cell embryos into three pseudopregnant hosts (Table 13.6). For the fourth attempt, the fertilized embryos that developed from zona-drilled B6C3 oocytes were allowed to develop to the blastocyst stage. At 4 DIV culture, it was observed that the embryos had properly developed to blastocysts, but they were trapped in the ZP during hatching. Thirty of these trapped blastocysts were treated with acid Tyrode's solution to remove the ZP, and these zona-free blastocysts were transferred to the uterus of two day 2.5 pseudopregnant SW hosts. Eleven births were obtained from these two surgeries, representing 37% of the transferred zona-free blastocysts (Table 13.6). This result indicates that the combination of ZD IVF and assisted hatching with acid Tyrode's prior to uterine surgery can indeed be an effective means of recovering lines from cryopreserved sperm. By alleviating the hatching problems that have been found with embryos that develop from zona-drilled oocytes, this method should allow investigators to recover lines from sperm that inefficiently fertilizes oocytes in standard IVF recovery experiments.

Table 13.6 Recovery of live mice from a knockout line using acid Tyrode's-treated trapped blastocysts

	1. ZI IVF	2. ZI IVF	3. ZD IVF	4. ZD IVF
Oocytes	269	114	80	175
Two-cell (%)[a]	19 (7%)	10 (9%)	75 (94%)	149 (85%)
Two-cell surgeries (# two-cell)[b]	1 (19)	ND	3 (75)	ND
Two-cell births (%)[c]	0 (0%)	NA	0 (0%)	NA
Zona-free blastocyst surgeries (#)[d]	ND	NA	ND	2 (30)
Zona-free blastocyst births (%)[e]	NA	NA	NA	11 (37%)

Four separate IVF experiments were performed with either zona-intact (ZI IVF, # 1 and #2) or zona-drilled (ZD IVF, #3 and #4) B6C3 oocytes.
IVF assays were prepared with thawed, cryopreserved sperm using VitroFert media.
ND, not done; NA, not applicable.
[a] Number in parentheses is the percentage of oocytes that developed to the two-cell stage after IVF with thawed, cryopreserved sperm.
[b] Number in parentheses is the number of two-cell stage embryos transferred into the oviduct of pseudopregnant day 0.5 SW female mice.
[c] Number in parentheses is the percent of transferred two-cell embryos that developed to term.
[d] Number in parentheses is the number of acid Tyrode's-treated (zona-free) trapped blastocysts that were transferred into the uterus of day 2.5 pseudopregnant SW hosts.
[e] Number in parentheses is the percent of transferred zona-free blastocysts that developed to term.

REFERENCES

Ahmad, T., Conover, J. C., Quigley, M. M., Collins, R. L., Thomas, A. J., and Gwatkin, R. B. (1989). Failure of spermatozoa from T/t mice to fertilize in vitro is overcome by zona drilling. *Gamete Res.* **22,** 369–373.

Bunge, R. G., and Sherman, J. K. (1953). Fertilizing capacity of frozen human spermatozoa. *Nature* **172,** 767–768.

Conover, J. C., and Gwatkin, R. B. (1988). Fertilization of zona-drilled mouse oocytes treated with a monoclonal antibody to the zona glycoprotein, ZP3. *J. Exp. Zool.* **247,** 113–118.

Germond, M., Nocera, D., Senn, A., Rink, K., Delacretaz, G., Pedrazzini, T., and Hornung, J. P. (1996). Improved fertilization and implantation rates after non-touch zona pellucida microdrilling of mouse oocytes with a 1.48 micron diode laser beam. *Hum. Reprod.* **11,** 1043–1048.

Gordon, J. W., and Talansky, B. E. (1986). Assisted fertilization by zona drilling: A mouse model for correction of oligospermia. *J. Exp. Zool.* **239,** 347–354.

Gordon, J. W., Scangos, G. A., Plotkin, D. J., Barbosa, J. A., and Ruddle, F. H. (1980). Genetic transformation of mouse embryos by microinjection of purified DNA. *Proc. Natl. Acad. Sci.* **77,** 7380–7384.

Goto, K., Kinoshita, A., Takuma, Y., and Ogawa, K. (1990). Fertilisation of bovine oocytes by the injection of immobilised, killed spermatozoa. *Vet. Rec.* **127,** 517–520.

Kawase, Y., Iwata, T., Ueda, O., Kamada, N., Tachibe, T., Aoki, Y., Jishage, K.-I., and Suzuki, H. (2002). Effect of partial incision of the zona pellucida by piezo-micromanipulator for in vitro fertilization using frozen-thawed mouse spermatozoa on the developmental rate of embryos transferred at the 2-cell stage. *Biol. Reprod.* **66,** 381–385.

Khalifa, E.-A. M., Tucker, M. J., and Hunt, P. (1992). Cruciate thinning of the zona pellucida for more successful enhancement of blastocyst hatching in the mouse. *Hum. Reprod.* **7,** 532–536.

Kimura, Y., and Yanagimachi, R. (1995). Intracytoplasmic sperm injection in the mouse. *Biol. Reprod.* **52,** 709–720.

Lanzendorf, S. E., Maloney, M. K., Veeck, L. L., Slusser, J., Hodgen, G. D., and Rosenwaks, Z. (1988). A preclinical evaluation of pronuclear formation by microinjection of human spermatozoa into human oocytes. *Fertil. Steril.* **49,** 835–842.

Liow, S. L., Bongso, A., and Ng, S. C. (1996). Fertilization, embryonic development and implantation of mouse oocytes with one or two laser-drilled holes in the zona, and inseminated at different sperm concentrations. *Hum. Reprod.* **11,** 1273–1280.

Nagy, A., Gertsenstein, M., Vintersten, K., and Behringer, R. (2003). *Manipulating the Mouse Embryo: A Laboratory Manual.* 3rd edn. Cold Spring Harbor Laboratory Press, Cold Spring Harbor, NY.

Nakagata, N., and Takeshima, T. (1993). Cryopreservation of mouse spermatozoa from inbred and F1 hybrid strains. *Exp. Anim.* **42,** 317–320.

Nakagata, N., Okamoto, M., Ueda, O., and Suzuki, H. (1997). Positive effect of partial zona-pellucida dissection on the in-vitro fertilizing capacity of cryopreserved C57Bl/6J transgenic mouse spermatozoa of low motility. *Bio. Reprod.* **57,** 1050–1055.

Ostermeier, G. C., Wiles, M. V., Farley, J. S., and Taft, R. A. (2008). Conserving, distributing and managing genetically modified mouse lines by sperm cryopreservation. *PLoS ONE* **3,** e2792.

Palermo, G., Joris, H., Devroey, P., and Van Steirteghem, A. C. (1992). Pregnancies after intracytoplasmic injection of single spermatozoon into an oocyte. *Lancet* **340,** 17–18.

Perona, D., and Wassarman, P. M. (1986). Mouse blastocysts hatch in vitro by using a trypsin-like proteinase associated with cells of the mural trophectoderm. *Dev. Biol.* **114,** 45–52.

Quinn, P. (1995). Enhanced results in mouse and human embryo culture using a modified human tubal fluid medium lacking glucose and phosphate. *J. Assist. Reprod. Genet.* **12,** 97–105.

Tada, N., Sato, M., Yamanoi, J., Mizorogi, T., Kasai, K., and Ogawa, S. (1990). Cryopreservation of mouse spermatozoa in the presence of raffinose and glycerol. *J. Reprod. Fertil.* **89,** 511–516.

Takeshima, T., Nakagata, N., and Ogawa, S. (1991). Cryopreservation of mouse spermatozoa. *Jikken Dobutsu* **40,** 493–497.

Uehara, T., and Yanagimachi, R. (1976). Microsurgical injection of spermatozoa into hamster eggs with subsequent transformation of sperm nuclei into male pronuclei. *Biol. Reprod.* **15,** 467–470.

Whittingham, D. G. (1971). Survival of mouse embryos after freezing and thawing. *Nature* **233,** 125–126.

Whittingham, D. G., Leibo, S., and Mazur, P. (1972). Survival of mouse embryos frozen to −196 degrees and −269 degrees C. *Science* **178,** 411–414.

CHAPTER FOURTEEN

ICSI in the Mouse

Paula Stein *and* Richard M. Schultz

Contents

Abstract

Fertilization by intracytoplasmic sperm injection (ICSI) is a powerful technique that can be used to understand better the biology of fertilization, in addition to a form of assisted reproduction both in humans and in endangered species. Mouse is often the model organism of choice to study mammalian fertilization. The ability to fertilize successfully mouse eggs by sperm injection, however, has been hard to achieve. The introduction of piezo-actuated injection made possible to perform mouse ICSI in an efficient fashion. Piezo-actuated ICSI is also currently used to fertilize eggs of other mammalian species, for injection of earlier spermatogenic cells into mouse eggs, as well as for somatic cell nuclear transfer, and cloning. This chapter provides detailed experimental procedures to perform ICSI in mouse eggs.

Department of Biology, University of Pennsylvania, Philadelphia, Pennsylvania, USA

Methods in Enzymology, Volume 476
ISSN 0076-6879, DOI: 10.1016/S0076-6879(10)76014-6

1. Introduction

Fertilization is the fusion of haploid sperm and egg to produce a diploid embryo. In mammals, the ovulated egg is arrested at the metaphase of the second meiotic division (MII). Upon sperm penetration, the egg completes meiosis, and enters interphase to give rise to a zygote that contains a maternal and a paternal pronucleus. Experiments dating back more than 50 years resulted in successful *in vitro* fertilization (IVF) in several mammalian species, including humans (reviewed in Chang, 1968). The first report of fertilization by intracytoplasmic sperm injection (ICSI) into mammalian eggs was published in 1976 (Uehara and Yanagimachi, 1976) and this technique was later employed to fertilize rabbit, cattle, and human eggs (Goto *et al.*, 1990; Iritani *et al.*, 1988; Lanzendorf *et al.*, 1988). Although ICSI using human eggs successfully led to live births in 1992 (Palermo *et al.*, 1992), attempts at fertilizing mouse eggs by ICSI were unsuccessful. The utilization of a piezo-actuated drill to penetrate the egg's extracellular coat, the *zona pellucida* (ZP), and break the egg's plasma membrane or oolemma ultimately led to the success of ICSI using mouse eggs (Kimura and Yanagimachi, 1995).

There are several factors that likely explain why mouse eggs are so difficult to inject without using a piezoinjector. The wound healing ability of the oolemma and the viscosity of the ooplasm are much lower, whereas the elasticity of the oolemma is higher in the mouse than in eggs from other species (Kimura and Yanagimachi, 1995). The piezo-actuated microinjection technique has become the standard for mouse ICSI and is also used for sperm injection in several other species, for injection of earlier stages of spermatogenic cells, as well as for somatic cell nuclear transfer and cloning (Wakayama, 2007; Yanagimachi, 2005). This chapter describes the use of ICSI in mouse eggs.

2. Experimental Procedures

2.1. Animals

Different strains of mice have been used as egg donors for ICSI (Kawase *et al.*, 2001). In general, hybrid and outbred strains are better with respect to survival after sperm injection and especially with respect to development to term (Kawase *et al.*, 2001; Kurokawa and Fissore, 2003). Furthermore, it is easier to visualize the meiotic spindle in eggs from outbred or hybrid strains than in those from inbred strains. We typically use 6-week-old CF-1 (Harlan, Indianapolis, IN) or B6D2F1/J (Jackson Laboratories, Bar Harbor, ME) female mice as the source of eggs. The male mouse strain, on the other hand, does not influence the outcome of ICSI (Kawase *et al.*, 2001).

2.2. Culture dish setup

Place several 50 μl drops of KSOM (Erbach *et al.*, 1994) on the bottom of a 60-mm sterile plastic tissue culture dish and cover the dish with light paraffin oil. Place the dish in a humidified incubator containing 5% CO_2 in air at 37 °C. Using the same procedure, set up a dish containing drops of Whitten's/PVA (Table 14.1). The KSOM dish is for culture of eggs after ICSI. The Whitten's/PVA dish is for keeping the eggs before sperm injection. Another dish containing a single 500-μl drop of Whitten's/PVA should be set up in the same way for sperm preparation. All these culture dishes should be prepared at least a few hours before use—it can be done the night before—to allow equilibration of the culture medium to the right pH.

2.3. Isolation of metaphase II eggs

Female mice are superovulated by an intraperitoneal (i.p.) injection of pregnant mare's serum gonadotropin (PMSG; 5 IU) followed by an i.p. injection of human chorionic gonadotropin (hCG; 5 IU) 48 h later. Metaphase II eggs are collected from ampullae of the oviducts 13–16 h later.

Table 14.1 Composition of Whitten's media used in this protocol

Component	Whitten's/PVA	Whitten's/HEPES/PVA[a]
NaCl (109.5 m*M*)	1600 mg	1600 mg
KCl (4.7 m*M*)	89 mg	89 mg
KH_2PO_4 (1.2 m*M*)	40.5 mg	40.5 mg
$MgSO_4 \cdot 7H_2O$ (1.2 m*M*)	73.5 mg	73.5 mg
Glucose (5.5 m*M*)	250 mg	250 mg
Pyruvic acid, sodium salt (0.23 m*M*)	6.25 mg	6.25 mg
L-(+)-Lactic acid, hemicalcium salt (4.8 m*M*)	131.75 mg	131.75 mg
Gentamycin (10 mg/ml solution)	0.25 ml	0.25 ml
Phenol Red (1% stock)	0.25 ml	0.25 ml
$NaHCO_3$	475 mg (22 m*M*)	147 mg (7 m*M*)
Polyvinyl alcohol (PVA; 0.01%)	25 mg	25 mg
HEPES (sodium salt)	–	975 mg (15 m*M*)

For both: bring to 250 ml with water. Filter in Nalgene Filterware (0.2 mm cellulose nitrate filter).
[a] For Whitten's/HEPES/PVA only: adjust pH to 7.4 with 6 *M* HCl.

Table 14.2 List of reagents used in this protocol[a]

Reagent	Source	Catalog number
PMSG	Calbiochem	367222
hCG	Sigma	CG5
KSOM	Millipore	MR-106-D
Polyvinyl alcohol (PVA)	Sigma	P8136
Hyaluronidase	Sigma	H3506
Mercury	Fisher	M-140
VacuTips	Eppendorf	930001015
Piezo drill tips (ICSI)	Eppendorf	930001091
Hexamethyldisilazane (HMDS)	Pierce	84770
Hamilton syringes:	Hamilton	
Gastight 1705		80900
Microliter #705		80500

[a] All other chemicals were obtained from Sigma-Aldrich, and were of tissue culture or embryo-tested grade where appropriate.

1. Sacrifice the females, remove the oviducts, and place them in a 35-mm plastic culture dish containing Whitten's/HEPES/PVA (Tables 14.1 and 14.2).
2. Transfer the individual oviducts to individual drops of Whitten's/HEPES/PVA containing 3 mg/ml hyaluronidase. Using a pair of forceps, hold the upper part of the oviduct (ampulla) and pierce it with a 27G½ needle to release the eggs into the medium. Let the eggs sit in the hyaluronidase solution for a few minutes at room temperature until the cumulus cells detach. It is important to watch carefully and move the eggs from the hyaluronidase solution as soon as the cumulus cells get loose to avoid damaging the eggs.
3. Wash the cumulus cell-free eggs through several drops of Whitten's/HEPES/PVA and then transfer them to a culture dish containing microdrops of Whitten's/PVA covered with light paraffin oil. Place the dish in a humidified incubator containing 5% CO_2 in air at 37 °C until use.

2.4. Sperm collection

Although ICSI can be successfully performed using motile intact sperm (Kimura and Yanagimachi, 1995), most laboratories utilize immobilized sperm or isolated sperm heads because sperm immobilization before injection improves the fertilization rate (Tateno *et al.*, 2000). Several methods can be used to immobilize sperm, including brief sonication, freeze-thawing, separation of head and tail by piezo pulses, detergent treatment, and rubbing the sperm tail with a micropipette (Tateno *et al.*, 2000). We use sonication, which immobilizes sperm and also results in a substantial percentage of sperm whose tails have been clipped off. Better survival rates are usually found when injecting isolated sperm heads instead of a whole spermatozoon because the volume of medium entering the egg is much smaller. In addition, sonication is not harmful as long as it is performed in appropriate media that contain low sodium and high potassium concentrations (Tateno *et al.*, 2000). We use a modification of the nucleus isolation medium (NIM) originally described by Yanagimachi's group (Sasagawa *et al.*, 1998) to sonicate our sperm preparation. NIM contains 123 m*M* KCl, 2.6 m*M* NaCl, 7.8 m*M* Na_2HPO_4, 1.4 m*M* KH_2PO_4, and 3 m*M* EDTA (disodium salt); the pH is adjusted to 7.2 by the addition of 1 *M* KOH. The addition of 1% PVA prevents sperm stickiness (NIM/PVA). When preparing NIM/PVA, stir for a long time (~48 h) after adding the PVA because the solubility of PVA is low.

1. Sacrifice the male and remove the epididymides and a portion of *vas deferens*. Place the isolated tissues into a 500-μl drop of Whitten's/PVA under oil.
2. Make several cuts using surgical scissors through the epididymides and *vas deferens* and place the dish in the incubator. Allow sperm to swim out for ~10 min; remove pieces of tissue.
3. Collect the 500 μl of sperm suspension into a 1.5-ml microcentrifuge tube. Put the tube on ice; the rest of the protocol is done on ice.
4. Centrifuge at 700×*g* for 5 min at 4 °C. Carefully remove the supernatant, as the pellet is loose.
5. Add 500 μl ice-cold NIM/PVA and invert to mix.
6. Sonicate the suspension four times for 15 s each time; invert the tube and put back on ice between sonications. These specifications are for a Branson Sonic Bath Model 1210 (Branson, Danbury, CT), and may need to be modified if using another type of sonicator.
7. After sonication, take a 5-μl aliquot of the sperm suspension and observe under a light microscope to make sure enough tails have been clipped off. It is not necessary to have a high percentage of isolated heads, as only a few will be used for injection. A typical preparation will have ~30% heads/70% intact sperm. If the treatment was not strong enough, perform additional sonication cycles.
8. Wash sperm twice with 500 μl ice-cold NIM/PVA. The preparation will be cloudy after sonication and a white layer of fat will appear on top of the solution after centrifugation. Make sure you remove it when you remove the supernatant after each wash.
9. Remove the supernatant, resuspend sperm in 500 μl of NIM/PVA containing 50% glycerol, and store at −20 °C until use. This preparation can be stored for several weeks.

10. On the day of injection, take 10 μl of the sperm suspension and wash twice by centrifugation with 250 μl of NIM/PVA to remove glycerol. Resuspend the sperm preparation in ~100 μl of NIM/PVA. Place a 5-μl drop of sperm suspension on a Petri dish and place on the microscope stage. Similarly, place 5 μl drops of a 1:10 and a 1:100 dilution of the sperm preparation in the same dish. Adjust the concentration as needed in order to have enough sperm heads to inject, but avoid a very crowded preparation.

2.5. Injection dish setup

Figure 14.1 shows the setup that we use for injections. Make a 5-μl flat drop of sperm suspension at the center of the lid of a 100-mm Petri dish. Make 3 × 5 μl drops of NIM/PVA at one end of the dish along the midline. Make 2 × 5 μl injection drops of Whitten's/HEPES/PVA at the other end of the dish along midline. Cover with mineral oil and place on the micromanipulator stage.

2.6. Pipette preparation

Injection pipettes are blunt-ended, with an internal diameter of 5–6 μm at the tip. They can be either straight or with an angle, depending on personal preferences or equipment requirements. These pipettes are commercially available from a few suppliers. We use Eppendorf's piezo drill tips (catalog # 930001091). The pipettes can also be made using a pipette puller and a microforge. In this case, the pipettes should be siliconized before use.

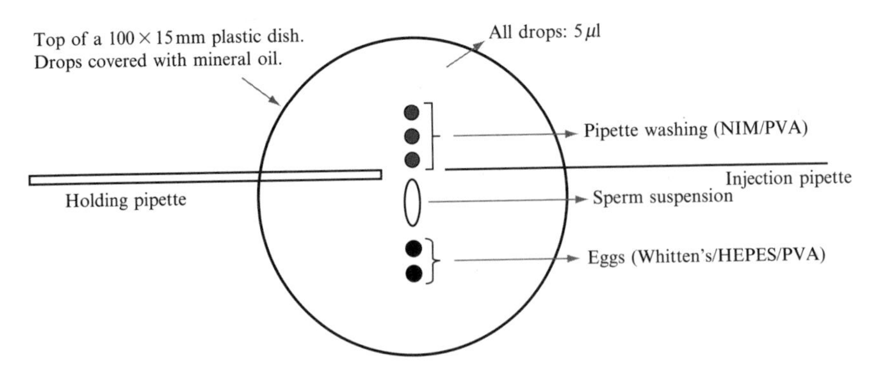

Figure 14.1 Injection dish setup. The injection dish is the lid of a 100 × 15 mm plastic Petri dish. Five-microliter drops of the different media (as indicated) are arranged along the midline, the dish is covered with mineral oil, and placed on the micromanipulator stage. Injection and holding pipettes are positioned as shown.

To siliconize the pipettes, put them in a Petri dish held by clay or tape so that they do not touch the bottom of the dish and place the dish in a dessicator with vacuum port. Add a few drops (~50 μl) of hexamethyldisilazane (Pierce, catalog # 84770) into the Petri dish and apply vacuum briefly (<1 min). Keep the pipettes under vacuum until use.

One of the keys for efficient injection using a piezo-driving unit is to add a substance heavier than water in the injection pipette; this improves the efficacy of drilling. Most investigators use mercury, which is very dense and hence provides excellent drilling efficiency, its only disadvantage being its toxicity. There are also nontoxic alternatives, such as Fluorinert.

1. To add mercury to the pipettes, under a fume hood, back-fill injection pipettes with ~0.5–1 cm of mercury using a Hamilton syringe (Fig. 14.2A). Place the mercury as close as possible to the tip of the

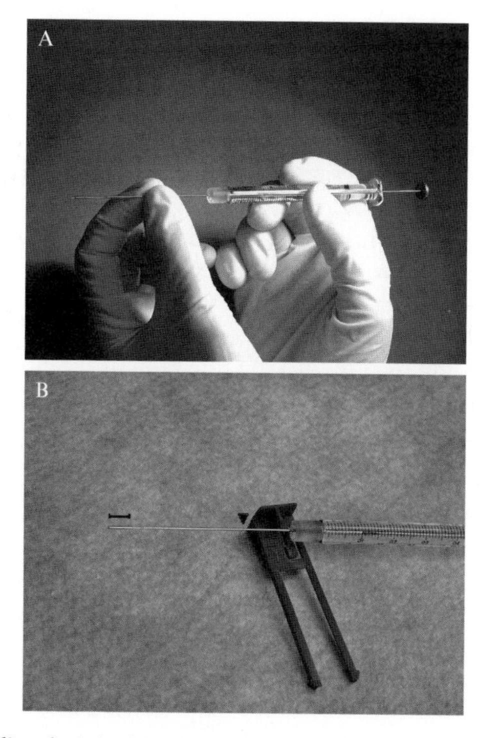

Figure 14.2 Loading the injection pipette with mercury to improve drilling efficiency. (A) A Hamilton syringe containing a 22-gauge needle is loaded with mercury. The syringe is inserted in the back end of the injection capillary and advanced forward as close to the tip as possible. A column of mercury (0.5–1 cm in length) is deposited. (B) The injection capillary with the Hamilton syringe inserted through its back end is shown. The arrowhead points to the end of the capillary. The bar indicates the length of the mercury column, which is deposited where the capillary narrows.

pipette (Fig. 14.2B). If this is not possible as the needle is not slim enough, the mercury can be moved forward once the pipette is inserted in the pipette holder.

2. Place the pipette into the piezoinjector holder and immerse the tip in one of the NIM/PVA drops.
3. Remove the air in the pipette by advancing the mercury all the way to the tip of the injection pipette until a couple of drops are released into the NIM/PVA. Draw in NIM/PVA. Repeat these two steps several times until all air and oil are expelled and a good control of the flow is achieved. It is also very important to make sure that no air bubbles are present inside the tubing connected to the injection pipette.
4. Move the injection pipette to a clean drop of NIM/PVA and draw in medium (the mercury column should be outside of the field of view under low magnification (4× or 10× objectives)).

2.7. Injection

1. Bring the injection pipette to the sperm drop and pick up a single sperm head. Place the head back in the pipette to a point where the pipette is wider than the sperm head turned sideways. Load another sperm head. Three to five sperm heads can be loaded into the pipette in this way keeping them apart to avoid injecting more than one sperm head.
2. Move the stage to place the pipette into the injection drop (Whitten's/HEPES/PVA).
3. Place the holding pipette into the injection drop. Holding pipettes can be made using a pipette puller and a microforge (Gordon, 1993), or they can be purchased from a few suppliers. We use Eppendorf's VacuTips (catalog # 930001015).
4. Place approximately five eggs in the injection drop.
5. Position an egg on the holding pipette so that the metaphase II spindle is at either 6 or 12 o'clock, and the egg is touching the dish (Fig. 14.3). Being able to see the spindle depends greatly on the mouse strain used to isolate the eggs and on the optics. It is recommended to use a microscope equipped with either differential interference contrast (DIC) or Hoffman modulation contrast (HMC). It is also desirable to find a position in which there is some space between the ZP and the oolemma at the site of injection (3 o'clock). If the ZP and plasma membrane are in very close proximity, the oolemma could be easily damaged when drilling the ZP.
6. Bring the injection pipette close to the egg and press against the ZP slightly to ensure that the pipette is in the same plane as the *zona* (Z-axis) (Fig. 14.4A).

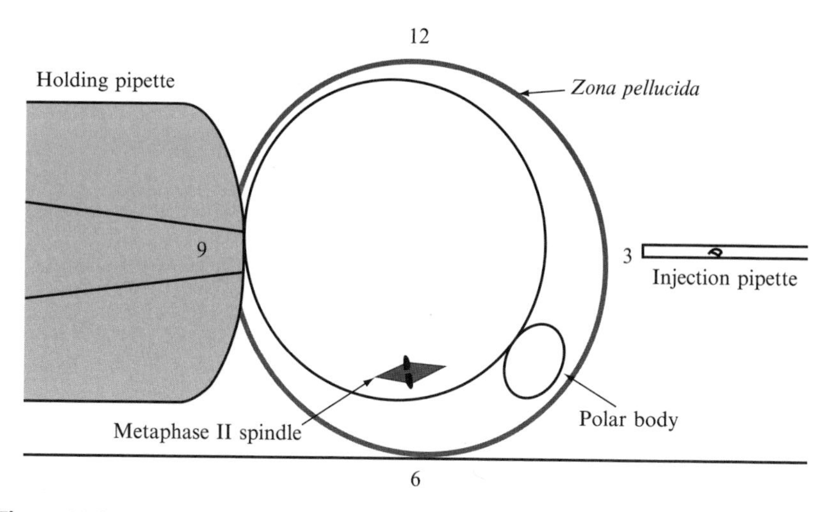

Figure 14.3 Position of the egg for injection. The egg is held by the holding pipette in such a way that it is touching the dish. The egg is rotated until the metaphase II spindle is at either the 12 o'clock or the 6 o'clock position and making sure there is some space between the oolemma and the *zona pellucida* (ZP) at the site of injection. The injection pipette, containing three to five sperm heads, is positioned at 3 o'clock and brought to contact the ZP at the equator of the egg.

7. Use the piezoinjector to make a hole in the ZP (speed 6, intensity 2–6 usually if using the Prime Tech Piezo device). We recommend using the lowest settings possible; start drilling with intensity = 2, if drilling is unsuccessful, increase to 3, and so on. While drilling, advance the pipette very gently, without pushing and as soon as the pipette has passed through the ZP stop drilling to avoid contact with the oolemma (Fig. 14.4B).
8. Advance one sperm head to the tip of the injection pipette (Fig. 14.4C). Once the sperm head is close to the tip, move the pipette forward into the cortex of the egg on the side opposite to the entry site (Fig. 14.4D). Apply slight negative pressure to the injection pipette to aid in breaking the membrane.
9. Apply a single piezo pulse (speed 2, intensity 1–3 usually) and watch for relaxation of the oolemma (Fig. 14.4E).
10. Inject the sperm head into the egg with a minimal volume of medium.
11. Withdraw the pipette applying a slight negative pressure (Fig. 14.4F).

Repeat the whole procedure with the remaining eggs. If the pipette becomes clogged or the flow control is not smooth, wash the pipette in a drop of NIM/PVA, advancing the mercury all the way to the tip and then drawing in fresh NIM/PVA several times. If this solves the problem, collect

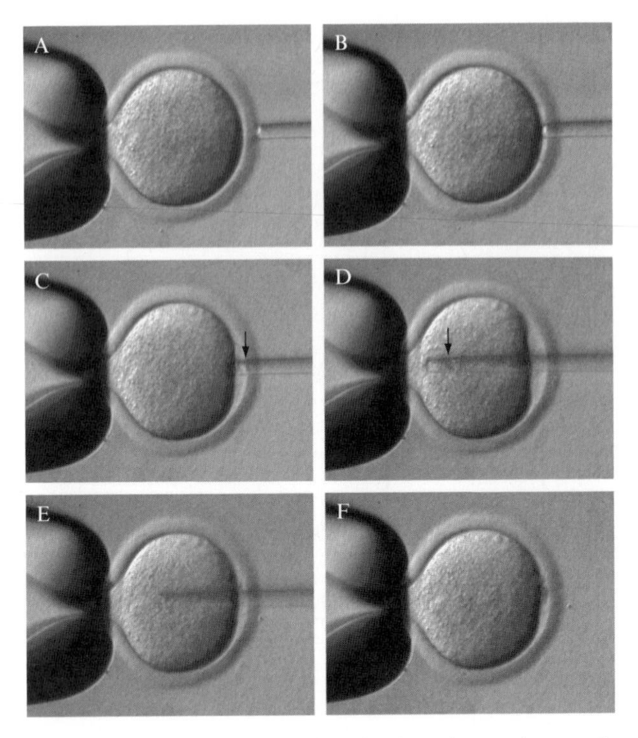

Figure 14.4 ICSI procedure. (A) Bring the injection pipette close to the egg until it makes contact with the *zona pellucida* (ZP) around the equator of the egg. (B) Drill through the ZP using the drill's lowest settings possible and being careful not to touch the oolemma. (C) Advance one sperm head all the way to the tip of the pipette. The arrow points to the sperm head. When the sperm head gets close to the tip, slightly advance the pipette until it contacts the oolemma to prevent the sperm head to come out of the pipette. (D) Advance the injection pipette as far as possible, near the cortex to the opposite side of the egg (9 o'clock). The arrow points to the sperm head. Tension of the oolemma at the injection site can be observed. (E) Apply a single piezo pulse to break the plasma membrane. Relaxation of the oolemma is a good indicator that the membrane has been pierced. (F) Withdraw the injection pipette.

more sperm heads and proceed with injections. If the flow is not good enough, discard the pipette and use a new injection pipette.

2.8. Disposal of mercury-containing material

Mercury must be handled with caution. All injection pipettes should be placed in designated containers for safe disposal. Similarly, injection dishes containing drops of mercury should be decontaminated before being thrown away. Using an automatic pipette, remove the drops of mercury and place them in the designated container. Contact your institution's safety office for proper disposal of these containers.

2.9. Embryo culture

After eggs have undergone ICSI, place them in KSOM under mineral oil in a humidified chamber containing 5% CO_2, 5% O_2, and 90% N_2 or in the incubator (5% CO_2 in air).

2.10. Embryo transfer

After reaching the two-cell stage, transfer the embryos to the oviducts of a pseudopregnant female (day 0.5 p.c.) or culture to the blastocyst stage in KSOM and transfer to the uteri of a pseudopregnant female mouse (day 2.5 p.c.). Detailed procedures can be found elsewhere (Nagy *et al.*, 2003).

ACKNOWLEDGMENTS

We thank Francesca E. Duncan and Karen Schindler for helpful discussions and critical reading of the manuscript, and the NIH (HD 22681 and HD 22732 to R. M. S.) for financial support of the research in this laboratory.

REFERENCES

Chang, M. C. (1968). In vitro fertilization of mammalian eggs. *J. Anim. Sci.* **27**(Suppl. 1), 15–26.

Erbach, G. T., Lawitts, J. A., Papaioannou, V. E., and Biggers, J. D. (1994). Differential growth of the mouse preimplantation embryo in chemically defined media. *Biol. Reprod.* **50,** 1027–1033.

Gordon, J. W. (1993). Micromanipulation of gametes and embryos. *Methods Enzymol.* **225,** 207–238.

Goto, K., Kinoshita, A., Takuma, Y., and Ogawa, K. (1990). Fertilisation of bovine oocytes by the injection of immobilised, killed spermatozoa. *Vet. Rec.* **127,** 517–520.

Iritani, A., Utsumi, K., Miyake, M., Hosoi, Y., and Saeki, K. (1988). In vitro fertilization by a routine method and by micromanipulation. *Ann. N. Y. Acad. Sci.* **541,** 583–590.

Kawase, Y., Iwata, T., Toyoda, Y., Wakayama, T., Yanagimachi, R., and Suzuki, H. (2001). Comparison of intracytoplasmic sperm injection for inbred and hybrid mice. *Mol. Reprod. Dev.* **60,** 74–78.

Kimura, Y., and Yanagimachi, R. (1995). Intracytoplasmic sperm injection in the mouse. *Biol. Reprod.* **52,** 709–720.

Kurokawa, M., and Fissore, R. A. (2003). ICSI-generated mouse zygotes exhibit altered calcium oscillations, inositol 1,4,5-trisphosphate receptor-1 down-regulation, and embryo development. *Mol. Hum. Reprod.* **9,** 523–533.

Lanzendorf, S. E., Maloney, M. K., Veeck, L. L., Slusser, J., Hodgen, G. D., and Rosenwaks, Z. (1988). A preclinical evaluation of pronuclear formation by microinjection of human spermatozoa into human oocytes. *Fertil. Steril.* **49,** 835–842.

Nagy, A., Gertsenstein, M., Vintersten, K., and Behringer, R. (2003). Manipulating the Mouse Embryo—A Laboratory Manual. 3rd edn. Cold Spring Harbor Laboratory Press, Cold Spring Harbor, NY.

Palermo, G., Joris, H., Devroey, P., and Van Steirteghem, A. C. (1992). Pregnancies after intracytoplasmic injection of single spermatozoon into an oocyte. *Lancet* **340,** 17–18.

Sasagawa, I., Kuretake, S., Eppig, J. J., and Yanagimachi, R. (1998). Mouse primary spermatocytes can complete two meiotic divisions within the oocyte cytoplasm. *Biol. Reprod.* **58,** 248–254.

Tateno, H., Kimura, Y., and Yanagimachi, R. (2000). Sonication per se is not as deleterious to sperm chromosomes as previously inferred. *Biol. Reprod.* **63,** 341–346.

Uehara, T., and Yanagimachi, R. (1976). Microsurgical injection of spermatozoa into hamster eggs with subsequent transformation of sperm nuclei into male pronuclei. *Biol. Reprod.* **15,** 467–470.

Wakayama, T. (2007). Production of cloned mice and ES cells from adult somatic cells by nuclear transfer: How to improve cloning efficiency? *J. Reprod. Dev.* **53,** 13–26.

Yanagimachi, R. (2005). Intracytoplasmic injection of spermatozoa and spermatogenic cells: Its biology and applications in humans and animals. *Reprod. Biomed. Online* **10,** 247–288.

SECTION FIVE

ES AND IPS CELLS

CHAPTER FIFTEEN

A Simple Procedure for the Efficient Derivation of Mouse ES Cells

Esther S. M. Wong,*[,1] Kenneth H. Ban,†[,1] Rafidah Mutalif,* Nancy A. Jenkins,† Neal G. Copeland,† *and* Colin L. Stewart*

Contents

Abstract

Embryonic stem (ES) cells were first derived from inner cell mass (ICM) explants of preimplantation stage mouse blastocysts some 30 years ago. ES cells are of primary interest as they are used to genetically modify the genome of mice via gene targeting. Although many founder ES lines have been established, there is still a need to obtain new ES lines or their derivatives, often from new mutant mouse lines, to study the function of a mutated gene in different cell types. Existing methods for isolating ES cell lines are inefficient. Here, we describe a

* Institute of Medical Biology, Immunos, Singapore
† Institute of Molecular and Cell Biology, Proteos, Singapore
[1] These authors contributed equally to the work.

Methods in Enzymology, Volume 476
ISSN 0076-6879, DOI: 10.1016/S0076-6879(10)76015-8

reproducible, efficient, and economical method to derive ES cells from different mouse strains using a defined serum-free, serum replacement (KO-SR) media, with 50–85% efficiency. We have derived over 100 ES lines, which when karyotyped >70% were euploid. Two of these lines, when tested, produced germ-line chimeras. We also present procedures for the routine maintenance and karyotyping of the ES cells.

1. Introduction

Embryonic stem (ES) cells are derived from the inner cell mass (ICM) of mouse and rat blastocysts. ES cells are pluripotent in that they can give rise to both the entire somatic and germ-line lineages of chimeric embryos, with the exception of some extraembryonic tissues, such as the trophectoderm and extraembryonic endoderm. Mouse ES cells were first derived in 1981 (Evans and Kaufman, 1981; Martin, 1981). Rat ES cells have only recently become available, primarily due to an increased understanding as to which signaling pathways govern rodent ES cell growth and differentiation (Buehr *et al.*, 2008; Li *et al.*, 2008; Ying *et al.*, 2008).

1.1. The need for improved methods of ES cell derivation

The primary interest in ES cells is that they are the principal route to genetically modifying the genome of mice. Specific mutations are introduced into genes in the ES cells by the procedure generally known as gene targeting. The genetically altered ES cells are then used to derive mice carrying the modified gene by making germ-line chimeras. From these chimeras, new mouse lines are established carrying the altered gene. In recent years, research on ES cells has broadened due to a greater interest in understanding the molecular basis of pluripotentiality or "stemness" (Chen *et al.*, 2008; Cole and Young, 2008; Takahashi and Yamanaka, 2006). In addition, extensive efforts are being made with mouse ES cells, to model tissue regeneration, and to develop protocols whereby differentiated cells derived from ES cells may be used in the therapeutic replacement of diseased or damaged tissues (Murry and Keller, 2008).

Despite the existence of a few mouse ES cell lines that have been widely distributed to many labs (Kontgen *et al.*, 1993; Simpson *et al.*, 1997), and having served as the workhorses in gene targeting, there is still a need to establish new ES lines—often from new mutant mouse lines, so that the function of the mutated gene can be studied in different cell types *in vitro*.

1.2. Overview of previous methods for deriving and culturing ES cells

Previous methods for isolating new ES cell lines were largely based on standard protocols established by Robertson and colleagues (Abbondanzo *et al.*, 1993; Bradley *et al.*, 1984; Robertson, 1987). These methods were generally inefficient, resulting in yields ranging between 0% to at best 50% of blastocysts giving rise to new ES lines; with mouse strain and genetic background being a significant influence on the frequency of derivation. However, with a deeper understanding of the factors that regulate ES growth and differentiation, more efficient procedures have recently been established.

Originally, ES cells were derived by isolating the ICM from explanted blastocysts, by first allowing the blastocysts to attach to a tissue culture (TC) dish. Within 4 days of isolation, the trophoblast would outgrow, with the ICM appearing as a clump of cells on top of the trophectoderm (Fig. 15.1A). The ICM would then be picked and disaggregated into individual or small clumps of cells by brief trypsinization. The disaggregated ICM was then placed onto a feeder layer of fibroblasts, either primary mouse embryonic fibroblasts (PMEFs) or immortalized lines such as STO cells. The explanted ICM cells on the feeder layer were cultured in standard culture medium (Dulbecco's modified Eagle's medium (DMEM) or alpha-medium) supplemented with 15% fetal calf serum (FCS), with the serum being carefully batch screened to support established ES cell growth. Under ideal conditions, colonies of ES cells would appear within a week, which could be expanded into new ES cell lines.

Undifferentiated growth of ES cells is dependent on PMEFs, since removal of the feeder layer results in the rapid differentiation of ES cells. The feeders produce the cytokine leukemia inhibitory factor (LIF)-which was the sole factor required to maintain the proliferation of ES cells in DMEM + FCS, in the absence of feeders (Smith *et al.*, 1988; Stewart *et al.*, 1992; Williams *et al.*, 1988).

Subsequent studies revealed that the FCS used to support ES cell growth was, paradoxically, a major factor promoting their differentiation. LIF treatment of the ES cells activates the JAK/STAT3 signaling pathway which inhibits the differentiation-inducing activities of FCS-primarily caused by FGF and BMP factors present in the serum, coupled to the autocrine production of FGF by the ES cells that drives ES differentiation via activation of the ERK pathway. By inhibiting, FGF receptor FGFr/ERK pathway signaling, using small molecule inhibitors, together with a highly specific GSK-3β inhibitor, whose precise role is still somewhat unclear, ICM explants can be converted into ES colonies. These colonies can then be expanded into cell lines and maintained in a defined serum-free medium (N2B27) with the attachment factor, laminin. Such a method for

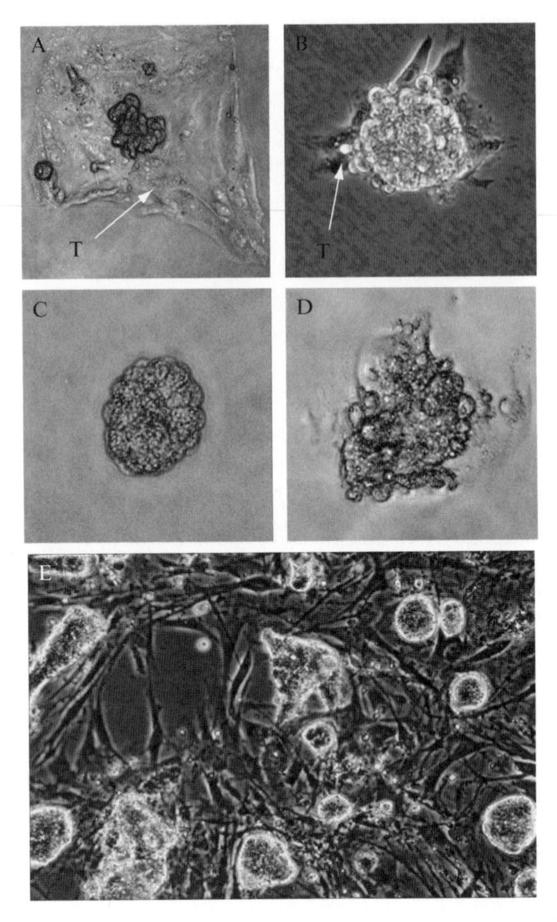

Figure 15.1 (A) Blastocysts cultured in DMEM + FCS show extensive trophoblast outgrowth, whereas in SR medium, (B–D), attachment to the dish surface is reduced with minimal trophoblast outgrowth. (E) ES colonies expanding on MEF feeders - derived from disaggregated ICM from blastocyst cultured in SR medium.

the derivation and maintenance of ES cells is referred to as the "3i procedure." Of the 3 inhibitors (i), two blocked FGFr-mediated activation of the ERK pathway: SU5402, an FGFr inhibitor and PD184352, inhibitor of MEK activation. The third inhibitor CHIR99021, blocks GSK-3β activity, which may enhance canonical Wnt signaling and potentially enhance global metabolomic and biosynthetic activity in ES cells (Ying *et al.*, 2008). More recently, the procedure has been refined to the "2i + LIF" procedure-in which the two FGFr/MEK inhibitors are replaced with a single more potent MEK inhibitor PD0325901, and is used in combination with the GSK-3β inhibitor and LIF (Silva *et al.*, 2008).

While the small molecule inhibitors and serum-free defined medium has improved the efficiency of establishing ES cell lines, even from previously recalcitrant strains such as NOD mice (Nichols *et al.*, 2009), there is concern that the dependence on a GSK-3β inhibitor may enhance chromosomal instability in the ES lines. GSK-3β inhibition results in significant chromosomal nondisjunction due to impaired chromosomal alignment at mitosis in both mouse preimplantation embryos and in cultured HeLa cells (Acevedo *et al.*, 2007; Tighe *et al.*, 2007). Secondly, the inhibitors, LIF, serum-free medium, attachment factors, and supplements needed to maintain the ES cells are relatively expensive. Also, immunosurgery is required to isolate ICMs from the trophectoderm prior to their culture in defined medium (Solter and Knowles, 1975).

As an alternative, we developed a simplified version of a previously described procedure, which we found to be very efficient at establishing ES cell lines from C57BL/6J and 129/J strains (Bryja *et al.*, 2006). Furthermore, it is also effective in deriving ES lines from other strains, such as FVBN and the outbred Swiss and CD1 lines, where ES cell establishment had proved to be difficult using previous methods.

This method involves the initial culture of blastocysts in the commercially available serum replacement (SR) medium. With this media, blastocyst attachment and outgrowth on TC dishes is reduced and seemingly inhibits trophoblast proliferation (Fig. 15.1B–D). The resulting embryonic balls are disaggregated and the embryonic cells plated onto PMEFs overnight in DMEM + FCS, with the transient exposure to FCS providing attachment factors, permitting ES cell adherence to the feeder layer. Within 24 h after attachment, FCS is replaced with the SR medium and the ES colonies are expanded into cell lines. We then karyotype each line we intend to use (as in our previous experience, 30–40% of primary ES lines maybe aneuploid).

This method has the advantage in that the media used are readily available. Additionally, the use of defined serum-free, SR medium reduces issues of batch variability of FCS, and any undefined, differentiation promoting factors that may impede ES cell derivation. Moreover, LIF is only transiently required during the initial stages of blastocyst culture and we dispense with its use once ES cell lines are established on feeders. Lastly, there is minimal handling of the embryos.

2. Methodology/Workflow

2.1. Source of E3.5 embryos

- C57BL/6J (or expt strain) female mice age 4–6 weeks, weighing 16–20 g
- Stud males 12–16 weeks of age

2.1.1. Superovulation

Day 1: Dissolve 10,000 IU of the powdered form of pregnant mare's serum (PMS; Sigma-Aldrich, G-4877) in 10 ml sterile 0.9% saline (store in aliquots at -20 °C); inject 100 μl (1000 IU) intraperitoneally per mouse, using a 30-gauge needle, at 16:00–18:00 h.

Day 3: Dissolve 10,000 IU of the powdered form of human chorionic gonadotrophin (hCG; Sigma-Aldrich, CG-5) in 10 ml sterile 0.9% saline (store in aliquots at -20 °C); inject 100 μl (1000 IU) intraperitoneally per mouse, using a 30-gauge needle, at 16:00–18:00 h (~48 h post-PMS injection). Set up paired matings immediately after hCG injection.

Day 4: Morning — check for copulation plug.

2.1.2. Dissecting mouse uteri

- *Essential equipment*:
  A good stereo dissection microscope with 40× magnification
  A TC microscope equipped with phase-contrast optics (10–40×)
- *Standard TC facilities*:
  A sterile/filtered air culture hood
  A 37 °C, 5% CO_2-gassed incubator
- Fill 60 mm TC dishes each with 4 ml PBS (Ca^{2+}/Mg^{2+}-free) at RT.
- Fill additional 60 mm TC dishes with 4 ml HEPES-buffered M2 media (EmbryoMax®, Millipore, MR-015D)-equilibrated at 37 °C.
- *Dissection tools*:
  Forceps-World Precision Instruments, #501985
  Spring scissors-SuperFine Vannas Scissors (WPI, #501778)
  Curved scissors-Iris Scissors, SuperCut (WPI, #14218)

Day 7: Flushing blastocysts from uteri of Day 4 (day of plug ≡ equivalent to Day 1 of pregnancy) between 09:00 and 11:00 h:

1. Euthanize the females.
2. Wipe a 70% ethanol (EtOH) solution over the abdomen to disinfect the skin.
3. Cut the abdominal skin with sharp scissors, grasp the skin at site of incision, and tear longitudinally to expose the underlying peritoneum.
4. Grasp the peritoneum with forceps and make a longitudinal incision.
5. Push the intestines to one side to expose the uteri and trim off as much fat tissue from the uteri.
6. Locate the ovary using forceps-noting whether redish corpora lutea are present, as they are indicative that the female had ovulated and that blastocysts are potentially present in the uteri. Grasp the uterus near the oviduct and separate from the ovary, using a pair of fine spring scissors.

7. Holding onto the ovarian end of the uterus with forceps, carefully trim away, using fine curved scissors, the mesentery containing the blood vessels supply to the uterus, downward toward the cervix.
8. Make a second cut at the cervix, and continue separating the mesenchyme upward, and away from the contralateral uterus toward the ovary. Remove the entire uterus by cutting through the oviduct near the remaining ovary.
9. Wash the uteri in two changes of Ca^{2+}/Mg^{2+}-free PBS to remove bits of fat, loose tissue, and wash away any blood. Transfer to a 60-mm dish of prewarmed M2 media.

2.1.3. Flushing and isolating blastocysts

- Prepare droplets (5–10 μl) of M2 media (EmbryoMax®, Millipore, MR-015D) (at 37 °C) in a 35-mm Petri dish. Flood the dish with mineral oil (Sigma-Aldrich, M8410) over the droplets to prevent evaporation.
- Prepare drops (10–20 μl) of Acid Tyrode's (EmbryoMax®, Millipore, MR-004-D) overlaid with mineral oil on a 35-mm dish and preequilibrated at 37 °C.
- Prefill a 3-ml syringe with M2 media, attach a 27-gauge needle.
- A mouth-controlled pipette (BioMedical Instruments)-for embryo transfer.
- Hard glass capillary tubes (BDH Ltd, #321242C).

  To pull capillary to the required diameter, the center of the tube is rotated over a 70% alcohol burner until it begins to melt. The tube is promptly withdrawn from the flame while pulling both ends sharply in opposite directions-so stretching center of the tube into a fine, narrow diameter, and causing a break (into two) near midsection. With practice, the correct degree of force can be appropriately applied to obtain any desired internal diameter of the pipette. Hard glass capillaries usually give a clean vertical break without any jagged edges-inspect under a dissecting scope to check for bad (nonvertical) ends. The capillary tip is polished by holding tip opening close to the edge of the cooler, base of the flame.
- Alternatively, ready-to-use glass pipettes of varying internal diameters are commercially available (BioMedical Instruments, www.pipettes.de)

1. Under a dissecting scope, cut away any remaining extraneous tissue/fat from the uteri in a 35-mm dish containing 1–2 ml of prewarmed M2 medium.
2. At both uterine–oviductal ends make a single 2 mm longitudinal cut, splaying the uterus. This prevents the ends from closing due to uterine smooth muscle contraction.
3. Holding one uterine horn with forceps close to the cervical end, insert the 27-gauge needle prefilled with M2 into the cervix and flush.

The uterus becomes elongated and dilates, with the embryos being expelled from the splayed oviductal end.

4. Repeat the flushing of the other uterine horn into the same dish. Inspect bottom of the dish for blastocysts, and mouth-pipet the blastocysts into a fresh 35 mm dish of M2 to remove any blood cells and uterine tissue fragments.
5. Transfer the washed blastocysts into a microdrop of prewarmed Acid Tyrode's to remove the zona pellucida. This can take between 15–60s for the zona to dissolve and should be monitored with the binocular microscope.
6. Mouth-pipet and wash the dezonulated blastocysts two to three times in drops of M2 to remove Acid Tyrode's.

2.2. Embryo attachment and outgrowth: Optimal culture

The effective growth of ES cells requires that all culture media be made with pure, deionized water. A variety of commercially available media are used to culture embryos and ES cells, namely, DMEM, Glasgow modified Eagle's medium, and DMEM/Ham's F12 mixture.

Mouse and rat ES cells are dependent on the cytokine LIF to maintain them as an undifferentiated proliferating population. The cytokine is supplied by growing the cells on a feeder layer of fibroblasts that produce LIF. Recombinant LIF is also commercially available. ES cells may also be established and maintained in the absence of feeders, either using the 3i or 2i + LIF protocols with defined media formulation and the attachment factor, laminin (Nichols *et al.*, 2009; Ying *et al.*, 2008).

- *ES-DMEM* media: KO-DMEM + 15% KO-SR + L-Glut + NEAA + P/S + βME
  KO™-DMEM, high glucose-4.5 g/l (Gibco® Invitrogen, #10829-018)
  KO-SR (Gibco® Invitrogen, #10828-028)
  L-Glutamine (Gibco® Invitrogen, #25030-081)-2 m*M* final concentration
  MEM nonessential amino acids (Gibco® Invitrogen, #11140-050)-0.1 m*M*
  P/S, penicillin/streptomycin (Gibco® Invitrogen, #15140-122)-50IU/ml
  β-Mercaptoethanol (Gibco® Invitrogen, #21985-023)-0.1 m*M*
- hLIF (ESGRO®, Millipore, #LIF1010) 1000 U/ml, to be added before use

1. Transfer blastocysts into a 35-mm dish with 2 ml ES-DMEM + hLIF preequilibrated at 37 °C, 5% CO_2.
2. Incubate at 37 °C, 5% CO_2. Do not disturb for ~48 h to allow the blastocysts to partially attach to the culture dish. One to two days later, that is, 3–4 days after explanting the blastocysts, prepare the embryos for trypsinization.

2.3. Embryo outgrowth/colony disaggregation: Initial culture

ES cells are sensitive to the type and quality of serum in which they are grown. As the quality of the FCS can vary from batch to batch, it is essential to test each batch to ensure that it is optimal for ES cell growth.

The procedure to identify an optimal batch of FCS is based on a "colony-forming assay"-where each batch is tested to determine the maximum number of ES colonies that grow in a fixed concentration of serum after plating. Single cell suspensions (3.0×10^3 ES cells) are plated onto 60 mm TC dishes precoated with a layer of feeder cells. The single cells will attach and proliferate to form ES colonies over the next 5 days, which are then scored. This assay is performed in duplicates for each batch of serum, at two different concentrations: 15%, the usual concentration used for ES cell growth; and at 30%, to determine if there is any residual toxicity associated with that batch. A good batch of serum should yield 30% or more of the ES cells forming colonies, at both 15% and 30% FCS concentrations.

- *M-DMEM* media: DMEM + 10% FCS + L-Glut + NEAA + P/S + βME
  DMEM, 4.5 g/l D-glucose w/o L-Glut (Gibco® Invitrogen, #31053-028)
  FCS (Gibco® Invitrogen, #10437-028)
  L-Glutamine (Gibco® Invitrogen, #25030-081)-2 m*M* final concentration
  MEM nonessential amino acids (Gibco® Invitrogen, #11140-050)-0.1 m*M*
  P/S, penicillin/streptomycin (Gibco® Invitrogen, #15140-122)-50 IU/m
  β-Mercaptoethanol (Gibco® Invitrogen, #21985-023)-0.1 m*M*
- Preseed 4-well TC plates (Nunc, #176740) with mitomycin-C (Mit-C) arrested PMEFs (iMEFs-see Section 2.4.2; at 37 °C, 5% CO_2) at 5.0×10^5 cells/well in M-DMEM
- ES-DMEM media: KO-DMEM + 15% KO-SR + L-Glut + NEAA + P/S + βME
- hLIF (ESGRO®, Millipore, #LIF1010)-1000 U/ml, added to the stock media.
- 0.25% trypsin (Gibco® Invitrogen, #25200-114)

- Sterile PBS, Ca^{2+}/Mg^{2+}-free (Gibco® Invitrogen, #14190-250)
- Two sets of micropipettes to pick attached blastocysts:
  1 wide bore 90–100 μm
  1 narrow bore 25–30 μm, flame-polished tip

1. Place 20 μl drops of trypsin onto a 60-mm TC dish-overlay with mineral oil.
2. Using a wide bore pipette prefilled with medium from the dish containing the blastocysts, gently lift off an embryo (Fig. 15.1B–D) and transfer to a drop of trypsin under mineral oil. Incubate for at least 5 min (up to 30 min) at 37 °C, 5% CO_2, till the embryos start to disaggregate.
3. Switch to a narrow bore pipette prefilled with M-DMEM media from the iMEF plate, repeatedly pipet the embryos/blastocyst up and down till the cell mass disaggregates into small clusters of some four to five cells.
4. Transfer the entire disaggregated embryo onto a 16-mm 4-well dish preseeded with iMEFs in M-DMEM + 15% FCS + hLIF. Incubate at 37 °C, 5% CO_2 for 24 h. After 24 h, change the medium to ES-DMEM. Colonies of ES cells should appear within 3–4 days. (To prepare iMEFs, see Section 2.4.2.)

2.4. ES cell expansion, growth and maintenance: Routine culture

- Seed a 60-mm dish with 1.0×10^6 cells iMEFs (M-DMEM) 1-day prior to cell expansion.

1. On the day of expansion, change iMEF media to fresh M-DMEM + hLIF.
2. To expand from a 16-mm 4-well to a 60-mm plate-wash each well with 500 μl PBS, add 200 μl 0.25% trypsin and incubate at 37 °C for 5 min.
3. Add 500 μl M-DMEM (+10% FCS) to neutralize the trypsin.
4. Using a sterile 1 ml Gilson tip and Gilson, pipet the trypsinized cultures up and down a few times to disaggregate the fibroblasts and ES colonies.
5. Transfer the entire contents of the well onto a 60-mm dish preplated of iMEFs (in M-DMEM + hLIF)-incubate at 37 °C, 5% CO_2 for overnight.
   NB: Switching back to M-DMEM (+FCS) is required for cell attachment (~24 h).

   The following day, ES cells would have attached-replace the M-DMEM to ES-DMEM + hLIF.

6. Upon confluency (~3–5 days), ES cells can be expanded onto a 100-mm dish, with LIF being no longer required as a supplement.

7. ES cells can be frozen down in freezing media (refer to Section 2.6) in 10% dimethyl sulfoxide (DMSO; Sigma-Aldrich, D2650) + 90% FCS at −80 °C.

2.4.1. Primary mouse embryonic fibroblast (PMEF) derivation

PMEFs can be made from any strain of mouse, with mouse lines being available that constitutively express many selectable marker genes. These mice are a source of feeders that are resistant to a variety of drugs used in the various selection schemes employed in gene targeting, such as neomycin, puromycin, etc. (Tucker *et al.*, 1997). We have also found, with 25 years of experience, that without addition of LIF to the culture medium is sufficient to maintain, clonally select for and expand ES cells on inactivated PMEFs (iMEFs). Supplementing the culture medium with recombinant LIF to maintain ES cells on an inactivated PMEF feeder layer has proven to be unnecessary and is a waste of resources.

Materials required-refer to Section 2.1.2.

Day 0: Setup paired matings, at 16:00–18:00 h.
Day 1: Check for copulation plug (equivalent to Day 1 of pregnancy). Remove plugged females to a separate cage.
Day 13: Sacrifice the pregnant females.

1. Open the peritoneal cavity and remove the uteri with the embryos.
2. Dissect embryos from uteri in Ca^{2+}/Mg^{2+}-free PBS, remove yolk sac, amnion, placenta. Wash the embryos twice in fresh Ca^{2+}/Mg^{2+}-free PBS to remove traces of blood.
3. Using forceps, pinch off the head, pinch out and remove the liver-discard both tissues.
4. Dissagregate the remaining embryonic tissues by first, placing the embryos in the barrel of a sterile 3 cc syringe with an 18-gauge needle. Break up the embryos in the M2 by repeatedly syringing the embryos through the needle about five times.
5. The disaggregated embryos are then incubated in 2 ml of 100 μg/ml DNAseI (Sigma, D-4527) + 500 μg/ml Collagenase IV (Sigma, C-9407) in DMEM/M2.
6. Leave the disaggregated embryos to incubate in a 37 °C water bath for 30 min, mixing occasionally to produce a single cell suspension.
7. Centrifuge at 1000 rpm for 5 min, and remove collagenase/DNAse solution from the cell pellet.
8. Resuspend the cell pellet well in 20 ml M-DMEM media. Add cell suspension to a 150-mm TC dish (or flask) and incubate at 37 °C, 5% CO_2. The cells will attach almost immediately and produce fibroblasts (PMEFs) over the next 2–3 days. The culture should reach confluency by the third day.

9. Wash the culture dish twice in Ca^{2+}/Mg^{2+}-free PBS, and add 3 ml of 0.25% trypsin + 1 m*M* EDTA in PBS + 1% (v/v) chicken serum (Gibco® Invitrogen, #16110-082). Trypsinize cells, spin down at 1200 rpm for 5 min, and transfer the resuspended single cells to three new 150 mm dishes.
10. After further culture, the PMEFs should be split at no more than a ratio of 1:3. PMEFs are only good as a feeder layer for 5–6 passages (P5–P6) under standard culture conditions in 5% CO_2 in air, since they cease to proliferate around P5 due to the PMEFs being sensitive to ambient O_2 tension that induces a cellular stress response and proliferative arrest (Parrinello *et al.*, 2003). Long-term cultures that continuously proliferate can however be maintained under hypobaric conditions using a hypobaric incubator.
11. Once sufficient numbers are established, PMEFs can be prepared as feeder layers (Section 2.4.2) or frozen for future use.

2.4.2. Inactivated PMEF (iMEF) feeders: Mitomycin-C treatment of PMEFs

As feeders, it is necessary to stop PMEF proliferation so that they do not become transformed or contaminate the ES cells. This is achieved in two ways: If available, a source of γ-radiation (600 rads) is used to inactivate PMEFs. More commonly, Mit-C is used to treat a proliferating population of PMEFs, by inhibiting mitosis—thus becoming growth inhibited PMEFs or iMEFs.

1. Mitomycin-C (Sigma-Aldrich, M0503)-2 mg powdered form dissolved in 5 ml PBS. Filter sterilize.
2. Prepare 10 μg/ml Mit-C working solution in 195 ml of M-DMEM + 10% FCS. Aliquot into 10 ml volumes, store frozen at −20 °C in the dark. *NB. Mit-C is light-sensitive-its activity decreases with exposure to light.*
3. Aspirate culture medium from a near confluent 150 mm dish of proliferating PMEFs. Replace with 10 ml of Mit-C solution and return dish to the incubator for 3 h, sufficient time for Mit-C to permanently arrest PMEF proliferation.
4. Following a minimal 3 h (up to 5 h) treatment, remove the Mit-C-containing medium-PMEFs may be trypsinized for reseeding onto smaller TC dishes, as feeders (iMEF) or iMEFs can be washed twice with fresh medium to remove traces of Mit-C, refed with 20 ml of M-DMEM + 10% FCS, and kept for up to 1 week before use.
5. For seeding into smaller culture dishes, iMEFs are washed twice in PBS, trypsinized, the cells are pelleted by centrifuging at 1000 rpm for 5 min at RT, resuspended in 10 ml of M-DMEM + 10% FCS, and counted using a hemocytometer or Coulter counter.

6. TC dishes on which the iMEFs are to be seeded at this stage are pretreated with 0.1% (w/v) gelatin solution (Sigma-Aldrich, G2500)-prepared by dissolving 0.5 g of swine skin type III collagen in 500 ml distilled water (dH_2O) and sterilized by autoclaving gelatin coats TC plating surface, thus enhancing the attachment of feeders and ES cells.
7. To gelatinize a culture dish, a small volume of sterile gelatin solution is added to adequately cover the plating surface-kept at 37 °C for 15–30 min. Excess solution is removed and the treated dishes can be stored at 4 °C.
8. Gelatin is removed from the gelatinized dishes and sufficient iMEFs are seeded onto the dish to form a uniform monolayer after attachment. The number of cells varies according to surface area of the culture dish:
   100 mm dish: 1.7–2.0 $\times$ 10^6 cells; 60 mm dish: 1.0 $\times$ 10^6cells;
   35 mm dish: 5.0–6.0 $\times$ 10^5 cells; 16 mm well: 1.0 $\times$ 10^5 cells.
9. The iMEFs are left to attach in M-DMEM for 6–12 h prior to addition of ES cells. The feeder layers prepared in this manner are good for up to 2 weeks, after which they are discarded.

2.4.3. Feeder-free culture of ES cells

Under certain circumstances (e.g., metabolic labeling of proteins in ES cells or preparation of mRNA for cDNA libraries), it is preferable to obtain ES cells without any fibroblast feeder layer contaminants. To do this, it is necessary to culture the ES cells in the absence of any feeders for 1–2 passages in order to dilute out residual fibroblasts. ES cells can be grown in absence of feeders on gelatinized dishes, provided that LIF is included in the medium. Substitution of feeder cells by LIF alone appears to work reasonably well for some ES cells that were originally maintained on feeder layers. However for other lines, it does not work as well, with the ES cells undergoing extensive differentiation.

An effective way to inhibit differentiation and grow almost 100% pure populations of stem cells is to use an extracellular matrix (ECM) deposited by the feeder layer as a substitute for gelatin. ES cells grown on an ECM in the presence of exogenous LIF retain their stem cell phenotype, with very little or no differentiation. Furthermore, the ES cells can be passaged through multiple rounds without differentiating, provided that they are replated onto such matrix with exogenous LIF.

Preparation of ECM for feeder-free culture of ES cells:

1. Seed appropriate numbers of proliferating PMEFs (not Mit-C treated) onto a gelatinized dish in M-DMEM + 10% FCS.
2. PMEFs are cultured to confluency over 3–4 days with the fibroblasts secreting and depositing an ECM as they proliferate.
3. To remove the cells without disrupting the matrix, wash the fibroblast layer twice with PBS, add lysis buffer just sufficient to cover the cells

~0.5–1.0 ml for a 60-mm dish. Lysis buffer is *freshly made* 0.5% (v/v) deoxycholic acid (Sigma-Aldrich, D6750) in PBS.

4. Incubate the dish at RT for 5–10 min. The cell membranes lyse with release of the cytoplasmic contents; the nuclear membranes remain intact, allowing the nuclei to be removed without releasing their DNA. The ECM remains attached to the culture dish and appears as fibrous strands.
5. Gently wash off the cellular contents-nuclei, detergent, etc., with six to eight changes of PBS (plus Ca^{2+}/Mg^{2+}). Then add DMEM—high glucose (4.5 g/l) with no supplemented FCS and incubate the ECM coated dish overnight.
6. The ECM dish can now be used to support the undifferentiated proliferation of ES cells in ES-DMEM + hLIF. The feeder-free ES culture should be refed every 2 days. ES cells can be passaged onto freshly prepared plates of lysed feeders (ECM) to dilute out remaining feeder contaminants, and expanded as an almost pure, undifferentiated population.

2.4.4. Passaging ES cells

The ES cells are maintained as an undifferentiated population by trypsinizing and replating the cells onto dishes containing fresh feeders, every 5–6 days if the cells are plated out at a sufficiently low density. A 60-mm dish at maximum density will contain about $1.0–2.0 \times 10^7$ ES cells, and a 150-mm dish contains up to $2.0–3.0 \times 10^8$ cells at maximal density. The cells will start to differentiate or die if they are maintained beyond the maximum density level, with the optimal period of time ES cells can be maintained before passaging being 5–7 days.

To maintain a line, trypsinize a semiconfluent dish and plate out single cell suspension at a 1:100–500 dilution. If the cells are replated at a reasonably low density, the culture medium needs to be changed every other day to keep cells under optimal conditions. If more cells and higher densities are required, then the cells should be refed every day. Under optimal conditions, the ES cells should grow as small clusters or mounds (Fig. 15.1E). If the conditions are suboptimal, differentiated derivatives will appear, and the mounds of ES cells will flatten out, with individual cells becoming more distinct. Under extreme conditions, the majority of the cells will have differentiated.

2.5. ES cell characterization: Karyotyping and C-banding

The mouse has 40 chromosomes: 19 pairs of autosomes, and 2 sex chromosomes. Unlike human counterparts, mouse chromosomes do not exhibit a great variation in size, making the identification of specific autosomes more difficult. In our experience a significant proportion (30–50%) of newly established ES cell lines do not contain a normal diploid (euploid) karyotype. If the intention is to use the ES cells for genetic manipulation of the mouse germ line, it is crucial to identify ES cell lines, preferably males, with a euploid karyotype (n=40). This is achieved by making *in situ* metaphase chromosome spreads of ES cells. Any Robertsonian translocation can be easily detected, as can loss or gain of a chromosome(s).

Determination of the sex is relatively straightforward using the C-banding technique, which highlights the regions of constitutive heterochromatin, principally the centromeres. All autosomes in the mouse have a darkly stained centromere, as does the X-chromosome. The centromere of the Y-chromosome does not stain as darkly, and a male karyotype can be recognized by lack of staining on one of the three smallest chromosomes visible in a typical spread. The other two chromosomes that stain are both copies of chromosome 19, which are about the same size as the Y-chromosome.

2.5.1. Karyotyping

1. Sterilize standard 22 × 22 mm microscope cover-slips by soaking in 100% EtOH. Dry by flaming over a burner in the TC hood.
2. Place a cover-slip into a 35-mm culture dish.
3. Trypsinize a 60-mm dish of ES cells in log growth phase-visible ES colonies.
4. Seed about 200–400 ES cells (as a single cell suspension) onto the cover-slip. Culture for 48 h in ES-DMEM + hLIF-allow ample time for ES cells to attach.
5. To arrest cells in mitosis, add 50 μl/ml Colcemid (Gibco® Invitrogen, #15212-012) to each dish and incubate at 37 °C, 5% CO_2 for 45 min.
6. Aspirate the medium. Carefully add-in dropwise, 2 ml of 0.8% sodium citrate (Sigma-Aldrich, S1804) hypotonic solution, incubate for 20 min at RT-this causes cells and nuclei to swell, thus releasing the mitotic chromosomes onto surface of the cover-slip. Gently add 1 ml of fresh fixative (3:1 MeOH:glacial acetic acid) to each dish.
7. Aspirate the hypotonic/fixative solution from the edge of each dish. Add 2 ml fixative to fix chromosomes onto the cover-slip, leave for 30 min at RT. Repeat the fixation twice, leaving for 20 min each time at RT.
8. Aspirate the fixative, and dry the cover-slip immediately over a spirit lamp.
9. Carefully remove cover-slip from the Petri dish, keeping it cell side up-label with a permanent marker, leave on slide warmer set to 60 °C for 18 – 24 h. The fixed ES cells can now be processed for cytogenetic analysis.

2.5.2. C-banding analysis

1. Bring 100 ml of distilled water (dH_2O) to boil.
2. Add 0.5 g of barium hydroxide (Sigma-Aldrich, #433373) to dissolve. Cool solution to 55 °C and remove any surface scum with filter paper.
3. Immerse the cover-slip (fixed ES cells) for 2–3 min in $Ba(OH)_2$ solution.
4. Rinse the cover-slip thrice with dH_2O to remove excess $Ba(OH)_2$.
5. Incubate the cover-slip at 50 °C for 2 h in 2× SSC (Promega, V4261); SSC: 30 m*M* NaCl, 30 m*M* trisodium citrate.
6. Rinse the cover-slip twice in dH_2O.
7. Stain with 5% Giemsa (Gibco® Invitrogen, #10092-013) for 5–10 min. Rinse in dH_2O to remove excess stain and allow the cover-slips to dry.
8. Mount cover-slips in DPX (Sigma-Aldrich, #44581).
9. Examine 30–40 chromosome spreads, count and record chromosome numbers, using a microscope with 200× magnification. This should be sufficient to give an accurate description of the ES line's karyotype, that is, what percentage of cells has a normal diploid chromosome number. Once the karyotype and sex have been verified, an ES cell line can then be tested for germ-line transmission.

2.6. ES storage: Freezing down cell lines

ES lines with a high percentage (80–95%) of cells with a normal diploid karyotype should be expanded, such that as many early passage cells can be frozen down in stock. This ensures an ample resource for future experiments, since early passage ES cells tend to form better chimeras, albeit at a higher frequency than if P15–P20 or later passages are used. However, there is no absolute correlation, since some lines such as E14 produce germ-line chimeras at P52 (C. L. Stewart, personal observation).

1. A culture of ES cells to be frozen down should be in the log growth phase, that is, not at maximal density-wash the dish twice in PBS and trypsinize.
2. Harvest the cells, resuspend in M-DMEM (with 10% FCS to neutralize trypsin), count cells with a hemocytometer.
3. Freezing media: FCS + DMSO (Sigma, D2650)-10% (v/v).
4. 1.0 ml freezing media containing 1.0–5.0 × 10^6 ES cells is aliquoted into a sterile 1.8 ml freezing vial (Nunc, #368632).
5. The vials are labeled with the ES line and passage number, placed in a slow freezing container or polystyrene holding rack, and left overnight in a −70 °C freezer.

6. The following day, transfer vials of frozen cells to liquid nitrogen for long-term storage.
7. To thaw out a vial of ES cells, preplate a 60-mm TC dish of iMEFs in M-DMEM.
8. Remove a stock vial of ES cells-place in a beaker of sterile dH_2O warmed to 37 °C until contents of the vial have visibly thawed.
9. Remove the vial, swab with 100% ethanol to sterilize the outside surface, pipet out the cell suspension using a sterile Pasteur pipette, and seed cells immediately onto the 60 mm feeder dish refed with ES-DMEM + LIF.
10. The next day, the culture medium is replaced with fresh ES-DMEM + LIF, to remove all traces of DMSO and any dead cells. If freezing and thawing of the ES cells were performed correctly, ES colonies should already be visible in the culture dish.

3. Concluding Remarks and Future Perspective

Here, we present a reproducible, efficient, and economical method to derive ES cells from different mouse strains (five listed) using a defined serum-free, serum replacement (KO-SR) media, with strain-dependent efficiency ranging from 50% to 85% (see Table 15.1). As evident from the good yield, euploid karyotypes and germ-line capabilities of the ES cell lines, this method in deriving ES cell lines may be generally applicable to a wide range of mouse strains.

Table 15.1 Summary of ES cell derivation from listed mouse strains

Mouse strain	No. of blastocysts explanted	No. of ES cell lines established	Efficiency of ES cell derivation (%)	Euploidy$^{n=20}$ (%)
129/B6 × C57BL/6J	17, 23, 20	9, 18, 12	52–78	74–100
129S3 × C57Bl6J	24, 48	13, 41	54–85	86–100
C57BL/6J	30, 25	18, 13	52–60	72–100
FVBN	16, 30	8, 17	50–57	53–100
C57BL/6J × C3H	20, 8	11, 6	55–75	65–100

REFERENCES

Abbondanzo, S. J., Gadi, I., and Stewart, C. L. (1993). Derivation of embryonic stem cell lines. *Methods Enzymol.* **225,** 803–823.

Acevedo, N., Wang, X., Dunn, R. L., and Smith, G. D. (2007). Glycogen synthase kinase-3 regulation of chromatin segregation and cytokinesis in mouse preimplantation embryos. *Mol. Reprod. Dev.* **74,** 178–188.

Bradley, A., Evans, M., Kaufman, M. H., and Robertson, E. (1984). Formation of germ-line chimaeras from embryo-derived teratocarcinoma cell lines. *Nature* **309,** 255–256.

Bryja, V., Bonilla, S., and Arenas, E. (2006). Derivation of mouse embryonic stem cells. *Nat. Protoc.* **1,** 2082–2087.

Buehr, M., Meek, S., Blair, K., Yang, J., Ure, J., Silva, J., McLay, R., Hall, J., Ying, Q. L., and Smith, A. (2008). Capture of authentic embryonic stem cells from rat blastocysts. *Cell* **135,** 1287–1298.

Chen, X., Vega, V. B., and Ng, H. H. (2008). Transcriptional regulatory networks in embryonic stem cells. *Cold Spring Harb. Symp. Quant. Biol.* **73,** 203–209.

Cole, M. F., and Young, R. A. (2008). Mapping key features of transcriptional regulatory circuitry in embryonic stem cells. *Cold Spring Harb. Symp. Quant. Biol.* **73,** 183–193.

Evans, M. J., and Kaufman, M. H. (1981). Establishment in culture of pluripotential cells from mouse embryos. *Nature* **292,** 154–156.

Kontgen, F., Suss, G., Stewart, C., Steinmetz, M., and Bluethmann, H. (1993). Targeted disruption of the MHC class II Aa gene in C57BL/6 mice. *Int. Immunol.* **5,** 957–964.

Li, P., Tong, C., Mehrian-Shai, R., Jia, L., Wu, N., Yan, Y., Maxson, R. E., Schulze, E. N., Song, H., Hsieh, C. L., *et al.* (2008). Germline competent embryonic stem cells derived from rat blastocysts. *Cell* **135,** 1299–1310.

Martin, G. R. (1981). Isolation of a pluripotent cell line from early mouse embryos cultured in medium conditioned by teratocarcinoma stem cells. *Proc. Natl. Acad. Sci. USA* **78,** 7634–7638.

Murry, C. E., and Keller, G. (2008). Differentiation of embryonic stem cells to clinically relevant populations: Lessons from embryonic development. *Cell* **132,** 661–680.

Nichols, J., Jones, K., Phillips, J. M., Newland, S. A., Roode, M., Mansfield, W., Smith, A., and Cooke, A. (2009). Validated germline-competent embryonic stem cell lines from nonobese diabetic mice. *Nat. Med.* **15,** 814–818.

Parrinello, S., Samper, E., Krtolica, A., Goldstein, J., Melov, S., and Campisi, J. (2003). Oxygen sensitivity severely limits the replicative lifespan of murine fibroblasts. *Nat. Cell Biol.* **5,** 741–747.

Robertson, E. J. (1987). Embryo-Derived Stem Cell Lines. IRL Press, Oxford.

Silva, J., Barrandon, O., Nichols, J., Kawaguchi, J., Theunissen, T. W., and Smith, A. (2008). Promotion of reprogramming to ground state pluripotency by signal inhibition. *PLoS Biol.* **6,** e253.

Simpson, E. M., Linder, C. C., Sargent, E. E., Davisson, M. T., Mobraaten, L. E., and Sharp, J. J. (1997). Genetic variation among 129 substrains and its importance for targeted mutagenesis in mice. *Nat. Genet.* **16,** 19–27.

Smith, A. G., Heath, J. K., Donaldson, D. D., Wong, G. G., Moreau, J., Stahl, M., and Rogers, D. (1988). Inhibition of pluripotential embryonic stem cell differentiation by purified polypeptides. *Nature* **336,** 688–690.

Solter, D., and Knowles, B. B. (1975). Immunosurgery of mouse blastocyst. *Proc. Natl. Acad. Sci. USA* **72,** 5099–5102.

Stewart, C. L., Kaspar, P., Brunet, L. J., Bhatt, H., Gadi, I., Kontgen, F., and Abbondanzo, S. J. (1992). Blastocyst implantation depends on maternal expression of leukaemia inhibitory factor. *Nature* **359,** 76–79.

Takahashi, K., and Yamanaka, S. (2006). Induction of pluripotent stem cells from mouse embryonic and adult fibroblast cultures by defined factors. *Cell* **126,** 663–676.

Tighe, A., Ray-Sinha, A., Staples, O. D., and Taylor, S. S. (2007). GSK-3 inhibitors induce chromosome instability. *BMC Cell Biol.* **8,** 34–40.

Tucker, K. L., Wang, Y., Dausman, J., and Jaenisch, R. (1997). A transgenic mouse strain expressing four drug-selectable marker genes. *Nucleic Acids Res.* **25,** 3745–3746.

Williams, R. L., Hilton, D. J., Pease, S., Willson, T. A., Stewart, C. L., Gearing, D. P., Wagner, E. F., Metcalf, D., Nicola, N. A., and Gough, N. M. (1988). Myeloid leukaemia inhibitory factor maintains the developmental potential of embryonic stem cells. *Nature* **336,** 684–687.

Ying, Q. L., Wray, J., Nichols, J., Batlle-Morera, L., Doble, B., Woodgett, J., Cohen, P., and Smith, A. (2008). The ground state of embryonic stem cell self-renewal. *Nature* **453,** 519–523.

CHAPTER SIXTEEN

Producing Fully ES Cell-Derived Mice from Eight-Cell Stage Embryo Injections

Thomas M. DeChiara, William T. Poueymirou, Wojtek Auerbach, David Frendewey, George D. Yancopoulos, *and* David M. Valenzuela

Contents

Abstract

In conventional methods for the generation of genetically modified mice, gene-targeted embryonic stem (ES) cells are injected into blastocyst-stage embryos or are aggregated with morula-stage embryos, which are then transferred to the uterus of a surrogate mother. F0 generation mice born from the embryos are chimeras composed of genetic contributions from both the modified ES cells and the recipient embryos. Obtaining a mouse strain that carries the gene-targeted mutation requires breeding the chimeras to transmit the ES cell genetic component through the germ line to the next (F1) generation (germ line transmission, GLT). To skip the chimera stage, we developed the VelociMouse® method, in which injection of genetically modified ES cells into eight-cell embryos followed by maturation to the blastocyst stage and transfer to a surrogate mother produces F0 generation mice that are fully derived from

VelociGene Division, Regeneron Pharmaceutical, Inc., Tarrytown, New York

Methods in Enzymology, Volume 476
ISSN 0076-6879, DOI: 10.1016/S0076-6879(10)76016-X

the injected ES cells and exhibit a 100% GLT efficiency. The method is simple and flexible. Both male and female ES cells can be introduced into the eight-cell embryo by any method of injection or aggregation and using all ES cell and host embryo combinations from inbred, hybrid, and outbred genetic backgrounds. The VelociMouse® method provides several unique opportunities for shortening project timelines and reducing mouse husbandry costs. First, as VelociMice® exhibit 100% GLT, there is no need to test cross chimeras to establish GLT. Second, because the VelociMouse method permits efficient production of ES cell-derived mice from female ES cells, XO ES cell subclones, identified by screening for spontaneous loss of the Y chromosome, can be used to generate F0 females that can be bred with isogenic F0 males derived from the original targeted ES cell clone to obtain homozygous mutant mice in the F1 generation. Third, as VelociMice are genetically identical to the ES cells from which they were derived, the VelociMouse method opens up myriad possibilities for creating mice with complex genotypes in a defined genetic background directly from engineered ES cells without the need for inefficient and lengthy breeding schemes. Examples include creation of F0 knockout mice from ES cells carrying a homozygous null mutation, and creation of a mouse with a tissue-specific gene inactivation by combining null and floxed conditional alleles for the target gene with a transgenic Cre recombinase allele controlled by a tissue-specific promoter. VelociMice with the combinatorial alleles are ready for immediate phenotypic studies, which greatly accelerates gene function assignment and the creation of valuable models of human disease.

1. Introduction

An established method for the elucidation of mammalian gene function *in vivo* is gene-targeted mutagenesis in embryonic stem (ES) cells by homologous recombination (Capecchi, 2001; Evans, 2001; Smithies, 2001), commonly referred to as knockout mouse technology because mutations are often designed to ablate gene function. Recently, large-scale functional genomic efforts, the Knockout Mouse Project (KOMP; Austin *et al.*, 2004) and the European Conditional Mouse Mutagenesis Project (Auwerx *et al.*, 2004), were begun to take full advantage of the wealth of new information revealed by the completion of the human and mouse genome sequences. To knockout and screen >3500 genes in a 5-year time horizon, we used our VelociGene® technology (Valenzuela *et al.*, 2003), an assembly line method for the creation of precise, gene-targeted heterozygous mutations in ES cells. VelociGene's unprecedented speed and throughput immediately shifted the bottleneck from the production of gene-targeted ES cells to the efficient use of these cells to generate genetically altered mice ready for phenotypic studies. To meet this challenge, we developed VelociMouse (Poueymirou *et al.*, 2007), a novel method for the generation of F0 generation mice that are fully derived from gene-targeted ES cells.

The most commonly used method for converting genetically altered ES cells into mice is to first inject into blastocyst stage embryos (Bradley and Robertson, 1986; Gardner, 1968), which develop *in utero* in a recipient female to produce chimeric mice (F0 generation). Chimeras are useful only as breeders to transmit the mutant allele through the germ line to heterozygous progeny in the F1 generation. The efficiency of germ-line transmission varies according to the extent of the ES cell contribution to the germ cells. Another round of breeding between F1 heterozygous mice produces homozygous F2 generation mice that are used for phenotypic studies. Assuming chimeras are efficient germ-line transmitters, multiple rounds of breeding are time consuming, requiring 9 months from the birth of F0 chimers to obtain F2 mice. In addition, the breeding required for the production of F2 study cohorts requires extensive husbandry efforts and mouse housing resources that are expensive in labor, materials, and often limiting. Moreover, it is often the case that combining more than one mutant or conditional allele by the crossing of individual mouse lines is very inefficient and can affect the genetic background of the study mice.

To eliminate these inefficiencies for producing mice for phenotypic study, we developed the VelociMouse method (Poueymirou *et al.*, 2007) as a universally applicable approach for directly producing mice in the F0 generation, which were genetically identical to and fully derived from genetically modified ES cells.

The original version of the VelociMouse method employed a laser to make a hole in the zona pellucida (ZP) of a diploid eight-cell stage host embryo to facilitate the injection of ES cells without damaging the embryo. To promote ES cell contribution to the epiblast of the developing embryo, the injected embryos are cultured overnight resulting in a preponderance of fully ES cell-derived F0 generation mice. We have used inbred and hybrid ES cells to efficiently generate normal, healthy, and fertile F0 mice that contain no detectable host embryo contribution (<0.1%). Historically, for making high-quality chimeric mice from gene-targeted ES cells injected into host blastocysts, it is necessary to use the desirable strain combinations of inbred ES cells and inbred blastocysts, despite the high cost of inbred mouse strains. The VelociMouse method can overcome the severe cost limitations because it allows for the injection of inbred ES cells into inexpensive outbred eight-cell host embryos to obtain fully ES cell-derived mice (Poueymirou *et al.*, 2007), whereas this challenging ES cell/host combination yields only poor quality chimeras from blastocyst injections. In addition to fully ES cell-derived mice, our method also produces some high-quality chimeras but the cost benefit of using eight-cell embryos from an inexpensive outbred strain (e.g., Swiss Webster, SW) more than outweighs the minor drawback of generating of some high quality chimeras when compared to injecting eight-cell embryos obtained from inbred lines.

The advantages of the VelociMouse method are: (1) inbred or hybrid ES cells can be injected into either inbred or less expensive outbred host embryos to obtain fully ES cell-derived mice that are 100% germ-line transmitters; (2) XY and isogenic XY-derived XO ES cells carrying a heterozygous mutation can be used to independently produce male and female F0 mice that are 100% germ-line transmitters, which can be intercrossed to obtain heterozygous, homozygous, and WT mice in the F1 generation; (3) ES cells engineered to carry homozygous mutations can be used to produce F0 mice that are ready for immediate phenotyping; (4) ES cells engineered to carry multiple genetic alterations (e.g., hemizygous knockout a conditional alleles and a transgene carrying a tissue-specific promoter driving an inducible Cre gene) can produce F0 mice identical in genotype to the ES cell, without the lengthy and inefficient breeding; (5) the production of F0 mice from gene-targeted or transgenic-inbred ES cells obviates the need for extensive backcrossing to obtain a pure inbred background; and (6) in addition to fully ES cell-derived F0 mice, the overall quality of the F0 chimeras that are sometimes produced from the injection of eight-cell stage embryos have better ES cell contribution than those obtained from blastocyst injections. The F0 mice that are fully ES cell-derived provide a faster route to the homozygous allele, and hence we have named them VelociMice.

2. Methods

2.1. ES cells and tissue culture

This chapter describes the injection of clones from two parental ES cell lines: VGF1 and C57BL/6N. The derivation of both cell lines and the culture methods for each has described previously (Auerbach *et al.*, 2000). ES cell clones for microinjection are grown to 50% confluence in a 24-well tissue culture plate (Falcon) containing a monolayer of γ-irradiated primary embryonic fibroblasts (Hogan *et al.*, 1994a). The medium is removed and the cells are washed with 1 ml Mg^{2+}/Ca^{2+}-free phosphate-buffered saline (Gibco, no. 20012-043). Following a 10-min incubation at 37 °C with 0.5 ml of 0.25% Trypsin–EDTA (Gibco, no. 25200), the ES cells are pipetted up and down to a single cell suspension using a presterilized RT-L1000F tip (Molecular BioProducts). The suspension is mixed with 1 ml of ES cell growth medium to inhibit the trypsin and the cells are collected by centrifugation at $2000\times g$ for 5 min. The cells harvested from one well of a 24-well plate are suspended in 1 ml of ES cell medium (LIF) and placed on ice with occasional gentle mixing for the length of the embryo injection period.

2.2. Mouse strains

SW and C57BL/6NTac (B6) mice were purchased from Taconic Farms, Germantown, NY, and were housed in semirigid isolator units (Plastic Design) in pathogen-free conditions maintained on a 10 h dark/14 h light cycle. SW females are used both as eight-cell embryo donors for either VGB6 or VGF1 ES cell clones and as recipient females for all injected embryos. B6 females are specifically used as embryo donors for VGF1 ES cell clones.

2.3. Producing eight-cell stage embryos

To obtain a maximum yield of eight-cell stage embryos from both SW and B6 strains, 3–4-week-old female mice were superovulated by hormone injection. Stud males are singly housed and are used twice weekly as breeders. Females are generated from our in-house colonies and are weaned at 21 days of age and either hormone injected immediately or housed for an additional week prior to use. Pregnant mare's serum gonadotrophin (PMSG; Calbiochem) is intraperitoneally administered (5 IU/mouse) using a 1-ml syringe fitted with a 27.5-gauge needle at 5–6 pm approximately 48 h prior to mating. Human chorionic gonadotrophin (hCG; Calbiochem) is similarly administered (7.5 IU/mouse) 2 h prior to mating. To obtain a yield of about 150 eight-cell stage embryos at 2.5 days post coitum (dpc), thirty females are mated at approximately 4–5 pm. The females are visually inspected for a copulation plug the following morning (0.5 dpc). At 2.5 dpc ($\sim$9 am) the females are sacrificed by CO_2 asphyxiation. The mice are placed on a clean draping ventral side up. The abdominal fur is wiped with 70% ethanol. An incision is made along the ventral midline from the pubis to the rib cage. The skin and mesentery are resected and the viscera are moved aside to expose the ovaries, oviducts, and uterine horns. The oviduct is carefully cut away from the ovary and the uterine horn and flushed with ES media (−LIF) into a 35-mm culture dish using a 1-ml syringe fitted with a bent 30.5-gauge needle inserted in to the infundibulum. After visual inspection under an MZ6 dissecting microscope (Leica), only the uncompacted eight-cell stage are selected for ES cell injection. To ensure the highest possible yield of fully ES cell-derived mice, the use of uncompacted eight-cell stage embryo is recommended. The embryos are placed into a new 35 mm dish containing a droplet of ES cell medium (−LIF) that has been overlaid with filtered mineral oil (no. M8410, Sigma-Aldrich). The embryos are incubated at 37 °C in a 7.5% CO_2 atmosphere prior to use.

2.4. Using frozen eight-cell stage embryos

Because of the very high throughput of our VelociGene pipeline and the large number of clones we microinject, we have successfully been using Swiss Webster (CFW) and B6 frozen uncompacted eight-cell stage embryos purchased from Charles River Laboratories. This eliminates the need to maintain a large number of mice, the effort required to superovulate the embryo donors, and the day-to-day variability in yield. The embryos come frozen (25/straw) in a liquid N_2 shipper and are transferred to a tank prior to use. Embryos (50 for each ES clone injected) are thawed at 11:00 am on the morning of injection as described in the CRL protocol for CryoTech^{TM} straws and are incubated at 37 °C in a 7.5% CO_2. Injections are begun at 1:00 pm.

2.5. Injection of eight-cell embryos

The equipment used for the injection of eight-cell embryos is identical to that used for standard blastocyst injections. Additionally, in the original version of the VelociMouse method (Poueymirou *et al.*, 2007), we used an XYClone Laser System (Hamilton-Thorne Biosciences) to assist in the injection process by "punching a hole" in the ZP and provide easier access to the embryo. This computer-controlled device utilizes an infrared laser (1480 nm) fired through a 50× objective to ablate a small portion of the ZP without damaging adjacent blastomeres or causing embryonic lethality. The laser intensity was fixed at 100% power with an 800-μs pulse duration to minimize heat conduction. To visualize injections we used an Inverted microscope (model IX50, Olympus America) with 4× and 20× objectives and phase contrast optics. ES cells and embryos are manipulated using glass microcapillary cell transfer pipettes (20° capillary angle and 15 μm inner diameter, Eppendorf) fitted on to micromanipulators (Leica) with manually controlled joysticks. All injections were performed in the top of a 35-mm dish, half-filled with ES cell medium (−LIF) and overlaid with filtered mineral oil. Only healthy-looking ES cells were drawn into the injection pipette. These are defined as rounded, refractive, and smooth. We try to avoid irregularly shaped cells and those containing abundant dark pigmentation, which is indicative of ES cells that are beginning to differentiate. Applying and releasing gentle suction through the Eppendorf holding pipette, the embryo is maneuvered to identify a region of the embryo with the greatest distance between a blastomere and laser punch site in the ZP to minimize heat transfer to the embryo. As shown in Fig. 16.1 the laser punch site is oriented at the "1 o'clock" position of the embryo for optimal ES cell deposition across at the "10 o'clock" position. The innermost isotherm ring (red) should be centered over the laser site on the ZP and a single laser pulse is used. The injection needle is introduced into the opening of ZP and inserted under and along the ZP to minimize damage to the blastomeres. To help minimize damage to the blastomeres during injection process, the tip of a new injection pipette can be dulled by gently striking against the holding pipette. Seven to nine ES cells are deposited into

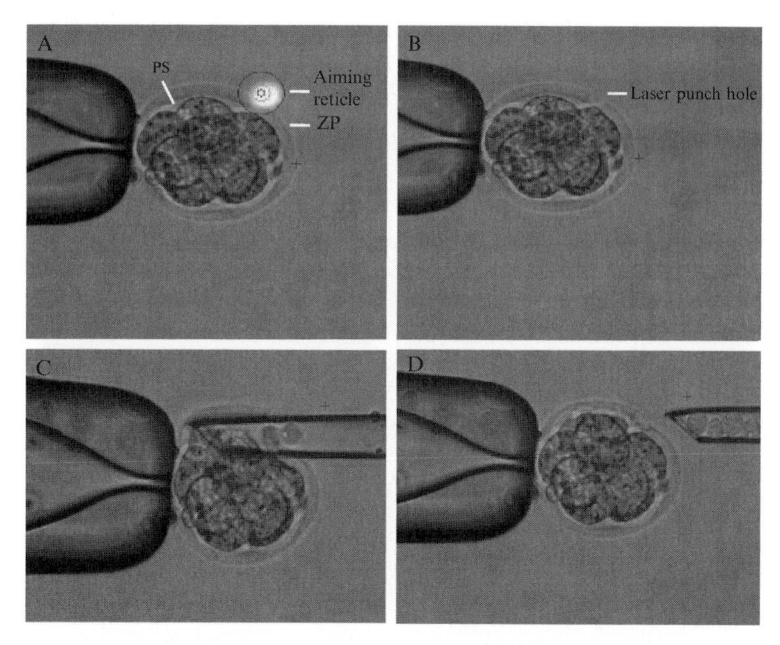

Figure 16.1 The eight-cell embryo injection process. (A) An uncompacted eight-cell embryo is held on the holding pipette with the laser aiming reticle positioned over the target site on the zona pellucida (ZP) at the "1 o'clock" position. ES cells are deposited in the perivitelline space (PS) between the blastomeres and the ZP. (B) Following one pulse of the laser, a hole in the ZP is easily visualized. (C) The injection pipette containing ES cells is gently pushed through the hole in the ZP and into the ps to deposit seven to nine ES cells at the "10 o'clock" position. (D) When the injection pipette is removed the ES cells remain in the ps. The injection of more than nine ES cells will normally result in them leaking out of the laser punch hole. (See Color Insert.)

each embryo. We have found that the injection of fewer than six ES cells can increase the production of chimeric F0 mice, while the injection of more than nine cells has no added benefit for making fully ES cell-derived F0 mice because the excess cells tend to leak out through the opening in the ZP. Depositing the ES cells at a "10 o'clock" location is preferred. More recently, after becoming proficient with eight-cell embryo injections, we have found that embryo injections by penetrating at the "3 o'clock" position using a piezo drill or simply a sharp injection needle would give results equal to those obtained with the use of laser-assisted injections. Thus, the key for generating VelociMice is not the method for introducing the ES cells, but the combination of ES cells and uncompacted blastomeres. Finally, we have also generated VelociMice from the injection of ES cells into four-cell stage embryos.

2.6. Culturing eight-cell embryos following ES cell injection

The postinjection culture of embryos is a critical step for generating fully ES cell-derived mice and takes advantage of ES cells, which have been adapted to grow *in vitro*, and embryos that normally develop *in vivo*. During the overnight culture, ES cells "outcompete" the blastomeres to form the inner cell mass of the developing embryo. The injected eight-cell embryos are cultured in KSOM (Millipore) drops overlaid with mineral oil, and incubated for 20–24 h at 37 °C in 7.5% CO_2. The following morning only those embryos that progressed to the morula or blastocyst stage were transferred into the uterine horns of recipient SW females. Following the injection, four-cell stage embryos were cultured for 36–40 h and similarly transferred into recipient females.

2.7. Transferring injected embryos to recipient females

Pseudopregnant SW females were used as recipient females. Females in natural estrus were mated with vasectomized SW males. The following morning, females with an observed copulation plug (0.5 dpc) are used as recipients on 2.5 dpc for the uterine horn transfer of injected embryos. Uterine horn transfer is an established procedure and we perform it as essentially described in Hogan *et al.* (1994b). Briefly, we anesthetize female mice with Avertin (2,2,2-tribromoethanol; Fluka 90710, Aldrich Chemical Co.) 2.5% in *tert*-amyl alcohol is administered i.p. at a dose of 0.1 ml/10 g body weight. This solution is stored at 4 °C in a foil wrapped 50 ml Falcon tube and used for up to 1 month. The proper plane of anesthesia is determined by the lack of a flinching response to a firm paw pinch with a blunt forceps. If there is a response, an additional 0.1 ml is administered. The surgical site is liberally treated with 70% alcohol and after the incision is made, the excess fur is wiped clear of the site with an alcohol pad. A 1-cm incision is made in the skin over the lateral lumbar region with a sharp scissors to create a "window" to visualize the underlying ovarian fat pad and attached ovary. A 0.5-cm incision is made in the body wall overlying the ovarian fat pad and a blunt forceps is used to gently remove the fat pad and the attached ovary and uterine horn. While holding the uterine horn nearly vertical, a 27.5-gauge needle is used to make a puncture into the lumen beginning at a point 2–3 mm distal to the oviduct. Prior to the embryo transfer procedure, transfer pipettes are prepared by quickly passing a glass pipette (no. 14672-380, 9 in., VWR Scientific) over a Bunsen burner flame, and drawing it to a diameter just small enough to fit into the uterine puncture. Using an aspirator tube assembly (no. A5177-SEA, Sigma-Aldrich) fitted onto the prepared glass pipette, six to seven embryos are transferred into the uterine lumen. Prior to collecting the embryos, a small air bubble is taken into the pipette to locate the position of the embryos as they are deposited into the uterine lumen.

The uterine horn, ovary, and fat pad are then gently returned into the body cavity and the skin incision is sealed with suture clips. The procedure is repeated on the other uterine horn. Following the surgery, recipient females are placed on a heating pad. The mice are observed for several minutes to ensure that their breathing appears normal for their anesthetized condition. They will usually stay down for 15–20 min following the surgical procedure. Following recovery from the uterine transfer surgery, the recipient females are housed two per cage.

2.8. Genotyping F0 mice

To determine if F0 generation mice are fully ES cell-derived, we perform a very sensitive PCR-based assay to look for genetic contribution from the host embryo. For example, when using SW host embryos, a PCR assay is performed to detect the SW albino locus. We have previously demonstrated that fully ES cell-derived F0 mice have $<0.1\%$ host embryo contamination. For the details of such an assay, refer to Chapter 17.

3. Summary

This chapter describes the VelociMouse® method of the injecting ES cells into eight-cell stage embryos to produce F0 generation mice that are fully ES cell-derived, which we call VelociMice (Poueymirou *et al.*, 2007). We have been using this technology at Regeneron for over 4 years and have found it to be very easy transition from conventional blastocyst injections and a technology that is more efficient and allows for more applications. When injecting nontargeted parental ES cells into eight-cell embryos, the resulting VelociMice are normal, healthy, and 100% germ-line transmitters. Because the method allows for ES cells to fully contribute to F0 mice, it is excellent for evaluating the pluripotency of newly established ES cell lines, as we have done for VGB6 that is being used to knockout 3500 genes for the KOMP initiative (Austin *et al.*, 2004).

The injection of heterozygous gene-targeted ES cells into eight-cell embryos produces 100% germ-line transmitting F0 breeders, which reduces the breeding burden for generating F1 heterozygous mice. A major advantage of the method is the ability to produce female F0 VelociMice. Thus, heterozygous male and female F0 VelociMouse breeders, generated from XY and isogenic XY-derived XO ES cell clones, respectively, can be generated efficiently. These F0 breeders are then crossed to produce WT, heterozygous, and homozygous (1:2:1) mice in the F1 generation, saving a full generation time and the associated husbandry resources. Moreover, genetically altered ES cells carrying homozygous mutations can be used to produce F0 VelociMice that are ready for phenotypic studies without the need for any breeding; a savings of two generation times. WT control littermates can be produced from separately injected eight-cell embryos with the nontargeted parental ES cells, which are cotransferred in recipient females with the embryos from homozygous ES cell injections. The ability to generate a fully ES cell-derived mouse with a genotype identical to that of the injected ES cells implies that combining mutant or transgenic alleles in ES cells to make VelociMice obviates the need for time-consuming and inefficient breeding strategies to achieve the desired allele combinations. Finally, we have demonstrated that the VelociMouse® method also works with inbred ES cells lines derived from 129' C57BL/6, and Balb/C mouse strains (Poueymirou *et al.*, 2007).

REFERENCES

Auerbach, W., Dunmore, J. H., Fairchild-Huntress, V., Fang, Q., Auerback, A. B., Huszar, D., and Joyner, A. L. (2000). Establishment and chimera analysis of 129SvEv- and C57BL/6-dervied mouse embryonic stem cell lines. *Biotechniques* **29,** 1024.

Austin, C. P., *et al.* (2004). The knockout mouse project. *Nat. Genet.* **36,** 921.

Auwerx, J., Avner, P., Baldock, R., Ballabio, A., Balling, R., Barbacid, M., Berns, A., Bradley, A., Brown, S., Carmeliet, P., Chambon, P., Cox, R., *et al.* (2004). The European dimension for the mouse genome mutagenesis program. *Nat. Genet.* **36,** 925.

Bradley, A., and Robertson, E. J. (1986). Injection of Cells into the Blastocyst. *In* "Manipulating the Mouse Embryo: A Laboratory Manual," (B. Hogan, F. Costantini, and E. Lacy, eds.), p. 188. Cold Spring Harbor Laboratory Press, Cold Spring Harbor, NY.

Capecchi, M. R. (2001). Generating mice with targeted mutations. *Nat. Med.* **7,** 1086.

Evans, M. J. (2001). The cultural mouse. *Nat. Med.* **7,** 1081.

Gardner, R. L. (1968). Mouse chimeras obtained by the injection of cells into the blastocyst. *Nature* **220,** 596.

Hogan, B., Beddington, R., Costantini, F., and Lacy, E. (1994a). Preparing STO or Mouse Embryo Fibroblast Feeder Dishes. *In* "Manipulating the Mouse Embryo: A Laboratory Manual", p. 261. Cold Spring Harbor Laboratory Press, Cold Spring Harbor, NY.

Hogan, B., Beddington, R., Costantini, F., and Lacy, E. (1994b). Uterine Transfer. *In* "Manipulating the Mouse Embryo: A Laboratory Manual", p. 178. Cold Spring Harbor Laboratory Press, Cold Spring Harbor, NY.

Poueymirou, W. T., Auerbach, W., Frendewey, D., Hickey, J. F., Escaravage, J. M., Esau, L., Doré, A. T., Stevens, S., Adams, N. C., Dominguez, M. G., Gale, N. W., Yancopoulos, G. D., *et al.* (2007). F0 generation mice fully derived from gene-targeted embryonic stem cells allowing immediate phenotypic analyses. *Nat. Biotechnol.* **25,** 91.

Smithies, O. (2001). Forty years with homologous recombination. *Nat. Med.* **7,** 1083.

Valenzuela, D. M., Murphy, A. J., Frendewey, D., Gale, N. W., Economides, A. N., Auerbach, W., Poueymirou, W. T., Adams, N. C., Rojas, J., Yasenchak, J., Chernomorsky, R., Boucher, M., *et al.* (2003). High-throughput engineering of the mouse genome coupled with high-resolution expression analysis. *Nat. Biotechnol.* **21,** 652.

CHAPTER SEVENTEEN

The Loss-of-Allele Assay for ES Cell Screening and Mouse Genotyping

David Frendewey, Rostislav Chernomorsky, Lakeisha Esau, Jinsop Om, Yingzi Xue, Andrew J. Murphy, George D. Yancopoulos, *and* David M. Valenzuela

Contents

Abstract

Targeting vectors used to create directed mutations in mouse embryonic stem (ES) cells consist, in their simplest form, of a gene for drug selection flanked by mouse genomic sequences, the so-called homology arms that promote site-directed homologous recombination between the vector and the target gene. The VelociGene® method for the creation of targeted mutations in ES cells employs targeting vectors, called BACVecs, that are based on bacterial artificial chromosomes. Compared with conventional short targeting vectors, BacVecs provide two major advantages: (1) their much larger homology arms promote high targeting efficiencies without the need for isogenicity or negative selection strategies; and (2) they enable deletions and insertions of up to 100 kb in a single targeting event, making possible gene-ablating definitive null alleles and other large-scale genomic modifications. Because of their large arm sizes, however, BACVecs do not permit screening by conventional assays, such as long-range PCR or Southern blotting, that link the inserted targeting vector to the targeted locus. To exploit the advantages of BACVecs for gene targeting, we inverted the conventional screening logic in developing the loss-of-allele (LOA) assay, which quantifies the number of copies of the native locus to which the mutation was directed. In a correctly targeted ES cell clone, the LOA assay detects one of the two native alleles (for genes not on the X or Y chromosome), the other allele being disrupted by the targeted modification. We apply the same principle in reverse as a gain-of-allele assay to quantify the copy number

VelociGene Division, Regeneron Pharmaceuticals, Inc., Tarrytown, New York

Methods in Enzymology, Volume 476
ISSN 0076-6879, DOI: 10.1016/S0076-6879(10)76017-1

of the inserted targeting vector. The LOA assay reveals a correctly targeted clone as having lost one copy of the native target gene and gained one copy of the drug resistance gene or other inserted marker. The combination of these quantitative assays makes LOA genotyping unequivocal and amenable to automated scoring. We use the quantitative polymerase chain reaction (qPCR) as our method of allele quantification, but any method that can reliably distinguish the difference between one and two copies of the target gene can be used to develop an LOA assay. We have designed qPCR LOA assays for deletions, insertions, point mutations, domain swaps, conditional, and humanized alleles and have used the insert assays to quantify the copy number of random insertion BAC transgenics. Because of its quantitative precision, specificity, and compatibility with high throughput robotic operations, the LOA assay eliminates bottlenecks in ES cell screening and mouse genotyping and facilitates maximal speed and throughput for knockout mouse production.

1. Introduction

The VelociGene method (Valenzuela *et al.*, 2003) of creating targeted genetic modifications in mouse embryonic stem (ES) cells uses large targeting vectors, termed BACVecs, which are based on bacterial artificial chromosomes (BACs). Because of their long homology arms, often greater than 50 kilobase pairs (kb), BACVecs enable very large genomic modifications not possible with conventional plasmid-based targeting vectors. For example, with a BACVec it is possible in a single ES cell targeting event to humanize a mouse gene by deleting 100 kb of mouse genomic DNA and replacing it with the homologous segment of the human genome. Besides such large-scale genomic engineering, BACVecs can be employed to generate the complete range of gene mutations—insertions, deletions, point mutations, or conditional alleles—in the modern mouse genetics tool kit. In addition to the precision and flexibility afforded by BACVecs, their long homology arms, which can be 10–50 times larger than those of conventional targeting vectors, promote high ES cell targeting efficiencies (5–10%) without the need for isogenic DNA or selection against nonhomologous insertion events. To take advantage of the benefits of BACVecs, however, requires a different approach to the screening of targeted mutations in ES cells.

Southern blotting or long-range polymerase chain reaction (PCR) assays are usually employed to screen for targeted mutations created with conventional targeting vectors. The goal of these methods is to obtain evidence for a linkage between the inserted targeting vector and the target genomic locus. For the Southern blotting assay, a probe that recognizes a sequence at the target locus that lies outside of the targeting vector's arms detects a predicted change in the size of a restriction enzyme cleavage fragment in a targeted ES cell clone as the result of correct targeting by double crossover homologous recombination. Because of the resolving power of conventional agarose gels, the Southern blotting assay restricts the length of the targeting vector's homology arms to under about 10 kb. Similarly, for the long-range PCR assay, in which one primer recognizes a sequence within the inserted DNA while the other recognizes a target locus sequence beyond the ends of the targeting vector's homology arms, the reliable efficiency of PCR makes it very difficult to perform the assay for vectors with homology arms larger than about 5 kb. Thus, by limiting the size of the homology arms, the common screening assays also limit the allele design possibilities of conventional targeting vectors.

To exploit the power and flexibility of genomic engineering afforded by BACVecs, we developed the loss-of-allele (LOA) assay (Valenzuela *et al.*, 2003) (also known as loss-of-native-allele, LONA), a quantitative assay that inverts the logic of conventional ES cell screening. Instead of a qualitative assay to detect the modified allele, LOA screening uses quantitative PCR (qPCR) to count the number of copies of the targeted (native) allele. LOA assays are designed to detect a sequence at the target locus that is altered by the intended modification, for example, a deletion, insertion, or replacement. For the targeting of a gene present in two copies (as for a gene on a somatic chromosome), the LOA assay will detect both copies in nontargeted cells or one copy in a targeted ES cell clone that has had one of its two alleles disrupted by the mutation. For a deletion, the LOA assay detects the native sequences targeted for deletion, while for an insertion, the LOA assay can be designed to detect the native sequence that is disrupted by the insertion event. Point mutations can be detected with an LOA assay that is essentially a quantitative screen for a single nucleotide polymorphism (SNP).

LOA assays are easy to design and have proven to be more reliable than long-range PCR in side-by-side comparisons (Gómez-Rodríguez *et al.*, 2008). In contrast to conventional assays that produce images that require visual interpretation, the output of an LOA assay is a number, which facilitates robotic automation and computational scoring and record keeping for increased speed and throughput. LOA screening readily accommodates multiple replicate assays for each sample to increase the confidence of the calls. The ES cell screening and mouse genotyping procedures described in this chapter combine an LOA assay for the targeting event with a gain-of-allele (GOA) assay that quantifies the copies of the inserted vector sequences such as a reporter or selection cassette or a human gene replacement. A correctly targeted clone in a knockout experiment should have lost one of its native alleles while acquiring one copy of the inserted sequence. The same assay can be used to quantify the copy number of randomly inserted BACVec transgenic clones. We have found that BAC-borne gene expression level correlates linearly with BACVec copy number both in ES cells and mice derived from them. The LOA and GOA assays that we use to screen for targeted mutations in ES cells or mice are essentially qPCR-based genomic copy number variation assays which we call QPCNV to distinguish them from other copy number variation methods. QPCNV assays have broad application for the detection and quantification of a wide array of genomic variations, whether they be directed mutations, random insertions, or naturally occurring differences between genomes. For example, LOA assays can serve as CNV assays to investigate genomic stability, while point mutation LOA assays can serve as SNP assays for allele-specific targeting in hybrid ES cell lines or for genotyping mice with a mixed strain background. Finally, LOA assays validated for ES cell screening are also ideal for genotyping mice derived from the targeted ES cells.

Since we first published the method (Valenzuela *et al.*, 2003), we have made many modifications and improvements to the original assay design, but the general principle remains the same. We use the TaqMan® method (Livak *et al.*, 1995) of qPCR for our LOA assays, but any method that screens ES cells by quantifying the targeted locus is an LOA assay. For the mouse researcher who intends to target one or a few genes or maintain a few mouse lines, the best approach may be to carefully design a series of thoroughly optimized LOA assays for each target. Our use of the LOA assay has always been to enable the screening of as many genes as possible. The methods described in this chapter have, therefore, been developed to maximize the throughput of ES cell screening and mouse genotyping for large-scale, multigene or whole genome knockout projects such as the U.S. National Institutes of Health Knockout Mouse Project (KOMP; http://www.knockoutmouse.org, http://www.komp.org). For the KOMP we are creating genetically modified ES cells at a pace of 1000 correctly targeted genes per year, with an average of 23 newly targeted genes per week. The throughputs for all projects passing through our facility are approximately 10,000 ES cell clones screened and 2000–3000 mice genotyped per week.

2. Methods

2.1. Equipment and materials

The one essential piece of equipment needed to perform LOA assays is a machine that can perform qPCR and collect fluorescence data in real time to determine a quantification cycle value (Cq; also commonly referred to as threshold cycle, Ct, or crossing point, Cp), the fractional PCR cycle at which the target is quantified, usually at a point where the accumulated fluorescence reaches a preset threshold. The system we use is the ABI Prism 7900HT (Applied Biosystems, http://www3.appliedbiosystems.com). All other processes can be performed manually with conventional laboratory pipetting devices, test tubes, and microtiter plates. The key items of equipment identified in the methods described below are important for our automated high-throughput LOA ES cell screening and mouse genotyping processes but are not essential for a lower throughput operation. In the methods we specifically note the key materials that we have used to optimize our high-throughput LOA assays. Similar materials from alternative suppliers may function as suitable substitutes.

2.1.1. LOA assay design

Our LOA assays employ TaqMan® qPCRs, which consist of two primers and an intervening fluorogenic probe, usually labeled with 5-carboxyfluorescein (FAM) at the 5′ end and a black hole quencher (BHQ) dye at the 3′ end (Biosearch Technologies, http://www.biosearchtech.com). Quantitative PCRs without a fluorogenic probe in which the real-time fluorescence is provided by the DNA-binding dye SYBR® Green also function in LOA assays, but for high-throughput screening of multiple projects, we prefer the confidence, reliability, and specificity imparted by the third oligonucleotide in the TaqMan format. LOA screening, especially when practiced in high-throughput production, can be particularly challenging and demands high-quality qPCR reagents. We have validated TaqMan qPCR assays provided by Biosearch Technologies, Applied Biosystems, Integrated DNA Technologies, and Roche for high-throughput LOA screening.

We typically design two TaqMan qPCR assays for each LOA assay of a targeted gene modification, regardless of the allele design. For deletions, we design qPCRs that assay for sequences at both ends of the deleted sequence that fall within 1 kb of the deletion boundaries. Ideally, we want to assay sequences at the extreme ends of the deletion, but this may not always be possible with every deleted sequence. An ES cell clone is scored as targeted only if both qPCRs detect a loss of one allele (a change in target gene copy number from 2 to 1 or from 1 to 0). We apply the same criterion to all LOA assays even if the allele is a small deletion, insertion, or point mutation in which the two qPCR assays may vary from one another by only a slight shift in the positions of the primers and probe. For insertions we design the qPCRs to assay the sequence at the insertion point such that the insertion would disrupt the binding of either the probe or one of the primers. For these assays and others in which the target of the assay is severely restricted, we often employ MGB probes (Applied Biosystems), which are shorter and have higher dissociation temperatures than conventional TaqMan probes of similar size, thereby providing more sensitivity to minor sequence variations.

2.1.1.1. Assay quality control TaqMan qPCR assays can be optimized by varying the primer and probe concentrations along with other components of the reaction to produce the lowest Cq value at a given template DNA concentration. For an ideally perfect qPCR, in which the amount of amplified DNA produced doubles with every cycle, every twofold difference in template DNA concentration will be reflected by a one-cycle difference in the Cq value. In a template DNA titration experiment, a plot of Cq as a function of the $\log_2$ of the DNA concentration will produce a straight line with a slope of -1 (-3.3 for a $\log_{10}$ plot). In practice, qPCRs can vary considerably from the theoretically perfect case. The important consideration for a relative quantification assay, such as an LOA assay, is that the efficiencies of the target-gene and reference-gene qPCRs be similar, that is, that the difference in the slopes of their DNA titration plots be minimal, with a ± 0.1 difference usually considered acceptable. The demands of high-throughput LOA screening do not permit careful optimization of every target-gene TaqMan assay and a perfect match with the reference-gene qPCR. As a surrogate quality control (QC) assay, prior to use in LOA screening we compare each target-gene TaqMan assay with our standard reference-gene assays to ensure that they produce similar Cq values on the same control DNA sample. Target-gene QC failure because of high Cq indicates poor amplification efficiency. Such assays often exhibit erratic performance that is more sensitive to variations in sample DNA concentration and minor contaminants that do not affect more robust TaqMan assays. A low target-gene Cq can be the result of an assay that is not completely specific for the target gene in that it recognizes, at least partially, other sequences in the genome. The logic of the LOA assay assumes a one-copy difference between a targeted clone and the wild-type (WT) sample. LOA scoring (see below) will not detect the loss of one allele when more than two are detected by the qPCR, which is an intentional feature of the LOA calculations designed to prevent interpretation of data from poorly performing assays. A final consideration in assessing the quality of target-gene assays is that they do not amplify on DNA samples, such as human DNA or nontemplate controls, that do not contain the target. This criterion is especially important for an LOA assay that detects a knockout of an X-linked gene in our male ES cells.

2.1.1.2. Reference gene considerations For high-throughput LOA ES cell screening, we prefer to use a set of standard reference-gene assays for all projects. We apply the following criteria when choosing a reference-gene assay: (1) an amplicon efficiency near 1 in DNA titration experiments; (2) compatibility with many different target-gene assays in ES cell screening and mouse genotyping; (3) stable performance in relative quantification assays with DNA samples whose concentrations vary over a 20-fold range; (4) robust performance with DNA samples of varying quality and purity from different types of biological samples; and (5) reliable and reproducible batch-to-batch quality from different suppliers.

2.1.1.3. Insert cassette assays Knockout alleles usually contain inserted sequences that are foreign to the mouse genome, such as those for genes that impart drug resistance or that report target-gene expression. Quantifying the copy of inserted sequences is an important part of targeted ES cell quality control. We have designed assays for bacterial selection and reporter cassettes, for promoters and polyadenylation signals, and for human replacement sequences in humanized alleles. Design considerations for good insert cassette assays are the same as those for general TaqMan assays and for the reference-gene assay. The amplicon efficiency of an insert cassette assay should closely match that of the reference gene, and the assay should reliably quantify the copy number of the inserted sequence in control DNA samples with known copies of the insert, such as DNA from genetically modified mice with well-defined genotypes.

2.1.2. ES cell screening

2.1.2.1. Purification of genomic DNA from ES cell clones We receive live cultures of ES cell clones from gene-targeting experiments that have been selected for drug resistance and split into two replicate plates: one for temporary storage at −80 °C and the other for LOA screening. It is best to allow the screening plate cultures to grow to confluency. Some differentiation of the ES cell clones prior to screening does not affect the LOA assay. Cultures that do not grow well and are not confluent produce low DNA yields and poor quality results in the LOA assay. Prior to DNA preparation, remove the growth medium from the cultures and wash the cells once with 0.2 ml of phosphate-buffered saline (e.g., DPBS, Invitrogen/Gibco no. 14190). This operation can be performed manually or with an automatic plate washer (e.g., SkanWasher 400, Molecular Devices, http://www.moleculardevices.com). The washed cells can be immediately processed for DNA purification or frozen at −20 °C for several weeks prior to DNA preparation.

We have performed successful LOA assays with genomic DNA purified by a variety of methods from careful manual spooling to high-throughput robotic methods. The method described below was optimized for speed and throughput and is performed on a Biomek® FX robotic liquid handling workstation (dual arm robotic platform with 96 and span-8 pipetting heads, Beckman-Coulter, http://www.beckmancoulter.com). In brief, the protocol consists of chemical and mechanical lysis of the cells in the presence of chaotropic salts; capture of the DNA on silica-coated magnetic particles ("beads"); washing of the beads to remove salts, proteins, and cell debris; elution of the bound DNA in a Tris–EDTA buffer followed by dilution with water.

1. Lyse the washed ES cells in 120 μl of RNA Lysis Buffer (Promega no. Z3051) by a combination of rotary shaking and up-and-down pipetting.
2. Transfer 60 μl of each cell lysate to a new 96-well plate containing 20 μl of MagneSil Red® beads in each well (Promega no. A1641) and mix as in step 1. (The remainder of the cell lysate can be stored at −20 °C for at least several months if a second DNA preparation is required.)

3. Collect the beads with bound DNA by placing the plate on a 24-prong magnetic platform and remove the supernatant.
4. Remove the plate from the magnet and wash the beads with 120 μl of 80% ethanol, mixing as in step 1.
5. Repeat step 3.
6. Repeat the ethanol washing (steps 4 and 5) two more times.
7. After the last wash, allow the beads with bound DNA to air-dry for 2 min.
8. Disengage the magnet and elute the bound DNA by adding 140 μl of TE (10 m*M* Tris–HCl, 0.1 m*M* ethylenediaminetetraacetic acid, EDTA, pH 8.0) and mixing thoroughly as in step 1.
9. Capture the beads on the magnet and transfer 110 μl of the DNA eluate into a new 96-well plate containing 100 μl of distilled water in each well.

The purity and quality of the DNA is sufficient for reliable, high-quality LOA assays in a high-throughput setting. The extensive mechanical agitation and pipetting undoubtedly cause significant shearing of the genomic DNA, a desirable property for the template of a qPCR, which usually has an amplicon length of less than 100 base pairs (bp). Another consideration in the development of the method was uniformity of DNA concentration across the plate. The average DNA concentration of the samples is usually between 0.5 and 1.0 ng/μl, with most of the samples differing in concentration by less than a factor of 3. The method recovers only a small fraction of the total genomic DNA in the ES cell cultures, but given the sensitivity of qPCR, total DNA yield was not as important a consideration for our purposes as were speed and throughput. As the LOA assay employs $\Delta\Delta$Cq relative quantification to determine gene copy number (explained below), we do not routinely determine the concentrations of the DNA samples nor do we normalize the concentrations prior qPCR setup.

2.1.2.2. qPCR setup We perform our qPCRs in a 384-well PCR plate (MicroAmp™ Optical 384-Well Reaction Plate, Applied Biosystems no. 4343370384) using TaqMan® Gene Expression Master Mix (Applied Biosystems no. 4370074) according to the manufacturer's suggestions in a reaction that is one part DNA and on part master mix containing the primers and probe. After distributing DNA to all the wells of the PCR plate, we then add the qPCR master mix with primer-probe sets for two independent target-gene assays (e.g., qPCRs that assay the upstream and downstream ends of a targeted deletion) and two reference-gene assays. We use single reaction qPCRs as opposed to duplex or multiplex reactions. Although duplex reactions, in which a target-gene-specific TaqMan assay is paired with a reference-gene assay in the same tube, often produce excellent LOA assay results, we have found that some target-gene assays are incompatible with some reference-gene assays; the two assays interfere with each other in that their Cq values are higher in the duplex reaction than in the individual single reactions. The frequency of interference in duplex reactions may be relatively low, but for a high-throughput ES cell screening and mouse genotyping operation, single reactions are more reliable and provide simplicity and flexibility of design, setup, and calculations. The reliability of the single qPCR format does, however, depend on accurate pipetting of the sample DNA that will be assayed in separate reactions for the target and reference genes. The accuracy of DNA pipetting for a given robotic or manual pipetting protocol can be very precisely determined by assaying the same DNA sample with a single primer-probe set in all the wells of a qPCR plate. Differences in Cq values reflect differences in pipetting accuracy and reproducibility.

2.1.2.3. Replicates and confirmation assays The PCR plate setup can be replicated in different formats. For example, the entire plate can be robotically "stamped" onto one or more additional plates as technical repeats to assure pipetting accuracy and reproducibility. (ES cell clones are normally screened as single entities; biological replicates are impractical.) Alternatively, different target-gene or reference-gene assays can be used to ensure the reliability of the qPCR assays or the compatibility of the target–reference pairs. However, we have found that the low rate of false positives—clones called as targeted with a one-plate screen that were not confirmed on the replicates—and false negatives—targeted clones missed with a one-plate screen that were confirmed as targeted on the replicates—does not justify the additional costs in materials and time of the replicate plates. A high false negative rate is wasteful of resources in that more ES cell clones must be screened to reveal the correctly targeted clones. Gómez-Rodríguez *et al.* (2008) demonstrated that LOA ES cell screening has a lower false negative rate than screening by long-range PCR, a common practice in the field, and was able to find more true, targeted clones for every 96 screened. Our retrospective analyses have indicated a false negative rate for single-plate screening compared with replicates of less than 2% (less than on correctly targeted clone missed for every 50 scored as positive).

False positives are a more serious concern. Such clones, if injected into embryos, would produce transgenic mice with a random insertion of the targeting vector rather than the desired targeted gene modification. To screen out false positives we use a three-tiered screening approach consisting of a primary screen and two confirmation assays. The primary screen is a single-plate LOA assay setup as described above. As a confirmation assay, candidate targeted clones from the primary screen are "cherry picked" from the ES cell DNA plate and redistributed as technical replicates on one half of new 384-well PCR plate (columns 1–12, two rows for each clone tested). Clones that confirm as positive in the six replicates of the confirmation assay are entered into our database as targeted clones. The ES cell culture lab will thaw the targeted clones from the replica ES cell plate stored at −80 °C and expand them for microinjection into embryos and for long-term storage in liquid nitrogen. Before the targeted clones are used in microinjection experiments to make knockout mice, we assay them for a third time to reconfirm that they are correctly targeted clones. Each ES cell clone is thawed, expanded in a 6-well culture plate, and then split and plated as six biological replicates on a 96-well gelatin-coated culture plate for LOA screening. We prepare DNA from the six replicates by our standard procedure (see above) and then distribute it into a PCR plate in the same pattern as that used for the confirmation assay of the primary screening candidates. A replica plate is also made. For one of the replicates we distribute master mixes for the target and reference genes. This PCR plate generates data for the LOA assay that reconfirms that the clones are correctly targeted. On the second replicate the target-gene master mixes are replaced by those containing TaqMan assays for markers in the inserted portion of the targeting vector, for example, the *lacZ* reporter gene, which encodes β-galactosidase, and the *neo*^{r} gene, which encodes neomycin phosphotransferase. This PCR plate generates data that quantifies the copy number of the inserted reporter and selection cassettes or other DNA sequences introduced in the gene-targeting event. One of the reference-gene assays we include on the reconfirmation plate is for the *Sry* gene on the Y chromosome. ES cells growing in culture tend to lose the Y chromosome at a fairly high rate; about 1% of selected clones will have no Y chromosome. The *Sry* gene assay is included to screen for the rare targeted clones that have lost their Y chromosome. Such clones can still be valuable. If injected into eight-cell embryos by the VelociMouse method (Poueymirou *et al.*, 2007; Chapter 16 of this volume) or into tetraploid blastocysts (Eggan *et al.*, 2002), they can produce fully ES-cell-derived X0 female mice in the F0 generation that are fertile and will transmit the targeted mutation through the germline. Only clones that reconfirm as correctly targeted and have a single copy of the inserted vector sequences are used for microinjection to create genetically modified mice or are shipped to the KOMP repository (https://www.komp.org) in the case of gene-targeting projects that we perform for the NIH KOMP.

2.1.2.4. PCR cycling and data collection After PCR setup is complete, we cover the plates with optical seals (Applied Biosystems no. 4311971) and affix a label with a barcode that identifies the plate and designates the file name for the qPCR data. Using the software provided with the 7900HT instrument (Sequence Detection System, SDS version 2.3), the plate bar codes are read and entered into a cycling queue in the computer that controls the 7900HT. We place the plates on a stacker platform attached to the 7900HT from which they are robotically fed into the instrument by a Zymark Twister arm. Plates are batched in groups of up to 25 and are cycled. For each PCR, the SDS software determines the Cq (threshold cycle, Ct, in Applied Biosystems terminology), the point in the PCR at which the fluorescence signal reaches the preset threshold. We use a default fluorescence threshold of 0.1 for all of our qPCRs.

2.1.2.5. Calculations and LOA scoring The LOA assay uses relative quantification to calculate target gene copy number by the so-called ΔΔCq method according to the methods provided in the manual for the Applied Biosystems 7900HT. For high-throughput production, we have developed a program in the Ruby programming language that examines the qPCR data files, performs calculations to score the targeted clones, and enters them in our database, but similar calculations can be performed in conventional spreadsheet or graphing applications.

Setting the stringency of acceptable copy number for LOA scoring is defined by the needs and methods of the particular screening operation. For example, a calculated copy number stringency of 1.0 ± 0.5 might be chosen to score a clone as a targeted candidate in a fist-pass primary screen, while in a confirmation assay the stringency might be increased to 1.0 ± 0.25. The best approach is to empirically determine a stringency that assures that a clone scored as targeted will produce mice with the intended modification after microinjection into embryos and transmission of the modified allele through the germline. The same applies to statistical tests of significance or confidence that compare the copy number of a candidate clone to that of the population of clones tested: production of targeted mice should be the final determinant of any LOA scoring algorithm.

We use a reconfirmation assay that includes a determination of the copy number of the inserted DNA, for example, the reporter and selection cassettes. We calculate copy number by the standard formula (Applied Biosytsems 7900HT manual):

$$\text{Copy number} = \text{Calibrator copy number} \times 2^{-\Delta\Delta\text{Cq}} \qquad (17.1)$$

We use both a one-copy and a two-copy calibrator for our standard cassette markers derived from a biological sample of known copy number, for example, heterozygous (Het) or homozygous (Hom) mice carrying one or two copies of a targeted allele. As quantitative screening and genotyping assays are rarely applied in standard gene-targeting methods, it has not been generally appreciated that ES cell clones that have been correctly targeted by double crossover homologous recombination can have multiple copies of the targeting vector inserted both at the target locus and elsewhere in the genome. In our experience, targeted clones with multicopy inserts (usually two) occur fairly frequently. In addition, certain targeted loci are more prone to yield multi-insert targeted clones, perhaps because of poor selection cassette expression at the insertion site. Our standard knockout/ knock-in alleles carry a *lacZ* gene followed by a drug selection cassette that is flanked by *loxP* sites. We have found that for this simple type of allele, multicopy insertions can be collapsed by the action of Cre recombinase to give alleles in which the selection cassettes have been removed while retaining one copy of the *lacZ* reporter gene. More complicated alleles, such as those that contain multiple functional parts, may not be amenable to this solution. In the absence of insert cassette quantification, a multicopy-targeted allele could be missed, with the result that interpretation of the phenotype, especially for complicated alleles, could be confusing or incorrect. For these reasons, we strongly recommend that quantitative assessment of insert copy number be a part of any ES cell screening or QC protocol.

2.1.3. Mouse genotyping

2.1.3.1. Purification of genomic DNA from mouse tail biopsies and qPCR setup We collect mouse tail biopsies in 1 ml 96-deep-well blocks (VWR no. 40002-009). The tails snips (2–3 mm) are digested in 0.25 ml of proteinase K lysis solution [0.1 *M* Tris–HCl, pH 7.6, 1 m*M* EDTA, 0.2% (w/v) sodium dodecyl sulfate (SDS), 0.2 *M* NaCl, ~0.04 mg/ml proteinase K (a 1/500 dilution of a buffered proteinase K solution, pH 7.5), whose concentration, 18 ± 4 mg/ml, varies with the batch, Roche no. 03 115 844 0010] for at least 16 h at 55 °C in an air incubator, after which DNA is purified, as for ES cells (see above), by capture on silica-coated magnetic particles. The qPCR setup, cycling, and data collection for mouse genotyping is similar to the format we use for ES cell screening except that for each sample we set up one target-gene assay, one assay for an insert marker gene (e.g., *lacZ* or neo^r), and two reference-gene assays.

2.1.3.2. Scoring genotypes Combining an LOA assay with a copy number assay for the insertion sequences makes calling genotypes unambiguous for a mouse that contains a single targeted allele. Our scoring algorithm first looks for samples in which no insert markers were detected and assigns them a provisional genotype of WT, that is, unmodified at both alleles ($Tge^{+/+}$, e.g., for a gene whose symbol is *Tge*). The program then calculates the average ΔCq value for the LOA assays (Cq of the target-gene assay − Cq of the reference-gene assay) to determine a two-copy calibrator. The program compares the LOA ΔCq values of each sample to the WT calibrator (ΔΔCq) and assigns a Het genotype ($Tge^{+/-}$) to a tail sample whose ΔΔCq value equals or approximately equals 1.0 (copy number = 1 by formula (17.1)). A WT ($Tge^{+/+}$) sample has a ΔΔCq value of 0 (copy number = 2), while a sample with a Hom genotype ($Tge^{-/-}$) is easy to score because it fails to amplify in the target-gene assay (copy number = 0). The WT, Het, and Hom genotypes called by the LOA assay are confirmed by the copy number assay for the inserted DNA: WT = 0, Het = 1, and Hom = 2.

In a large mouse operation in which many different lines are being genotyped at any given time, we find that LOA genotyping provides higher confidence in the results and eliminates rare mix-ups. For a lab that handles only one or a few lines at a time, once germline transmission has been established for a targeted ES cell clone by the production of Het mice in several F1 litters, a strictly quantitative assay may no longer be necessary. Genotypes can be called with a simple qualitative matrix: WT mice are positive for the target gene and negative for the insert assays; Het mice are positive for both the target gene and the insert assays; and Hom mice are negative for the target gene but are positive for the insert assays.

Genotypes can be determined by LOA assays for mice that carry different alleles for the same gene (e.g., a deletion and a conditional) or that combine modified alleles for multiple genes, but the calling of these genotypes may be difficult for a simple genotyping algorithm, especially if insert markers are shared among the various alleles. Such cases usually require some deductive reasoning by the genotyper to arrive at genotype calls that are consistent with all the data.

REFERENCES

Eggan, K., Rode, A., Jentsch, I., Samuel, C., Hennek, T., Tintrup, H., Zevnik, B., Erwin, J., Loring, J., Jackson-Grusby, L., Speicher, M. R., Kuehn, R., *et al.* (2002). Male and female mice derived from the same embryonic stem cell clone by tetraploid embryo complementation. *Nat. Biotechnol.* **20,** 455.

Gómez-Rodríguez, J., Washington, V., Cheng, J., Dutra, A., Pak, E., Liu, P., McVicar, D. W., and Schwartzberg, P. L. (2008). Advantages of q-PCR as a method of screening for gene targeting in mammalian cells using conventional and whole BAC-based constructs. *Nucleic Acids Res.* **36,** e117.

Livak, K. J., Flood, S. J., Marmaro, J., Giusti, W., and Deetz, K. (1995). Oligonucleotides with fluorescent dyes at opposite ends provide a quenched probe system useful for detecting PCR product and nucleic acid hybridization. *PCR Methods Appl.* **4,** 357.

Poueymirou, W. T., Auerbach, W., Frendewey, D., Hickey, J. F., Escaravage, J. M., Esau, L., Doré, A. T., Stevens, S., Adams, N. C., Dominguez, M. G., Gale, N. W., Yancopoulos, G. D., *et al.* (2007). F0 generation mice fully derived from gene-targeted embryonic stem cells allowing immediate phenotypic analyses. *Nat. Biotechnol.* **25,** 91.

Valenzuela, D. M., Murphy, A. J., Frendewey, D., Gale, N. W., Economides, A. N., Auerbach, W., Poueymirou, W. T., Adams, N. C., Rojas, J., Yasenchak, J., Chernomorsky, R., Boucher, M., *et al.* (2003). High-throughput engineering of the mouse genome coupled with high-resolution expression analysis. *Nat. Biotechnol.* **21,** 652.

CHAPTER EIGHTEEN

Induced Pluripotent Stem Cells

Holm Zaehres,* Jeong Beom Kim,*,† *and* Hans R. Schöler*

Contents

Abstract

Reprogramming of mouse and human somatic cells into induced pluripotent stem (iPS) cells has been possible with retroviral expression of the pluripotency-associated transcription factors Oct4, Sox2, Nanog, and Lin28 as well as Klf4 and c-Myc. iPS cells hold great potential as a model for diseases from the perspective of the individual patient and as an alternative source of pluripotent stem cells for therapeutic applications. In this chapter, we discuss how the use of retroviruses as well as other expression vectors, protein transduction, and small molecules can effectively and efficiently induce pluripotent stem cells from a variety of mouse and human starting somatic cell populations.

1. Generation of iPS Cells

The derivation of embryonic stem (ES) cells from mouse blastocysts (Evans and Kaufman, 1981; Martin, 1981) has revolutionized biomedical research. Together with recombinant DNA technology used for gene transfer, gene knockdown, or reporter gene expression, it has supported

* Department of Cell and Developmental Biology, Max Planck Institute for Molecular Biomedicine, Münster, NRW, Germany
† UNIST, School of Nano-Biotechnology and Chemical Engineering, Ulsan, Korea

Methods in Enzymology, Volume 476
ISSN 0076-6879, DOI: 10.1016/S0076-6879(10)76018-3

the generation of more than 500 mouse models to study embryonic development, adult physiology, various diseases, as well as the aging process. ES cells are *pluripotent*—they have the ability to differentiate along all embryonic and adult developmental cell lineages. Discovery of human embryonic stem cells (hESCs) (Thomson *et al.*, 1998) afforded the opportunity to create human models of development and diseases and, moreover, a potential source of cells for gene and cell-based therapies. The enormous potential benefit of hESCs has been offset by ethical concerns regarding their derivation, spurring a huge debate on moral, religious, and political considerations associated with this research.

Nuclear reprogramming of somatic cells to a pluripotent state holds potential for creating pluripotent cells specific to individual patients, which can be used to model diseases or generate an autologous source of cells for gene and cell-based therapies. Nuclear reprogramming, which has been studied in frogs for more than five decades (Gurdon *et al.*, 1958), has culminated in the birth of Dolly, the first application of somatic cell nuclear transfer in the cloning of mammals (Wilmut *et al.*, 1997). Fusion of somatic cells with pluripotent cells has provided alternative means to transfer pluripotency to somatic cells (Cowan *et al.*, 2005; Do and Schöler, 2004; Tada *et al.*, 2001). Since the discovery and functional description of the pluripotency transcription factors Oct4 (Nichols *et al.*, 1998; Schöler *et al.*, 1990), Sox2 (Yuan *et al.*, 1995), and Nanog (Chambers *et al.*, 2003; Mitsui *et al.*, 2003), it was envisioned that the delivery of a cocktail of factors could induce somatic cells to a state of pluripotency. Takahashi and Yamanaka (2006) conclusively demonstrated the achievement of this goal by retroviral transduction of mouse fibroblasts with the transcription factors Oct4, Sox2, Klf4, and c-Myc. Originally starting with 24 candidate genes, they reduced the cocktail to these four factors. The resultant cells, termed induced pluripotent stem (iPS) cells, were found to be remarkably similar to ES cells. The study was reproduced by other laboratories within a year, providing proof for the developmental similarity of mouse iPS cells to mouse embryonic stem cells (mESCs) (Maherali *et al.*, 2007; Okita *et al.*, 2007; Wernig *et al.*, 2007).

Researchers have proposed that there is remarkable similarity in the regulatory network of transcription factors in mouse and human ES cells and analogous functions of the reprogramming factors in the mouse and human systems (Boyer *et al.*, 2005; Yamanaka, 2007). Takahashi *et al.* (2007) transferred their results from mouse to the human system transducing human fibroblast cultures with Oct4, Sox2, Klf4, and c-Myc. They cultured cells under hESC culture conditions until the appearance of hESC-like colonies that could be further propagated and expanded. In parallel, Yu *et al.* (2007) reported the generation of human iPS cells transducing the pluripotency transcription factors Oct4, Sox2, Nanog, and Lin28 in human fibroblasts.

We were able to demonstrate that ectopic expression of Oct4 alone or of Oct4 together with either Klf4 or c-Myc is sufficient to generate iPS cells from mouse and human neural stem cells (NSCs), which endogenously express Sox2, Klf4, and c-Myc (Kim *et al.*, 2008, 2009b,c). These one or two-factor (1F, 2F) iPS cells are similar to ES cells at the molecular level, can be efficiently differentiated *in vitro*, and are capable of teratoma formation *in vivo*. Mouse NSC iPS cells are also capable of germline contribution and transmission. We propose that the number of reprogramming factors required to generate pluripotent stem cells can be reduced when starting with somatic cells that endogenously express appropriate levels of complementing factors.

The iPS cell generation process has been diversified by the use of different starting somatic cell populations in mouse (fibroblasts, blood, liver, hepatocytes, pancreas, and neural progenitor and stem cells) and human systems (fibroblasts, blood, keratinocytes, NSCs), different transcription factor combinations, different gene transfer systems for the reprogramming factors, the emerging use of proteins and small chemical compounds, and different iPS cell expansion and characterization conditions.

2. Expression Vectors for iPS Cell Generation

Recombinant DNA technology has been applied to redesign retroviruses to be used as vector systems for the efficient gene transfer and stable expression of genomically integrated transgenes in a broad range of mammalian cells for basic research and gene therapeutic applications. Retroviral vector expression in mESCs has been studied since their derivation, but it has proved problematic, as the provirus can be transcriptionally inactivated through the process of silencing (Cherry *et al.*, 2000). Initial transcriptional repression of proviruses has been attributed to the effect of trans-acting repressors or the lack of transactivators to start transcription from the regulatory elements in the retroviral long terminal repeat (LTR) and leader region. DNA methylation is the main mechanism for the silencing of retroviral control elements in mESCs undergoing differentiation. This mechanism has been described in detail for murine leukemia virus (MLV)/gammaretrovirus-based vectors. The lack of significant provirus transcription under nonselective conditions has hampered the use of simple retroviral vectors in transgenic *in vivo* experiments. Unlike its counterproductive role in gene transfer applications, silencing is beneficial for iPS cell generation. A key feature in the induction of a pluripotent cell state from a somatic cell state during the process of reprogramming is that expression of the reprogramming factors leads to initiation of the reprogramming of the starting cell population, yet once the iPS cell state is achieved, it extinguishes any further reprogramming. Fully reprogrammed cells therefore no longer require transgene expression.

In their first description of iPS cells, Takahashi and Yamanaka (2006) used the MX vector system to induce reprogramming (Fig. 18.1). The Moloney murine leukemia virus (MoMuLV)-LTR drives expression of the Oct4, Sox2, Klf4, and c-Myc transgenes in mouse fibroblasts, and it is basically silenced in the iPS cell state. For human iPS cell generation, Takahashi *et al.* (2007) used the MX retroviral vector with MoMuLV-LTR to drive Oct4, Sox2, Klf4, and c-Myc transgene expression, which is silenced in iPS cells. Park *et al.* (2008b) used mouse embryonic stem cell virus (MSCV)-based retroviral vectors to accomplish the same goal.

The advent of more complex vectors based on lentivirus (e.g., human immunodeficiency virus (HIV)) has considerably enhanced the prospect of successful retroviral gene transfer (Naldini *et al.*, 1996). Lentiviral vectors derived from HIV-1 with the chicken beta-actin/cytomegalovirus enhancer (CAG) promoter (Pfeifer *et al.*, 2002) and the ubiquitin-C-promoter (Lois *et al.*, 2002) have been used successfully to express transgenes in mESCs and *in vitro* differentiated progeny. Transfer of lentivector-transduced ESCs into

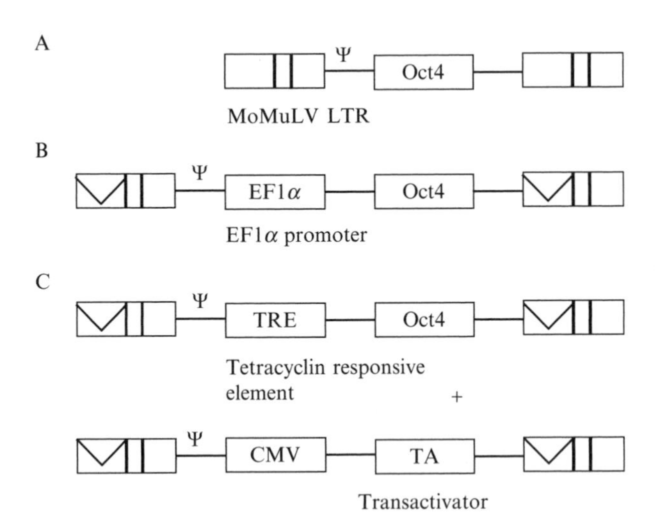

Figure 18.1 Retroviral and lentiviral vectors for iPS cell induction. (A) Retroviral vectors constitutively express transgenes from the LTR (long terminal repeat). The MX retroviral vector incorporates an MoMuLV LTR, which mediates silencing of gene expression in iPS cells (Takahashi and Yamanaka, 2006; Takahashi *et al.*, 2007). (Ψ: retroviral packaging signal). (B) Lentiviral vectors constitutively express transgenes from constitutive internal promoters, such as the EF1α (elongation factor 1 alpha) promoter (Yu *et al.*, 2007). (C) Inducible lentiviral vector systems express transgenes from a tetracycline responsive element (TRE) and require a vector expressing the drug-inducible transactivator (TA) from the CMV (cytomegalovirus) promoter.

blastocysts can lead to the generation of chimeric mice that express the transgene in multiple tissues. Thus, germline transmission of lentiviral vector-driven transgenes has supported the efficient generation of novel transgenic mouse models. Lentivirus-mediated gene transfer and expression of the green fluorescent protein (GFP) in hESCs has been described using vectors incorporating the EF1α (elongation factor 1 alpha), the PGK (phosphoglycerate kinase), and the CMV (cytomegalovirus) promoters (Gropp *et al.*, 2003; Ma *et al.*, 2003; Zaehres and Daley, 2006).

When inducing iPS cells from human fibroblasts, constitutively lentiviral EF1α promoter driven transgenes of Oct4, Sox2, Nanog, and Lin28 also appear to be silenced in the pluripotent state (Haase *et al.*, 2009; Yu *et al.*, 2007) (Fig. 18.1). Inducible lentiviral vectors provide the opportunity to "actively" silence reprogramming factor transgene expression upon drug administration, which has been demonstrated in the mouse (Brambrink *et al.*, 2008; Stadtfeld *et al.*, 2008a) and human systems (Hockemeyer *et al.*, 2008; Stadtfeld *et al.*, 2008a) (Fig. 18.1). Inducible lentiviral vectors have paved the way for "secondary iPS cell" generation systems, whereby cells derived from "primary iPS cells" could be reprogrammed again using homogeneous (re)expression from the same proviral integration sides (Eminli *et al.*, 2009; Wernig *et al.*, 2008).

Retroviral and lentiviral gene transfer incurs the risk of insertional mutagenesis associated with every transgene integration into a host genome. This makes iPS cell generation with retroviral vectors more suitable in the context of *in vitro* applications, such as disease modeling, and less suitable in the context of *in vivo* applications, such as cell-based therapies. iPS cell generation with fewer factors, such as Oct4 and Klf4 or Oct4 alone as outlined above, reduces the risk for insertional mutagenesis. However, the efficiency of programming declines and the duration of the reprogramming process increases when using fewer factors, during which time more cellular changes could accumulate.

Episomal expression vectors include plasmids and nonintegrating viral vectors such as adenoviruses, herpes viruses, Sendai virus, and Epstein-Barr-virus-based vector systems. Transgene expression of reprogramming factors is "silenced" by the simple dilution of episomal DNA during cell replication. For mouse iPS cell generation, the use of plasmids (Okita *et al.*, 2008) and adenoviruses (Stadtfeld *et al.*, 2008b) has been successful at low frequency, thereby requiring multiple rounds of application. For human iPS cell generation, the use of an Epstein-Barr-virus-based plasmid system (Yu *et al.*, 2009) and Sendai RNA virus (Fusaki *et al.*, 2009) has been described. It should be noted that episomal DNA can integrate randomly into the genome at very low frequency.

Integration-free reprogramming can also be achieved with transiently integrating vectors. The piggyBac transposition system has allowed the removal of (four) reprogramming factors upon transient expression of the transposase enzyme in mouse and human iPS cells (Woltjen *et al.*, 2009). Cre-recombinase-mediated excisions of loxP-flanked lentiviruses have also been utilized to generate factor-free human iPS cells (Soldner *et al.*, 2009).

3. Retrovirus Production

Retroviral and lentiviral vectors are key instruments in the transfer of reprogramming factors for the efficient generation of iPS cells in the mouse and human systems. Retroviruses can be produced either as ecotropic viruses (Naviaux *et al.*, 1996), which can only infect mouse cells, or viruses pseudotyped with the vesicular stomatitis virus surface protein (VSV-G) (Burns *et al.*, 1993), which can infect mammalian cells of all species. An advantage of using ecotropic virus is the potential to handle infections under safety level 1, whereas the VSV-G pseudotyping allows efficient infection with a small volume of concentrated virus. Takahashi *et al.* (2007) also used ecotropic retrovirus to transduce human fibroblasts, in which the ecotropic receptor (mouse Slc7a1 gene) was expressed by lentiviral gene transfer as previously described (Koch *et al.*, 2006). This procedure obviates the use of oncogene-expressing (e.g., Klf4 and c-Myc) retrovirus, which can infect human cells.

Retroviral and lentiviral vectors pseudotyped with VSV-G can infect all mammalian cells, and the titers produced upon ultracentrifugation are within the range of virus stocks used in human gene therapy applications (10^8–10^9 infectious particles/ml). Therefore, refer to the local biosafety guidelines of your institution and always handle virus production under BL2/BL2+ conditions. Bleach and autoclave all used plastic materials that have been in contact with virus.

1. Plate 293T cells (maintained in Dulbecco's modified Eagle's medium (DMEM) supplemented with 10% fetal bovine serum (FBS)) in 10 100-mm tissue culture dishes 24 h before transfection to attain cells with ~80% confluency for transfection.
2. Cotransfect 293T cells with 3 μg retroviral/lentiviral vector, 2 μg packaging plasmid, 1 μg VSV-G expression vector (for a three-plasmid system), and 18 μl *FuGENE 6* (Roche) per plate according to the supplier's conditions. As *FuGENE 6* is not toxic to 293T cells, no further medium changes are required.
3. Collect virus supernatant from all plates 48 or 72 h after transfection by using plastic pipettes and filter supernatant through a 0.45-μM filter.
4. Transfer filtered supernatants (35 ml each) to centrifuge tubes (Beckman Polyallomer, 25 × 89 mm) and spin at 23,000 rpm for 90 min at 4 °C in a Beckman SW28 swinging bucket rotor.

5. Remove supernatant carefully. You typically see a pellet that is opaque at the bottom center of the tube. Add 1 ml DMEM to the tube and keep the tubes wrapped in paraffin overnight at 4 °C.
6. Resuspend and mix all tubes containing the same virus; aliquot into 10 freezing vials and store virus stocks at −80 °C until use. Avoid multiple freeze–thaw cycles.
7. For titer determination, plate 1×10^5 293T cells/well in a 6-well plate and add 1, 10, and 100 μl of concentrated virus 6 h later; 24 h later, wash twice with PBS and supplement with fresh media. Analyze cells for GFP expression 48 h after infection by fluorescence-activated cell sorter (FACS). The titer can be calculated as follows: (percentage of GFP + cells) $\times 10^5 \times$ (dilution factor), and it is represented as IU (infectious units)/ml of concentrated vector. 100× concentrated stocks of the Lentilox vector have titers in range of 1×10^8–1×10^9 IU/ml. For vectors without a reporter gene, quantitative real-time PCR can be used to determine the number of vector DNA molecules/ml. Use 0.4% paraformaldehyde in your transduced cell solutions to inactivate remaining virus during titer determination.
8. Virus stocks should be assayed for replication-competent retrovirus (RCR). Supernatant collected from infected cells (e.g., 293T) should itself be tested for presence of virus 48 h after infection (by transferring supernatant onto another dish of 293T cells). This supernatant should be virus-free. The literature on retroviruses describes more detailed protocols for ruling out the generation of RCR during viral passage.

4. iPS Cell Generation with Retroviral Vectors

iPS cell generation can be technically reduced to the expression/delivery of reprogramming factors in/to the reprogramming starting cell population and further cultivation of the transduced cells under mouse or hESC culture conditions until iPS cell colonies appear.

Most studies in the mouse system have used mouse embryonic fibroblasts as the cell population to be reprogrammed. The key advantage to using fibroblasts is that they grow on their own feeder layer. They can simply be grown under ES cell conditions. Other mouse or human cells require replating to mouse embryonic fibroblasts.

1. Plate 1×10^5 cells of your target cell population on a 6-well tissue culture plate in 2 ml of medium.
2. Add virus supernatant 6 h later. You can use a GFP control virus to measure the transduction efficiency or infect with a calculated multiplicity of infection (MOI). The MOI is defined as the number of infectious units

per cell to be infected. For reprogramming induction, an MOI of 10 (1×10^6 IU on 1×10^5 cells) is reasonable. Evaluate different transduction efficiencies by infecting with different MOIs. For mouse iPS cell generation, use of ecotropic virus is sufficient; for human iPS cell generation, VSV-G pseudotyped virus can be utilized.

3. Add 6 μg/ml of protamine sulfate (Sigma) to your culture to enhance virus binding to the cells.
4. Culture cells with the virus for 24 h, wash three times with PBS, and then add fresh media.
5. On day 3 after infection, measure transduction efficiency by FACS analysis of the GFP control virus. *Optional*: Determine the provirus copy number by Southern-blot or QRT-PCR analysis of genomic DNA from the transduced cells.
6. On day 5 after infection, plate 5×10^4 transduced cells on a 6-well tissue culture plate preseeded with mouse embryonic feeder cells (MEFs) in MEF medium (DMEM/10% FCS).
7. On day 6 after infection, change to hESC medium and continue the culture with daily changes of medium. *Optional*: Supplement hESC medium with 1 m*M* valproic acid (VPA) to enhance iPS cell generation efficiency (Huangfu *et al.*, 2008).
8. mESC-like colonies will appear within 20 days; hESC-like colonies (Fig. 18.2) will appear within 30 days (transduction with four factors; Oct4, Sox2, Klf4, and c-Myc).
9. Mouse iPS cell cultures can be propagated like mESCs on MEFs with trypsinization.
10. Human iPS cell colonies should initially be passaged by mechanical splitting. Cut the colonies in smaller pieces and suck them up with a 20 μl pipette tip to transfer them on a 12-well plate preseeded with MEFs. The standard culture can be maintained on 6-well plates.
11. We cultivate our human iPS cell cultures on MEFs (Thomson *et al.*, 1998) (CF1 strain) or on Matrigel in basic fibroblast growth factor (bFGF)-supplemented MEF-conditioned medium (Xu *et al.*, 2001). Refer to the numerous protocols describing hESC culture.

iPS cell generation is greatly facilitated when using reporter genes for pluripotency gene reactivation as reprogramming marker systems. For generating mouse iPS cells, approaches involving Oct4, Nanog knockin reporter alleles (Maherali *et al.*, 2007; Okita *et al.*, 2007; Wernig *et al.*, 2007), transgenic Oct4-GFP reporters (Kim *et al.*, 2008), as well as no reporter genes (Meissner *et al.*, 2007) have been described. For human iPS cell generation, lentiviral vectors incorporating GFP or other reporter genes under the control of human Oct4 promoter elements could be utilized (see below).

Additional applications of transgenes can lead to increased iPS cell generation efficiency. Coexpression of Lin28 (Viswanathan *et al.*, 2008) with Oct4, Sox2, Klf4, and c-Myc has been shown to significantly increase

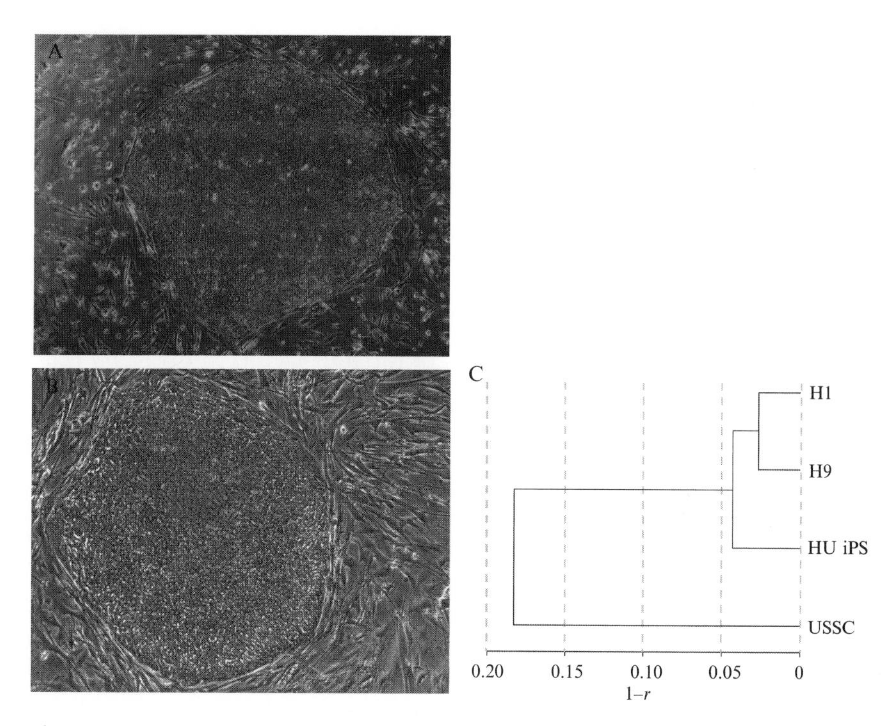

Figure 18.2 Human induced pluripotent stem cells. (A) Human OCT4 induced pluripotent stem (iPS) cells derived from human neural stem cells (Kim *et al.*, 2009b) grown under human embryonic stem (ES) cell culture conditions on a mouse embryonic feeder layer. (B) Human iPS cell colony generated from human cord blood–derived USSC cells (Kögler *et al.*, J Exp Med, 2004) by retroviral transduction with OCT4, SOX2, KLF4, and c-MYC. (C) Hierarchical cluster analysis of global gene expression comparing cord blood USSC (starting population), induced pluripotent stem cells (HUiPS), and the human ES cell lines H1/WA01 and H9/WA09 (microarray data from an Illumina platform, analysis by B. Greber).

colony formation (Liao *et al.*, 2008). Knockdown of certain genes can be used, all with similar effects. Knockdown of p53 by small interfering RNA (siRNA) applications during iPS cell induction has increased iPS cell generation in mouse and human systems by more than one order of magnitude (Hong *et al.*, 2009; Zhao *et al.*, 2008).

5. iPS Cell Generation with Proteins and Small Molecules

All the methodology described above relies on gene transfer and expression of the reprogramming factors. Two other methods for inducing pluripotency are available: direct delivery of the reprogramming factors as proteins or activation of the endogenous reprogramming factors by small molecules. Protein transduction, that is, crossing of the cell membrane and translocation into the nucleus, is feasible by the intrinsic ability of some proteins, as shown for the Cre-recombinase (Will *et al.*, 2002), or by conjugation of the proteins with short peptides, such as the HIV tat or a polyarginine tract. Zhou *et al.* (2009) fused a polyarginine protein transduction domain to the four reprogramming factors Oct4, Sox2, Klf4, and c-Myc and expressed them in *Escherichia coli*-inclusion bodies. Four rounds of transduction over a period of 8 days resulted in the generation of iPS cells from a starting population of murine OG2 MEFs. Kim *et al.* (2009a) expressed polyarginine-fused versions of the four factors in HEK293 cells. Human neonatal fibroblasts were transduced with six cycles over a period of 42 days, resulting in outgrowths of hESC-like colonies after 50 days. Since gene transfer is not involved with these methods, no risk of insertional mutagenesis exists, making these protein iPS cells in theory suitable for cell-based therapy applications.

A defined chemical reprogramming cocktail comprises small molecules instead of expression vectors or transducing proteins. Currently, different small molecules that can substitute for Oct4, Sox2, or Klf4—c-Myc can be omitted (Nakagawa *et al.*, 2008)—have been described; however, the presence or expression of other factors is required. Use of the histone methyltransferase inhibitor BIX01294 can increase the efficiency of iPS cell colony formation from neural progenitor cells and facilitates iPS cell generation from these cells in the presence of Sox2, Klf4, and c-Myc, thereby substituting for Oct4 (Shi *et al.*, 2008). Induction of pluripotency from human fibroblasts has been possible with Oct4 and Sox2 in the presence of the histone deacetylase inhibitor VPA (Huangfu *et al.*, 2008). This molecule inherits cell toxicity at required doses, thereby just allowing survival of cell subpopulations. Murine fibroblasts can also be generated without Klf4 using chemical complementation with the small molecule *kenpaullone* (Lyssiotis *et al.*, 2009). A Tgf-β signaling inhibitor and a GSK-3 inhibitor were used to substitute Sox2 in the presence of the other factors (Ichida *et al.*, 2009; Li *et al.*, 2009). Mikkelsen *et al.* (2008) have found iPS cell generation from mouse fibroblasts to increase when applying the classic DNA demethylating agent 5-azacytidine. Reprogramming of somatic cells with small molecules should support high-throughput production of iPS cells for *in vivo* applications, a feat yet to be achieved, however. Some of the molecules described to date are cytotoxic, pending safety testing, such as their risk for carcinogenesis. Most small molecules are not yet approved by the United States Food and Drug Administration.

Protein transduction does not involve transgene integration into the host genome, which occurs with the use of retrovirus vectors but also to a lower degree with episomal expression vectors. The efficiency of generating iPS cells with proteins is significantly lower and protein production is more tedious. Multiple rounds of application are currently necessary to generate iPS cells using this method. In addition, the potential for batch-to-batch variation in protein preparations makes high-throughput iPS cell production unlikely. Currently, the oncogenic potential of applying high doses of c-Myc and Klf4 proteins to cells remains unknown. The use of retroviruses is the most efficient method for generating iPS cells, but it is associated with risk for insertional mutagenesis. Nonintegrating expression vectors or proteins lead to the generation of iPS cells that are safer, but the process by which this is accomplished is much less efficient. Small molecules are most likely to be used in combination with other approaches to increase the efficiency of iPS cell generation with expression vectors or transducing proteins.

6. iPS Cell Characterization

The characterization of mouse and human iPS cell lines embraces a molecular and a developmental evaluation. The first criterion for characterizing reprogrammed cells is their morphology—mouse iPS cells resemble mESCs and human iPS cells resemble hESCs in their appearance. Mouse iPS cells can be passaged as single cells with trypsin treatment; human iPS cells are propagated mechanically as small cell clumps.

Analysis of pluripotency marker expression includes immunostaining for Oct4, Sox2, Nanog, as well as for SSEA-1 (mouse) or SSEA-3/-4 and TRA-1-60/-81 (human), and RT-PCR analysis. DNA methylation analysis of the Oct4 (Sox2, Nanog) promoter regions of the starting cell populations as well as of the corresponding iPS cell lines using bisulfite sequencing should distinguish the different epigenetic states of the cells. As outlined earlier, iPS cells should express the pluripotency genes from the endogenous loci and no longer rely on exogenous transgene expression. Therefore, PCR analysis should demonstrate silencing of transgene expression for retro/lentiviral vectors with virus-specific primers.

In vitro differentiation with induction of embryoid bodies (EBs) resulting in either random or directed differentiation into cells of all three germ layers using protocols described for ES cells can be applied to characterize the developmental potential of mouse and human iPS cells. Genetically, marked mouse iPS cells can be further injected into blastocysts to generate chimeric embryos, demonstrating the contribution of iPS cell-derived tissues in all germ layers (Kim *et al.*, 2008; Okita *et al.*, 2007; Wernig *et al.*, 2007). To assess germline contribution and transmission, chimeras have to be mated with mice, and the proviruses or other genetic iPS cell marks must be detected in the F1 generation. The most stringent test for developmental potency in the mouse system is tetraploid (4N) embryo aggregation. Since tetraploid cells have no full-term developmental potential, the progenitors of the entire developing animal must be the injected iPS cells, thereby demonstrating the cells' full-term developmental potential. Zhao *et al.* (2009) reported evaluation of several iPS cell lines used to generate viable, live-born progeny (27 live pups out of 848 injected blastocysts) by tetraploid complementation, at an efficiency comparable with that of ES cells generated from *in vivo* embryos.

The formation of teratomas upon injection of mouse or human iPS cells under the skin of immunodeficient mice, resulting in tumors consisting of cells representing all three germ layers, is a classic assay system widely used in the past to assess the *in vivo* pluripotency of mouse or human ES cells (Lensch *et al.*, 2007). However, the significance of this assay system is a concern. To distinguish real teratomas from teratocarcinoma-like cells with predominantly neuroectodermal elements, specific histology experience is required (Daley *et al.*, 2009). On the other hand, iPS cell lines with reduced or no potential for teratoma formation might be a preferable source for cell transplantation strategies (Ellis *et al.*, 2009).

Transcriptional profiling using microarrays is one method to compare starting cell populations, reprogrammed cells, and ES cells (Fig. 18.2). Since no other *in vivo* developmental assays, apart from teratoma formation, are feasible in the human system, transcriptional profiling is even more significant for comparisons of human iPS cells with hESC lines.

To select pluripotent cell populations or to derive and propagate pure populations of specific cell types from a complex culture of differentiating embryonic or pluripotent stem cells, lineage-selection methodologies must be applied to the culture systems. Apart from gene targeting by classic homologous recombination, essentially all strategies employ vectors with positive or negative selectable markers, whose expression is directed by stem cell or lineage-specific promoter/enhancer elements. A lentiviral vector with Oct4 and Sox2 promoter and enhancer elements (EOS) driving a GFP transgene was used to mark and purify emerging mouse and human iPS cell colonies (Hotta *et al.*, 2009). A correlation between GFP expression and pluripotency status was found in these cells.

The concept of lineage selection from iPS-derived cells using lentiviral vector technology is widely applicable. The strategy can be applied as follows:

1. Clone a positive or negative selectable marker under the control of a lineage-specific promoter/enhancer element in a (SIN-) lentiviral vector backbone.
2. Transduce iPS cells with lentivirus at a low MOI.
3. Sort single-cell clones and expand the undifferentiated stem cell population.
4. Evaluate the cell population for the selectable marker upon differentiation.

Gain-of-function and loss-of-function analyses can be achieved by direct translation of genetic modification approaches originally developed for ES cells to iPS cells (Zaehres and Daley, 2006).

7. Biotechnological and Medical Applications

hESCs have the ability to differentiate along all embryonic and adult developmental lineages, which makes them a valuable model to study early developmental processes and provide a potential source of cells for gene and cell-based therapies (Thomson *et al.*, 1998). However, the derivation of hESCs is fraught with ethical considerations. The discovery of human iPS cells obviates these ethical concerns and paves the way for generating pluripotent cells from somatic cells of individual patients for disease modeling, drug discovery, and cell-based therapeutic applications (Park *et al.*, 2008a; Takahashi *et al.*, 2007; Yu *et al.*, 2007; Zaehres and Schöler, 2007).

Engineering of cardiomyocytes, blood, neurons, pancreatic cells, or a multitude of other cell types appears feasible by transferring established protocols from ES cells to iPS cells. Transgene expression and gene knockdown technologies are crucial to the development of these *in vitro* differentiation protocols, requiring lineage-selection methodologies for clinical applications. Transgenic mice with constitutive, inducible, conditional, and cell type-specific gain-of-function or loss-of-function phenotypes have proven invaluable as advanced models for human diseases and for evaluation of drug, protein, gene, and cell-based therapies. Human iPS cells and their *in vitro* differentiation into defined cell lineages should have an even greater impact as model systems for human diseases, as they are human cells. The largest impact of iPS cell technology over the short term will be on drug screening. The long-term goal of iPS cell technology is to provide a source of stem cell-derived allogeneic or autologous cell grafts for regenerative medicine approaches, which may take a decade or longer to come to fruition. The suitability of reprogramming human cord blood-derived cells to iPS cells (Eminli *et al.*, 2009; Giorgetti *et al.*, 2009; Haase *et al.*, 2009; Loh *et al.*, 2009) (Fig. 18.2) opens the perspective to generate HLA-matched pluripotent stem cell banks based on existing cord blood banks. The lessons learned from reprogramming somatic cells to iPS cells might be further applied to discover and modulate endogenous regenerative cell programs in the human body and to directly reprogram cell lineages *in vivo* to other cell lineages.

Acknowledgments

This work was supported by grants and financial support from the German Ministry of Research and Education (BMBF), the Deutsche Forschungsgemeinschaft (DFG), the German State of North Rhine Westphalia (NRW), and the Max Planck Society (MPG).

REFERENCES

Boyer, L. A., Lee, T. I., Cole, M. F., Johnstone, S. E., Levine, S. S., Zucker, J. P., Guenther, M. G., Kumar, R. M., Murray, H. L., Jenner, R. G., *et al.* (2005). Core transcriptional regulatory circuitry in human embryonic stem cells. *Cell* **122,** 947–956.

Brambrink, T., Foreman, R., Welstead, G. G., Lengner, C. J., Wernig, M., Suh, H., and Jaenisch, R. (2008). Sequential expression of pluripotency markers during direct reprogramming of mouse somatic cells. *Cell Stem Cell* **2,** 151–159.

Burns, J. C., Friedmann, T., Driever, W., Burrascano, M., and Yee, J. K. (1993). Vesicular stomatitis virus G glycoprotein pseudotyped retroviral vectors: Concentration to very high titer and efficient gene transfer into mammalian and nonmammalian cells. *Proc. Natl. Acad. Sci. USA* **90,** 8033–8037.

Chambers, I., Colby, D., Robertson, M., Nichols, J., Lee, S., Tweedie, S., and Smith, A. (2003). Functional expression cloning of Nanog, a pluripotency sustaining factor in embryonic stem cells. *Cell* **113,** 643–655.

Cherry, S. R., Biniszkiewicz, D., van Parijs, L., Baltimore, D., and Jaenisch, R. (2000). Retroviral expression in embryonic stem cells and hematopoietic stem cells. *Mol. Cell. Biol.* **20,** 7419–7426.

Cowan, C. A., Atienza, J., Melton, D. A., and Eggan, K. (2005). Nuclear reprogramming of somatic cells after fusion with human embryonic stem cells. *Science* **309,** 1369–1373.

Daley, G. Q., Lensch, M. W., Jaenisch, R., Meissner, A., Plath, K., and Yamanaka, S. (2009). Broader implications of defining standards for the pluripotency of iPSCs. *Cell Stem Cell* **4,** 200–201. (Author reply 202).

Do, J. T., and Schöler, H. R. (2004). Nuclei of embryonic stem cells reprogram somatic cells. *Stem Cells* **22,** 941–949.

Ellis, J., Bruneau, B. G., Keller, G., Lemischka, I. R., Nagy, A., Rossant, J., Srivastava, D., Zandstra, P. W., and Stanford, W. L. (2009). Alternative induced pluripotent stem cell characterization criteria for in vitro applications. *Cell Stem Cell* **4,** 198–199. (Author reply 202).

Eminli, S., Foudi, A., Stadtfeld, M., Maherali, N., Ahfeldt, T., Mostoslavsky, G., Hock, H., and Hochedlinger, K. (2009). Differentiation stage determines potential of hematopoietic cells for reprogramming into induced pluripotent stem cells. *Nat. Genet.* **41,** 968–976.

Evans, M. J., and Kaufman, M. H. (1981). Establishment in culture of pluripotential cells from mouse embryos. *Nature* **292,** 154–156.

Fusaki, N., Ban, H., Nishiyama, A., Saeki, K., and Hasegawa, M. (2009). Efficient induction of transgene-free human pluripotent stem cells using a vector based on Sendai virus, an RNA virus that does not integrate into the host genome. *Proc. Jpn. Acad., Ser. B, Phys. Biol. Sci.* **85,** 348–362.

Giorgetti, A., Montserrat, N., Aasen, T., Gonzalez, F., Rodriguez-Piza, I., Vassena, R., Raya, A., Boue, S., Barrero, M. J., Corbella, B. A., *et al.* (2009). Generation of induced pluripotent stem cells from human cord blood using OCT4 and SOX2. *Cell Stem Cell* **5,** 353–357.

Gropp, M., Itsykson, P., Singer, O., Ben-Hur, T., Reinhartz, E., Galun, E., and Reubinoff, B. E. (2003). Stable genetic modification of human embryonic stem cells by lentiviral vectors. *Mol. Ther.* **7,** 281–287.

Gurdon, J. B., Elsdale, T. R., and Fischberg, M. (1958). Sexually mature individuals of *Xenopus laevis* from the transplantation of single somatic nuclei. *Nature* **182,** 64–65.

Haase, A., Olmer, R., Schwanke, K., Wunderlich, S., Merkert, S., Hess, C., Zweigerdt, R., Gruh, I., Meyer, J., Wagner, S., *et al.* (2009). Generation of induced pluripotent stem cells from human cord blood. *Cell Stem Cell* **5,** 434–441.

Hockemeyer, D., Soldner, F., Cook, E. G., Gao, Q., Mitalipova, M., and Jaenisch, R. (2008). A drug-inducible system for direct reprogramming of human somatic cells to pluripotency. *Cell Stem Cell* **3,** 346–353.

Hong, H., Takahashi, K., Ichisaka, T., Aoi, T., Kanagawa, O., Nakagawa, M., Okita, K., and Yamanaka, S. (2009). Suppression of induced pluripotent stem cell generation by the p53-p21 pathway. *Nature* **460,** 1132–1135.

Hotta, A., Cheung, A. Y., Farra, N., Vijayaragavan, K., Seguin, C. A., Draper, J. S., Pasceri, P., Maksakova, I. A., Mager, D. L., Rossant, J., *et al.* (2009). Isolation of human iPS cells using EOS lentiviral vectors to select for pluripotency. *Nat. Methods* **6,** 370–376.

Huangfu, D., Osafune, K., Maehr, R., Guo, W., Eijkelenboom, A., Chen, S., Muhlestein, W., and Melton, D. A. (2008). Induction of pluripotent stem cells from primary human fibroblasts with only Oct4 and Sox2. *Nat. Biotechnol.* **26,** 1269–1275.

Ichida, J. K., Blanchard, J., Lam, K., Son, E. Y., Chung, J. E., Egli, D., Loh, K. M., Carter, A. C., Di Giorgio, F. P., Koszka, K., *et al.* (2009). A small-molecule inhibitor of tgf-Beta signaling replaces sox2 in reprogramming by inducing nanog. *Cell Stem Cell* **5,** 491–503.

Kim, J. B., Zaehres, H., Wu, G., Gentile, L., Ko, K., Sebastiano, V., Arauzo-Bravo, M. J., Ruau, D., Han, D. W., Zenke, M., *et al.* (2008). Pluripotent stem cells induced from adult neural stem cells by reprogramming with two factors. *Nature* **454,** 646–650.

Kim, D., Kim, C. H., Moon, J. I., Chung, Y. G., Chang, M. Y., Han, B. S., Ko, S., Yang, E., Cha, K. Y., Lanza, R., *et al.* (2009a). Generation of human induced pluripotent stem cells by direct delivery of reprogramming proteins. *Cell Stem Cell* **4,** 472–476.

Kim, J. B., Greber, B., Arauzo-Bravo, M. J., Meyer, J., Park, K. I., Zaehres, H., and Schöler, H. R. (2009b). Direct reprogramming of human neural stem cells by OCT4. *Nature* **461,** 649–653.

Kim, J. B., Sebastiano, V., Wu, G., Arauzo-Bravo, M. J., Sasse, P., Gentile, L., Ko, K., Ruau, D., Ehrich, M., van den Boom, D., *et al.* (2009c). Oct4-induced pluripotency in adult neural stem cells. *Cell* **136,** 411–419.

Koch, P., Siemen, H., Biegler, A., Itskovitz-Eldor, J., and Brustle, O. (2006). Transduction of human embryonic stem cells by ecotropic retroviral vectors. *Nucleic Acids Res.* **34,** e120.

Lensch, M. W., Schlaeger, T. M., Zon, L. I., and Daley, G. Q. (2007). Teratoma formation assays with human embryonic stem cells: A rationale for one type of human-animal chimera. *Cell Stem Cell* **1,** 253–258.

Li, W., Zhou, H., Abujarour, R., Zhu, S., Young Joo, J., Lin, T., Hao, E., Schöler, H. R., Hayek, A., and Ding, S. (2009). Generation of human-induced pluripotent stem cells in the absence of exogenous Sox2. *Stem Cells* **27,** 2992–3000.

Liao, J., Wu, Z., Wang, Y., Cheng, L., Cui, C., Gao, Y., Chen, T., Rao, L., Chen, S., Jia, N., *et al.* (2008). Enhanced efficiency of generating induced pluripotent stem (iPS) cells from human somatic cells by a combination of six transcription factors. *Cell Res.* **18,** 600–603.

Loh, Y. H., Agarwal, S., Park, I. H., Urbach, A., Huo, H., Heffner, G. C., Kim, K., Miller, J. D., Ng, K., and Daley, G. Q. (2009). Generation of induced pluripotent stem cells from human blood. *Blood* **113,** 5476–5479.

Lois, C., Hong, E. J., Pease, S., Brown, E. J., and Baltimore, D. (2002). Germline transmission and tissue-specific expression of transgenes delivered by lentiviral vectors. *Science* **295,** 868–872.

Lyssiotis, C. A., Foreman, R. K., Staerk, J., Garcia, M., Mathur, D., Markoulaki, S., Hanna, J., Lairson, L. L., Charette, B. D., Bouchez, L. C., *et al.* (2009). Reprogramming of murine fibroblasts to induced pluripotent stem cells with chemical complementation of Klf4. *Proc. Natl. Acad. Sci. USA* **106,** 8912–8917.

Ma, Y., Ramezani, A., Lewis, R., Hawley, R. G., and Thomson, J. A. (2003). High-level sustained transgene expression in human embryonic stem cells using lentiviral vectors. *Stem Cells* **21,** 111–117.

Maherali, N., Sridharan, R., Xie, W., Utikal, J., Eminli, S., Arnold, K., Stadtfeld, M., Yachechko, R., Tchieu, J., Jaenisch, R., *et al.* (2007). Directly reprogrammed fibroblasts show global epigenetic remodeling and widespread tissue contribution. *Cell Stem Cell* **1,** 55–70.

Martin, G. R. (1981). Isolation of a pluripotent cell line from early mouse embryos cultured in medium conditioned by teratocarcinoma stem cells. *Proc. Natl. Acad. Sci. USA* **78,** 7634–7638.

Meissner, A., Wernig, M., and Jaenisch, R. (2007). Direct reprogramming of genetically unmodified fibroblasts into pluripotent stem cells. *Nat. Biotechnol.* **25,** 1177–1181.

Mikkelsen, T. S., Hanna, J., Zhang, X., Ku, M., Wernig, M., Schorderet, P., Bernstein, B. E., Jaenisch, R., Lander, E. S., and Meissner, A. (2008). Dissecting direct reprogramming through integrative genomic analysis. *Nature* **454,** 49–55.

Mitsui, K., Tokuzawa, Y., Itoh, H., Segawa, K., Murakami, M., Takahashi, K., Maruyama, M., Maeda, M., and Yamanaka, S. (2003). The homeoprotein Nanog is required for maintenance of pluripotency in mouse epiblast and ES cells. *Cell* **113,** 631–642.

Nakagawa, M., Koyanagi, M., Tanabe, K., Takahashi, K., Ichisaka, T., Aoi, T., Okita, K., Mochiduki, Y., Takizawa, N., and Yamanaka, S. (2008). Generation of induced pluripotent stem cells without Myc from mouse and human fibroblasts. *Nat. Biotechnol.* **26,** 101–106.

Naldini, L., Blomer, U., Gallay, P., Ory, D., Mulligan, R., Gage, F. H., Verma, I. M., and Trono, D. (1996). In vivo gene delivery and stable transduction of nondividing cells by a lentiviral vector. *Science* **272,** 263–267.

Naviaux, R. K., Costanzi, E., Haas, M., and Verma, I. M. (1996). The pCL vector system: Rapid production of helper-free, high-titer, recombinant retroviruses. *J. Virol.* **70,** 5701–5705.

Nichols, J., Zevnik, B., Anastassiadis, K., Niwa, H., Klewe-Nebenius, D., Chambers, I., Schöler, H., and Smith, A. (1998). Formation of pluripotent stem cells in the mammalian embryo depends on the POU transcription factor Oct4. *Cell* **95,** 379–391.

Okita, K., Ichisaka, T., and Yamanaka, S. (2007). Generation of germline-competent induced pluripotent stem cells. *Nature* **448,** 313–317.

Okita, K., Nakagawa, M., Hyenjong, H., Ichisaka, T., and Yamanaka, S. (2008). Generation of mouse induced pluripotent stem cells without viral vectors. *Science* **322,** 949–953.

Park, I. H., Arora, N., Huo, H., Maherali, N., Ahfeldt, T., Shimamura, A., Lensch, M. W., Cowan, C., Hochedlinger, K., and Daley, G. Q. (2008a). Disease-specific induced pluripotent stem cells. *Cell* **134,** 877–886.

Park, I. H., Zhao, R., West, J. A., Yabuuchi, A., Huo, H., Ince, T. A., Lerou, P. H., Lensch, M. W., and Daley, G. Q. (2008b). Reprogramming of human somatic cells to pluripotency with defined factors. *Nature* **451,** 141–146.

Pfeifer, A., Ikawa, M., Dayn, Y., and Verma, I. M. (2002). Transgenesis by lentiviral vectors: Lack of gene silencing in mammalian embryonic stem cells and preimplantation embryos. *Proc. Natl. Acad. Sci. USA* **99,** 2140–2145.

Schöler, H. R., Ruppert, S., Suzuki, N., Chowdhury, K., and Gruss, P. (1990). New type of POU domain in germ line-specific protein Oct-4. *Nature* **344,** 435–439.

Shi, Y., Do, J. T., Desponts, C., Hahm, H. S., Schöler, H. R., and Ding, S. (2008). A combined chemical and genetic approach for the generation of induced pluripotent stem cells. *Cell Stem Cell* **2,** 525–528.

Soldner, F., Hockemeyer, D., Beard, C., Gao, Q., Bell, G. W., Cook, E. G., Hargus, G., Blak, A., Cooper, O., Mitalipova, M., *et al.* (2009). Parkinson's disease patient-derived induced pluripotent stem cells free of viral reprogramming factors. *Cell* **136,** 964–977.

Stadtfeld, M., Maherali, N., Breault, D. T., and Hochedlinger, K. (2008a). Defining molecular cornerstones during fibroblast to iPS cell reprogramming in mouse. *Cell Stem Cell* **2,** 230–240.

Stadtfeld, M., Nagaya, M., Utikal, J., Weir, G., and Hochedlinger, K. (2008b). Induced pluripotent stem cells generated without viral integration. *Science* **322,** 945–949.

Tada, M., Takahama, Y., Abe, K., Nakatsuji, N., and Tada, T. (2001). Nuclear reprogramming of somatic cells by in vitro hybridization with ES cells. *Curr. Biol.* **11,** 1553–1558.

Takahashi, K., and Yamanaka, S. (2006). Induction of pluripotent stem cells from mouse embryonic and adult fibroblast cultures by defined factors. *Cell* **126,** 663–676.

Takahashi, K., Tanabe, K., Ohnuki, M., Narita, M., Ichisaka, T., Tomoda, K., and Yamanaka, S. (2007). Induction of pluripotent stem cells from adult human fibroblasts by defined factors. *Cell* **131,** 861–872.

Thomson, J. A., Itskovitz-Eldor, J., Shapiro, S. S., Waknitz, M. A., Swiergiel, J. J., Marshall, V. S., and Jones, J. M. (1998). Embryonic stem cell lines derived from human blastocysts. *Science* **282,** 1145–1147.

Viswanathan, S. R., Daley, G. Q., and Gregory, R. I. (2008). Selective blockade of microRNA processing by Lin28. *Science* **320,** 97–100.

Wernig, M., Meissner, A., Foreman, R., Brambrink, T., Ku, M., Hochedlinger, K., Bernstein, B. E., and Jaenisch, R. (2007). In vitro reprogramming of fibroblasts into a pluripotent ES-cell-like state. *Nature* **448,** 318–324.

Wernig, M., Lengner, C. J., Hanna, J., Lodato, M. A., Steine, E., Foreman, R., Staerk, J., Markoulaki, S., and Jaenisch, R. (2008). A drug-inducible transgenic system for direct reprogramming of multiple somatic cell types. *Nat. Biotechnol.* **26,** 916–924.

Will, E., Klump, H., Heffner, N., Schwieger, M., Schiedlmeier, B., Ostertag, W., Baum, C., and Stocking, C. (2002). Unmodified Cre recombinase crosses the membrane. *Nucleic Acids Res.* **30,** e59.

Wilmut, I., Schnieke, A. E., McWhir, J., Kind, A. J., and Campbell, K. H. (1997). Viable offspring derived from fetal and adult mammalian cells. *Nature* **385,** 810–813.

Woltjen, K., Michael, I. P., Mohseni, P., Desai, R., Mileikovsky, M., Hamalainen, R., Cowling, R., Wang, W., Liu, P., Gertsenstein, M., *et al.* (2009). piggyBac transposition reprograms fibroblasts to induced pluripotent stem cells. *Nature* **458,** 766–770.

Xu, C., Inokuma, M. S., Denham, J., Golds, K., Kundu, P., Gold, J. D., and Carpenter, M. K. (2001). Feeder-free growth of undifferentiated human embryonic stem cells. *Nat. Biotechnol.* **19,** 971–974.

Yamanaka, S. (2007). Strategies and new developments in the generation of patient-specific pluripotent stem cells. *Cell Stem Cell* **1,** 39–49.

Yu, J., Vodyanik, M. A., Smuga-Otto, K., Antosiewicz-Bourget, J., Frane, J. L., Tian, S., Nie, J., Jonsdottir, G. A., Ruotti, V., Stewart, R., *et al.* (2007). Induced pluripotent stem cell lines derived from human somatic cells. *Science* **318,** 1917–1920.

Yu, J., Hu, K., Smuga-Otto, K., Tian, S., Stewart, R., Slukvin, I. I., and Thomson, J. A. (2009). Human induced pluripotent stem cells free of vector and transgene sequences. *Science* **324,** 797–801.

Yuan, H., Corbi, N., Basilico, C., and Dailey, L. (1995). Developmental-specific activity of the FGF-4 enhancer requires the synergistic action of Sox2 and Oct-3. *Genes Dev.* **9,** 2635–2645.

Zaehres, H., and Daley, G. Q. (2006). Transgene expression and RNA interference in embryonic stem cells. *Methods Enzymol.* **420,** 49–64.

Zaehres, H., and Schöler, H. R. (2007). Induction of pluripotency: From mouse to human. *Cell* **131,** 834–835.

Zhao, Y., Yin, X., Qin, H., Zhu, F., Liu, H., Yang, W., Zhang, Q., Xiang, C., Hou, P., Song, Z., *et al.* (2008). Two supporting factors greatly improve the efficiency of human iPSC generation. *Cell Stem Cell* **3,** 475–479.

Zhao, X. Y., Li, W., Lv, Z., Liu, L., Tong, M., Hai, T., Hao, J., Guo, C. L., Ma, Q. W., Wang, L., *et al.* (2009). iPS cells produce viable mice through tetraploid complementation. *Nature* **461,** 86–90.

Zhou, H., Wu, S., Joo, J. Y., Zhu, S., Han, D. W., Lin, T., Trauger, S., Bien, G., Yao, S., Zhu, Y., *et al.* (2009). Generation of induced pluripotent stem cells using recombinant proteins. *Cell Stem Cell* **4,** 381–384.

SECTION SIX

IMAGING MOUSE DEVELOPMENT

CHAPTER NINETEEN

Imaging Mouse Embryonic Development

Ryan S. Udan*,† *and* Mary E. Dickinson*,†

Contents

Abstract

For the past three decades, methods for culturing mouse embryos *ex vivo* have been optimized in order to improve embryo viability and physiology throughout critical stages of embryogenesis. Combining advances made in the production of transgenic animals and in the development of different varieties of fluorescent proteins (FPs), time-lapse imaging is becoming more and more popular in the analysis of dynamic events during mouse development. Targeting FPs to specific cell types or subcellular compartments has enabled researchers to

* Department of Molecular Physiology and Biophysics, Baylor College of Medicine, Houston, Texas, USA
† Program in Development Biology, Baylor College of Medicine, Houston, Texas, USA

Methods in Enzymology, Volume 476
ISSN 0076-6879, DOI: 10.1016/S0076-6879(10)76019-5

study cell proliferation, apoptosis, migration, and changes in cell morphology in living mouse embryos in real time. Here we provide a guide for time-lapse imaging of early stages of mouse embryo development.

1. Introduction

During development, cell–cell interactions, divisions, and coordinated movements underlie the changes in shape known as morphogenesis. How cells and tissues form and rearrange during morphogenesis has fascinated scientists for generations, yet the ability to directly observe such profound and fundamental changes has only recently been attained. Before the 1950s (when video cameras were combined with microscopes), knowledge about morphogenetic events was stored only in the minds of the observers and could only be communicated in drawings (Inoue and Gliksman, 2003). In more recent times, video microscopy and continued advances in technology have provided new tools for cell and developmental biologists to discern closer and clearer views of morphogenesis in action. The advent of confocal and multiphoton microscopes in the 1980s and 1990s together with the development of fluorescent proteins (FPs) that can be used to vitally label cells of interest (also emerging in the 1990s) has transformed what we can directly visualize in a living embryo, and provides us with exciting new insights into both cellular and subcellular dynamics during development. Moreover, advances in digital media now make it possible not only to record the dynamic events of development but to publish and share movies with the world with the click of a button.

Despite advances in imaging technology, probe development, and digital media, there are still limitations imposed by the samples themselves as microscopic analysis of the embryo requires that the organism can be grown on a microscope stage. Thus, most imaging studies in vertebrate embryos have been carried out using zebrafish, avian (chick and quail), and *Xenopus* embryos which can be readily maintained in culture. Imaging mammalian model systems, such as mouse or rat embryos, has been significantly more difficult as uterine implantation, which supports growth and development of the embryo, does not permit direct visualization of embryonic development at all stages. To observe rodent embryos *in utero*, approaches such as MRI and ultrasound have been used (Dickinson, 2006). These methods have excellent depth penetration but with relatively poor resolution (50–200 μm) and relatively few specific labels exist to visualize cells or subcellular details of interest. Fluorescence microscopy methods offer cellular and subcellular resolution and a rapidly growing list of promoters and enhancers to label cells with FPs, but fluorescence microscopy cannot be performed through the thick and light scattering uterine wall. However, it is possible to maintain mouse and rat embryos in culture during pre- and early postimplantation stages, prior to forming a maternal–placental connection (Copp and Cockroft, 1990; Hsu, 1979; Jones *et al.*, 2002; New and Cockroft, 1979; Tam, 1998; Wiley *et al.*, 1978). We and others have adapted these protocols for static culture on the microscope stage to allow time-lapse imaging of mouse embryo development (Aulehla *et al.*, 2008; Fraser *et al.*, 2005; Kwon *et al.*, 2008; Plusa *et al.*, 2008; Srinivas *et al.*, 2004). In this chapter, we review the methods that have been used successfully to enable direct imaging of mouse embryos. We have highlighted some of the key concerns and criteria for successful embryo maintenance in static culture on the microscope stage and have also included insight into imaging and image analysis to aid researchers in obtaining the most from live imaging experiments.

2. Overview of Key Parameters

For imaging mouse embryonic development, optimization of three important criteria have been shown to be essential—culture conditions, cell/tissue labeling, and imaging tools.

2.1. Culture conditions

To grow embryos *ex vivo* in static culture, culture media must be carefully maintained at the appropriate temperature, pH, and humidity and must be supplied with the right balance of gasses. Carefully controlling these parameters allow mouse embryos to survive and develop comparably to embryos grown *in utero*. Mouse embryos can be cultured successfully during two different stages of development—preimplantation stages up until the early somite stages (Hsu, 1979; Sadler and New, 1981; Sherbahn *et al.*, 1996; Tam, 1998) and prestreak stages up until embryonic day 10.5 (E10.5) (Hsu, 1979; Jones *et al.*, 2002). The protocols discussed here have been optimized for E6.5–E10.5 embryos. Embryos at these stages are sensitive to changes in environmental conditions and should be imaged using an environmental stage unit which can be used to control these parameters such as the one shown in Fig. 19.1. Detailed culture conditions are given below.

2.2. Cell/tissue labeling

Over the past 10 years, there have been tremendous advances in vital cell labeling with improvements both in dyes as well as the development of bright and stable FPs. While much can be said about fluorescent labels and further details are provided below, the best labels are bright and nontoxic. Bright fluorescent molecules or proteins are generally those that have both a

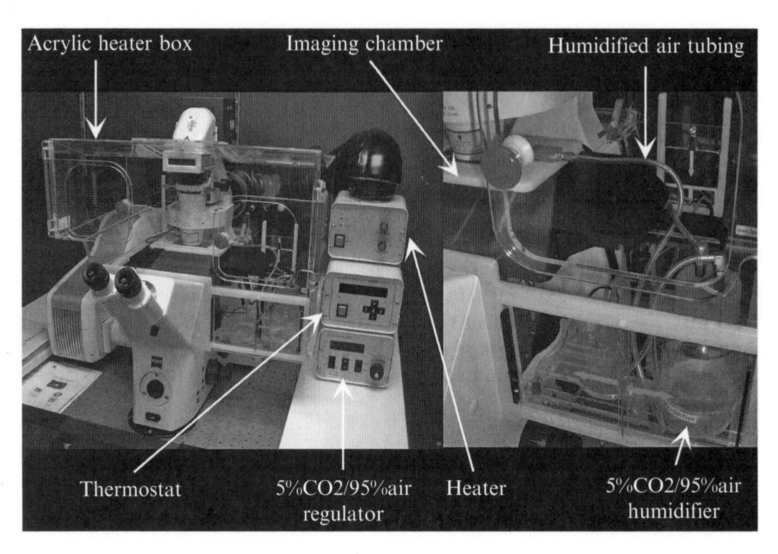

Figure 19.1 Zeiss LSM 5 LIVE confocal laser scanning microscope setup. The microscope is equipped with an acrylic heater box which surrounds the microscope stage. Directly on the microscope stage is the imaging chamber which consists of a lower platform in which the culture dish rests on, and a top platform that encloses the imaging chamber and has a clear acrylic portion to allow light to pass through for bright-field imaging. The lower platform has an open circular area at the bottom that allows the objective to collect light. Two thermosensors are placed in the imaging chamber and within the heater box to control the thermostat. The thermostat regulates the temperature (37 °C) by controlling the heater, which is attached to the back of the heater box. For humidified air, 5% CO_2/95% air flows through the regulator and humidifier, and it is sent through tubing to the imaging chamber.

high extinction coefficient and a high quantum yield. This means that the label readily absorbs photons and emits photons as the excited electron relaxes back to the ground state. Fluorochromes with a poor quantum yield exhibit nonradiative decay which can result in local heating or increases in free radicals. Also, dim fluorochromes require higher power illumination which can damage the tissue directly. The fluorescent-labeling strategy should be optimized as much as possible to reduce excess illumination. The choice of fluorescent label will also depend somewhat on the imaging system being used, so take care to match these choices with the illumination wavelengths and emission filters that are available (see below).

2.3. Imaging tools

The protocols that have been established to grow mouse embryos on the microscope stage can be used in conjunction with a number of imaging strategies or microscopies (for instance, using optical coherence tomography; Larina *et al.*, 2008, 2009), but fluorescence microscopy offers the most utility for imaging cellular and subcellular events with the widest range of available cell labels. Both wide-field fluorescence microscopy and confocal laser scanning microscopy (CLSM) have been used for time-lapse microscopy of developing mouse embryos, but for many applications confocal microscopy can produce considerably better images with enhanced resolution and signal-to-noise over wide-field fluorescence approaches. Most modern confocal microscopes have time-lapse imaging capabilities built into their acquisition software for automated collection and offer considerable flexibility in scan strategies to enable efficient and sensitive collection of fluorescence signal while minimizing damage. Movies generated from these experiments can be further processed with image analysis tools to quantify changes in cell proliferation, apoptosis, morphology, and migration.

There are many examples in which live imaging of cultured mouse embryos or embryonic tissue explants, in combination with transgenic reporters, has revealed insights into the mechanisms behind various cellular and developmental events. For example, by labeling precursors to specific cell types in early development, a more accurate understanding of the specific morphogenetic events driving germ layer formation has been determined (Burtscher and Lickert, 2009; Kwon *et al.*, 2008; Plusa *et al.*, 2008; Srinivas *et al.*, 2004). Imaging has revealed new information about notochord formation (Yamanaka *et al.*, 2007), peripheral nerve outgrowth (Brachmann *et al.*, 2007), neurogenesis from basal neuroepithelium (Haubensak *et al.*, 2004), ureteric bud branching (Srinivas *et al.*, 1999; Watanabe and Costantini, 2004), and FP fusions have provided insights into specific cellular events such as cell migration and motion (Anderson *et al.*, 2000; Druckenbrod and Epstein, 2005; Jones *et al.*, 2002, 2004; Molyneaux *et al.*, 2001; Young *et al.*, 2004), cell mitosis and G1 to S-phase transitions (Fraser *et al.*, 2005; Sakaue-Sawano *et al.*, 2008), nodal cilia dynamics (Nonaka *et al.*, 1998; Okada *et al.*, 1999), and cell death in primitive endoderm-fated cells (Hadjantonakis and Papaioannou, 2004; Plusa *et al.*, 2008).

This chapter details an approach for live imaging the mouse embryonic yolk sac. However, similar culturing and imaging methods can be adapted to study other embryonic structures. Here, we will discuss culturing conditions, dissection strategies, use of FP transgenics, confocal fluorescence microscopy, time-lapse imaging, and image analyses.

3. Whole Embryo Culture

3.1. Whole embryo culture staging

Although there are well-described roller bottle protocols for culturing postimplantation mouse embryos (Downs and Gardner, 1995; Lawson *et al.*, 1986; New and Cockroft, 1979; Sadler and New, 1981), static culture is required for microscopy. For early postimplantation stage embryos (starting at E6.5) grown in static culture, cultures can begin between early streak formation to E9.5 (Jones *et al.*, 2002; Nagy, 2003; Tam, 1998). They can be grown for 18–24 h until E9.5, and with limited success to E10.5 (Jones *et al.*, 2002). By E10.5 there is limited diffusion through thicker, more complex embryos and the lack of placental support prevents extensive normal development. For imaging dynamic events in later stage embryos, explant culture has been used successfully to image salivary glands, lungs, kidneys, ovaries, testes, and heart (epicardium) (Coveney *et al.*, 2008; Nel-Themaat *et al.*, 2009; Rhee *et al.*, 2009; Sakai and Onodera, 2008; Watanabe and Costantini, 2004) but with the caveat that the tissues are not perfused by blood flow as in live, intact embryos.

To verify success of the static cultures, developmental timing can be compared to embryos grown *in vivo* at similar time points by assessing morphological changes. For instance, for embryo cultures starting at E8.5, the formation of the neural plate, appearance of the first somite, formation of a linear heart tube, appearance of the first heart beat, commencement of head fold closure, initiation of heart looping, fusion of head folds, and the onset and completion of axial rotation or turning are the various developmental changes that can be observed (Jones *et al.*, 2002). In embryo cultures starting at E9.5, changes in embryo size and maturation of head features can be compared. In addition, heart rate and the rate of somite formation can be used as metrics of continuing development (Jones *et al.*, 2002; Nagy, 2003) and more sophisticated approaches to measure blood flow (Jones *et al.*, 2004) can provide detailed information about cardiovascular physiology.

3.2. Culture conditions

For whole embryo culture of postimplantation embryos, two types of media are prepared fresh on the day of use in a sterile tissue culture hood (see Section 3.4): dissection media and culture media. Dissection media is comprised of DMEM/F12 (Invitrogen, Cat. # 11330) which contains a mixture of salts, buffers, amino acids, nutrients, vitamins, and pH indicator. To prepare dissection media, 90% (v/v) DMEM/F12 is supplemented with 10% (v/v) fetal bovine serum (Gibco) and penicillin (1 unit/ml)/streptomycin (1 μg/ml) antibiotics. For culturing embryos at E7.5 and beyond, culture media is comprised of a 1:1 ratio of DMEM/F12 to homemade rat serum and with penicillin (1 unit/ml)/streptomycin (1 μg/ml) antibiotics (see Section 3.3); however, we have also had success in using a 6:1 ratio of dissection media to homemade rat serum. It is important to note that the homemade rat serum (on the day of use) should be incubated for a minimum of 1 h at 37 °C (5% CO_2/95% air, for E7.5–E9.5 cultures) with an open cap in a sterile 50 ml Falcon® tube (BD Biosciences), to allow for excess ether to evaporate (too much ether can reduce cardiac contractility and blood flow).

After incubating the rat serum, the culture media is prepared and placed through a 0.45 μm syringe filter (Nalgene®) for sterilization. Both dissection media and culture media are warmed to 37 °C prior to use.

An important parameter for a successful whole embryo culture is maintenance of the pH of the media. For bicarbonate or HEPES-based buffers, exposure to 5% CO_2 can maintain the pH of the media to ~7.2, the appropriate physiological pH. Culture media should be exposed to CO_2 by placing the media in a gas incubator (which is fed 5% CO_2/95% air) in a 50 ml Falcon® tube with the cap partially unscrewed. If pH levels are not sufficiently maintained, extra HEPES buffer can be added to the culture media to bring the final concentration of HEPES from 7.5 to 15 m*M*.

Evaporation can also affect the overall health of the embryo. In postimplantation embryos with an intact yolk sac, the yolk sac becomes wrinkled upon excessive evaporation which can subsequently impede blood flow throughout the yolk sac and embryo (Jones *et al.*, 2002). Thus, media must be kept in a humidified environment. In static culture, embryos can be kept in a gas incubator with humidifier pan or when being imaged they can be kept in an imaging chamber that is fed 5% CO_2/95% air at a low flow rate sent through a bubbler to keep the air humidified. Evaporation rates are dependent upon the percent humidity present in the laboratory environment. Some regions of the country are drier than others. In these cases, extra measures can also be taken to prevent excess evaporation including the use of a layer of sterile mineral oil on top of the culture media, silicon grease to seal the edges of the culture dish, or the addition of Teflon® tape to seal the edges of the imaging chamber (Jones *et al.*, 2002).

Temperature must also be maintained at 37 °C. Thus, the incubator can be kept at this temperature and the confocal microscope stage must have an enclosure that maintains this temperature. Keeping both the temperature of the stage and the objective can help to prevent drift caused by changes in the glass so an environmental chamber that encases the stage and nosepiece is recommended. Many microscope systems offer environmental control chambers (such as the Zeiss XL systems) that are custom fitted to that microscope base and that surrounds a portion of the microscope (Fig. 19.1). The box has a thermostat that can detect the temperature both within the box and the imaging chamber, and the amount of heat is adjusted to maintain the temperature. Alternatively, a homemade box can be made out of cardboard, insulating material, a heater, and a temperature regulated power outlet (Jones *et al.*, 2002, 2005a). Embryos can be very sensitive to fluctuations in temperature which can sometimes affect the timing of development and rate of cardiac contraction; thus, appropriate temperature regulation is critical (Nishii and Shibata, 2006).

For culturing late-stage embryos (E9.5–E11.5), growth of the embryo can limit O_2 diffusion rates. Thus, the concentration of O_2 should be adjusted depending on the developmental stage at the beginning of a culture.

For example, cultures starting at E9.5 should be cultured in 20% O_2; whereas cultures starting at E10.5 should be cultured in 95% O_2 (Nagy, 2003).

Overall, dissection conditions should be as sterile as possible. Media is prepared using sterile containers and pipette tips in a sterile tissue culture hood. Dissection tools are cleaned with distilled water and ethanol. The tools should never be placed in contact with detergents or fixative, as this can adversely affect the health of the embryos.

3.3. Preparation of rat serum

For our culturing purposes, we have utilized many different types of commercially available rat serum from several different companies, but this was met with limited success, as normal cardiovascular physiology was impaired (Jones *et al.*, 2002). Thus, homemade rat serum is prepared essentially according to previously established protocols (Fraser *et al.*, 2005; Hogan, 1994; Jones *et al.*, 2005a,b). This protocol requires two people to perform. To collect rat serum, adult male Sprague-Dawley rats at around 12 weeks of age weighing approximately 300 g (Charles River) are anesthetized (as approved by animal protocols) by exposure to ether. Once rats are unresponsive, they are laid down in a supine position, and the Rat's nose and mouth is placed inside a 50-ml Falcon® tube containing an ether-soaked paper towel to maintain anesthetization. The abdomen is prepared by spraying with 70% ethanol and a V-shaped incision is made in the peritoneum to open up the abdominal cavity. The internal organs are moved aside and excess fascia is wiped away with a Kimwipe®, leaving the dorsal aorta exposed. A beveled butterfly needle (BD Biosciences) is inserted into the dorsal aorta. Directly after insertion (when blood is present in the needle), the opposite end of the needle is inserted into a Vacutainer® blood collection tube with anticoagulants (BD Biosciences). When blood flows into the collection tube, the tube is inverted to mix the anticoagulants with the blood. The collection tube is then placed on ice, and rats are euthanized as per animal protocols (decapitation using a guillotine). Carcasses are placed in a fume hood for several hours or overnight to allow ether to evaporate before disposing and should be stored in an explosion proof freezer prior to disposal. *Note: working with ether is dangerous and all safe practices should be observed according to the regulations of your home institution.*

Once blood collection is complete, the blood is centrifuged in the collection tubes at 1300×*g* for 20 min to separate the blood cells from the serum. After centrifugation, the sera (top layer) quality is evaluated. Low-quality sera with a pinkish hue, as compared to a high-quality and less-turbid/clearer sera, is discarded because the pinkish color represents components from lysed red blood cells which impede embryo growth when present in media. Only high-quality sera is collected and transferred to a 15 ml Falcon® tube and centrifuged again at 1300×*g* for 10 min. The serum is pooled into a 50 ml Falcon® tube and is heat inactivated by placing in a 56 °C water bath for 30 min in a tissue culture hood with the lid partially unscrewed to allow the ether to evaporate. After incubation, the sera is sterile filtered through a 0.45 μm syringe filter (Nalgene®), and in a sterile hood dispensed into microcentrifuge tubes which then are stored at -80 °C for up to 1 year. Typically, we collect sera from about 30 rats, which produce about 100 ml of rat serum.

3.4. Dissection and isolation of postimplantation mouse embryos

To isolate E8.5 embryos, timed matings are performed. The presence of a vaginal plug in the morning after mating signifies a potential pregnancy. The resulting embryos are E0.5 on the afternoon of the vaginal plug. Eight days later, embryos are harvested from the mothers. Dissections are performed on a dissection stereomicroscope surrounded by a homemade heater box regulated to 37 °C to avoid interruptions in cardiac activity. The heater box is made from insulated cardboard and is heated by a space heater connected to a thermostat (Fisher Biosciences) (Jones *et al.*, 2005a,b). Embryos are isolated by humanely sacrificing mothers (using euthanization procedures approved by animal protocols). A V-shaped incision is made in the peritoneum, starting from the posterior and working anteriorly, to expose the abdominal cavity. The uterus is removed by cutting the uterus at the uterotubal/ovary junction for each horn of the uterus. The uterus is subsequently cut at the cervix and it is then placed in dissection media in a 35 mm culture dish, and further dissected in the heated box. A sterile transfer pipette is used to wash away some of the blood with dissection media in order to more clearly see the uterus. Using dissection scissors, incisions are made perpendicular to the uterus to separate each individual embryo still surrounded by the uterus (Fig. 19.2). The embryos with the uterus are then transferred (by forceps or transfer pipette) to a new 35 mm culture dish with fresh dissection media. To prevent excess nutrient expenditure, it is critical that each embryo has at least 1 ml of dissection or culture media. Typically, we place three embryos per culture dish in about 3–4 ml of media. Embryos that are not immediately being dissected are kept in dissection media and moved to the incubator until ready to dissect. Using two pairs of forceps, remove the rest of the uterus by cutting away the uterine tissue that surrounds the decidua. The embryo proper is surrounded by visceral yolk sac, parietal yolk sac, and the decidua. The decidua is shaped with the distal region being more pointed and the proximal region being wider at the ectoplacental cone (Fig. 19.2). At the base of the ectoplacental cone, carefully use both forceps to remove the decidua by cutting around this base, and slightly pulling off the decidua proximally. Then, remove the parietal endoderm cutting in the same manner. Finally, remove the clear and thin membrane, Reichert's membrane, from the yolk sac to allow for better nutrient or media exchange. For future immobilization of an intact embryo during an imaging session, keep a portion of the ectoplacental cone attached to the embryo

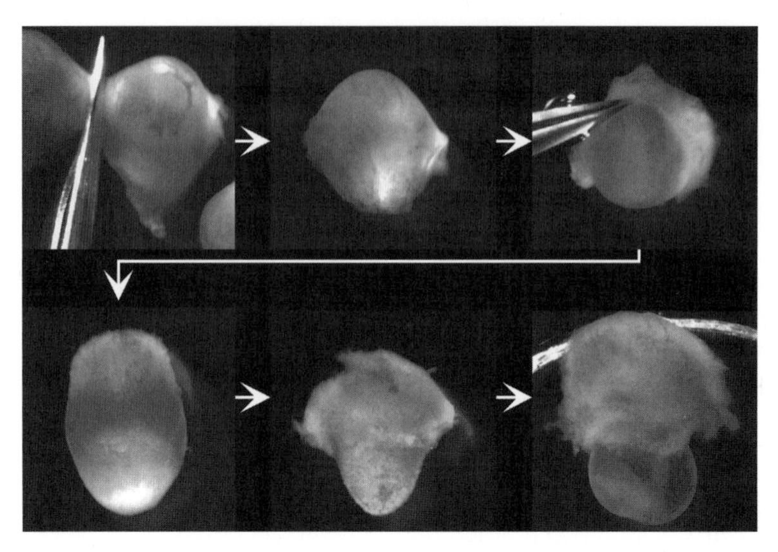

Figure 19.2 Example for dissecting E8.25–E8.75 embryos for static culture (bottom right embryo ~16 somites). Uteri are removed and placed in dissection medium in a culture dish. The embryos are dissected out by first cutting perpendicular to the uterus between implanted embryos using dissection scissors. Using forceps, the uterus is removed from the decidua, and the distal portion of the uterus is removed by holding one portion of the decidua with one pair of forceps, and gently pulling of the other part of the decidua away from the embryo with another pair of forceps. This is done around the embryo, and what remains is the embryo attached to the ectoplacental cone. The parietal endoderm and Reichert's membrane are removed, and a portion of the ectoplacental cone can be pared down and placed adjacent to the sticky sides of the Mat-Tek® culture dish glass bottom microwell to prevent drift.

(see Section 5.2 and Fig. 19.2). Embryos that are freshly dissected are transferred to fresh dissection media using a sterile transfer pipette cut off at the base, and then moved to the incubator. After all embryos are dissected, healthy embryos with strong heart beats are chosen and transferred to a 35 mm culture dish with a 10 mm glass bottom microwell (Mat-Tek®) containing 3 ml of culture media, or they can be transferred to Lab-Tek™ culture chambers (Lab-Tek™ II chambered coverglass) with 2 ml per chamber. A maximum of three embryos per dish/chamber can be cultured. Embryos are then allowed to recover in a 37 °C tissue culture incubator for 15–30 min before they are imaged.

4. Labeling Cells of Interest

As mentioned above, there are two categories of fluorescent labels that are used in embryos for time-lapse analysis, dyes and FPs. Molecular Probes (Invitrogen) is an excellent source of dyes and dye conjugates that can be used to label tissues within embryos. Some examples of these probes that have been used for vital imaging are fluorescent dextrans for labeling blood flow or iontophoretic injection into single cells for lineage tracing, lipophilic dyes (i.e., DiI and DiO) for labeling clusters of cells for tracking migration and movement or axonal connections, CellTracker dyes which are taken up into the cytoplasm marking cells of interest, BODIPY-ceramide which labels cell membranes and can be used to outline tissue structures and organization, and SYTO dyes (cell-permeant cyanine nucleic acid stains) which label nuclei in order to follow mitosis, apoptosis, or single cell migration. In cases where the investigator wishes to label a cluster of cells within a tissue, vital lipophilic fluorescent dyes such as carbocyanine lipophilic dye (DiI/DiO) can be used. While application of exogenous dyes can be very effective and convenient, these need to be injected or applied, requiring additional manipulation of the embryo which can lead to impaired viability or abnormal development and suffer from the drawback of having limited specific control over the cells that are labeled.

Genetically encoded markers such as FPs provide bright, stable markers and have a number of advantages over traditional chemical dyes. FPs can be introduced as stable elements in the genome either via viruses or by the production of transgenic mice. In transgenic mice, FPs can be driven by particular promoter sequences that have either part of a gene expression construct introduced via pronuclear injection or by "knocking in" the FP into a specific loci (Gordon and Ruddle, 1981; Hadjantonakis *et al.*, 2003; Megason *et al.*, 2006). While the knock-in approach has the advantage that specific promoter elements need not be defined, only a single FP gene per locus is expressed which can result in weak expression and low fluorescence signal unless the targeted gene is expressed at high levels. This is less of a problem with transgenes introduced via pronuclear injection since multiple copies of the transgene integrate as an array into the genome. The mutation and optimization of FPs has now resulted in dozens of available colors with many bright, stable choices for live cell imaging (Davidson and Campbell, 2009; Shaner *et al.*, 2005). Currently, there are 29 common FPs that are excitable and can emit light in the visible spectrum, where at least three spectra can be easily separated with standard filter sets and laser sources (Nowotschin *et al.*, 2009).

FPs also offer the advantage of being able to direct the fluorescence to subcellular compartments such as the cell membrane, cytoplasm, and nucleus; thus permitting direct visualization of membrane dynamics and cell migration, cell division, or changes in cell shape (Hadjantonakis *et al.*, 2003; Nowotschin and Hadjantonakis, 2009a; Passamaneck *et al.*, 2006; Rizzo *et al.*, 2009). For example, myristoylation (myr) or glycosylphosphotidylinosital (gpi) FP fusions can be used to localize FPs at the cell membrane (Hadjantonakis *et al.*, 2003; Nowotschin and Hadjantonakis, 2009b). Histone H2B::FP fusions can target FPs to the cell nucleus. Lack of these tags causes FPs to distribute throughout the cell including the cytoplasm and the nucleus. The use of these tags not only provides information about overall morphology of subcellular, cellular, and tissue structures, but it also allows researchers to visualize many dynamic processes such as cell membrane dynamics, mitosis, apoptosis, migration, nuclear import/export, and vesicular trafficking. In addition, concentrating FPs within specific domains within the cell can produce a brighter signal, allowing for less illumination light to be used to image fluorescence. There is a great deal of interest in both continuing to improve the properties of FPs as well as to adapt FPs for new purposes, as in the development of photactivateable/photoconvertible FPs and sensors for intracellular signaling pathways as well as in facilitating very high resolution microscopy (Davidson and Campbell, 2009; Nowotschin and Hadjantonakis, 2009b). The combination of different cell-specific promoters driving expression of FPs proves to be very popular markers for live cell imaging in many systems, especially the mouse.

5. Time-Lapse Imaging of Early Mouse Embryos

5.1. Confocal fluorescence microscopy setup

As discussed above, CLSM is a popular tool for *in vivo* time-lapse imaging of embryos. However, since these systems can use fairly powerful lasers, care must be taken to limit the exposure of embryos to excessive laser illumination which can cause cell damage and cell death. The amount of light exposure can be reduced by altering several aspects: scanning time intervals, transmission of the laser to the sample, optimizing the balance of illumination versus efficiency of detection specific filters, use of objectives or zooming features, and the number of optical slices or images acquired. Here are some key issues to remember when optimizing these conditions.

5.1.1. Start by knowing the excitation and emission spectrum of your FP or dye

The wavelength of the excitation laser should be as close to the excitation peak as possible and the emission filters that you choose should encompass as much of the emission spectra as possible. For instance, a long-pass 505 filter will allow you to collect more emission signal than a 505–530 band-pass filter which means you will be able to use a lower laser power setting and expose the specimen to less light for the same amount of emission signal. Thus, using single fluorochromes or well-separated pairs is advantageous over labels that have close or overlapping spectra. Narrower emission filters wastes valuable photons since they are blocked from the detector. To avoid using narrow band-pass filters, another option is to collect different color signals sequentially to avoid overlap. For instance, excite the green dye and collect the signal with a long-pass or broad band-pass filter, then excite the red dye in the second image and collect with a long-pass filter. In many cases this will produce the same amount of emission signal for less overall laser power.

5.1.2. Maximize the gain

Make sure when you are adjusting the power of the laser that you do this with the gain set to the maximum. Many people make the mistake of optimizing the gain and leaving the laser power at some default level. Set the laser at zero and the gain at its maximum, then bring up the laser power in small increments until you get the signal level that you need. If there is still considerable noise from the detector, lower the gain gradually, paying close attention to how much laser power needs to be increased to compensate.

5.1.3. Embrace the ugly

The settings that are the safest and most informative for your imaging studies may not always yield cover photo ready images but may provide plenty of information. Averaging or slow scanning can both reduce noise and improve the signal-to-noise ratio, but if viability is a concern, consider scanning faster without averaging. Faster scan times not only enable faster processes to be tracked but also minimizes laser exposure to the sample resulting in less damage. However, these benefits come with increased noise. Longer pixel dwell times and averaging reduce noise, but require longer or repeated laser exposure and can damage tissue, so beauty has its price. Consider aiming to make the images pretty enough for what you need to answer your question of interest.

5.1.4. Open the pinhole

For many live imaging experiments using thick samples, *Z*-stacks are necessary to follow cells migrating along the *Z*-axis or to image other changes in 3-D. That said, it is not always the case that thin optical sections are needed. Unless the highest possible resolution is important to the experiment (such as in imaging submicron subcellular domains), the pinhole can be opened up to allow for thicker optical sections and more signal to reach the detector and this may be sufficient to follow cell translocations, mitosis, etc. Also, opening the pinhole even a little will reduce the number of optical sections that will need to be acquired in the *Z*-axis which will reduce the amount of scans used to generate 3-D datasets and reduce the chance of damage to the embryo. Many confocal microscopes provide an estimation of the size of the optical slice thickness and this will change from lens to lens so think about how much resolution you really need. Moreover, if resolution is very important, you may also want to consider using deconvolution after you have acquired the data to make further improvements without endangering the sample. Take some stacks through a sample and try out different pinhole settings to evaluate these settings for your particular question.

5.1.5. Choose the best objective

Although most people think about the magnification needed to image a particular sample, the numerical aperture (NA) is actually most important. This number relates to the axial and lateral resolution that can be achieved as well as the amount of light that can be collected. Higher values are more desirable but higher NA lenses usually also have a shorter working distance (w.d.) which may limit the depth along the *Z*-axis that can be imaged. Also, if fluorescence imaging is being used, choose a lens that is optimized for fluorescence transmission, not for DIC or phase contrast (PH). Lenses that have DIC or PH on the lens itself are designed specifically for these types of contrast and are optimized for aberration correction and preservation of polarization but at the expense of fluorescence transmission. These lenses can contain more glass elements which limit transmission. Before you start, you should ask your local microscope representative or core director about the best lenses available for live imaging. Many manufacturers now offer lenses that are optimized for live imaging with improved transmission, longer working distances for imaging thicker specimens, temperature stability for use at 37 °C and for immersion into biological media. The right lens can make all the difference, so compare those that are available to you to find the best lens for your sample.

5.2. Transferring embryos to the microscope stage

Embryos are dissected following appropriate procedures (see Section 3.4). For imaging, embryos are placed in culture media in either Mat-Tek® culture dishes or Lab-Tek™ culture chambers. After allowing the embryos to recover in the incubator for a minimum of 15 min, they are quickly removed and placed in the imaging chamber. It is important that the imaging chamber be equilibrated to the appropriate temperature, gaseous phase and humidity for at least 1 h prior to imaging. To prevent drift of the embryos on the *XY* and *Z* positions, embryos with an intact yolk sac can be stabilized by placing the ectoplacental cone adjacent to the sticky edge of the glass microwell. If using the Lab-Tek™ culture chambers, a human hair can be tied around the ectoplacental cone in a manner which can prop up the embryo in place at the bottom of the dish (Jones *et al.*, 2005a). Alternatively, embryos can also be positioned using a holding pipette attached to a micromanipulator. In some cases, drift can still occur. This drift can be in both the *XY* plane (often caused by microcurrents in the media) and in the *Z* plane (often caused by the expansion of the yolk sac or growth of the embryo that occurs during the culturing period). To adjust for this drift, periodic evaluation of the culture should be performed, and appropriate readjustments of the *XY* and *Z* planes should be made if necessary. Time-lapse images that are readjusted in the middle of an imaging session can be realigned by adjusting for drift computationally using the Imaris software program (see Section 6).

The appropriate amount of light, magnification, Z-stack number and scanning intervals are empirically chosen as to not disrupt tissue viability. To assess tissue viability, there are three major parameters that should be performed to assess health of the embryo being imaged. First, cells labeled with a nuclear-localized FP can be used to assess whether abnormal apoptosis occurs by visualizing nuclear fragmentation. For example, time-lapse imaging of Flk1-H2B::eYFP labeled vessels can clearly show nuclear fragmentation events which can take anywhere between 0.5 and 1.5 h to observe before the fragments are cleared (Fig. 19.3). The duration of this process ensures that all apoptotic events can be captured by the 5–10 min scan time intervals. Cell death can also be assessed in nonlabeled embryos after the imaging session to assess viability. Second, normal cardiovascular physiology can be assessed by performing periodic evaluations of blood flow (visual inspection of blood flow through the eye piece or by performing bright-field imaging to detect streaks of dark cells within vessels that represent moving blood cells) during the imaging

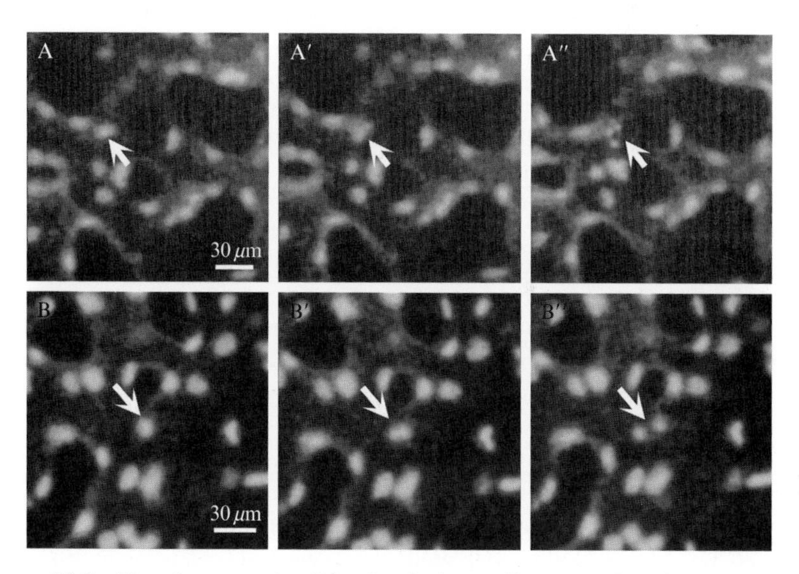

Figure 19.3 Time-lapse movie of the developing Yolk sac vessels (cultures starting at E8.5). Yolk sac vessels of intact Flk1-myr::mCherry$^{tg/tg}$ (to visualize endothelial cell membrane); Flk1-H2B::eYFP$^{tg/tg}$ (to visualize endothelial cell nuclei) embryos are imaged on the LSM 5 LIVE confocal with a 25× objective (NA of 0.45). Images are taken every 6 min at three Z-planes (~30 μM) each with the 488 nm (0.3% power) and 561 nm (2.0% power) lasers to image eYFP and mCherry, respectively. Endothelial cell apoptosis (panels A-A″) and mitosis (panels B-B″) can be captured in movies by observing fragmentation of nuclei and division of a single nucleus into two nuclei, respectively. Beginning of chromatin condensation is observed in A′ (6 minutes after A), and nuclear fragmentation is observed in A″ (1 hour after A). Beginning of mitosis is observed in B′ (12 minutes after B) and formation of two separate nuclei is apparent in B″ (18 minutes after B). (See Color Insert.)

session. Also, normal yolk sac vessel remodeling in embryos cultured from E8.5–E9.5 is a good sign of normal cardiovascular physiology. Lastly, cultured embryos should form similar morphological structures, at similar time points, to embryos grown *in utero* as shown in (Jones *et al.*, 2002). These structures can thus be used as hallmarks to compare health of the embryo being imaged.

6. Image Analysis

There are many different parameters that can be analyzed for different types of morphogenetic events. The rate of mitosis (Fig. 19.3), apoptosis, cell/tissue morphology changes, and cell migration rate/directionality of movement are among the parameters that can be quantified by computational means. Tools for image analysis are available from several sources. Many convenient tools can be found within the confocal software. For instance, in the Zeiss LSM software there are many tools such as those for adding scale bars and time stamps, creating 3-D or 4-D reconstructions, analyzing signal colocalization, applying look up tables, cropping an image series, joining different image series and measuring changes in signal intensity. In addition to the LSM software, the Bitplane Imaris software has additional tools for image analysis and is used routinely by our lab for tracking cells, 3-D reconstructions and image postprocessing for denoising and drift. Other tools are also available from MatLab, Volocity, Slidebook, and other programs as well as for free in ImageJ. In this section, we will highlight some of the routine methods that we use for image analysis and data analysis but note that there are many available tools for many applications. Talk to your local microscope experts or company representatives if you are looking for a particular image analysis tool.

6.1. Image processing for a typical 3-D, time-lapse sequence

To begin processing the time-lapse images, entire movies are first reconstructed by performing an extended-depth view (projection) in the Zeiss LSM software. This projects the 3-D data from optical slices into a single image, so the resulting movie will be a simple time sequence. Sequential movies are combined together by concatenating them in order using the concatenation macro in the macros menu of the Zeiss LSM software. Image contrast and brightness can be adjusted using this software; however, it is recommended to do this in the Imaris software program where there are more tools for denoising. To use Imaris, we input the LSM-derived concatenated image sequences into Imaris, and open the image as a Surpass scene using the volume function. Imaris has a plug-in to recognize LSM images from Zeiss confocals to make this relatively easy. If needed, the image sequence is further processed by correcting the levels of each individual color (e.g., eYFP and mCherry). If the signal-to-noise ratio is low, Gaussian filters are then applied to denoise the images. Care is always taken in postprocessing to be sure that important features of the images are preserved and additional features are not created. Files are ultimately saved as Audio Video Interleaved (AVI) movies, and playback time is adjusted depending on the speed of morphogenesis and on the particular events to be displayed.

6.2. Adjusting for specimen drift using Imaris

Specimen drift is common in time-lapse imaging sessions. In order to correct for drift in the *XY* plane using Imaris, the user assigns several reference points on the specimen that can be followed in each image over time. The reference points represent a center of mass which the software will use to stabilize the structure. In our samples containing Flk1-H2B:: eYFP labeled nuclei, we use this signal as our reference spots but other markers or regions of reproducible contrast can be used. Nuclei are identified by selecting "objects" and "adding new spots." The appropriate channel (YFP) is selected, and the diameter of the spots to be detected is set ($\sim$10 μm for endothelial cell nuclei). To account for most nuclei, the threshold is appropriately set until all or most nuclei are detected and this is confirmed by visual inspection from the operator. The next step is to track spots by selecting the Tracking folder and choosing the "Autoregressive motion gap close 3 function." The motion of each spot is then tracked throughout the movie, and information is gathered in the spot track group1 folder. The last step is to depress the "correct for drift" icon in the Tracks folder (make sure the spot track group 1 folder is selected). This produces an image series that has been corrected for drift. Regions that are not common to all the corrected images are cropped so there is some sacrifice of the field of view encountered by drift correction. Also, structures that leave the plane of the imaging view during the middle of the time-lapse movie cannot be compared and are thus subsequently cropped off.

6.3. Tracking cell migration using Imaris

Another benefit of using nuclear markers is to analyze cell migration. The ability to easily identify each nucleus of an individual cell makes it possible to easily track the movement of cells with these markers. In instances where labeled cells invade a nonlabeled region, cytoplasmic or membrane-localized markers can be easily distinguished. However, in similarly labeled tissues, uniform expression of a cytoplasmic- or membrane-localized FP makes it difficult to distinguish between neighboring cells. Thus, nuclear-localized FPs can be used to define the nucleus of each cell in order to discern the migratory abilities of individual cells in similarly labeled tissues.

There are several parameters that can be quantified from a migrating cell. For instance, the overall speed, distance traveled, displacement, and directionality of cell migration can be determined. In the Imaris software program, each individual cell can be traced by tracking spots (as shown above). Spots are tracked by Imaris by plotting the location of a nucleus at a particular *XY* coordinate from one time point to the next. The program can take the sum of the distance traveled at all time points (total distance traveled), can compare the distance traveled at the beginning to the end of the time-lapse session (distance displaced), and determine the average speed of migration. Tracking information can be displayed in a Microsoft® Excel spreadsheet by depressing the "tracking folder" and then "statistics" tabs. After collecting the data, results can be averaged to determine the overall affect of migration in the sample of interest.

7. Summary

Here we have provided some details and examples into the procedures used by our lab to study cell dynamics in early mouse embryos using time-lapse confocal microscopy to guide new scientists toward the use of imaging to study mouse embryogenesis. We are grateful to all the investigators that have established protocols that we have modified for imaging strategies in our lab and we encourage more people to adapt and extend these protocols to generate new approaches to image developmental events.

REFERENCES

Anderson, R., Copeland, T. K., Scholer, H., Heasman, J., and Wylie, C. (2000). The onset of germ cell migration in the mouse embryo. *Mech. Dev.* **91,** 61–68.

Aulehla, A., Wiegraebe, W., Baubet, V., Wahl, M. B., Deng, C., Taketo, M., Lewandoski, M., and Pourquie, O. (2008). A beta-catenin gradient links the clock and wavefront systems in mouse embryo segmentation. *Nat. Cell Biol.* **10,** 186–193.

Brachmann, I., Jakubick, V. C., Shaked, M., Unsicker, K., and Tucker, K. L. (2007). A simple slice culture system for the imaging of nerve development in embryonic mouse. *Dev. Dyn.* **236,** 3514–3523.

Burtscher, I., and Lickert, H. (2009). Foxa2 regulates polarity and epithelialization in the endoderm germ layer of the mouse embryo. *Development* **136,** 1029–1038.

Copp, A. J., and Cockroft, D. L. (1990). Postimplantation Mammalian Embryos: A Practical Approach. IRL Press, Oxford, UK; New York, USA.

Coveney, D., Cool, J., Oliver, T., and Capel, B. (2008). Four-dimensional analysis of vascularization during primary development of an organ, the gonad. *Proc. Natl. Acad. Sci. USA* **105,** 7212–7217.

Davidson, M. W., and Campbell, R. E. (2009). Engineered fluorescent proteins: Innovations and applications. *Nat. Methods* **6,** 713–717.

Dickinson, M. E. (2006). Multimodal imaging of mouse development: Tools for the postgenomic era. *Dev. Dyn.* **235,** 2386–2400.

Downs, K. M., and Gardner, R. L. (1995). An investigation into early placental ontogeny: Allantoic attachment to the chorion is selective and developmentally regulated. *Development* **121,** 407–416.

Druckenbrod, N. R., and Epstein, M. L. (2005). The pattern of neural crest advance in the cecum and colon. *Dev. Biol.* **287,** 125–133.

Fraser, S. T., Hadjantonakis, A. K., Sahr, K. E., Willey, S., Kelly, O. G., Jones, E. A., Dickinson, M. E., and Baron, M. H. (2005). Using a histone yellow fluorescent protein fusion for tagging and tracking endothelial cells in ES cells and mice. *Genesis* **42,** 162–171.

Gordon, J. W., and Ruddle, F. H. (1981). Integration and stable germ line transmission of genes injected into mouse pronuclei. *Science* **214,** 1244–1246.

Hadjantonakis, A. K., and Papaioannou, V. E. (2004). Dynamic in vivo imaging and cell tracking using a histone fluorescent protein fusion in mice. *BMC Biotechnol.* **4,** 33.

Hadjantonakis, A. K., Dickinson, M. E., Fraser, S. E., and Papaioannou, V. E. (2003). Technicolour transgenics: Imaging tools for functional genomics in the mouse. *Nat. Rev. Genet.* **4,** 613–625.

Haubensak, W., Attardo, A., Denk, W., and Huttner, W. B. (2004). Neurons arise in the basal neuroepithelium of the early mammalian telencephalon: A major site of neurogenesis. *Proc. Natl. Acad. Sci. USA* **101,** 3196–3201.

Hogan, B. (1994). Manipulating the Mouse Embryo: A Laboratory Manual. Cold Spring Harbor Laboratory Press, Cold Spring Harbor, NY.

Hsu, Y. C. (1979). In vitro development of individually cultured whole mouse embryos from blastocyst to early somite stage. *Dev. Biol.* **68,** 453–461.

Inoue, T., and Gliksman, N. (2003). Techniques for optimizing microscopy and analysis through digital image processing. *Methods Cell Biol.* **72,** 243–270.

Jones, E. A., Crotty, D., Kulesa, P. M., Waters, C. W., Baron, M. H., Fraser, S. E., and Dickinson, M. E. (2002). Dynamic in vivo imaging of postimplantation mammalian embryos using whole embryo culture. *Genesis* **34,** 228–235.

Jones, E. A., Baron, M. H., Fraser, S. E., and Dickinson, M. E. (2004). Measuring hemodynamic changes during mammalian development. *Am. J. Physiol. Heart Circ. Physiol.* **287,** H1561–H1569.

Jones, E. A., Baron, M. H., Fraser, S. E., and Dickinson, M. E. (2005a). Dynamic in vivo imaging of mammalian hematovascular development using whole embryo culture. *Methods Mol. Med.* **105,** 381–394.

Jones, E. A., Hadjantonakis, A. K., and Dickinson, M. E. (2005b). Imaging mouse embryonic development. Imaging in Neuroscience and Development: A Laboratory Manual. Cold Spring Harbor Laboratory Press, Cold Spring Harbor, NY.

Kwon, G. S., Viotti, M., and Hadjantonakis, A. K. (2008). The endoderm of the mouse embryo arises by dynamic widespread intercalation of embryonic and extraembryonic lineages. *Dev. Cell* **15,** 509–520.

Larina, I. V., Sudheendran, N., Ghosn, M., Jiang, J., Cable, A., Larin, K. V., and Dickinson, M. E. (2008). Live imaging of blood flow in mammalian embryos using Doppler swept-source optical coherence tomography. *J. Biomed. Opt.* **13,** 060506.

Larina, I. V., Ivers, S., Syed, S., Dickinson, M. E., and Larin, K. V. (2009). Hemodynamic measurements from individual blood cells in early mammalian embryos with Doppler swept source OCT. *Opt. Lett.* **34,** 986–988.

Lawson, K. A., Meneses, J. J., and Pedersen, R. A. (1986). Cell fate and cell lineage in the endoderm of the presomite mouse embryo, studied with an intracellular tracer. *Dev. Biol.* **115,** 325–339.

Megason, S., Amsterdam, A., Hopkins, N., and Lin, S. (2006). Uses of GFP in transgenic vertebrates. *Methods Biochem. Anal.* **47,** 285–303.

Molyneaux, K. A., Stallock, J., Schaible, K., and Wylie, C. (2001). Time-lapse analysis of living mouse germ cell migration. *Dev. Biol.* **240,** 488–498.

Nagy, A. (2003). Manipulating the Mouse Embryo: A Laboratory Manual. Cold Spring Harbor Laboratory Press, Cold Spring Harbor, NY.

Nel-Themaat, L., Vadakkan, T. J., Wang, Y., Dickinson, M. E., Akiyama, H., and Behringer, R. R. (2009). Morphometric analysis of testis cord formation in Sox9-EGFP mice. *Dev. Dyn.* **238,** 1100–1110.

New, D. A., and Cockroft, D. L. (1979). A rotating bottle culture method with continuous replacement of the gas phase. *Experientia* **35,** 138–140.

Nishii, K., and Shibata, Y. (2006). Mode and determination of the initial contraction stage in the mouse embryo heart. *Anat. Embryol. (Berl.)* **211,** 95–100.

Nonaka, S., Tanaka, Y., Okada, Y., Takeda, S., Harada, A., Kanai, Y., Kido, M., and Hirokawa, N. (1998). Randomization of left-right asymmetry due to loss of nodal cilia generating leftward flow of extraembryonic fluid in mice lacking KIF3B motor protein. *Cell* **95,** 829–837.

Nowotschin, S., and Hadjantonakis, A. K. (2009a). Photomodulatable fluorescent proteins for imaging cell dynamics and cell fate. *Organogenesis* **5,** 135–144.

Nowotschin, S., and Hadjantonakis, A. K. (2009b). Use of KikGR a photoconvertible green-to-red fluorescent protein for cell labeling and lineage analysis in ES cells and mouse embryos. *BMC Dev. Biol.* **9,** 49.

Nowotschin, S., Eakin, G. S., and Hadjantonakis, A. K. (2009). Live-imaging fluorescent proteins in mouse embryos: Multi-dimensional, multi-spectral perspectives. *Trends Biotechnol.* **27,** 266–276.

Okada, Y., Nonaka, S., Tanaka, Y., Saijoh, Y., Hamada, H., and Hirokawa, N. (1999). Abnormal nodal flow precedes situs inversus in iv and inv mice. *Mol. Cell* **4,** 459–468.

Passamaneck, Y. J., Di Gregorio, A., Papaioannou, V. E., and Hadjantonakis, A. K. (2006). Live imaging of fluorescent proteins in chordate embryos: From ascidians to mice. *Microsc. Res. Tech.* **69,** 160–167.

Plusa, B., Piliszek, A., Frankenberg, S., Artus, J., and Hadjantonakis, A. K. (2008). Distinct sequential cell behaviours direct primitive endoderm formation in the mouse blastocyst. *Development* **135,** 3081–3091.

Rhee, D. Y., Zhao, X. Q., Francis, R. J., Huang, G. Y., Mably, J. D., and Lo, C. W. (2009). Connexin 43 regulates epicardial cell polarity and migration in coronary vascular development. *Development* **136,** 3185–3193.

Rizzo, M. A., Davidson, M. W., and Piston, D. W. (2009). Fluorescent protein tracking and detection: Applications using fluorescent proteins in living cells. *CSH Protoc.* **2009,** pdb top64.

Sadler, T. W., and New, D. A. (1981). Culture of mouse embryos during neurulation. *J. Embryol. Exp. Morphol.* **66,** 109–116.

Sakai, T., and Onodera, T. (2008). Embryonic organ culture. *Curr. Protoc. Cell Biol.* Chapter 19, Unit 19 8.

Sakaue-Sawano, A., Kurokawa, H., Morimura, T., Hanyu, A., Hama, H., Osawa, H., Kashiwagi, S., Fukami, K., Miyata, T., Miyoshi, H., Imamura, T., Ogawa, M., *et al.* (2008). Visualizing spatiotemporal dynamics of multicellular cell-cycle progression. *Cell* **132,** 487–498.

Shaner, N. C., Steinbach, P. A., and Tsien, R. Y. (2005). A guide to choosing fluorescent proteins. *Nat. Methods* **2,** 905–909.

Sherbahn, R., Frasor, J., Radwanska, E., Binor, Z., Wood-Molo, M., Hibner, M., Mack, S., and Rawlins, R. G. (1996). Comparison of mouse embryo development in open and microdrop co-culture systems. *Hum. Reprod.* **11,** 2223–2229.

Srinivas, S., Goldberg, M. R., Watanabe, T., D'Agati, V., al-Awqati, Q., and Costantini, F. (1999). Expression of green fluorescent protein in the ureteric bud of transgenic mice: A new tool for the analysis of ureteric bud morphogenesis. *Dev. Genet.* **24,** 241–251.

Srinivas, S., Rodriguez, T., Clements, M., Smith, J. C., and Beddington, R. S. (2004). Active cell migration drives the unilateral movements of the anterior visceral endoderm. *Development* **131,** 1157–1164.

Tam, P. P. (1998). Postimplantation mouse development: Whole embryo culture and micro-manipulation. *Int. J. Dev. Biol.* **42,** 895–902.

Watanabe, T., and Costantini, F. (2004). Real-time analysis of ureteric bud branching morphogenesis in vitro. *Dev. Biol.* **271,** 98–108.

Wiley, L. M., Spindle, A. I., and Pedersen, R. A. (1978). Morphology of isolated mouse inner cell masses developing in vitro. *Dev. Biol.* **63,** 1–10.

Yamanaka, Y., Tamplin, O. J., Beckers, A., Gossler, A., and Rossant, J. (2007). Live imaging and genetic analysis of mouse notochord formation reveals regional morphogenetic mechanisms. *Dev. Cell* **13,** 884–896.

Young, H. M., Bergner, A. J., Anderson, R. B., Enomoto, H., Milbrandt, J., Newgreen, D. F., and Whitington, P. M. (2004). Dynamics of neural crest-derived cell migration in the embryonic mouse gut. *Dev. Biol.* **270,** 455–473.

CHAPTER TWENTY

Imaging Mouse Development with Confocal Time-Lapse Microscopy

Sonja Nowotschin, Anna Ferrer-Vaquer, *and* Anna-Katerina Hadjantonakis

Contents

Abstract

The gene expression, signaling, and cellular dynamics driving mouse embryo development have emerged through embryology and genetic studies. However, since mouse development is a temporally regulated three-dimensional process, any insight needs to be placed in this context of real-time visualization. Live imaging using genetically encoded fluorescent protein reporters is pushing the envelope of our understanding by uncovering unprecedented insights into mouse development and leading to the formulation of quantitative accurate models.

Developmental Biology Program, Sloan-Kettering Institute, New York, USA

Methods in Enzymology, Volume 476
ISSN 0076-6879, DOI: 10.1016/S0076-6879(10)76020-1

1. Introduction

The goal of developmental biology is to understand how embryos develop from simple fertilized eggs into complex animals of diverse shapes, and elaborate but stereotypical internal architectures. A carefully orchestrated series of events takes the zygote to the generation and organization of groups of cells with distinct developmental fates. As cells acquire distinct fates, they adopt specific behaviors that drive growth and shape changes. Thus, once specified, groups of cells will become organized into the tissue layers that go on to form the organs of the fetus and adult organism. Therefore, to accomplish embryonic development, cells must execute and coordinate a variety of diverse behaviors including proliferation, apoptosis, migration, and differentiation. Key to unraveling these events is the observation of embryos as they develop, by microscopic imaging. Today live imaging provides unbiased, quantitative, digitized data that can be used to formulate experimentally testable models of how biological circuits regulate developmental processes.

The mouse is currently the preferred genetically tractable mammalian model organism. Thus, experimentally investigating mouse embryonic development provides an important framework for understanding mammalian embryonic development. Mouse, and hence mammalian embryonic development, is a highly dynamic and complex process, which we are only beginning to understand with respect to the intrinsic molecular control and cellular dynamics. However, if we are to understand mouse development at the molecular and cellular level, a great deal more needs to be learned about the dynamics of gene expression, cell behaviors, and fate, and how they are molecularly determined and integrated. Most importantly, given the dynamic nature of embryonic development, any insights will ultimately need to be placed in a spatiotemporal framework which can only be achieved by live imaging.

Recent advances in live imaging and consequently our understanding of the cellular dynamics driving mouse embryo and tissue morphogenesis have been spearheaded by the convergence of four key technologies. These include: (1) the availability of mouse mutant strains in which a process of interest is defective, (2) the availability of mouse reporter strains for live imaging populations of cells or tissues of interest, (3) methods for culturing specimens (embryos or tissue explants) of interest for live imaging, and (4) improved modalities for image data acquisition and methods for data processing.

Transgenic and gene targeting techniques for random and directed mutagenesis of the mouse genome are routine, and offer the possibility to study gene function *in vivo*. Genetically modified mice often serve as excellent models of human congenital diseases. Over the past decade, forward genetic screens using chemical (e.g., ENU) or genetic agents (e.g., transposons) to induce random mutations into the genome have provided an additional unbiased source of mutants and helped advance our understanding of gene function during the morphogenetic processes.

Microscopy has always been an essential tool in developmental biology both for determining the normal course of events and for contrasting with the effects of experimental perturbations. Staining tissues or labeling specific groups of cells in tissues has extended the information content of microscopic observation. Various labeling methods have been used to deliver readily visualized reporters into cells in mouse embryos or organ explants, and have been widely used to study cellular dynamics and fate. Traditionally, a variety of approaches have been used to label single or groups of cells in mouse embryos. Embryological methods include grafting of genetically distinct cells or tissues (Kinder *et al.*, 1999, 2001), injection of vital dyes (e.g., DiI or DiO), or the electroporation of nucleic acids or proteins (Bronner-Fraser and Fraser, 1988; Haas *et al.*, 2001). Over the past two decades, transgenic methods have been used to express reporter genes in cells or tissues of interest. Additional temporal control of reporter expression is afforded by binary transgenic systems, such as the genetic-induced fate mapping (GIFM) approach (Joyner and Zervas, 2006; Nagy, 2000).

The GIFM method relies on characterized *cis*-regulatory elements to drive tissue-specific expression of a genetically encoded reporter. By contrast, many of the embryological techniques are invasive, or are only applicable when tissues are accessible for manipulation. As a consequence, cells which lie in deeper tissue layers of an embryo may be difficult to label. Furthermore, it still remains a challenge to label single cells or groups of cells, which can then be visualized at single-cell resolution, by many of these methods. Most importantly, however, the majority of these methods only provide a static picture of the highly dynamic three-dimensional processes of mouse development. In this way, dynamics are inferred through the analysis of multiple samples, as opposed to the reiterative analysis of a single sample, which is the goal of any live imaging experiment.

1.1. Genetically encoded reporters for live imaging cellular dynamics during mouse development

Researchers have long exploited molecular fluorescence to observe the localization and dynamics of proteins, organelles, and cells. The main classes of fluorophores in use today are small organic vital dyes (such as DiI or DiO), inorganic nanocrystals which are known as quantum dots (QDs), and genetically encoded fluorescent proteins (FPs) (Giepmans *et al.*, 2006). Even though the smaller size of vital dyes and QDs makes them attractive over FPs, they need to be conjugated to protein-targeting molecules, such as antibodies, which serve to restrict their application for live imaging experiments. By contrast, FPs can, in principle, be fused to any protein of interest making them minimally invasive and ideal reporters for live imaging.

Recent advances in the isolation and availability of genetically encoded FP reporters (Nowotschin *et al.*, 2009b; Shaner *et al.*, 2005) in combination with increasingly sophisticated imaging acquisition and analysis methodologies have propelled the field to investigate mouse development in living specimens and in three dimensions (3D) through live imaging (Hadjantonakis *et al.*, 2003). Besides cytosolic reporters represented by native FPs, subcellularly localized FP fusions such as fusions of FPs to human histone H2B that label active chromatin (Fraser *et al.*, 2005; Hadjantonakis and Papaioannou, 2004), or fusions directed to the plasma membrane proteins by, for example, glycosylphosphatidylinositol (GPI) tagging (Larina *et al.*, 2009; Rhee *et al.*, 2006), facilitate the visualization of different aspects of a cell, and of individual cells in complex tissues. By segmenting nuclei, single cells can be indentified and tracked in complex populations. Indeed, nuclei are easier to track than whole cells since they have a simple spherical morphology, are separated in 3D space, so making them easier to segment. Importantly, H2B fusions have a distinct advantage over standard nuclear localization sequences. They facilitate tracking of daughter cells, since they remain associated with chromatin during cell division (Nowotschin *et al.*, 2009b). By combining nuclear and membrane labeling yields, live information equivalent to routine histology, such that dynamic cell behaviors including cell division, death, or morphology, can be acquired and quantified at high resolution in living specimens. Cell-type-specific resolution can be achieved when expressing these FP reporters under defined *cis*-regulatory elements. For example, our laboratory has recently generated mouse strains that express GFP and RFP under visceral endoderm-specific *cis*-regulatory elements, from the mouse *Afp* (*Alpha-fetoprotein*) and *Ttr* (*Transthyretin*) loci, respectively, where all visceral endoderm cells are labeled with a green or red fluorescent reporter, respectively (Kwon and Hadjantonakis, 2009; Kwon *et al.*, 2006).

Recently, a powerful tool for highlighting specific pools of molecules has emerged with the engineering of photomodulatable FPs (PM-FPs). PM-FPs offer an increase in spatiotemporal resolution of labeling for visualizing dynamic cell behaviors and mapping cell fate. PM-FPs undergo a conformational change leading to a change in FP color when illuminated briefly with light of a certain wavelength that can be used to label and track subpopulations of cells or even single cells in embryos or tissue explants in a noninvasive and selective manner (Chudakov *et al.*, 2004; Gurskaya *et al.*, 2006; Lippincott-Schwartz and Patterson, 2008; Nowotschin and Hadjantonakis, 2009; Patterson, 2008; Stark and Kulesa, 2005, 2007; Wiedenmann *et al.*, 2004).

To distinguish between, and simultaneously image, different cell populations, spectrally distinct reporters can be used (Nowotschin *et al.*, 2009a). Multiplexing of reporters often necessitates microscope systems capable of spectral separation. Instruments with these capabilities are quite common and likely available at most institutions.

Live imaging of other organisms such as *Drosophila* (Murray and Saint, 2007), zebrafish (Scherz *et al.*, 2008), and *Xenopus* (Davidson *et al.*, 2008; Woolner *et al.*, 2009) has shown to deliver massive data sets about the dynamic relationship between cells during embryonic development. By contrast, the mouse embryo poses a significant challenge for live imaging due to its in utero development. The mouse embryo is not readily accessible for visualization and therefore needs to be cultured *in vitro*. Therefore, it is imperative, when live imaging embryos, to create *ex utero* culture conditions that most resemble *in utero* development and to use genetically encoded fluorescent reporters that exhibit a minimum brightness at the levels they are expressed in the embryo to be able to be detected with low laser power.

1.2. Live imaging mouse development

Among the most important technical advances in optical imaging is the capability for optical sectioning, which is routinely facilitated by the use of a confocal microscope. Confocal imaging allows the observer to look deep inside a sample without its physical destruction and by cutting out interference from out-of-focus light and scatter. Improvements and affordability in confocal instruments have partnered the development of increasingly sensitive detectors, leading to acquisition of data of unprecedented spatial and temporal resolution. In this chapter, we provide a detailed description of the methods and conditions that we routinely use for culturing and time-lapse imaging of mouse embryos. We discuss requirements for both pre- and postimplantation stages that are routinely used to obtain high-resolution 3D time-lapse data revealing the gene expression, cellular and tissue dynamics of mouse embryonic development.

Once a reporter strain labeling cells of interest has been identified and imported from another investigator or a repository, or else generated within the lab, the steps of any live imaging experiment are quite simple. Animals of are timed mated to produce embryos of and stages. Then the actual imaging experiment can be divided into embryo dissection and mounting, followed by image data acquisition and analysis.

2. Materials

2.1. Media

Media for recovering and culturing embryos are commercially available from several vendors. Catalog numbers and vendor names are given for the media routinely used in our laboratory. However, alternative products will likely also suffice. In some cases, we also provide protocols for the home-made preparation of media compositions.

2.1.1. Media for recovering preimplantation embryos

M2: Commercially available from Millipore, Specialty Media (Cat. no. MR-015-D) or can be made up from constituent solutions (see Recipe for the preparation of 100ml of M2 on next page, and table of media compositions below).

2.1.2. Media for culturing preimplantation embryos

KSOM + amino acids: Commercially available by Millipore, Specialty Media (Cat. no. MR-121_D) or can be made up from constituent solutions (see Recipe for the preparation of 100ml of KSOM on next page, and table of media compositions below).

Important: KSOM should be stored frozen (−20 °C) for no more than 3 months, and once thawed, aliquoted and kept at (4 °C) for no more than 2 weeks.

Media compositions for preparing M2 and KSOM:

	Component	g/100 ml	Source	Cat. no.
Stock A (10×)	NaCl	5.534	Sigma	S3014
	KCl	0.356	Sigma	P9541
	KH_2PO_4	0.162	VWR	BDH0268
	$MgSO_4 \cdot 7H_2O$	0.293	Sigma	M5921
	Sodium lactate 60% syrup	2.610 or 4.349 g	Fisher	S326-500
	Glucose	1.000	Sigma	G8270
	Penicillin G	0.060	Sigma	P4687
	Streptomycin	0.050	Sigma	S1277
Stock B (10×)	$NaHCO_3$	2.101	VWR	BDH0280
	Phenol red	0.001	Sigma	P3532
Stock E (10×)	HEPES	5.958	Invitrogen	11344-041
	Phenol red	0.001	Sigma	P3532
Stock C (100×)	Sodium pyruvate	0.036	Sigma	P2256
Stock D (100×)	$CaCl_2 \cdot 2H_2O$	0.252	VWR	BDH0224
Stock F (10,000×)	$Na_2EDTA \cdot 2H_2O$	0.0.372	Sigma	E5134
Stock G (100×)	Glutamine	200 m*M*	Invitrogen	25030-081

Weigh solids into media bottles and add an appropriate quantity of water. If sodium lactate syrup is used (for Stock A), the weigh boat should be rinsed well and the wash added to the media flask. For Stock E, add half the required volume of water, then adjust the pH to 7.4 with 1 N NaOH. Make up to final volume using measuring cylinder. Filter all stock solutions through a 0.45 μm (Millipore) filter, aliquot into sterile tubes.

Preparation of 100 ml of M2 and 100 ml of KSOM from concentrated stocks:

Stock	M2 (ml)	KSOM (ml)
A (10×)	10.0	10.0
B (10×)	1.6	10.0
C (100×)	1.0	1.0
D (100×)	1.0	1.0
E (10×)	8.4	–
F (100×)	–	1
G (200×)	–	0.5
Water	78.0	76.5
BSA	400 mg	400 mg

BSA (bovine serum albumin), embryo tested (Sigma A3311).

Rinse all pipettes and tubes thoroughly into the final flask. When preparing BSA, allow to dissolve and gently swirl the medium without excessive frothing using a magnetic stirrer.

- *M2*: If necessary, readjust the pH to 7.2–7.4 with 1 N NaOH. Make up to final volume. Filter sterilize through a 0.45 μm (Millipore) filter and aliquot into polypropylene tubes. Store at +4 °C.
- *KSOM*: Gas with 5% CO_2 in air to adjust pH to 7.4, alternatively incubate the medium for several hours (or overnight) at 37 °C, 5% CO_2 (in a tissue culture incubator) with the cap loosened. The pH will not usually require adjustment, but it may vary with different batches of BSA. Make up to volume. Filter sterilize and aliquot into polypropylene tubes.

2.1.3. Media for dissection of postimplantation embryos

We routinely use two alternative media compositions for embryo dissection. We find both equally good at preserving embryo viability during dissection.

- 95% DMEM/F12 (1:1) (Invitrogen, 11330-057) + 5% newborn calf serum (Lonza, 14-416F).
- Modified PB1 with 10% fetal calf serum (Papaioannou and West, 1981; Whittingham and Wales, 1969).

Media composition for preparing modified PB-1 medium with 10% fetal calf serum:

1. Make up stock solutions and mix the indicated volumes.

	g/100 ml	ml
NaCl	0.9	68.96
KCl	1.148	1.84
$Na_2HPO_4 \cdot 12H_2O$	5.5101	5.44
KH_2PO_4	2.096	0.96
$CaCl_2 \cdot 2H_2O$	1.1617	0.88
$MgCl_2$	3.131	0.32
Na-pyruvate	0.02 in stock NaCl	22.40

2. Add the solutions below to make up 104 ml of medium.

Penicillin/streptomycin (100×)	1 ml
Glucose	104 mg
Distilled water	1.16 ml
Phenol red (1%)	0.1 ml

3. Add heat-inactivated fetal calf serum (56 °C for 30 min) to a final concentration of 10%.
4. Filter sterilize and aliquot.

2.1.4. Media for culturing postimplantation embryos

- DMEM/F12 (1:1) with GLUTAMAX (Invitrogen, 10565-018), 1% penicillin/streptomycin (Invitrogen/Gibco, 15140)m and rat serum.

The percentage of rat serum versus DMEM (Dulbecco's modified Eagles medium) varies depending upon the embryo stage. Commonly, we use 50% rat serum and 50% DMEM for E5.5–E8.25, 75% rat serum for stages E8.5–E9.5 and 100% for stages older than E9.5.

2.1.5. Rat serum

We routinely use rat serum that is specifically collected and available from commercial vendors such as Harlan Bioproducts (Cat. no. 4520). However, we find that the most consistent quality, but more labor intensive, is a homemade rat serum preparation, which we provide below.

2.1.5.1. Preparation of rat serum

1. Anesthetize adult male rat with ether (fume hood!) or isofluorane, according to institutional IACUC policies. Note that all procedures involving mice or rats should be written up in laboratory animal protocols, and be compliant with institutional policies and federal regulations.

2. Make a V-shaped incision into the skin and peritoneum of the lower abdomen of the anesthetized rat. Expose the dorsal aorta by pushing aside the internal organs.
3. Puncture the aorta using a beveled butterfly needle (Vacutainer blood collection set, BD Bioscience) and collect the blood by pressing the outlet needle into a vacutainer blood collection tube.
4. When the rat is completely exsanguinated, place blood on ice and euthanize rat using a guillotine or use the triple kill method.
5. Take all collected blood samples and centrifuge them for 20 min at $1300 \times g$.
6. Remove and collect the supernatant (serum) into a new tube and discard the pellet.
7. To remove remaining debris, centrifuge again for 10 min at $1300 \times g$ and keep supernatant.
8. Heat inactivate the serum at 56 °C for 30 min.
9. Filter serum under sterile conditions using a 0.45 μm filter.
10. Make 1–2 ml aliquots of the serum, freeze, and store at −80 °C. Aliquots can be stored frozen for up to 1–2 years.

2.2. Equipment

2.2.1. Embryo culture

- Two pairs of watchmaker's forceps #5 (Roboz, RS-4978) and small surgical scissors (Roboz, RS-5910) for embryo dissection
- 3.5, 6, and 10 cm plastic Petri dishes (Falcon, 351001; 351007 and 351029)
- Organ culture dishes (Falcon, 353037)
- 3.5 cm glass-bottom dishes (MatTek, P35G-1.5-14-C) or LabTek coverslip bottom chambers (NUNC, 155360; 155379; 155382)
- CoverWell perfusion chamber gaskets, 9 mm diameter; 1.0 mm deep (Invitrogen/Molecular Probes, C18140) or 2.0 mm deep (Invitrogen/Molecular Probes, C18141)
- Mouth pipette (homemade) consistent of a mouthpiece (HPI Hospital Products Med. Tech., 1501P-B4036-2), latex tubing (latex 1/8 in. ID, 1/32 in. wall, Fisherbrand, 22362772), and very fine pulled glass Pasteur pipettes using a 1000 μl pipette tip connector
- Plastic transfer pipettes (Fisherbrand, 13-711-7M) to transfer embryos stages E7.5 and older
- Pulled Pasteur pipettes
- Suction holding pipette (optional; Eppendorf CellTram Air, 5176000.017)
- 1 ml syringe and 26-gauge needle (Becton Dickinson, 309623), 30-gauge needle with a blunt end (Becton Dickinson, 305106). Use sandpaper or sharpening stone for blunting the needle.

- Embryo tested light weight mineral oil (Sigma, M8410)
- Incubator providing a humid atmosphere and constant level of 5% CO_2
- Roller apparatus (BTC Engineering, Cambridge, UK)
- Industrial gas supply containing gas mixtures of 5% CO_2/95% O_2; 5% CO_2/20% O_2, or 5% CO_2/5% O_2
- Microscope with an environmental chamber to keep temperature and gas levels stable throughout the culture
- Human eyelashes, or cat whiskers, sterilized with 70% ethanol. (*Note*: To prevent animal cruelty, cats should not be harmed during whisker collection. We therefore recommend the use of cat whiskers which have been naturally shed by the animal.)

2.2.2. Microscopes

- Stereomicroscopes for dissecting and immobilizing embryos with both incident and transmitting light, and a magnification range from 6.3× up to 100×. Though not providing high-resolution information, these systems can also be used for time-lapse imaging to visualize changes in the gross morphology of embryos. Software controlling CCD cameras routinely used on stereo dissecting microscopes often allows time-lapse acquisition. To set up a culture for live imaging, a Petri dish can be covered with a watch glass or glass lid to present evaporation, and an incident air stream heater can be used for heating the sample. In such a system, media containing embryos is gassed intermittently.
- *Single point laser scanning confocal systems*: comparable to Zeiss LSM700 or 710. This is our system of choice for imaging mouse embryos in whole mount from preimplantation through to early somite stages or in explant at various stages.
- *Slit scanning confocal systems*: comparable to Zeiss LSM 5 LIVE systems. This type of system facilitates high-speed imaging, and can be used for visualizing blood flow, calcium dynamics, or the movement of organelles such as cilia.
- *Spinning disk laser confocal systems*: comparable to Perkin-Elmer UltraView RS5 systems. These systems work well for imaging preimplantation or very early postimplantation stage embryos, for reasonably high-speed imaging or for high-resolution cellular imaging. In our experience, spinning disk systems are excellent for imaging a single fluorophore, GFP or RFP in our experience, but when multiplexing, many systems are prone to substantial of cross talk, especially when levels of reporter expression are low.
- *Two photon excitation systems*: can be used for live imaging of mouse embryos, but only offer benefits if deeper tissue imaging is required. TPLSM systems are usually more expensive to purchase and maintain. They often also require more time to set up imaging parameters in an

experiment, and it is sometimes difficult to maximally excite multiple reporters due to the single tunable laser line present on most systems.

- For live imaging mouse embryos, we prefer to have the confocal system fitted onto an inverted microscope. When using systems available through an Institutional core facility, one may not have a choice of upright versus inverted systems. Though not optimal, we have also had some success with homemade chambers fitted onto the stage of upright microscopes. Two benefits of using an upright microscope are the ability to use water immersion (or dipping) objectives, that can be physically placed into culture media just above the specimen, and the ease of positioning or manipulating the sample. The shortcoming of an upright microscope-based system is the difficulty of maintaining a closed, gassed, humidified, and heated environment which is critical for successfully live imaging mouse embryos or tissues.
- The microscope and optics carrier are enclosed in an incubator providing a heated, humidified, and gassed environment for on-stage cell and embryo culture (Fig. 20.6). These are commercially available from microscope manufacturers (usually the most costly option), specialist manufacturers of incubators for microscopes, or can be made in-house (usually the least costly option). They can be fitted onto both upright and inverted microscopes, though the latter is our system of choice.
- The microscope should be equipped with a 5× dry objective to scan the field of view, and position the embryo. A 10× dry objective can be used to image low-magnification 3D time-lapses. A 20× dry or 25× multi-immersion, as well as a 40× multi- or oil-immersion objectives, respectively, are used for high-magnification 3D time-lapses and acquisition of static images of fixed embryos. We rarely use 63× or higher magnification oil-immersion objectives for the acquisition of *z*-stacks of mouse embryos.
- Computer workstation running software for image data acquisition (Zeiss AIM—http://www.zeiss.com/, Perkin-Elmer Volocity—http://cellular-imaging.com/products/Volocity/, MetaMorph—http://www.molecular-devices.com/pages/software/metamorph.html) and image analysis (Amira—http://www.amiravis.com/, Volocity, Imaris—http://www.bitplane.com/, Metamorph).

3. Methods

3.1. Setting up timed matings for recovery of staged embryos

Mice are usually kept on a daily cycle of artificial light and darkness with the dark cycle starting usually at 6 pm and lasting 12 h. Mating is assumed to occur in the middle of the dark cycle. Breeding pairs or triangles of mice are set up using one male with one or two females, for pairs or triangles, respectively. It is preferable and usually more efficient for females that are in estrus to be preselected for matings. The morning after mating is set up, pregnant females are identified by the presence of a vaginal plug. By convention, females are aged as 0.5 days pregnant at midday of the next day. Embryos of the age of interest can be dissected accordingly. However, after dissection, embryos may be of the same age but are present as a variety of stages in any one litter, and so are usually staged by morphology, for example, using the nomenclature of Downs and Davies (1993) for early postimplantation embryos, or Theiler (1989).

3.2. Microscope setup and environmental control

Routinely, inverted microscope systems (Fig. 20.6) are used to live image cultured embryos. Unfortunately, it is not as easy to manipulate and position an embryo on an inverted system as it is on an upright one. The success of the *ex utero* culture of the embryo and the live imaging experiment depend on strict control of its environment that closely should resemble *in utero* development. Since mammalian embryos are very sensitive to variation in temperature, the culture condition should be able to accommodate a constant temperature of 37 °C. This can be achieved through an environmental chamber around the microscope stage (Fig. 20.6), which will keep the temperature and the gas content stable. Such a setup can either be built to personal specifications or commercially bought. The specific gas requirements vary for different embryonic stages. In general, the gas composition consists of 5% CO_2, a variable oxygen concentration and balanced nitrogen. Commonly, 95% O_2 is used for preimplantation embryos, 5% O_2 for early to mid postimplantation embryos (E6.5–E9.5), 20% O_2 for postimplantation embryos of stages E9.5–E10.5, 40% and 95% for late E10.5–11.5 (Nagy *et al.*, 2003). The gas is bubbled through a gas-washing bottle and directly supplied to the culture dish through tubing into a little plastic chamber placed on top of the dish.

Evaporation of the culture media needs to be kept to a minimum. For on-stage culture experiments, this is usually achieved by covering the culture dish with embryo-tested light mineral oil or, in cases when this is not possible, with water soaked paper towels placed around the culture dish. In roller culture experiments, using the intermittent gassing apparatus, tubes are sealed with silicone grease between periods of gassing.

When using an inverted confocal microscope system, embryos must be imaged through glass coverslips or dishes containing coverglass bottoms (e.g., MatTek dishes or LabTek 2-well glass coverslip chambers).

3.3. Culture and imaging of preimplantation mouse embryos

After fertilization, early mouse embryos float freely in the oviduct, which makes it easy to recover them from the mouse's oviduct. Culture conditions for *in vitro* culture of preimplantation mouse embryos have been very well established. Culture conditions should reflect the *in vivo* environment and therefore ask for appropriate media, temperature, and gas conditions. Preimplantation embryos are recovered in M2 medium and cultured in KSOM medium at 37 °C and 5% CO_2. The following section gives instruction on how to isolate, culture, and image preimplantation mouse embryos (Fig. 20.1).

1. Prewarm M2 medium and prepare and prewarm microdrop cultures of KSOM culture medium in a 35-mm culture dish covered with light mineral oil. The culture dish containing KSOM should be placed in a humidified incubator at 37 °C gassed with 5% CO_2 for at least 30 min for equilibration. The oil reduces the evaporation and pH and temperature changes when the dish is outside the incubator, though times of keeping embryos outside of proper atmospheric conditions should be minimized.

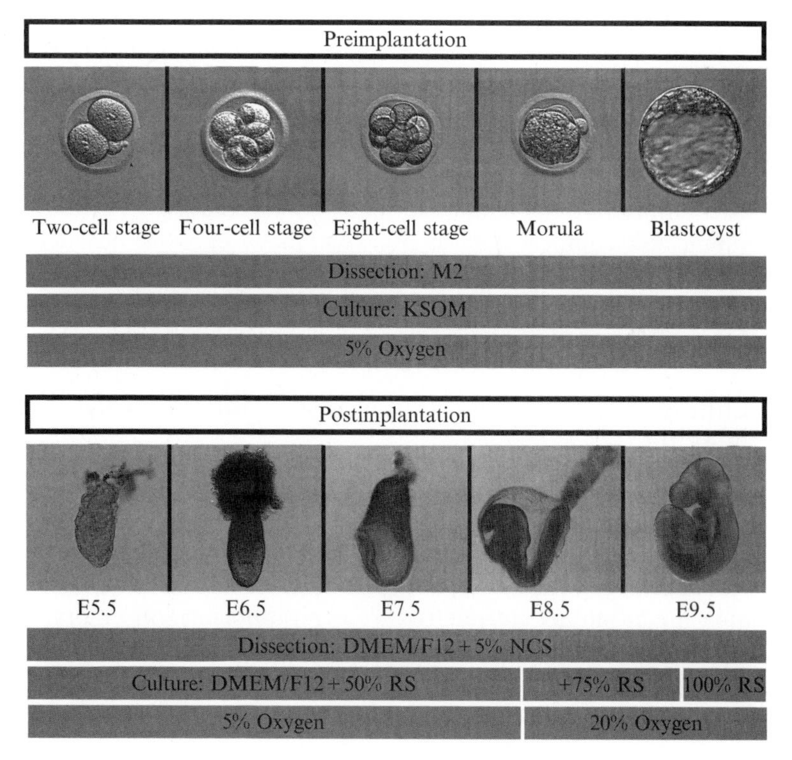

Figure 20.1 Overview of mouse embryonic development to midgestation. Preimplantation and early postimplantation stage mouse embryos are depicted. Stage-specific dissection, culture media, and gas composition are provided for each particular stage. Embryo images are not shown to scale. NCS, newborn calf serum; RS, rat serum; E, embryonic day.

2. Sacrifice pregnant female by cervical dislocation. Make an incision into the lower abdomen and then dissect out:
   - the oviduct when looking for E0.5–E2.5 embryos. When dissecting out the oviduct, the distal part of the uterus should be left.
   - the entire uterus to obtain E3.5–E4.5 embryos.
3. Place oviduct or uterus in a drop of prewarmed M2 media.
4. Flush the oviduct with prewarmed M2 media using a 1 ml syringe with a blunt 30-gauge needle by inserting the needle in the infundibulum of the oviduct or the uterus with 26-gauge needle, respectively.
5. Collect the embryos using a mouth pipette attached to a pulled Pasteur pipette and transfer them through several microdrops of KSOM under oil to rinse off the M2 media completely. Then transfer them into the prepared culture dish containing prewarmed KSOM culture medium. *Note*: Transfer multiple embryos together since it has been shown that groups of embryos cultured in a small amount of medium increases blastocyst development (Lane and Gardner, 1992).
6. For live imaging place the preimplantation embryos in a glass-bottom dish with a drop of prewarmed KSOM covered with light mineral oil, equilibrated at 37 °C in a humidified incubator gassed with 5% CO_2 for at least 30 min. (*Note*: Prewarm the on-stage incubator to 37 °C before setting up the culture for imaging.)
7. Place dish on the microscope stage and provide gassing with 5% CO_2. Image embryos using either a single point confocal system or the spinning disk confocal system. *Note*: Minimize exposure to laser light to avoid phototoxic effect on cells of the embryo by using low laser power, minimal number of scans, increasing scan speed and the size of the optical section.

3.4. Culture of postimplantation mouse embryos for live imaging

As preimplantation embryos make their way down the oviduct to the uterus, they undergo several cell divisions and develop into blastocysts. The blastocysts hatch from the zona pellucida and implant into the uterus, where they undergo the next steps of their development. Embryos that have implanted in the uterus are referred to as postimplantation embryos and are much more difficult to manipulate than preimplantation embryos. The following sections describe methods of dissecting, culturing, and imaging of postimplantation embryos (Fig. 20.1).

3.4.1. Collecting postimplantation embryos

1. Prewarm culture media in an organ culture dish by placing it into a humidified incubator at 37 °C and 5% CO_2 for at least 1 h before dissecting the embryos.

2. Sacrifice the female by cervical dislocation. Make an incision into the lower abdomen and then open up the peritoneum to remove the two horns of the uterus with the embryos and place them in a dish with dissecting media (DMEM/F12 + 5% NCS).
3. Dissect the decidua out from the uterus under a stereomicroscope. Then carefully remove the decidua and subsequent Reichert's membrane from the embryo using watchmaker's forceps (Fig. 20.2). *Important*: Dissection of embryos should be done in a timely manner – speed is of the essence! The yolk sac and the ectoplacental cone (EPC) of the embryo must not be damaged during dissection. Defective embryos will not develop properly in culture and should not be used for time-lapse imaging.

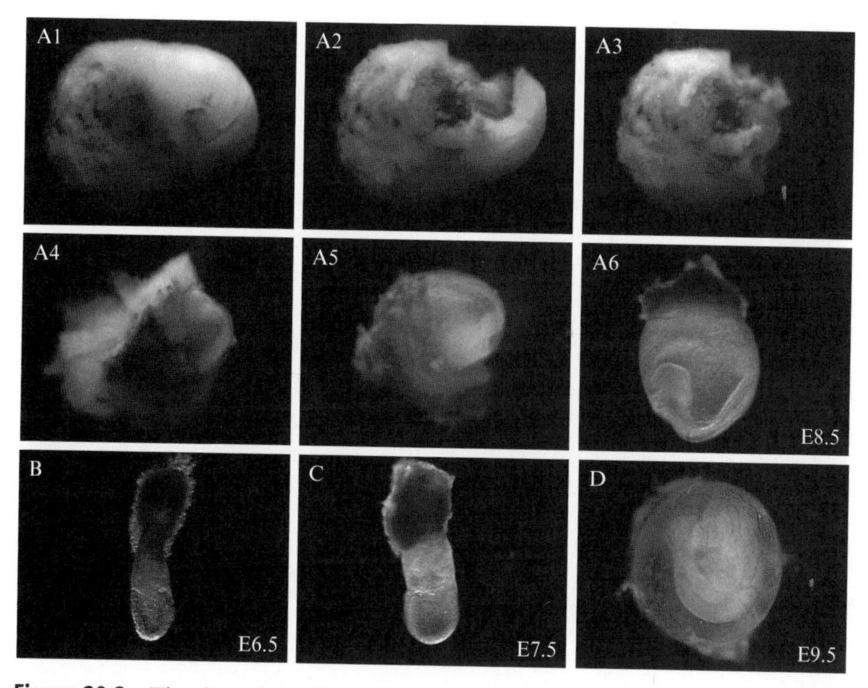

Figure 20.2 The dissection of postimplantation embryos for *ex utero* culture. Images show the sequence of dissection of a postimplantation embryo at stage E8.5: (A1) A deciduum, removed from the uterus, containing an embryo. (A2–A5) Panels show the careful, sequential peeling away of the decidual tissue (with watchmaker's forceps) leaving the yolk sac, which surrounds the embryo, intact. (A6) A dissected E8.5 embryo enclosed in its yolk sac. Note, small portion of ectoplacental cone (EPC) remaining on the top of the embryo. EPC should be left on to prevent rupturing of the yolk sac and for facilitating the immobilization of the embryo for culture. (B–D) Embryos of various stages dissected as shown in A1–A5: (B) E6.5 embryo, (C) E7.5 embryo, (D) E9.5 embryo.

4. After dissection, immediately move the embryos using a transfer pipette into a dish with prewarmed culture media and incubate them in an incubator at 37 °C and 5% CO_2 for 15–20 min before setting them for on-stage culture (see also Section 3.4.3) or when doing roller culture immediately transfer embryos in prewarmed culture media in roller culture bottles (see also Section 3.4.2).

3.4.2. Roller culture of postimplantation mouse embryos

The most preferred culture method for mouse embryos, when not live imaged, is a roller culture system. Constant motion of the embryos in rotating bottles/tubes, together with the constant temperature and gassing provide most optimal *ex utero* culture conditions. Two alternative roller culture system designs are commonly used. One uses intermittent gassing of the culture, whereas the other one provides gassing throughout the culture period (Fig. 20.3). Both types of system are commercially available from BTC Engineering. Our laboratory has experience with both systems; however, the latter has become our preferred method for culturing embryos since continuous gassing and a constant level of humidity provides a homogenous environment for the duration of the experiment, and is most favorable for the normal development of embryos.

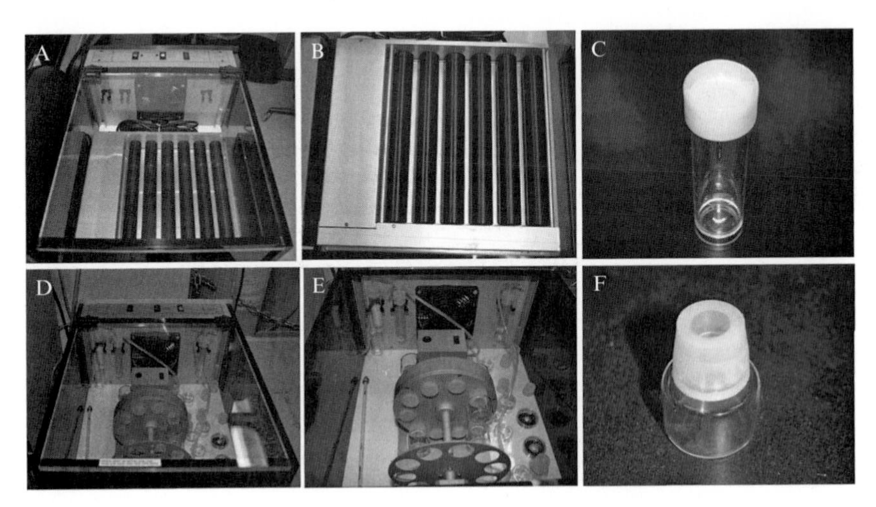

Figure 20.3 Two roller systems used for *ex utero* culture of mouse embryos. (A–C) Equipment for roller culture of mouse embryos with intermittent gassing: (A) incubator with roller apparatus inside, (B) close-up view of the roller apparatus, (C) tube for culturing embryos. For culture tube will be sealed with silicone grease. (D–F) Equipment for roller culture of mouse embryos with constant gassing: (D) incubator with drum inside, (E) close-up of drum, (F) bottle for culturing embryos. Bottle will be attached to drum and is connected to the gas flow inside the drum through opening on the top.

3.4.2.1. Roller culture with intermittent gassing

1. Prewarm the incubator and the culture media to 37 °C.
2. Culture media must be equilibrated with the appropriate gas combination at least 1 h before setting up the culture.
3. Dissect embryos as described in Section 3.4.1.
4. Transfer embryos into a tube with the appropriate culture media using 1 ml media per 1 embryo, seal tube tightly, and place them on the rotator that consists of several horizontal rollers.
5. Gas the culture media every 6 h with the appropriate gas combination for the stage of the embryo until the end of the culture.

3.4.2.2. Roller culture with constant gassing

1. Once you have prepared culture media and dissected the embryos, you are ready to set them up in the roller culture incubator.
2. Transfer dissected embryos into glass bottles, seal glass bottle with a silicone rubber stopper that has an opening to provide a constant supply of gas.
3. Attach glass bottles to the gassing system, a gas filled drum rotating around a horizontal axis. *Note*: Glass bottles should be rinsed with water or 70% ethanol and/or sterilized after each use.

3.4.3. Static culture (on-stage culture) and imaging of postimplantation mouse embryos

Static or on-stage culture, though less ideal for the *ex utero* development of the embryo, is the preferred method to culture embryos for time-lapse imaging of postimplantation mouse embryos. Embryos can be cultured up to 18–36 h on-stage. (The mentioned time span does not only depend on the culture conditions but also very much on the intensity and frequency of laser light the embryos are exposed to!)

1. Before setting up the embryo for a time-lapse experiment, prewarm the on-stage incubator at the microscope to 37 °C.
2. Dissect embryos as described before and incubate them for 15 min at least in prewarmed and pregassed culture media.
3. Prepare glass-bottom dish (MatTek, P35GC-1.5-14-C) for the culture and imaging of the embryo for stages E5.5–E8.5. Place a drop of prewarmed culture media to fill up the glass part of the dish and cover it with embryo tested light mineral oil. If embryo needs to be immobilized during culture, modifications according to the stage of the embryo are applied to the glass-bottom dish. See Section 3.4.4 for a detailed description.

4. Set up the culture dish with the embryo on the microscope stage. Gas the embryo by putting a small square plastic cover with an outlet for 5% CO_2 creating an appropriate culture atmosphere.
5. Set up imaging experiment. Note that the success of the culture and the imaging experiment depends on a good balance between exposure to laser power and brightness of the image. In general, laser intensity and frequency of scans should be kept as low as possible, otherwise laser light will be phototoxic to embryonic cells. Scan speed and adjustment of the thickness of optical sections can also minimize laser light exposure. However, this is very much dependent on the brightness of the fluorophore used, microscope setup used, and the biological question asked. The optimal setup has to be established empirically using wild-type embryos for each microscope system.

3.4.4. Methods for immobilizing postimplantation mouse embryos for time-lapse imaging

Immobilization of postimplantation embryos is necessary when a specific region of the embryo needs to be imaged repetitively. Various possibilities exist for immobilization of embryos. Suction pipettes, for example, can be used to hold the embryo while it is suspended in the culture. However, our preferred method of immobilizing embryos between E7.5 and E8.25 stages is to use modified culture dishes with inserts for mounting, for example, CoverWell chamber gaskets. These 'press-to-seal' gaskets are made of a plastic surface containing a hole (for inserting the embryo) and a rubber section that can be stuck onto the coverglass bottom of the culture dish. These gaskets are manually cut to fit the glass bottom of the imaging dish (Fig. 20.4). Different heights of these chamber gaskets are available to accommodate embryos of different sizes. Human eyelashes or cat whiskers inserted through the EPC of the embryo are used to suspend the embryo through the plastic hole and fix the embryo's position (Fig. 20.5A–C). For E8.5 embryos or older, we use a modified culture dish with a miniature crane that will hold the embryo during on-stage culture (Fig. 20.5D and E). The embryo crane consists of a thin platinum wire extending from a cantilever, which is mounted on top of a one-coil spring. The spring is fixed onto a washer sitting inside the Petri dish, and is regulated in height by an adjustment screw coupled to the washer. The wire from the cantilever extends down over the middle of the Petri dish and into a medium-filled ultra high molecular weight polyethylene ring that is fixed in the center of the dish. The bottom tip of the wire is formed into a hook from which the embryo hangs suspended in the medium, and over the coverslip. The latter two methods are described below in greater detail.

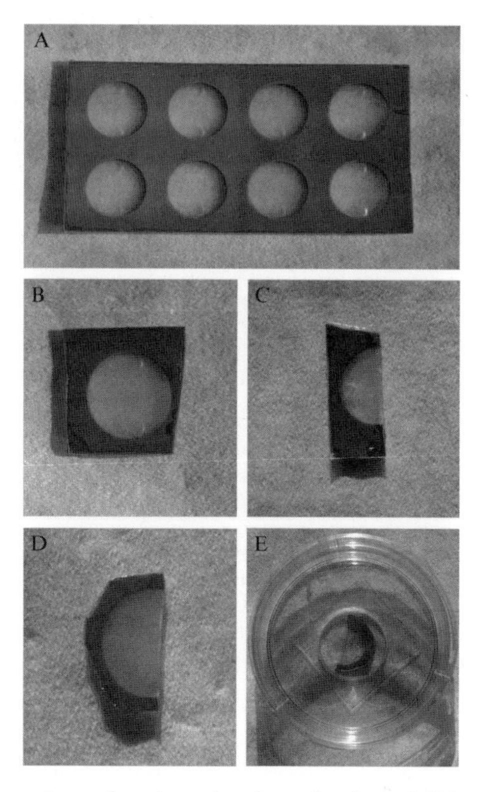

Figure 20.4 Preparation of gasket chambers for immobilizing postimplantation embryos for static culture and time-lapse imaging. (A–E) Preparation of inserts for immobilization of early postimplantation stages: (A) CoverWell (Invitrogen, USA) gasket chambers; (B, C) chambers are cut so to obtain a plastic insert containing one hole for the insertion of the embryo; (D) pieces of plastic are cut away so to fit insert into an imaging dish; (E) adhesive side of the insert is stuck to bottom of the glass-bottom imaging dish.

3.4.4.1. Embryo immobilization using CoverWell chamber gaskets

1. Cut chamber gaskets to fit the glass bottom of your culture dish and adhere the rubber part to the bottom. To facilitate adhesion to the glass, place dish on a hot surface (i.e., heating block).
2. Add prewarmed and equilibrated DMEM/F12 culture media to the dish covering the entire gasket and cover it with light mineral oil.
3. Transfer embryo and eyelash/whisker to the dish.
4. Under a stereo dissecting microscope, pierce the EPC of the embryo and position the embryo in such a way that it hangs through the plastic hole of the dish and the eyelash/whisker balances the embryo at the top of the plastic (Fig. 20.5C).

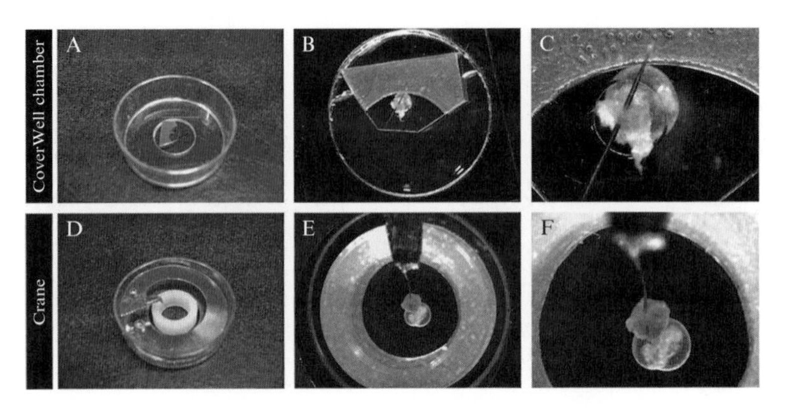

Figure 20.5 Two alternative methods for immobilization of postimplantation mouse embryos for static culture and time-lapse imaging. (A–C) Immobilizing a postimplantation embryo using a CoverWell gasket chamber: (A) glass-bottom imaging dish with a CoverWell gasket chamber adhered to the glass bottom; (B, C) close-ups of embryo after culture hanging through the hole in the plastic via an eyelash. (D–F) Immobilizing a postimplantation embryo using an embryo 'crane': (D) Embryo 'crane,' made of stainless steel, inserted into the glass bottom of the imaging dish. A plastic ring is inserted into the middle of the dish above the glass bottom that holds the culture medium, where the embryo is hung into via a platinum wire extending from a cantilever. (E, F) Close-up view of an embryo after culture hanging from the platinum wire of the embryo 'crane.'

5. Place dish on the microscope stage with environmental chamber using conditions for embryo culture described previously.
6. Note that the orientation of the embryo depends on the precise angle at which the eyelash/whisker is inserted into the EPC. If the embryo is not in the correct orientation, one can usually carefully remove and reinsert the eyelash/whisker. However, this process cannot be performed repeatedly until the desired orientation is achieved.
7. Image embryo (Section 3.5).

3.4.4.2. Embryo immobilization using an embryo "crane"

1. Before setting up the embryo for a time-lapse experiment prewarm the crane to 37 °C by placing in a tissue culture or on-stage incubator.
2. Dissect embryos as described previously and incubate them for at least 15 min in prewarmed and pregassed culture media in a tissue culture incubator.
3. Add prewarmed and equilibrated DMEM/F12 culture media to the inner ring.
4. Transfer embryo to the ring.
5. Using watchmaker's forceps, hang the embryo from the hook by piercing the EPC of the embryo and position the embryo (Fig. 20.5F).

6. Place dish on the microscope stage with an environmental chamber using the conditions for embryo culture described previously.
7. Image embryo (Section 3.5).

3.5. Live imaging cultured mouse embryos

While imaging, the frequency of acquisition, as well as the laser power used, should be kept to a minimum since laser light is phototoxic to cells and can impair proper development. Control cultures without exposure to laser light should be set up to work out the proper imaging conditions, especially when morphogenetic events in mutant embryos should be imaged.

3.5.1. Setting up the image data acquisition

1. Prewarm the microscope system to 37 °C for 2 h prior to starting the experiment. In the meantime the embryos are dissected, placed in a tissue culture incubator and then mounted. Of note, we prefer an incubator system (Fig. 20.6) that encloses the optics carrier and microscope stage over an on-stage incubator system. The former system is preferable as it provides more robust temperature control with fewer fluctuations, so preferable for sustained normal embryonic development, as well as preventing focal drift created by thermal expansion or contraction of the system's mechanical components.
2. Place coverglass-bottomed culture dishes containing immobilized embryo onto the stage insert so that it is securely in place. Since in many systems gassing occurs within the chamber above the stage, a secure fit of the dish within the stage insert is desired so as to limit gas leakage.
3. Find the embryo using a low-magnification (5×) scanning objective. This is usually done by visualizing the specimen in bright field and manually moving the stage using the *x*–*y* control.
4. Sequentially move through a series of dry objectives of increasing magnification (10× and 20×) refocusing and repositioning the stage to visualize the region of the embryo that you wish to image. We routinely use the 10× objective for low-magnification 3D time-lapse data acquisition to visualize global cell rearrangements (Fig. 20.7), and the 20× for higher magnification data acquisition where our aim is to be able to track individual cells. Occasionally, we will imaging using an immersion objective (25× or 40×), if so we add the oil, glycerol, or water-based immersion liquid onto the objective, move it out of position temporarily as we place the dish on the stage and scan it to identify the embryo, and then bring it back into position. Of note, since it readily evaporates, water cannot be used as an immersion medium for time-lapse imaging of

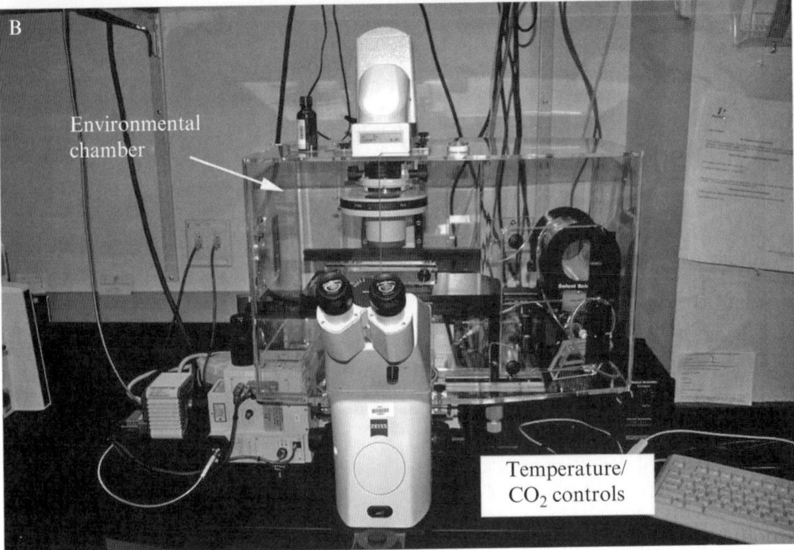

Figure 20.6 Enclosed incubator containing microscope systems for live imaging static cultured mouse embryos. (A) Single point laser scanning confocal inverted microscope with scan head placed on a base port. This arrangement facilitates unobstructed access to the microscope and placement of an environmental chamber enclosing the stage, objectives, and condenser. The microscope is positioned on an air table to buffer vibration and movement to the specimen during imaging. (B) Spinning disk confocal on an inverted microscope with laser module and Nipkow disk scan head on the left side of the microscope and custom-made environmental chamber. The microscope is positioned on a table top antivibration platform to buffer movement to the specimen during imaging.

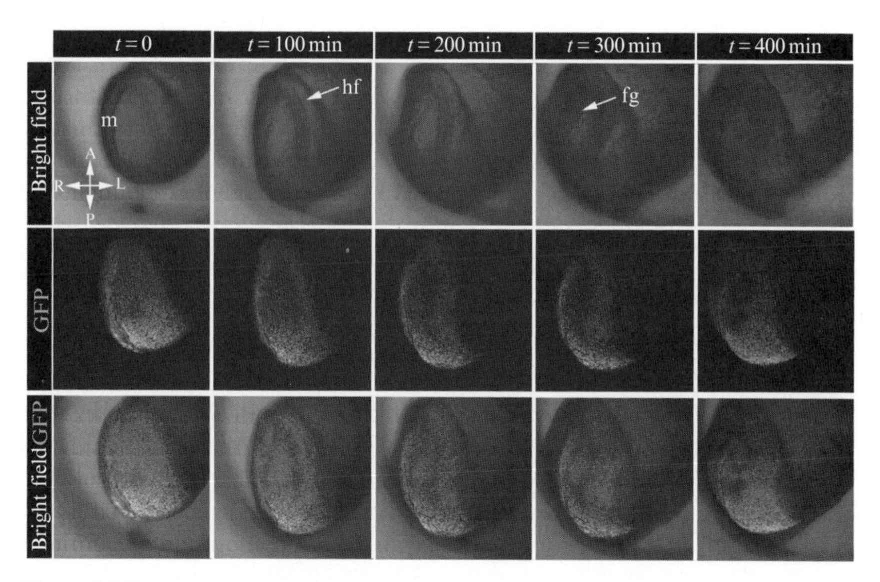

Figure 20.7 Live imaging of postimplantation mouse development, visualization of mesoderm cell dynamics during foregut invagination. Sequence of stills from a laser scanning confocal time-lapse experiment of an embryo expressing a histone H2B–YFP reporter in all mesoderm cells. Normal morphogenesis of the headfolds (hf) and invagination of the foregut (fg) are evident in the bright field panels of the time series. Embryo was imaged for a total of 8.5 h. Optical sections were taken every 4 μm, and z-stacks were acquired at 20 min intervals. Upper row shows a time series of bright field images. Middle row shows 3D rendered z-stack of the GFP channel from individual time points of the same time series. Bottom row shows the overlay of the GFP channel onto the bright field images. m, midline. (See Color Insert.)

embryos cultured at 37 °C. To overcome this issue, many water-based immersion compounds are available for use as substitutes.

5. Through the microscope software, set up the image acquisition parameters. The laser settings will vary based on the microscope system, and also on the reporter being visualized. Quite some effort needs to be invested in empirically evaluating each reporter strain to determine the optimal parameters for visualization without photobleaching of the reporter or phototoxicity to the sample. For example, we routinely use a Zeiss LSM 510 system with the following 25 mW 488 nm Argon laser (for GFP) output: running at 6.1 A with power set to <5% but preferably 2–3%. To increase the speed of data acquisition on a laser point scanning confocal we will routinely bidirectionally scan in 'single track' mode, where all channels are acquired at once. The alternative mode of 'multitracking,' where all channels are captured separately, may need to be used with overlapping fluorophores, but results in extended exposure of the specimen to laser light.

6. Through the microscope software, set the top and bottom limits of the *z*-stack as well as the *z*-spacing. When inputting these settings, make sure you account for inherent focal drift, embryo movement as well as slight drift and embryo growth by extending the *z*-stack by several micrometers at either end. We routinely acquire *z*-stacks of up to 150 μm with a spacing of 0.5–6 μm depending on the objective and experiment. For example, a 0.5 μm *z*-interval is not necessary when using a 10× objective, which inherently produces optical sections of several micrometers.
7. Through the microscope software, set the time interval and the number of *z*-stacks that will be acquired in, or total duration of, the experiment.
8. Click START.
9. Once a 3D time-lapse is started, we routinely return to the microscope to check on the data as it is being acquired and to make sure that the embryo is not drifting. If we do find embryo drift, we stop and save any useable *z*-stacks. We then reposition the embryo, reset the top and bottom limits of the *z*-stack, and restart time-lapse acquisition. Furthermore, if the experiment is taking place during the day, one can always return to the microscope and readjust any settings, for example, the positioning in *x*/*y*/*z* for embryo drift, growth or focal drift of the system, laser power or gain for reporter photobleaching, degradation or changes in the level of expression.
10. At the end of the experiment, save the data and score the embryo, usually by placing onto a stereo dissecting microscope, for normal development.
11. Once the experiment is terminated, clean the objective used to image, even if it was a dry one. Then move the objective turret to an empty position or a position with a low magnification (5×), long working distance objective.
12. If imaging embryos are suspended with the 'crane,' clean the 'crane' by wiping it with 70% ethanol.

3.5.2. Choosing reporters

- An increasing number of FP reporter strains are available from public repositories such as the JAX IMR (http://www.jax.org/imr/index.html).
- It should be borne in mind that any particular reporter strains demonstrated to be readily detectable on someone else's system may not necessarily work on every system.
- Fusing a protein of interest to an FP can affect its native behavior, and can in some cases subtly or significantly affect a developmental process being studied. Our test for developmental neutrality of any fusion that we generate is to determine if we can establish strains of transgenic mice exhibiting widespread expression of the fusion protein. Proof of

developmental neutrality comes if germline transmission and expression of the transgene is achieved, transgenics are represented in Mendelian ratios, and are viable, fertile, and indistinguishable from nontransgenic littermates.

- A knowledge of the physical and optical properties of FPs is crucial to determine if and to what extent FPs might affect the localization, function, and spatiotemporal dynamics of fusion proteins. *Aequeoria victoria* GFP variants have been shown to weakly dimerize. By contrast, FPs from *Anthozoa* species form obligate tetramers, a feature that precluded the use of first generation native proteins in mouse reporter strains, and has hampered the use of even newer generation variants in many fusion proteins. So although oligomerization does not preclude the use of newer generation cytosolic FPs as reporters for gene expression, it cautions their use in fusions. In our experience, even though both human histone H2B–GFP and H2B–mCherry FP fusions are bright and localize to active chromatin, we have been unable to generate a strain of mice exhibiting widespread expression of an H2B–mCherry fusion. By contrast, we have developed several strains expressing the H2B–GFP fusion (Nowotschin *et al.*, 2009b). The brightness of an FP (the product of its extinction coefficient and its fluorescence quantum yield) determines the intensity of the fluorescence signal that can be visualized, and thus captured. Bright FPs require low-intensity illumination or short scan times for acquisition, both of which are preferable for live imaging where they serve to minimize phototoxicity and photodamage to the tissue, as well as to reduce photobleaching (excitation-induced photodestruction) of the fluorophore. We therefore recommend that the latest generation FP variants be used in the generation of genetically modified mouse strains. A minimal investment of time at the start of an experiment, to generate transgenic or gene targeting DNA constructs incorporating newer FP variants, is later rewarded when strains are available for imaging. Even though improved FP variants are continually being developed which are optimized for faster folding and chromophore maturation, increased brightness and photostability and minimal self-association, in our experience not all mouse transgenic strains will express reporters at robust enough levels for live imaging. Unfortunately, little can be done to improve the fluorescent signal in a reporter strain exhibiting low levels of expression.

ACKNOWLEDGMENTS

We are indebted to Sneaky, Iffy, and Phoebe Nowotschin, as well as Arthur Soriano, for assistance with embryo immobilization. We thank Stefan Kirov and Ricardo Toledo-Crow for design and development of the embryo 'crane.' Work in our laboratory is supported by the National Institutes of Health and NYSTEM. S. N. is supported by an American Heart Association postdoctoral fellowship.

REFERENCES

Bronner-Fraser, M., and Fraser, S. E. (1988). Cell lineage analysis reveals multipotency of some avian neural crest cells. *Nature* **335,** 161–164.

Chudakov, D. M., Verkhusha, V. V., Staroverov, D. B., Souslova, E. A., Lukyanov, S., and Lukyanov, K. A. (2004). Photoswitchable cyan fluorescent protein for protein tracking. *Nat. Biotechnol.* **22,** 1435–1439.

Davidson, L. A., Dzamba, B. D., Keller, R., and Desimone, D. W. (2008). Live imaging of cell protrusive activity, and extracellular matrix assembly and remodeling during morphogenesis in the frog, *Xenopus laevis. Dev. Dyn.* **237,** 2684–2692.

Downs, K. M., and Davies, T. (1993). Staging of gastrulating mouse embryos by morphological landmarks in the dissecting microscope. *Development* **118,** 1255–1266.

Fraser, S. T., Hadjantonakis, A. K., Sahr, K. E., Willey, S., Kelly, O. G., Jones, E. A., Dickinson, M. E., and Baron, M. H. (2005). Using a histone yellow fluorescent protein fusion for tagging and tracking endothelial cells in ES cells and mice. *Genesis* **42,** 162–171.

Giepmans, B. N., Adams, S. R., Ellisman, M. H., and Tsien, R. Y. (2006). The fluorescent toolbox for assessing protein location and function. *Science* **312,** 217–224.

Gurskaya, N. G., Verkhusha, V. V., Shcheglov, A. S., Staroverov, D. B., Chepurnykh, T. V., Fradkov, A. F., Lukyanov, S., and Lukyanov, K. A. (2006). Engineering of a monomeric green-to-red photoactivatable fluorescent protein induced by blue light. *Nat. Biotechnol.* **24,** 461–465.

Haas, K., Sin, W. C., Javaherian, A., Li, Z., and Cline, H. T. (2001). Single-cell electroporation for gene transfer in vivo. *Neuron* **29,** 583–591.

Hadjantonakis, A. K., and Papaioannou, V. E. (2004). Dynamic in vivo imaging and cell tracking using a histone fluorescent protein fusion in mice. *BMC Biotechnol.* **4,** 33.

Hadjantonakis, A. K., Dickinson, M. E., Fraser, S. E., and Papaioannou, V. E. (2003). Technicolour transgenics: Imaging tools for functional genomics in the mouse. *Nat. Rev. Genet.* **4,** 613–625.

Joyner, A. L., and Zervas, M. (2006). Genetic inducible fate mapping in mouse: Establishing genetic lineages and defining genetic neuroanatomy in the nervous system. *Dev. Dyn.* **235,** 2376–2385.

Kinder, S. J., Tsang, T. E., Quinlan, G. A., Hadjantonakis, A. K., Nagy, A., and Tam, P. P. (1999). The orderly allocation of mesodermal cells to the extraembryonic structures and the anteroposterior axis during gastrulation of the mouse embryo. *Development* **126,** 4691–4701.

Kinder, S. J., Tsang, T. E., Wakamiya, M., Sasaki, H., Behringer, R. R., Nagy, A., and Tam, P. P. (2001). The organizer of the mouse gastrula is composed of a dynamic population of progenitor cells for the axial mesoderm. *Development* **128,** 3623–3634.

Kwon, G. S., and Hadjantonakis, A. K. (2009). Transthyretin mouse transgenes direct RFP expression or Cre-mediated recombination throughout the visceral endoderm. *Genesis* **47,** 447–455.

Kwon, G. S., Fraser, S. T., Eakin, G. S., Mangano, M., Isern, J., Sahr, K. E., Hadjantonakis, A. K., and Baron, M. H. (2006). Tg(Afp-GFP) expression marks primitive and definitive endoderm lineages during mouse development. *Dev. Dyn.* **235,** 2549–2558.

Lane, M., and Gardner, D. K. (1992). Effect of incubation volume and embryo density on the development and viability of mouse embryos in vitro. *Hum. Reprod.* **7,** 558–562.

Larina, I. V., Shen, W., Kelly, O. G., Hadjantonakis, A. K., Baron, M. H., and Dickinson, M. E. (2009). A membrane associated mCherry fluorescent reporter line for studying vascular remodeling and cardiac function during murine embryonic development. *Anat. Rec. (Hoboken)* **292,** 333–341.

Lippincott-Schwartz, J., and Patterson, G. H. (2008). Fluorescent proteins for photoactivation experiments. *Methods Cell Biol.* **85,** 45–61.

Murray, M. J., and Saint, R. (2007). Photoactivatable GFP resolves *Drosophila* mesoderm migration behaviour. *Development* **134,** 3975–3983.

Nagy, A. (2000). Cre recombinase: The universal reagent for genome tailoring. *Genesis* **26,** 99–109.

Nagy, A., Gertsenstein, M., Vintersten, K., and Behringer, R. (2003). Manipulating the Mouse Embryo. A Laboratory Manual. Cold Spring Harbor Laboratory Press, Cold Spring Harbor, NY.

Nowotschin, S., and Hadjantonakis, A. K. (2009). Use of KikGR a photoconvertible green-to-red fluorescent protein for cell labeling and lineage analysis in ES cells and mouse embryos. *BMC Dev. Biol.* **9,** 49.

Nowotschin, S., Eakin, G. S., and Hadjantonakis, A. K. (2009a). Dual transgene strategy for live visualization of chromatin and plasma membrane dynamics in murine embryonic stem cells and embryonic tissues. *Genesis* **47,** 330–336.

Nowotschin, S., Eakin, G. S., and Hadjantonakis, A. K. (2009b). Live-imaging fluorescent proteins in mouse embryos: Multi-dimensional, multi-spectral perspectives. *Trends Biotechnol.* **27,** 266–276.

Papaioannou, V. E., and West, J. D. (1981). Relationship between the parental origin of the X chromosomes, embryonic cell lineage and X chromosome expression in mice. *Genet. Res.* **37,** 183–197.

Patterson, G. H. (2008). Photoactivation and imaging of photoactivatable fluorescent proteins. *Curr. Protoc. Cell Biol.* Chapter 21, Unit 21.6.

Rhee, J. M., Pirity, M. K., Lackan, C. S., Long, J. Z., Kondoh, G., Takeda, J., and Hadjantonakis, A. K. (2006). In vivo imaging and differential localization of lipid-modified GFP-variant fusions in embryonic stem cells and mice. *Genesis* **44,** 202–218.

Scherz, P. J., Huisken, J., Sahai-Hernandez, P., and Stainier, D. Y. (2008). High-speed imaging of developing heart valves reveals interplay of morphogenesis and function. *Development* **135,** 1179–1187.

Shaner, N. C., Steinbach, P. A., and Tsien, R. Y. (2005). A guide to choosing fluorescent proteins. *Nat. Methods* **2,** 905–909.

Stark, D. A., and Kulesa, P. M. (2005). Photoactivatable green fluorescent protein as a single-cell marker in living embryos. *Dev. Dyn.* **233,** 983–992.

Stark, D. A., and Kulesa, P. M. (2007). An in vivo comparison of photoactivatable fluorescent proteins in an avian embryo model. *Dev. Dyn.* **236,** 1583–1594.

Theiler, K. (1989). The House Mouse. Atlas of Embryonic Development. Springer-Verlag, New York, NY.

Whittingham, D. G., and Wales, R. G. (1969). Storage of two-cell mouse embryos in vitro. *Aust. J. Biol. Sci.* **22,** 1065–1068.

Wiedenmann, J., Ivanchenko, S., Oswald, F., Schmitt, F., Rocker, C., Salih, A., Spindler, K. D., and Nienhaus, G. U. (2004). EosFP, a fluorescent marker protein with UV-inducible green-to-red fluorescence conversion. *Proc. Natl. Acad. Sci. USA* **101,** 15905–15910.

Woolner, S., Miller, A. L., and Bement, W. M. (2009). Imaging the cytoskeleton in live *Xenopus laevis* embryos. *Methods Mol. Biol.* **586,** 23–39.

CHAPTER TWENTY-ONE

Ultrasound and Magnetic Resonance Microimaging of Mouse Development

Brian J. Nieman*,† *and* Daniel H. Turnbull‡,§

Contents

Abstract

Ultrasound biomicroscopy (UBM) and magnetic resonance microimaging (micro-MRI) provide noninvasive, high-resolution images in mouse embryos and neonates, enabling volumetric and functional analyses of phenotypes, including longitudinal imaging of individual mice over critical stages of *in utero* and early-postnatal development. In this chapter, we describe the underlying principles of UBM and micro-MRI, including the advantages and limitations of these approaches for studies of mouse development, and providing a number of examples to illustrate their use. To date, most imaging studies have focused on the developing nervous and cardiovascular systems, which are also reflected in the examples shown in this chapter, but we also discuss the future application of these methods to other organ systems.

* Mouse Imaging Centre, Hospital for Sick Children, Toronto, Canada
† Department of Medical Biophysics, University of Toronto, Toronto, Canada
‡ Kimmel Center for Biology and Medicine at the Skirball Institute of Biomolecular Medicine, New York University School of Medicine, New York, USA
§ Departments of Radiology and Pathology, New York University School of Medicine, New York, USA

Methods in Enzymology, Volume 476
ISSN 0076-6879, DOI: 10.1016/S0076-6879(10)76021-3

1. Introduction

Compared to developmental biology studies in lower organisms, such as *Caenorhabditis elegans* and zebrafish, the mouse presents significant challenges for direct visualization and analysis of volumetric and dynamic changes in embryos and their developing organ systems. Despite advances in optical microscopy and the availability of mouse reporter lines expressing fluorescent proteins, *in vivo* optical imaging is generally restricted to tissue explants and *exo utero* imaging of early-stage embryos that are amenable to whole embryo culture, and imaging studies cover relatively short time windows over which normal development can be maintained. Ultrasound and magnetic resonance imaging (MRI) are widely used for human fetal and pediatric imaging, and can be scaled to provide effective microimaging tools for application in mice. Although the spatial resolution of these methods (typically 50–100 μm) is lower than optical microscopy, they offer the advantages of much greater penetration, enabling whole body imaging, and the ability to perform three-dimensional (3D) anatomical and functional phenotype analyses, including noninvasive longitudinal imaging over periods of days to weeks, both *in utero* in mouse embryos, and extending to neonatal through adult stages of organ development. A major challenge for imaging methods, including ultrasound and MRI, is the need for high image throughput necessary to match the requirements for efficient phenotypic screening of mutant and transgenic embryos and postnatal mice. In this context, both ultrasound and MRI offer significant advantages in terms of real-time imaging capability (ultrasound) and the recent development of multiple-mouse imaging systems (MRI). For phenotype screening with either method, the image analysis process is critically important, but is not discussed in detail in this chapter. It is worth noting that MRI data are particularly well-suited to computational 3D analysis approaches, providing the potential to greatly improve phenotyping throughput and detection of more subtle changes than can be evaluated by simple inspection of images. Such methods have produced impressive results in the adult mouse brain (Lerch *et al.*, 2008; Nieman *et al.*, 2006), but have not been employed extensively to date for analysis during development or in other tissues. In principle, however, these and emerging methods can be extended to provide embryonic and neonatal phenotype analysis. Development of such automated phenotyping methods are likely to continue, particularly in the context of embryo imaging, as large-scale efforts to generate mutants of every gene motivate improved high-throughput phenotyping methods. It should also be noted that ultrasound and MRI are two technologies among others that provide similar, often complementary information. Alternative techniques not described here include X-ray computed tomography after iodine staining or perfusion of X-ray opaque vascular agents (Marxen *et al.*, 2004; Metscher, 2009) and optical projection tomography, which advantageously may also permit the use of immunohistochemical fluorescent markers (Sharpe *et al.*, 2002; Walls *et al.*, 2008). Ultrasound and MRI are notable for their ability to provide *in vivo* data in mouse embryos.

With the recent increase in multimodality small-animal imaging facilities in many research centers, the ability to utilize ultrasound and MRI microimaging is now a reality for many mouse developmental biologists. In this chapter, we describe a variety of ultrasound and MRI methods that are now available for studies of mouse development.

2. Ultrasound Biomicroscopy

Ultrasound is the most common approach for fetal imaging in the clinic. Likewise, ultrasound biomicroscopy (UBM) is now a well-established method for *in utero* imaging of mouse embryos (Srinivasan *et al.*, 1998; Turnbull *et al.*, 1995; reviewed in Turnbull and Foster, 2002). UBM is a high-frequency (30–100 MHz) form of pulse-echo imaging, providing high-resolution (30–100 μm) images in real time. High-frequency Doppler ultrasound has also been incorporated into UBM scanners to measure blood velocity, originally using separate transducers for imaging and Doppler (Fig. 21.1) (Aristizabal *et al.*, 1998), and more recently with both functions provided by the same transducer in commercial scanners (Foster *et al.*, 2009; Zhou *et al.*, 2002). Originally, UBM systems were based on single, mechanically scanned transducers, but more recently array transducers have been developed to improve focusing (Aristizabal *et al.*, 2006) and increase image frame rates using electronic beam forming without the need for mechanical scanning (Foster *et al.*, 2009). *In utero* UBM of mouse embryos is most commonly performed at frequencies between 40 and 50 MHz, which allows sufficient penetration to image mouse embryos throughout gestation with high spatial resolution, starting from early postimplantation stages (Zhou *et al.*, 2002).

2.1. UBM of mouse embryos and neonates

Since its introduction for *in utero* mouse embryo imaging over 15 years ago, UBM has found numerous applications, mostly for brain and cardiovascular imaging (reviewed in Turnbull and Foster, 2002). Similar to clinical ultrasound, Doppler approaches have enabled analysis of blood flow properties in the embryo that provide new insights into development of cardiovascular function (reviewed in Phoon and Turnbull, 2003). The general approach taken for *in utero* UBM studies is to anesthetize the

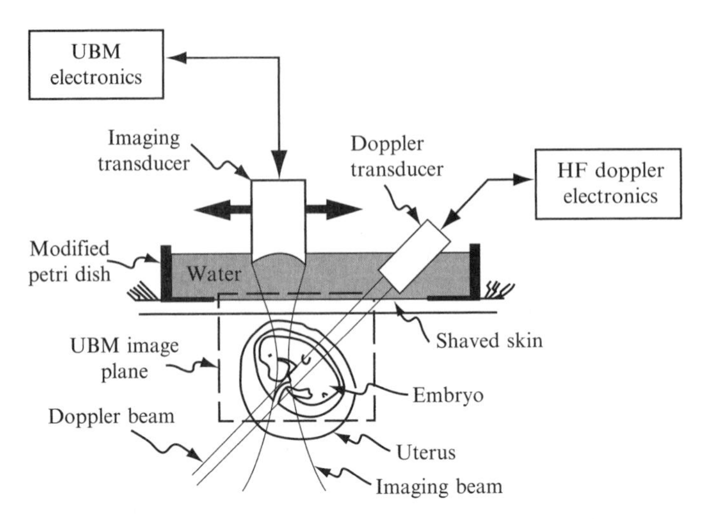

Figure 21.1 Schematic of setup for UBM analysis of mouse embryos. A pregnant mouse is anesthetized, the lower abdomen shaved, and the mouse is laid in the lower level of a two-level stage. A 100 mm plastic Petri dish with a 25 mm hole punched in the center is attached to the upper level and filled with water. The UBM and Doppler transducers are then scanned in the resulting water bath to acquire images and Doppler blood velocity waveforms. The temperature of the water and mouse are maintained at 37 °C with a feedback temperature controller for all physiological measurements. Reprinted with permission from Aristizabal *et al.* (1998).

pregnant mouse, remove the hair on the skin overlying the embryos, and couple the UBM transducer to the mouse using a commercially available ultrasound gel, or a holding system that incorporates a water bath between the transducer and the skin (Fig. 21.1). Two-dimensional (2D) UBM images are acquired in real time ($\geq$100 images/s with current technology), and 3D imaging can be accomplished by acquiring a stack of 2D UBM images (Aristizabal *et al.*, 2006). Recently, commercial UBM scanners have also included color flow imaging, in which Doppler signals are analyzed in real time to produce a color-coded map of blood velocities (Foster *et al.*, 2009).

To date, UBM has found greatest application in the mouse cardiovascular system, although similar techniques should provide data relevant to many organ systems. UBM and UBM–Doppler methods can be used over a very wide range of cardiovascular developmental stages (Fig. 21.2), from the onset of heart beat at E8.0 (Ji *et al.*, 2003), through the critical early stages of chamber formation between E10.5 and E14.5 (Phoon *et al.*, 2000, 2002; Srinivasan *et al.*, 1998; Zhou *et al.*, 2003), and into neonatal stages when cardiomyopathy and heart failure are first manifested in many mutants (Fatkin *et al.*, 1999). Interestingly, the first UBM studies for

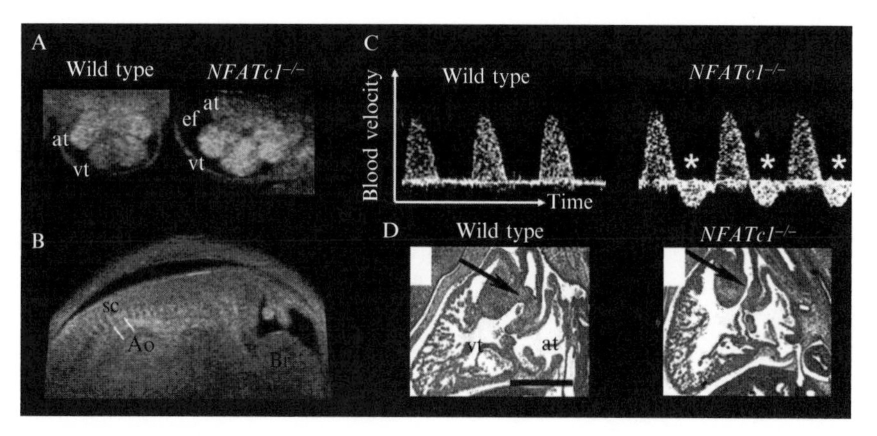

Figure 21.2 UBM analysis of cardiovascular defects in mouse embryos. (A) UBM provides *in utero* images of the E12.5 embryonic heart, enabling identification and dynamic analysis of the cardiac atria (at) and ventricles (vt), including detection of pericardial effusions in a subset of *NFATc1*$^{-/-}$ mutants. (B) UBM images were used to place the Doppler sample volume (hatch marks) over the aorta (Ao), to acquire blood velocity waveforms in E13.5 wild type and *NFATc1*$^{-/-}$ mutants (C). Doppler analysis showed clear evidence of regurgitant flow patterns (*) in the mutants resulting from defects in aortic valve formation (arrows; D). Other labels: Br, brain; sc; spinal cord. Panels (A), (C), and (D) reprinted with permission from Phoon *et al.* (2004).

phenotyping *NFATc1*$^{-/-}$ mutants, which lack aortic and pulmonary cardiac valves and die *in utero*, revealed unusual mechanisms of embryonic heart failure (Phoon *et al.*, 2004). Longitudinal studies of each embryo in a pregnant mouse, required for effective *in utero* phenotype analysis with UBM, is challenging but can be achieved through careful mapping of extra- and intraembryonic anatomical landmarks to enable accurate identification of individual embryos over a period of several days (Ji and Phoon, 2005).

2.2. *In utero* UBM-guided injections

For over a decade, UBM has provided a unique and powerful approach for direct *in utero* image-guided manipulation of mouse embryos (Liu *et al.*, 1998; Olsson *et al.*, 1997). This has been most utilized in the developing embryonic brain, where UBM-guided injections have enabled *in utero* neural cell transplantation (Butt *et al.*, 2005; Olsson *et al.*, 1997; Wichterle *et al.*, 2001), cell lineage tracing (Kimmel *et al.*, 2000), and gain-of-function studies with retroviruses or electroporation (Gaiano *et al.*, 1999; Punzo and Cepko, 2008; Weiner *et al.*, 2002). For UBM-guided injections, timed pregnant mice are anesthetized with nembutol or isoflurane, and the uterus exposed after laparotomy (Fig. 21.3). Hair is

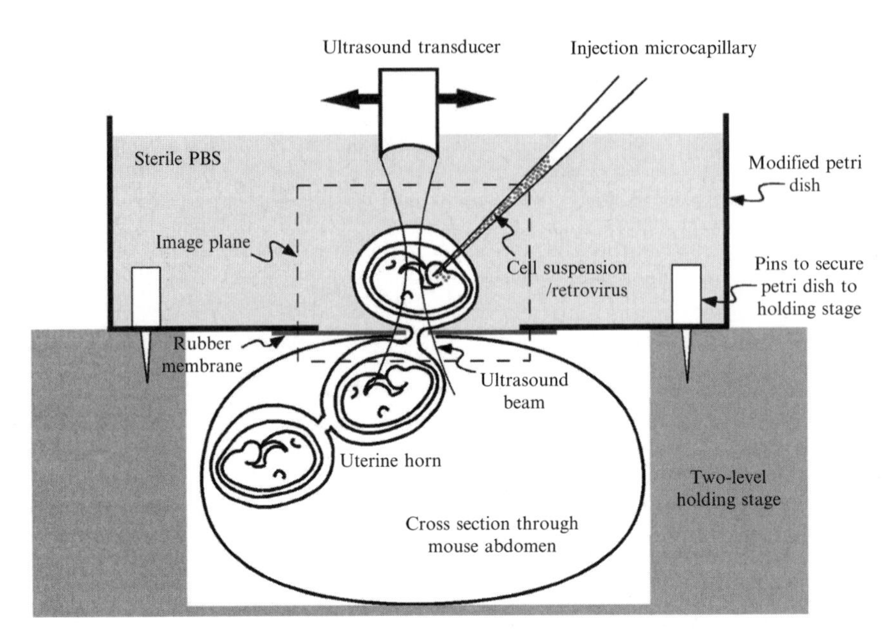

Figure 21.3 Schematic of setup for UBM-guided injections. A pregnant mouse is anesthetized, the abdomen shaved, and a midline incision made through the skin and muscle to gain access to the uterine horn. The mouse is laid in the lower level of a two-level stage, with a modified Petri dish attached to the upper level. Part of the uterus containing a single embryo is gently pulled through a slot in a thin rubber membrane, into a bath of sterile PBS for injection under real-time UBM image guidance. Reprinted with permission from Liu *et al.* (1998).

removed from the lower abdominal skin, and a midline incision made through the skin and peritoneum. The pregnant mouse is laid supine in the lower section of a two-level stage, and a plastic Petri dish, modified by punching a ~25 mm diameter central hole, is secured over her abdomen. Part of the uterus containing one to two embryos (depending on embryonic stage) is gently pulled through a slot, cut into a thin rubber membrane stretched across the hole in the Petri dish, into sterile PBS. The UBM transducer is then lowered into the PBS bath, providing real-time imaging of the mouse brain and image-guidance of a pulled and sharpened 50-μm diameter glass microcapillary injection needle, mounted on a 3D micromanipulator and inserted through the uterine wall and into the embryonic target region (Fig. 21.4). After injection, one to two new embryos are gently pulled into the PBS, and the injected embryos gently placed back through the slot into the abdominal cavity. In this way, an entire litter (8–10 embryos) can be injected in ~1 h, after which the muscle and skin are sutured or clipped and the pregnant mouse recovers in a warming chamber until regaining consciousness. Like most surgical

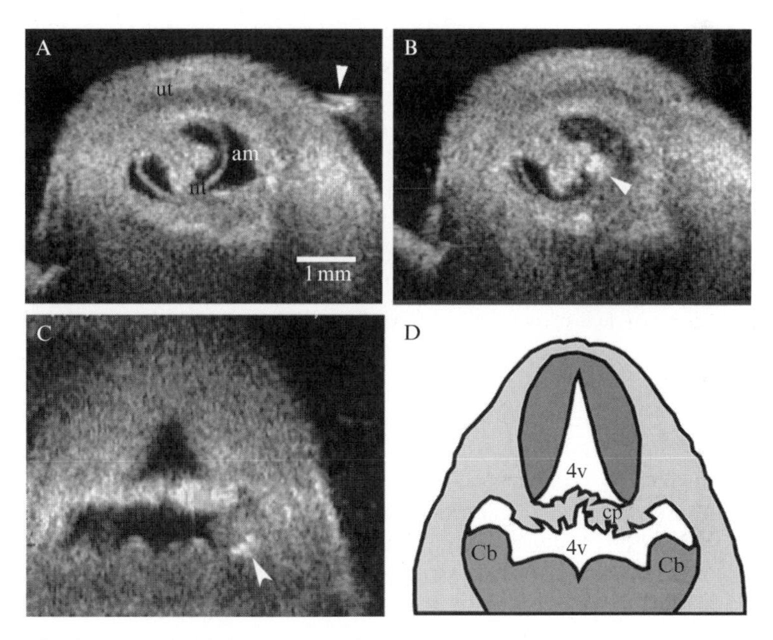

Figure 21.4 UBM-guided injections of the embryonic mouse nervous system. (A, B) At E8.5 the needle tip (arrows) is shown before (A) and after (B) insertion through the uterus (ut) and amniotic membrane (am) to inject a retrovirus adjacent to the (open) neural tube (nt). (C, D) At E13.5 a similar approach is used to inject a *Shh*-expressing retrovirus into the cerebellar anlage (Cb), which is easily identified from its lateral position in the mid-hindbrain, near the fourth ventricle (4v) and choroid plexus (cp). Panels (A) and (B) reprinted with permission from Gaiano *et al.* (1999); panels (C) and (D) reprinted with permission from Weiner *et al.* (2002).

procedures, survival of embryos is operator-dependent, but most people can achieve at least 50–60% embryonic survival after practice, and many can reach 80% survival or better for *in utero* brain injections. In addition to the applications described above, UBM-guided injection can also provide a unique method for precise targeting of cell-labeling agents in mouse embryos, for example, with MRI contrast agents that label cells that can then be followed with longitudinal MRI imaging.

3. Magnetic Resonance Microimaging

MRI is well-established in clinical radiology as an imaging tool that provides excellent soft tissue contrast for isolating pathologies or characterizing anatomy. Increasingly, MRI is also used for assessment of functional parameters, such as blood perfusion, oxygenation status, or cardiac dynamics, and in larger population studies to identify regions altered by disease or development (Aljabar *et al.*, 2008; Davatzikos *et al.*, 2005; Janke *et al.*, 2001). MRI, like ultrasound, can be scaled down for magnetic resonance microimaging (micro-MRI), providing a powerful quantitative approach for assessing changes in 3D anatomy and physiology during development, including isolating phenotypes in genetically modified mice.

Compared to UBM, micro-MRI image contrast is remarkably flexible, allowing optimization of image acquisition to highlight anatomy or pathology of interest. The MRI signal is ultimately derived from the perturbed magnetic state of water protons within the tissue; however, variations in the acquisition can emphasize image contrast from different physical processes affecting the proton signal. Traditional images are referred to as proton-density-, T_1-, or T_2-weighted images, according to the mechanism most reflected in the image contrast. However, diffusion, perfusion, blood flow or macromolecular content can also provide a basis for image contrast, and may be preferred in many applications in developmental biology studies. Administration of contrast agents that produce hyper- or hypointensity in an image is also prevalent, and particularly common in mouse imaging, both for *in vivo* and *ex vivo* imaging studies. This contrast flexibility provides opportunities in research to highlight physiology, anatomy, and even molecular/cellular regions of interest.

Typical resolutions achieved with *in vivo* micro-MRI are on the order of 100 μm in adult mice. The reduced size of the embryo, particularly during early developmental stages, requires higher resolution for visualization of anatomical features. This limitation has meant that MRI imaging has been most feasible in mid- to late-embryonic stages, starting around E12. Likewise, there has been a preference for studying embryos with *ex vivo* micro-MRI, where specimen imaging hardware and longer acquisition times can be tailored to achieve the necessary resolution increase without the concerns for anesthesia and animal maintenance necessary *in vivo*. Nevertheless, *in vivo* micro-MRI methods are particularly beneficial for examining early-postnatal mice, when the phenotype permits, and continued development of *in utero* imaging methods shows potential for more widespread *in utero* studies in the future.

Below we describe methods for both *ex vivo* and *in vivo* micro-MRI of mouse development (Turnbull and Mori, 2007). Developmental MRI studies to date have focused primarily on the heart, brain, or vascular systems and so these areas are highlighted in this chapter. This is, however, in part a reflection of the developmental biology community at large and should not be considered an inherent limitation of the technology. Indeed, with a high demand for phenotyping mice, and a large number of embryonic lethal phenotypes, continued development of embryo phenotyping tools will remain a priority.

3.1. *Ex vivo* anatomical micro-MRI

Although one of the frequently cited benefits of MRI is its *in vivo* potential, *ex vivo* images provide exquisite spatial resolution for detection of phenotypes at defined time points. This permits a more detailed anatomical comparison between mutant and wild-type groups for detection of subtle phenotypes. *Ex vivo* images generally run for many hours in an "overnight" scan session, achieving spatial resolutions between 20 and 50 μm. These images can be acquired on a routine basis, and are highly appropriate for screening new mutants and performing volumetric structural analyses of phenotypes (Fig. 21.5; Petiet *et al.*, 2008).

Ex vivo imaging requires fixation and mounting of embryonic specimens for micro-MRI. Mouse embryos are harvested at a defined stage, where by convention noon of the day after mating is defined as embryonic day 0.5 (E0.5). After extraction from the uterus, embryos are immersed overnight or longer in 4% paraformaldehyde (PFA) or equivalent for fixation at 4 °C. For mid- to late-stage embryos, embryos can also be perfusion-fixed for improved fixation, and for vascular imaging, as described below. Prior to imaging, specimens are embedded in 1–3% agar or immersed in a proton-free fluid, such as Fomblin (Ausimount) or Fluorinert (3 *M*). Some investigators also report "doping" the agar either proton-free fluid or an iron-oxide contrast agent in order to reduce the background agar signal in the image (Dhenain *et al.*, 2001).

For imaging, prepared sample tubes are placed within radiofrequency coils that collect the proton signal used to reconstruct images. Small solenoid coils (~10 mm diameter, ~15 mm length) that fit closely around individual sample tubes provide optimal detection of the water proton signal and signal-to-noise ratio, a common metric of image quality. However, high-resolution *ex vivo* scans commonly require many hours and are run in "overnight" scan sessions, so imaging one embryo per evening can be prohibitively slow. In the aim of increasing throughput, multiple samples can be embedded in larger tubes and placed in a single larger coil, and then imaged collectively (Schneider *et al.*, 2004). This results in one large image, from which individual, 3D images containing single specimens can be extracted. An alternative method for increasing throughput is offered in specialized MRI systems equipped for multiple-mouse MRI (Bock *et al.*, 2003). This technology enables many imaging experiments to be run in parallel, utilizing several single-specimen solenoid coils with one specimen each and avoiding any compromise in image quality—an unavoidable consequence of using a single larger coil.

With either hardware configuration, image acquisition must be prescribed to achieve the desired contrast. It is common to perform a refinement of imaging parameters to achieve the contrast optimal for a particular application. For general anatomical discrimination at high-field ($\geq$7 T),

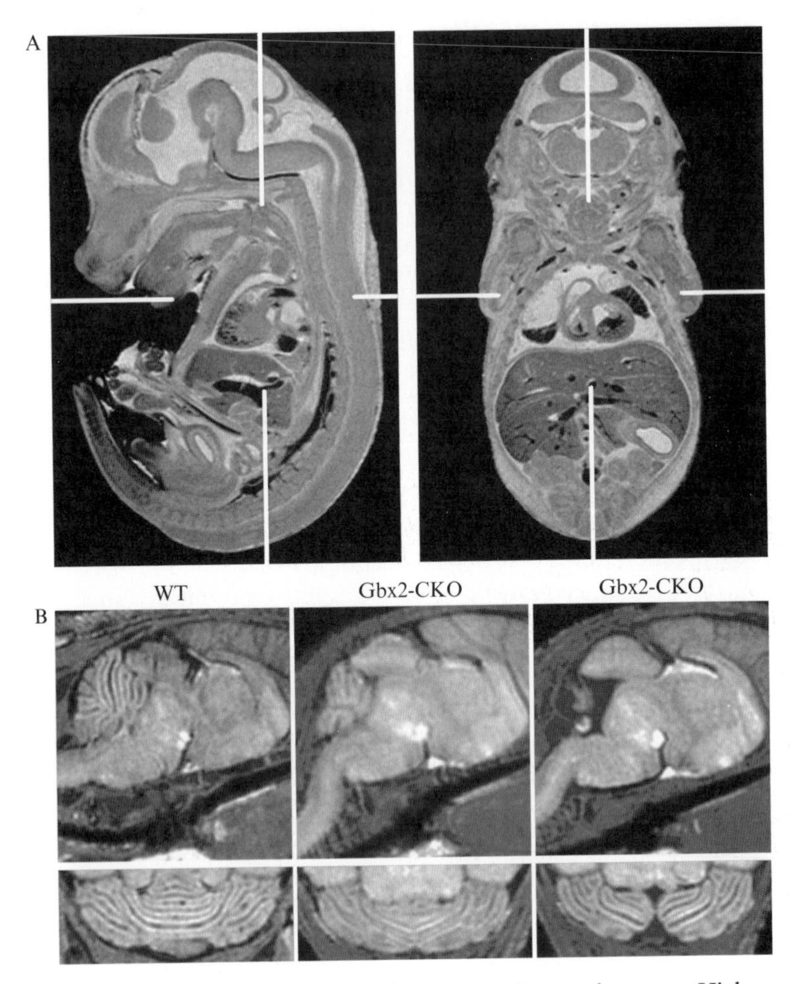

Figure 21.5 Anatomical micro-MRI in the mouse embryo and neonate. High-resolution, 3D images, such as the example in (A) in an E15.5 fixed mouse embryo, provide excellent anatomical detail for phenotyping. *In vivo* anatomical phenotyping is also possible, especially during early-postnatal development. One example is provided in (B), showing variable cerebellar defects in *Gbx2* conditional knockout mice on postnatal day P11. At left, is shown a wild-type mouse in sagittal (top) and horizontal (bottom) views. The second and third columns show a mild and severe example of the cerebellar defect. Panels (A) and (B) reprinted from Petiet *et al.* (2008) and Wadghiri *et al.* (2004), respectively with permission.

T_2-weighted imaging is generally preferred (Dhenain *et al.*, 2001), and can be achieved in fixed samples using repetition time (TR) $\geq$2000 ms and echo time (TE) $\sim$35 ms. Improved acquisition efficiency can be achieved by adding modest concentrations of MRI contrast agent—most commonly gadolinium compounds such as DTPA-Gd or DOTA-Gd—to the fixative and storage solutions. This serves to increase the rate of water proton signal dynamics, thereby allowing for either shorter imaging sessions or higher resolution images. A concentration of 2 m*M* Gd-chelate in the fixative, for instance, in combination with TR $\sim$ 300 ms and TE $\sim$ 35 ms retains excellent T_2-weighted contrast in a fraction of the acquisition time. Alternatively, more highly doped samples (with $\geq$30 m*M* Gd-chelate, for instance) may allow more rapid imaging (Johnson *et al.*, 2007), with TR and TE reduced to $\sim$50 and $\sim$5 ms. In the latter case, contrast is heavily dependent on contrast agent distribution and diligence ensuring consistent specimen preparation is paramount. Modest concentrations of contrast agent ($\leq$3 m*M* Gd-chelate) are also beneficial for diffusion tensor imaging (described below), but higher concentrations confound measurements of the diffusion related signal effects.

Methods for evaluating mouse embryo development with *ex vivo* micro-MRI have been available for over a decade (Smith *et al.*, 1996), with continued improvements resulting in atlases of mouse development from as early as E6.5, and with resolutions $\sim$20 μm (Dhenain *et al.*, 2001; Petiet *et al.*, 2008). Micro-MRI shows potential for phenotyping of embryos in developmental studies (Schneider *et al.*, 2003b), and has been particularly successful in studying cardiac morphology and structure (Smith, 2001). The heart, with its several chambers and intertwining inflow and outflow tracts, is a structure that can be very difficult to appreciate using traditional histological sections and is best-visualized with 3D imaging methods. These features have been helpful in phenotyping of the *Cited2* and *Cx43* mutants, identifying atrial and ventricular septal defects, inflow and outflow tract abnormalities, and aortic arch malformations (Schneider *et al.*, 2003a; Wadghiri *et al.*, 2007). Phenotyping has also been performed in the brain, revealing changes associated with ethanol or retinoic acid administration during pregnancy (Parnell *et al.*, 2009), and a preliminary study investigating study of bone development has also been reported (Ichikawa *et al.*, 2004).

Despite these successes, micro-MRI for study in development has not been used as widely as might be expected. MR scanner availability and challenges associated with analyzing large 3D image data sets in a systematic and efficient way have likely been among the historical barriers. In this regard, analysis of the resulting images for possible phenotypes is a crucial step, one that cannot be treated fully here. The processing steps depend largely on the phenotypes in question. Anatomical phenotypes remain the most commonly investigated by micro-MRI and are appropriate for initial investigations and general characterizations of new mutants. Gross phenotypes can be detected by simple inspection of images, and are of sufficient severity that further analysis provides little additional insight. In many instances, more subtle anatomical differences are present. In these cases, segmentation of structural volumes or computational analysis of local volume/shape changes is necessary to describe the anatomical phenotype. These analysis methods, while employed extensively in adult mouse brain, have had more limited application in developmental studies. However, the tools are not inherently limited to adults, and are beginning to be applied more widely in the embryo and neonate as imaging methods are refined. In general, anatomical phenotypes are very common in genetic mutants (Nieman *et al.*, 2007), so results are likely to be plentiful and micro-MRI provides an important starting point for further investigation.

3.2. *Ex vivo* diffusion tensor imaging

Some anatomical features in embryos, particularly in the central nervous system, are difficult to visualize with traditional MRI contrast weighting. White matter, for instance, which exhibits prominent contrast in later stages, is still under development in embryonic stages so that brain structures myelinated in the adult can be difficult to visualize in the embryonic brain. Novel contrast mechanisms based on water diffusion properties enable enhanced contrast of these structures, and further—through a method called diffusion tensor imaging (DTI)—provide data for computation of the preferential diffusion directions associated with the presence of axons or other anisotropic structures. DTI methods can be used in the postnatal human brain, and show great potential for MR-based phenotyping in the embryonic mouse brain.

DTI data is generally computed based on a set of six or more diffusion-weighted images in addition to a reference, minimally diffusion-weighted (e.g., T_2-weighted) image. The degree of diffusion weighting is controlled in the scan by adjusting a parameter called the *b*-value. Diffusion-weighted images are typically acquired with $b = 1000$–1500 s/mm^2. The low diffusion reference image is generally acquired with a nominal value of $b = 0$ s/mm^2, although higher values ($b \sim 200$ s/mm^2) have been used in fixed samples where some suppression of background signal from the embedding medium is desirable (Mori *et al.*, 2001). As each separate diffusion-weighted image requires as long or longer than an individual anatomical image (such as a T_2-weighted image), acquisition time in DTI studies is often a limiting factor.

The fixation of embryos for DTI–MRI may be performed as described above for standard *ex vivo* imaging. The use of contrast agents during fixation can also be beneficial for speeding image acquisition, but only modest concentrations (≤ 3 m*M* Gd-chelate) are appropriate as high concentrations of contrast agents may confound measurements of the diffusion related signal effects. While absolute measures of diffusion measured in fixed specimens differ from measurements in the *in vivo* case (Shepherd *et al.*, 2009; Sun *et al.*, 2009), relative measures including the principal diffusion directions are not altered (Kim *et al.*, 2009; Sun *et al.*, 2003).

After collection of the necessary images, it is common to process the images to produce a map representing diffusion properties in the brain. In the simplest case, a map showing the degree of anisotropy—the extent to which there is a preferred diffusion direction—can be produced and shown in a simple gray scale image, providing a novel anatomical map emphasizing structures altering water diffusion. To show the principle diffusion direction on DTI images, a color map convention for visualization of DTI data sets has emerged in which the medial–lateral, dorsal–ventral, and rostral–caudal directions are each represented as one of the red, green, or blue channels in an RGB color space. The intensity of the colors can then be modulated by the degree of anisotropy, fading to black for isotropic diffusion and appearing at full intensity for directional diffusion. Both MRI system manufacturer and independent software is available for the computation of the DTI maps from the raw diffusion-weighted imaging data.

DTI maps are particularly powerful for micro-MRI of mouse brain development (Fig. 21.6). Defects in the formation of key white matter structures including the corpus callosum and the hippocampal commissure, as in *Robo1* and *Netrin1* mutants, can be analyzed (Andrews *et al.*, 2006; Zhang *et al.*, 2003). Similarly, apparent degeneration of white matter structures—the fimbria and commissures—has been reported during postnatal development (Zhang *et al.*, 2005a). In addition to enhancing white matter conspicuity, DTI maps also provide surprising contrast in the neocortex of developing embryos (Mori *et al.*, 2001). Methods for automated analysis based on diffeomorphic mapping in DTI-derived maps have also been reported (Zhang *et al.*, 2005b), suggesting routine automated screening and analysis could be developed.

3.3. *Ex vivo* vascular imaging

A sensitive method for vascular imaging in *ex vivo* embryo specimens requires perfusion with an intravascular contrast agent, such as DTPA-Gd conjugated to albumen. For this, the umbilical vessels are isolated for cannulation and perfusion–fixation after exteriorizing the embryo from the uterus and yolk sac (Smith *et al.*, 1994, 1996). The intravascular contrast agent is mixed in gelatin solution and perfused into the embryo after flushing the blood and fixative solutions, after which the umbilical vessels are tied off and the embryo is further immersion fixed as described above (without contrast agent). Imaging then proceeds with a T_1-weighted imaging sequence (TR $\leq$ 50 ms; TE $\leq$ 5 ms), producing bright blood vessels in the presence of the intravascular contrast agent (Fig. 21.7).

The benefit of 3D imaging is particularly important for analysis of the vascular system. The branching, tree-like structure of the vasculature cannot be captured in 2D histological sections, making 3D methods a necessity. In some cases, important findings can be overlooked in traditional histology

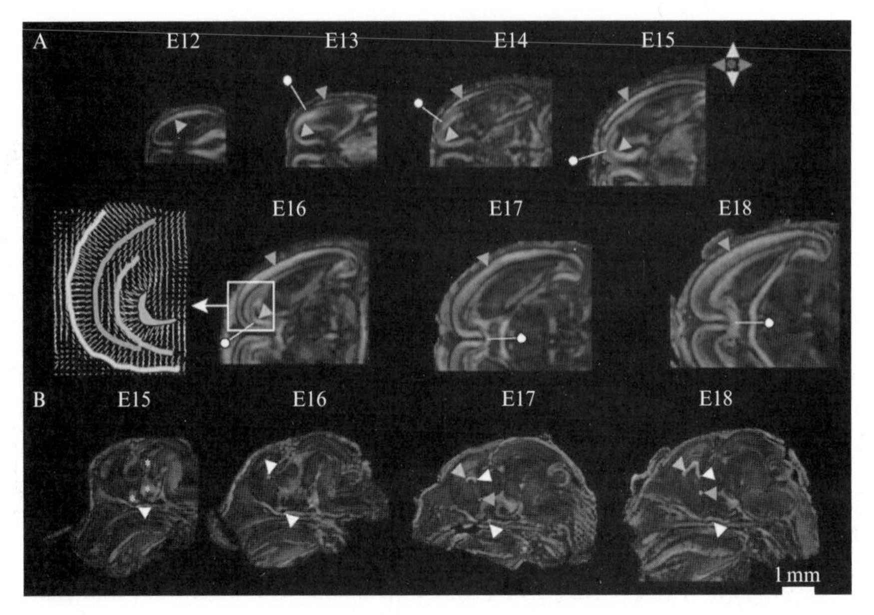

Figure 21.6 Diffusion tensor imaging in the developing mouse brain. Diffusion tensor data is shown with a color map, where red, green, and blue each represent diffusion in an orthogonal direction as indicated by the arrows. In panel (A), horizontal images from E12 to E18 are provided. Pink and blue arrowheads represent the cortical plate and neuroepithelium, respectively. Yellow pins indicate the leading edge of the growing intermediate zone. A white box is expanded at E16 to show the fiber orientation in a vector picture. In panel (B), the equivalent sagittal images are provided for E15–E18. In this case, white, yellow, pink, and blue arrowheads indicate the optic chiasm, hippocampal commissure, anterior commissure, and corpus callosum, respectively. Asterisks (*) mark the neuroepithelium around the third ventricle. All data are from full three-dimensional data sets. Reprinted from Zhang *et al.* (2003) with permission. (See Color Insert.)

or pathology due to the difficulty in assessing the entire vascular tree. For example, the deletion of basilar artery in *Gli2*$^{-/-}$ mutants was only noticed during micro-MRI analysis (Fig. 21.7; Berrios-Otero *et al.*, 2009). Similarly, micro-MRI of *Connexin43* mutant mice revealed abnormal development of both the right ventricle and major outflow tracts, difficult to appreciate from histological analysis (Huang *et al.*, 1998; Wadghiri *et al.*, 2007).

3.4. *In vivo* micro-MRI of mouse embryos and neonates

In vivo micro-MRI offers unique opportunities to assess longitudinal development and functional parameters such as blood perfusion, heart motion, or other physiological measures. *In vivo* studies can also be convenient in the context of comprehensive phenotyping studies, in which several additional

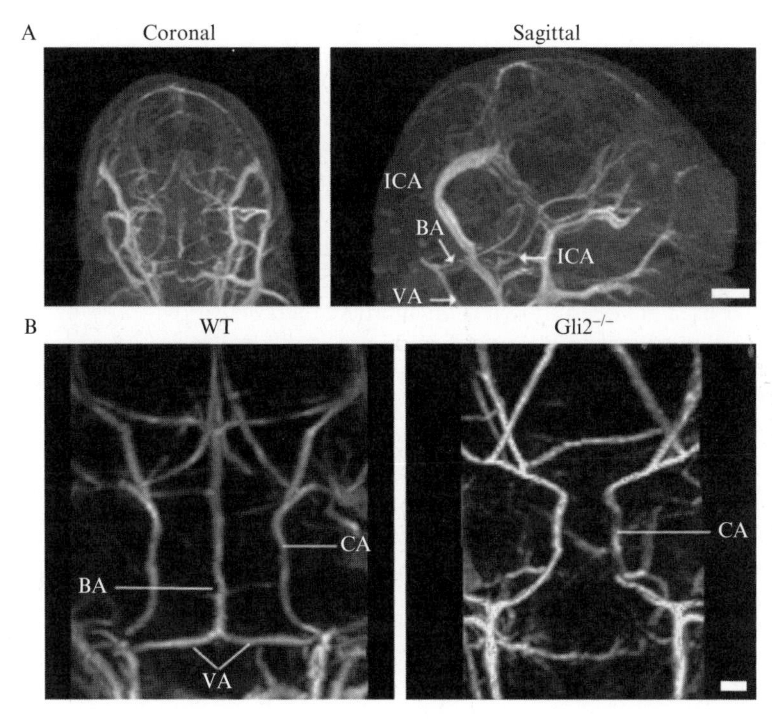

Figure 21.7 *Ex vivo* imaging of embryo vasculature. In panel (A), maximum intensity projections through an E15.5 embryo data set are shown in two orthogonal planes. The internal carotid arteries (ICA), vertebral arteries (VA), and basilar artery (BA) are labeled. In panel (B), comparison of wild type and $Gli2^{-/-}$ embryos at E17.5 reveals a missing basilar artery in the mutant. Reprinted from Berrios-Otero *et al.* (2009) with permission.

assays are required following initial imaging assessment, or in the case of conditional mutants, where variably penetrant phenotypes motivate detailed, time course imaging of individual animals. However, *in vivo* micro-MRI during development has been historically challenging. The small size of embryonic and neonatal mice requires high-resolution images, and consequently long scan times. The mice must be anesthetized throughout this period, and physiological monitoring and peripheral heating must be employed to ensure proper maintenance of their health. Physiological and other motion during the scans—detrimental to image quality—must be eliminated or compensated.

In vivo micro-MRI has been applied successfully in neonatal mice. Although the small size of neonates makes positioning and restraint challenging, good results can be achieved provided dedicated neonate cradles and setups similar to those used for adult mouse imaging. Isoflurane gas is the preferred method of anesthesia. After induction at 4–5%, animals can be maintained at ~1% isoflurane concentration for extended time periods (up to ~3 h). During this time, body temperature should be maintained with external heat sources and physiological signs must be monitored. Small ECG electrodes have been employed in mice as early as 3 days after birth, timing the acquisition of MRI data to the phase of the heart cycle to allow excellent visualization of the heart via prospectively gated cine-MRI methods (Wiesmann *et al.*, 2000). Despite these successes, physiological monitoring can be a challenging aspect of working with neonates. Detection of respiratory or cardiac events using additional data acquired from the MRI scanner itself has been shown as an alternate method for detection of physiological events, and in our experience can greatly improve the efficiency of setup for neonate imaging, without compromising monitoring capability during the 3D imaging session (Nieman *et al.*, 2009). Repeated *in vivo* imaging of individual mice postnatally provides a measure of brain development, mapping regions of the most rapid growth quantitatively (Fig. 21.8). Longitudinal growth maps through several time points can provide a quantitative picture of the growth process from birth through adulthood. Comparisons of *in vivo* images of neonatal mice with different genotypes have also demonstrated potential for phenotyping, visualizing, for instance, abnormalities in the *Gbx2* mutant cerebellum (Fig. 21.5; Wadghiri *et al.*, 2004).

Imaging the embryo *in utero* has been more limited, largely because it is not possible to reliably keep the embryo from moving inside the maternal abdomen. While neonates—with extra care and dedicated hardware—can be handled with similar procedures as adult animals, embryos *in utero* cannot be

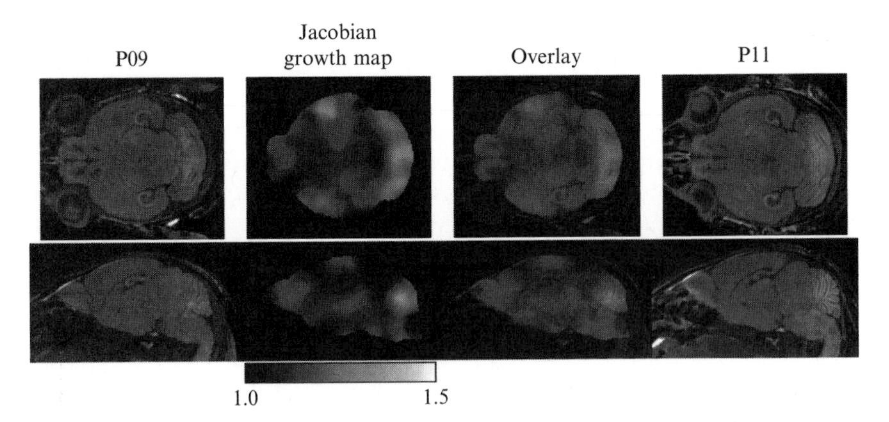

Figure 21.8 Computational mapping of growth in the developing mouse brain. *In vivo* Mn-enhanced MRI images at days 9 and 11 after birth (left and rightmost columns, respectively) provide a visual representation of growth-related changes. Computational image processing can provide a quantitative growth map (second column). An overlay of the growth map on the outline of the P09 mouse shows that the most significant growth is occurring in the cerebellum and regions of the cortex. (See Color Insert.)

restrained or monitored in the same fashion. One solution to this challenge is to image more quickly, using rapid acquisition of 2D slices rather than acquisition of high-resolution 3D volumes. This method has been applied successfully by several groups (Chapon *et al.*, 2002; Hogers *et al.*, 2000) using a T_2-weighted spin-echo imaging sequence, including one example in which *in utero* imaging could distinguish between genotypes in embryos transgenically expressing human ferritin (Cohen *et al.*, 2007). However, the 2D rapid MR images provide very limited image detail and lack sufficient anatomical data to investigate any but the most obvious phenotypes.

Imaging methods that permit acquisition of high-resolution 3D volumes in the presence of motion should permit much improved study of the live embryo *in utero*. In late-embryonic stages, where the resolution capabilities of MRI may be satisfactory, movement is somewhat restricted relative to earlier stages, so only a small amount of motion need be accounted for. We have found that gating on maternal respiratory motion, in combination with manganese (Mn) enhancement via maternal i.p. injection of $MnCl_2$, can provide high-quality *in utero* T_1-weighted images, enabling volumetric analysis of ventral forebrain defects in *Nkx2.1*$^{-/-}$ mutant embryos (Deans *et al.*, 2008). For additional motion compensation, prescription of a series of rapid 3D volume acquisitions (with 3–5 min acquisition time per image) provides images with sufficient quality to detect image motion between serially acquired images. In postprocessing, therefore, a set of serial images can be corrected for motion and then combined to produce high-quality image reconstructions. In combination with detection of cardiac motion, we have shown that this method can even be used to produce images of the beating embryonic heart (Fig. 21.9; Nieman *et al.*, 2009).

Further improvements for *in utero* imaging will broaden the potential application of these emerging methods. Most notably, dedicated hardware configurations for *in utero* imaging will be necessary to enhance imaging results sufficiently for routine application. Dedicated coil arrays, for instance, may improve the imaging outcome, allowing large field-of-view pilots to isolate individual embryos and yet still offer the sensitivity of a small surface coil for small field-of-view embryo imaging. Isotropic resolution close to 70 μm would likely be sufficient for many developmental studies, commencing in mid- to late-embryonic stages. Nonetheless, the complications associated with *in utero* imaging will mean it is not likely to serve the same screening role as *ex vivo* imaging, but will rather be used for assessing particular phenotypes requiring longitudinal examination. Important applications for *in utero* imaging include the evaluation of developmental changes that occur between late-embryonic and early-postnatal stages, functional measurements (such as cardiac performance or blood perfusion) in embryos with mutations lethal in late-embryonic stages, and time course studies of developmental abnormalities during maternal exposure to toxins.

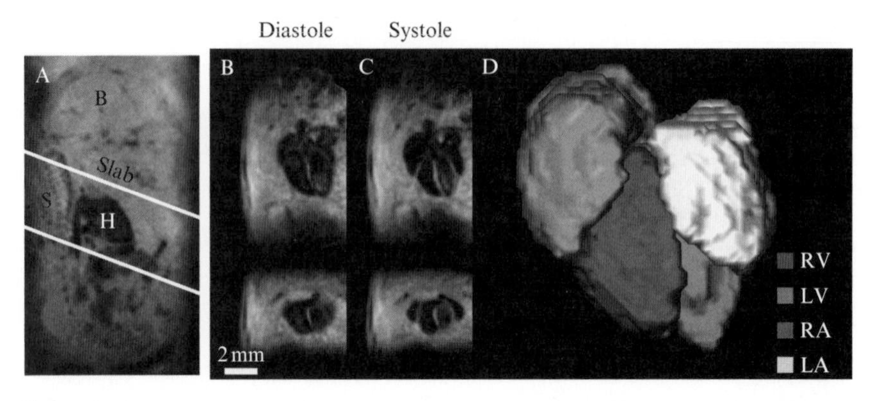

Figure 21.9 *In utero* MRI of the beating embryonic heart. With appropriate motion-correction, it is possible to image a volume (depicted in as an imaging slab in (A)) including the embryonic heart (day E17). Orthogonal long- and short-axis views are shown at diastole in (B) and (C). A volumetric rendering of the four heart chambers in (D) emphasizes the three-dimensional character of the data and shows the similarity in the four chamber volumes. Reprinted from Nieman *et al.* (2009) with permission. (See Color Insert.)

4. Summary

UBM and micro-MRI are microimaging techniques based on ultrasound and magnetic resonance, respectively, that provide powerful new approaches for anatomical and functional phenotype analysis in developing mouse embryos and neonates. UBM provides real-time image acquisition and Doppler blood velocity measurements, which has been widely used for studies of brain and cardiovascular development, and as a method for *in utero* image-guided injections. Micro-MRI has more flexibility for image contrast compared to UBM, but requires longer acquisition times. Micro-MRI of multiple fixed mouse embryos can be applied in a relatively high-throughput manner, with and without contrast agents, to analyze a wide range of phenotypes. In addition to these more conventional MRI methods, diffusion-weighted MRI and DTI have been demonstrated for 3D analysis of tissue microstructure and connectivity that is difficult to appreciate with standard histological analysis. Finally, recent advances have shown the feasibility of using micro-MRI for *in utero* imaging in live mouse embryos, providing the potential for future longitudinal studies of individual mice from embryonic to adult stages.

Acknowledgments

We thank Drs. Allan Johnson (Duke University), Susumu Mori (Johns Hopkins University), and Collin Phoon (New York University School of Medicine) for permission to reprint figures from their published work. We are grateful to all our current and past students, postdocs, and colleagues at the Mouse Imaging Centre and the Skirball Institute of Biomolecular Medicine who contributed to the work described in this chapter. We especially thank Kamila Szulc who provided the micro-MRI data used to compute the developmental growth map shown in Fig. 21.8. Some of the research described in this chapter was supported by grants from the National Institutes of Health (R01 NS038461, R01 HL078665) and contracts from the New York State Department of Health (C022053, C020926).

REFERENCES

Aljabar, P., Bhatia, K. K., Murgasova, M., Hajnal, J. V., Boardman, J. P., Srinivasan, L., Rutherford, M. A., Dyet, L. E., Edwards, A. D., and Rueckert, D. (2008). Assessment of brain growth in early childhood using deformation-based morphometry. *Neuroimage* **39,** 348–358.

Andrews, W., Liapi, A., Plachez, C., Camurri, L., Zhang, J., Mori, S., Murakami, F., Parnavelas, J. G., Sundaresan, V., and Richards, L. J. (2006). Robo1 regulates the development of major axon tracts and interneuron migration in the forebrain. *Development* **133,** 2243–2252.

Aristizabal, O., Christopher, D. A., Foster, F. S., and Turnbull, D. H. (1998). 40-MHz echocardiography scanner for cardiovascular assessment of mouse embryos. *Ultrasound Med. Biol.* **24,** 1407–1417.

Aristizabal, O., Ketterling, J. A., and Turnbull, D. H. (2006). 40-MHz annular array imaging of mouse embryos. *Ultrasound Med. Biol.* **32,** 1631–1637.

Berrios-Otero, C. A., Wadghiri, Y. Z., Nieman, B. J., Joyner, A. L., and Turnbull, D. H. (2009). Three-dimensional micro-MRI analysis of cerebral artery development in mouse embryos. *Magn. Reson. Med.* **62,** 1431–1439.

Bock, N. A., Konyer, N. B., and Henkelman, R. M. (2003). Multiple-mouse MRI. *Magn. Reson. Med.* **49,** 158–167.

Butt, S. J., Fuccillo, M., Nery, S., Noctor, S., Kriegstein, A., Corbin, J. G., and Fishell, G. (2005). The temporal and spatial origins of cortical interneurons predict their physiological subtype. *Neuron* **48,** 591–604.

Chapon, C., Franconi, F., Roux, J., Marescaux, L., Le Jeune, J. J., and Lemaire, L. (2002). In utero time-course assessment of mouse embryo development using high resolution magnetic resonance imaging. *Anat. Embryol. (Berl.)* **206,** 131–137.

Cohen, B., Ziv, K., Plaks, V., Israely, T., Kalchenko, V., Harmelin, A., Benjamin, L. E., and Neeman, M. (2007). MRI detection of transcriptional regulation of gene expression in transgenic mice. *Nat. Med.* **13,** 498–503.

Davatzikos, C., Shen, D., Gur, R. C., Wu, X., Liu, D., Fan, Y., Hughett, P., Turetsky, B. I., and Gur, R. E. (2005). Whole-brain morphometric study of schizophrenia revealing a spatially complex set of focal abnormalities. *Arch. Gen. Psychiatry* **62,** 1218–1227.

Deans, A. E., Wadghiri, Y. Z., Berrios-Otero, C. A., and Turnbull, D. H. (2008). Mn enhancement and respiratory gating for in utero MRI of the embryonic mouse central nervous system. *Magn. Reson. Med.* **59,** 1320–1328.

Dhenain, M., Ruffins, S. W., and Jacobs, R. E. (2001). Three-dimensional digital mouse atlas using high-resolution MRI. *Dev. Biol.* **232,** 458–470.

Fatkin, D., Christe, M. E., Aristizabal, O., McConnell, B. K., Srinivasan, S., Schoen, F. J., Seidman, C. E., Turnbull, D. H., and Seidman, J. G. (1999). Neonatal cardiomyopathy in mice homozygous for the Arg403Gln mutation in the alpha cardiac myosin heavy chain gene. *J. Clin. Invest.* **103,** 147–153.

Foster, F. S., Mehi, J., Lukacs, M., Hirson, D., White, C., Chaggares, C., and Needles, A. (2009). A new 15–50 MHz array-based micro-ultrasound scanner for preclinical imaging. *Ultrasound Med. Biol.* **35,** 1700–1708.

Gaiano, N., Kohtz, J. D., Turnbull, D. H., and Fishell, G. (1999). A method for rapid gain-of-function studies in the mouse embryonic nervous system. *Nat. Neurosci.* **2,** 812–819.

Hogers, B., Gross, D., Lehmann, V., Zick, K., De Groot, H. J., Gittenberger-De Groot, A. C., and Poelmann, R. E. (2000). Magnetic resonance microscopy of mouse embryos in utero. *Anat. Rec.* **260,** 373–377.

Huang, G. Y., Wessels, A., Smith, B. R., Linask, K. K., Ewart, J. L., and Lo, C. W. (1998). Alteration in connexin 43 gap junction gene dosage impairs conotruncal heart development. *Dev. Biol.* **198,** 32–44.

Ichikawa, Y., Sumi, M., Ohwatari, N., Komori, T., Sumi, T., Shibata, H., Furuichi, T., Yamaguchi, A., and Nakamura, T. (2004). Evaluation of 9.4-T MR microimaging in assessing normal and defective fetal bone development: Comparison of MR imaging and histological findings. *Bone* **34,** 619–628.

Janke, A. L., de Zubicaray, G., Rose, S. E., Griffin, M., Chalk, J. B., and Galloway, G. J. (2001). 4D deformation modeling of cortical disease progression in Alzheimer's dementia. *Magn. Reson. Med.* **46,** 661–666.

Ji, R. P., and Phoon, C. K. (2005). Noninvasive localization of nuclear factor of activated T cells c1−/− mouse embryos by ultrasound biomicroscopy-Doppler allows genotype-phenotype correlation. *J. Am. Soc. Echocardiogr.* **18,** 1415–1421.

Ji, R. P., Phoon, C. K., Aristizabal, O., McGrath, K. E., Palis, J., and Turnbull, D. H. (2003). Onset of cardiac function during early mouse embryogenesis coincides with entry of primitive erythroblasts into the embryo proper. *Circ. Res.* **92,** 133–135.

Johnson, G. A., Ali-Sharief, A., Badea, A., Brandenburg, J., Cofer, G., Fubara, B., Gewalt, S., Hedlund, L. W., and Upchurch, L. (2007). High-throughput morphologic phenotyping of the mouse brain with magnetic resonance histology. *Neuroimage* **37,** 82–89.

Kim, T. H., Zollinger, L., Shi, X. F., Rose, J., and Jeong, E. K. (2009). Diffusion tensor imaging of ex vivo cervical spinal cord specimens: The immediate and long-term effects of fixation on diffusivity. *Anat. Rec. (Hoboken)* **292,** 234–241.

Kimmel, R. A., Turnbull, D. H., Blanquet, V., Wurst, W., Loomis, C. A., and Joyner, A. L. (2000). Two lineage boundaries coordinate vertebrate apical ectodermal ridge formation. *Genes Dev.* **14,** 1377–1389.

Lerch, J. P., Carroll, J. B., Spring, S., Bertram, L. N., Schwab, C., Hayden, M. R., and Henkelman, R. M. (2008). Automated deformation analysis in the YAC128 Huntington disease mouse model. *Neuroimage* **39,** 32–39.

Liu, A., Joyner, A. L., and Turnbull, D. H. (1998). Alteration of limb and brain patterning in early mouse embryos by ultrasound-guided injection of Shh-expressing cells. *Mech. Dev.* **75,** 107–115.

Marxen, M., Thornton, M. M., Chiarot, C. B., Klement, G., Koprivnikar, J., Sled, J. G., and Henkelman, R. M. (2004). MicroCT scanner performance and considerations for vascular specimen imaging. *Med. Phys.* **31,** 305–313.

Metscher, B. D. (2009). MicroCT for developmental biology: A versatile tool for high-contrast 3D imaging at histological resolutions. *Dev. Dyn.* **238,** 632–640.

Mori, S., Itoh, R., Zhang, J., Kaufmann, W. E., van Zijl, P. C., Solaiyappan, M., and Yarowsky, P. (2001). Diffusion tensor imaging of the developing mouse brain. *Magn. Reson. Med.* **46,** 18–23.

Nieman, B. J., Flenniken, A. M., Adamson, S. L., Henkelman, R. M., and Sled, J. G. (2006). Anatomical phenotyping in the brain and skull of a mutant mouse by magnetic resonance imaging and computed tomography. *Physiol. Genomics* **24,** 154–162.

Nieman, B. J., Lerch, J. P., Bock, N. A., Chen, X. J., Sled, J. G., and Henkelman, R. M. (2007). Mouse behavioral mutants have neuroimaging abnormalities. *Hum. Brain Mapp.* **28,** 567–575.

Nieman, B. J., Szulc, K. U., and Turnbull, D. H. (2009). Three-dimensional, in vivo MRI with self-gating and image coregistration in the mouse. *Magn. Reson. Med.* **61,** 1148–1157.

Olsson, M., Campbell, K., and Turnbull, D. H. (1997). Specification of mouse telencephalic and mid-hindbrain progenitors following heterotopic ultrasound-guided embryonic transplantation. *Neuron* **19,** 761–772.

Parnell, S. E., O'Leary-Moore, S. K., Godin, E. A., Dehart, D. B., Johnson, B. W., Allan Johnson, G., Styner, M. A., and Sulik, K. K. (2009). Magnetic resonance microscopy defines ethanol-induced brain abnormalities in prenatal mice: Effects of acute insult on gestational day 8. *Alcohol. Clin. Exp. Res.* **33,** 1001–1011.

Petiet, A. E., Kaufman, M. H., Goddeeris, M. M., Brandenburg, J., Elmore, S. A., and Johnson, G. A. (2008). High-resolution magnetic resonance histology of the embryonic and neonatal mouse: A 4D atlas and morphologic database. *Proc. Natl. Acad. Sci. USA* **105,** 12331–12336.

Phoon, C. K., and Turnbull, D. H. (2003). Ultrasound biomicroscopy-Doppler in mouse cardiovascular development. *Physiol. Genomics* **14,** 3–15.

Phoon, C. K., Aristizabal, O., and Turnbull, D. H. (2000). 40 MHz Doppler characterization of umbilical and dorsal aortic blood flow in the early mouse embryo. *Ultrasound Med. Biol.* **26,** 1275–1283.

Phoon, C. K., Aristizabal, O., and Turnbull, D. H. (2002). Spatial velocity profile in mouse embryonic aorta and Doppler-derived volumetric flow: A preliminary model. *Am. J. Physiol. Heart Circ. Physiol.* **283,** H908–H916.

Phoon, C. K., Ji, R. P., Aristizabal, O., Worrad, D. M., Zhou, B., Baldwin, H. S., and Turnbull, D. H. (2004). Embryonic heart failure in NFATc1−/− mice: Novel mechanistic insights from in utero ultrasound biomicroscopy. *Circ. Res.* **95,** 92–99.

Punzo, C., and Cepko, C. L. (2008). Ultrasound-guided in utero injections allow studies of the development and function of the eye. *Dev. Dyn.* **237,** 1034–1042.

Schneider, J. E., Bamforth, S. D., Farthing, C. R., Clarke, K., Neubauer, S., and Bhattacharya, S. (2003a). Rapid identification and 3D reconstruction of complex cardiac malformations in transgenic mouse embryos using fast gradient echo sequence magnetic resonance imaging. *J. Mol. Cell. Cardiol.* **35,** 217–222.

Schneider, J. E., Bamforth, S. D., Grieve, S. M., Clarke, K., Bhattacharya, S., and Neubauer, S. (2003b). High-resolution, high-throughput magnetic resonance imaging of mouse embryonic anatomy using a fast gradient-echo sequence. *MAGMA* **16,** 43–51.

Schneider, J. E., Bose, J., Bamforth, S. D., Gruber, A. D., Broadbent, C., Clarke, K., Neubauer, S., Lengeling, A., and Bhattacharya, S. (2004). Identification of cardiac malformations in mice lacking Ptdsr using a novel high-throughput magnetic resonance imaging technique. *BMC Dev. Biol.* **4,** 16.

Sharpe, J., Ahlgren, U., Perry, P., Hill, B., Ross, A., Hecksher-Sorensen, J., Baldock, R., and Davidson, D. (2002). Optical projection tomography as a tool for 3D microscopy and gene expression studies. *Science* **296,** 541–545.

Shepherd, T. M., Thelwall, P. E., Stanisz, G. J., and Blackband, S. J. (2009). Aldehyde fixative solutions alter the water relaxation and diffusion properties of nervous tissue. *Magn. Reson. Med.* **62,** 26–34.

Smith, B. R. (2001). Magnetic resonance microscopy in cardiac development. *Microsc. Res. Tech.* **52,** 323–330.

Smith, B. R., Johnson, G. A., Groman, E. V., and Linney, E. (1994). Magnetic resonance microscopy of mouse embryos. *Proc. Natl. Acad. Sci. USA* **91,** 3530–3533.

Smith, B. R., Linney, E., Huff, D. S., and Johnson, G. A. (1996). Magnetic resonance microscopy of embryos. *Comput. Med. Imaging Graph.* **20,** 483–490.

Srinivasan, S., Baldwin, H. S., Aristizabal, O., Kwee, L., Labow, M., Artman, M., and Turnbull, D. H. (1998). Noninvasive, in utero imaging of mouse embryonic heart development with 40-MHz echocardiography. *Circulation* **98,** 912–918.

Sun, S. W., Neil, J. J., and Song, S. K. (2003). Relative indices of water diffusion anisotropy are equivalent in live and formalin-fixed mouse brains. *Magn. Reson. Med.* **50,** 743–748.

Sun, S. W., Liang, H. F., Xie, M., Oyoyo, U., and Lee, A. (2009). Fixation, not death, reduces sensitivity of DTI in detecting optic nerve damage. *Neuroimage* **44,** 611–619.

Turnbull, D. H., and Foster, F. S. (2002). In vivo ultrasound biomicroscopy in developmental biology. *Trends Biotechnol.* **20,** S29–S33.

Turnbull, D. H., and Mori, S. (2007). MRI in mouse developmental biology. *NMR Biomed.* **20,** 265–274.

Turnbull, D. H., Bloomfield, T. S., Baldwin, H. S., Foster, F. S., and Joyner, A. L. (1995). Ultrasound backscatter microscope analysis of early mouse embryonic brain development. *Proc. Natl. Acad. Sci. USA* **92,** 2239–2243.

Wadghiri, Y. Z., Blind, J. A., Duan, X., Moreno, C., Yu, X., Joyner, A. L., and Turnbull, D. H. (2004). Manganese-enhanced magnetic resonance imaging (MEMRI) of mouse brain development. *NMR Biomed.* **17,** 613–619.

Wadghiri, Y. Z., Schneider, A. E., Gray, E. N., Aristizabal, O., Berrios, C., Turnbull, D. H., and Gutstein, D. E. (2007). Contrast-enhanced MRI of right ventricular abnormalities in Cx43 mutant mouse embryos. *NMR Biomed.* **20,** 366–374.

Walls, J. R., Coultas, L., Rossant, J., and Henkelman, R. M. (2008). Three-dimensional analysis of vascular development in the mouse embryo. *PLoS ONE* **3,** e2853.

Weiner, H. L., Bakst, R., Hurlbert, M. S., Ruggiero, J., Ahn, E., Lee, W. S., Stephen, D., Zagzag, D., Joyner, A. L., and Turnbull, D. H. (2002). Induction of medulloblastomas in mice by sonic hedgehog, independent of Gli1. *Cancer Res.* **62,** 6385–6389.

Wichterle, H., Turnbull, D. H., Nery, S., Fishell, G., and Alvarez-Buylla, A. (2001). In utero fate mapping reveals distinct migratory pathways and fates of neurons born in the mammalian basal forebrain. *Development* **128,** 3759–3771.

Wiesmann, F., Ruff, J., Hiller, K. H., Rommel, E., Haase, A., and Neubauer, S. (2000). Developmental changes of cardiac function and mass assessed with MRI in neonatal, juvenile, and adult mice. *Am. J. Physiol. Heart Circ. Physiol.* **278,** H652–H657.

Zhang, J., Richards, L. J., Yarowsky, P., Huang, H., van Zijl, P. C., and Mori, S. (2003). Three-dimensional anatomical characterization of the developing mouse brain by diffusion tensor microimaging. *Neuroimage* **20,** 1639–1648.

Zhang, J., Chen, Y. B., Hardwick, J. M., Miller, M. I., Plachez, C., Richards, L. J., Yarowsky, P., van Zijl, P., and Mori, S. (2005a). Magnetic resonance diffusion tensor microimaging reveals a role for Bcl-x in brain development and homeostasis. *J. Neurosci.* **25,** 1881–1888.

Zhang, J., Miller, M. I., Plachez, C., Richards, L. J., Yarowsky, P., van Zijl, P., and Mori, S. (2005b). Mapping postnatal mouse brain development with diffusion tensor microimaging. *Neuroimage* **26,** 1042–1051.

Zhou, Y. Q., Foster, F. S., Qu, D. W., Zhang, M., Harasiewicz, K. A., and Adamson, S. L. (2002). Applications for multifrequency ultrasound biomicroscopy in mice from implantation to adulthood. *Physiol. Genomics* **10,** 113–126.

Zhou, Y. Q., Foster, F. S., Parkes, R., and Adamson, S. L. (2003). Developmental changes in left and right ventricular diastolic filling patterns in mice. *Am. J. Physiol. Heart Circ. Physiol.* **285,** H1563–H1575.

SECTION SEVEN

HEMATOPOIESIS

CHAPTER TWENTY-TWO

Use of Transgenic Fluorescent Reporter Mouse Lines to Monitor Hematopoietic and Erythroid Development During Embryogenesis

Stuart T. Fraser,[*,†,‡,1] Joan Isern,[*,†] *and* Margaret H. Baron[*,†,‡]

Contents

* Division of Hematology and Medical Oncology, Department of Medicine, Mount Sinai School of Medicine, New York, USA
† Tisch Cancer Institute, Mount Sinai School of Medicine, New York, USA
‡ Black Family Stem Cell Institute, Mount Sinai School of Medicine, New York, USA
1 Current address: Disciple of Physiology, School of Medical Sciences, NSW, Australia

Methods in Enzymology, Volume 476
ISSN 0076-6879, DOI: 10.1016/S0076-6879(10)76022-5

Abstract

The use of fluorescent reporter proteins such as GFP, RFP, and their variants to tag and track cells within the embryo has revolutionized developmental biology. Expression of these proteins within restricted populations has been achieved through the use of lineage-specific regulatory elements. This approach has proven especially powerful in the hematopoietic system, where it has been possible to monitor the generation, expansion, maturation, and migration of primitive erythroid cells, macrophages, and megakaryocytes during embryogenesis at unprecedented resolution. Such analyses have provided novel insights into the development of these lineages. In this chapter, we discuss the design considerations and methodologies involved in the production and analysis of transgenic mouse lines in which fluorescent reporters are expressed in the hematopoietic system of the mouse embryo.

1. Introduction

1.1. Development of the hematopoietic system in the mouse embryo

The ontogeny of the mammalian hematopoietic system is a complex and precisely orchestrated process that results in the rapid production of erythroid, myeloid, hematopoietic stem and lymphoid cells in the developing embryo. A detailed description of the pathways involved in the regulation of embryonic hematopoiesis is beyond the scope of this chapter but is available in a number of reviews (Baron and Fraser, 2005; Dzierzak and Speck, 2008; McGrath and Palis, 2008; Orkin and Zon, 2008).

Primitive erythroid cells (EryP), the first hematopoietic lineage to develop in the embryo, arise in the yolk sac (YS) shortly after gastrulation, expand rapidly, and fill the early embryonic bloodstream, prior to the appearance of hematopoietic stem cells (HSCs) and their definitive (adult type) progeny (Palis *et al.*, 1999). From midgestation onwards, HSC activity emerges within the YS, the aorta–gonad–mesonephros (AGM) region of the embryo, the placenta, large arteries, and then expands in the fetal liver (FL) (discussed in Dzierzak and Speck, 2008). Within the FL, definitive erythroid cells (EryD) are generated from HSCs in enormous numbers and rapidly outnumber EryP (McGrath and Palis, 2008). In addition, the myeloid and lymphoid lineages expand dramatically (reviewed in Laiosa *et al.*, 2006; Luc *et al.*, 2008). Shortly before birth, the site of HSC activity shifts, first to the spleen, then permanently to the bone marrow (reviewed in Laiosa *et al.*, 2006; Orkin and Zon, 2008). Lymphoid progenitors feed the developing thymus and other peripheral lymphoid organs, where they mature into functional T and B lymphocytes (reviewed in Laiosa *et al.*, 2006). The wide range of hematopoietic lineages and the dynamic shifts in their generation and expansion have challenged both hematologists and immunologists alike. The expression of fluorescent reporter proteins has provided a powerful approach for monitoring and isolating distinct hematopoietic cell types.

1.2. Considerations in designing a transgenic mouse line that expresses a fluorescent reporter in hematopoietic lineages

Transgenic mouse lines expressing fluorescent reporter proteins in specific lineages have provided useful models for monitoring hematopoietic development (Table 22.1). The value and versatility of such lines will be maximized if their design is carefully thought through in advance, with consideration given to the following points:

1. Regulatory elements for driving reporter gene expression

    Lineage-restricted expression of a fluorescent protein (FP) requires tight control by tissue- or cell type-specific promoters/enhancers. Such control regions have been used to drive the transcription of genes that are widely expressed in hematopoietic cells (e.g., *CD45*; Yang *et al.*, 2008), tightly regulated within a narrow developmental window (such as c-kit (Cairns *et al.*, 2003) or Bmi-1 (Hosen *et al.*, 2007)), or restricted to specific hematopoietic lineages such as the embryonic erythroid lineage (Fraser *et al.*, 2007), macrophages (Sasmono *et al.*, 2003), or regulatory T lymphocytes (Fontenot *et al.*, 2005). Fluorescent reporter mice currently available for analysis of hematopoietic cells are listed in Table 22.1.

    Two methods are commonly used to place a fluorescent reporter cDNA under the control of the desired regulatory elements: transgenesis, in which DNA sequences are integrated at random sites within the genome (Nagy *et al.*, 2003) or a "knock-in" approach in which the cDNA is targeted to a specific locus within the genome (Joyner, 1995; Nagy *et al.*, 2003) or within a bacterial artificial chromosome (BAC) (Yang *et al.*, 1997; Zhang *et al.*, 1998) and, in principle, will be tightly controlled by all regulatory elements of that specific gene. Once suitable regulatory elements have been identified, the generation of transgenic mouse lines is relatively straightforward and rapid. However, each line must be characterized carefully to ensure that the transgene is expressed at sufficiently high levels and in the expected spatiotemporal pattern to be useful for the intended studies. Expression of the integrated transgene

Table 22.1 Transgenic mouse lines expressing fluorescent reporter proteins in the hematopoietic system

Gene	Reporter	Lineage labeled	Reference
Gata1	GFP (Tg)	Hemangioblast, EryP, EryD, Megs	Nishimura *et al.* (2000)
Gata2	GFP (Ki)	HSC	Suzuki *et al.* (2006)
Gfi1B	GFP (Ki)	HSC, erythroid, and myeloid progenitors	Vassen *et al.* (2007)
Ly6.1	GFP (Tg)	HSC, lymphoid, and myeloid cells	Ma *et al.* (2002)
CD41	Farnesyl-YFP (Tg)	Megs, platelets, HSC, progenitors	Zhang *et al.* (2007)
CD45	YFP (Ki)	Widespread hematopoietic	Yang *et al.* (2008)
Eklf	GFP (Tg)	EryD	Lohmann and Bieker (2008)
ε-globin	KGFP (Tg); H2B-GFP (Tg)	EryP	Dyer *et al.* (2001), Isern *et al.* (2008)
β-globin	Farnesyl-CFP (Tg)	MEP, EryD	Faust *et al.* (2000)
EpoR	GFP-Cre (Ki)	Erythroid, endothelial	Heinrich *et al.* (2004)
c-kit	GFP (Tg)	HSC, progenitors	Cairns *et al.* (2003)
Bmi1	GFP (Ki)	Highest in HSC, decreasing with maturation	Hosen *et al.* (2007)
Runx1	IRES-GFP (Ki)	Lymphoid, myeloid, lower levels in erythroid	Lorsbach *et al.* (2004)
Abcg2	IRES-GFP (Ki)	Erythroid, HSC	Tadjali *et al.* (2006)
Pu.1	IRES-GFP (Ki)	Highest in HSC, lower levels in CMP	Nutt *et al.* (2005)
Lysozyme M	EGFP (Tg)	Macrophages, granulocytes	Faust *et al.* (2000)
c-fms	GFP (Tg)	Macrophages, dendritic cells, myeloid cells	Sasmono *et al.* (2003)

Table 22.1 (*continued*)

Gene	Reporter	Lineage labeled	Reference
CX3CR1	GFP (Ki)	Macrophages, monocytes, NK cells, dendritic cells, microglia	Jung *et al.* (2000)
MafB	GFP (Tg)	Myeloid cells	Hamada *et al.* (2003)
TCRβ	GFP (Tg)	Lymphoid progenitors	Norris *et al.* (2007)
CD2	GFP (Tg)	T lymphoid cells	Singbartl *et al.* (2001)
FoxP3	GFP (Ki)	T regulatory lymphoid cells	Fontenot *et al.* (2005)
Ror(gT)	EGFP (Ki)	T helper 17 lymphoid cells	Lochner *et al.* (2008)
Pax5	EGFP (Ki)	Pre-B, B lymphoid cells	Fuxa and Busslinger (2007)
Rag1	GFP (Ki)	Lymphoid cells	Kuwata *et al.* (1999)
Rag2	GFP (Ki)	B and T lymphoid cells	Monroe *et al.* (1999)
Blimp1	IRES-EGFP (Ki)	B lymphoid cells, plasma cells	Kallies *et al.* (2004)
Langerin	IRES-EGFP (Ki)	Langerhans cells	Kissenpfennig *et al.* (2005)

Abbreviations: Tg, transgenic; Ki, knock-in; GFP, green fluorescent protein; IRES, internal ribosomal entry site; EGFP, enhanced GFP; YFP, yellow fluorescent protein; CFP, cyan fluorescent protein; HSC, hematopoietic stem cell; EryP, primitive erythroid; EryD, definitive erythroid; MEP, megakaryocyte–erythroid progenitor; Meg, megakaryocyte; NK, natural killer cells.

may be driven ectopically, under the influence of genomic sequences neighboring the insertion site, or it may be silenced from the time of integration or even months to years later (Garrick *et al.*, 1998; Nagy *et al.*, 2003). Silencing may be the result of genomic imprinting (Preis *et al.*, 2003). An important advantage of the knock-in approach is that reporter expression is driven by endogenous regulatory elements for the gene whose expression pattern is to be recapitulated by the FP transgene. However, a significant investment of time, effort, and expertise is required to generate knock-in reporter mouse lines. A potential limitation of the knock-in approach is that the targeted allele may not be developmentally neutral, in heterozygous and/or homozygous form, for example, loss of one allele of *Runx1* alters HSC development (North *et al.*, 2002). Forced or ectopic expression of a protein may result in abnormal embryonic development (Nagy *et al.*, 2003). Knock-in of reporter cDNAs into dosage-sensitive alleles can also lead to abnormal embryonic development (Schedl *et al.*, 1996). To address this issue, a number of investigators have generated knock-in vectors in which an internal ribosomal entry sequence (IRES) is linked to the fluorescent reporter as used in Kallies *et al.* (2004) and Nutt *et al.* (2005). This strategy results in the production of normal transcripts from the targeted gene as well as the expression of the fluorescent reporter. Transgenic or knock-in mouse lines can be generated either by microinjection of DNA into the male pronucleus of a fertilized egg or by blastocyst injection of embryonic stem (ES) cells that have been genetically manipulated to contain the desired sequences to yield chimeras (Pirity *et al.*, 1998).

2. Choice of fluorescent reporter

   The choice of FP is critical if multicolor analyses will be undertaken, for example, if the intended transgenic line will later be crossed with another FP-expressing line and/or if immunofluorescence studies are planned. For example, two or more hematopoietic lineages can be distinguished within the circulation or within a tissue such as fetal liver by judicious choice of reporter lines (Heck *et al.*, 2003; Stadtfeld *et al.*, 2005). For multicolor analysis of FPs, the spectral overlap of each FP must be considered. For microscopy, the combinations GFP/CFP and GFP/YFP exhibit significant emission overlap, so that it is difficult to detect each individual signal (discussed in Stadtfeld *et al.*, 2005). CFP and YFP are a useful combination of fluorophores for microscopy, as they show insignificant overlap and excellent contrast (Shaner *et al.*, 2005). For flow cytometry, the overlap between GFP and CFP is not problematic, as these FPs are excited by the violet and blue lasers, respectively (Heck *et al.*, 2003; Stadtfeld *et al.*, 2005) (S. T. Fraser, J. Isern, and M. H. Baron, unpublished observations). The range of FPs currently available is extensive. FPs can emit signals from the blue range (e.g., CFP, cerulean) through green (GFP), yellow (YFP, Venus), and red-shifted regions of the spectrum. The latter include orange/red (dsRed and variants such as tomato and cherry), far-red (mPlum) and even, as recently reported, into the infrared (IFP) (Shu *et al.*, 2009). A number of excellent reviews present the characteristics and utilities of the different FPs (Giepmans *et al.*, 2006; Nowotschin *et al.*, 2009b; Shaner *et al.*, 2005).

3. Detection of fluorescence

   The two approaches used most commonly for monitoring expression of FPs are microscopy and flow cytometry. Conventional fluorescence microscopes have a more limited range of detection than confocal microscopes. It is essential to confirm that appropriate filters are available for the instruments to be used for the analyses (discussed in Stadtfeld *et al.*, 2005). For flow cytometry, the lasers and filter sets required for excitation and for detection must be considered (Pruitt *et al.*, 2004).

While most flow cytometers can excite and detect GFP, YFP, and RFP variants, detection of CFP requires a violet laser.

4. Cellular localization of fluorescent reporter

   FPs can be targeted to specific regions of the cell, offering another dimension for imaging of different lineages. The nucleus has been successfully targeted by incorporating a nuclear localization signal into the construct or by using histone H2B–FP fusions to localize the FP to chromatin (reviewed in Hadjantonakis and Papaioannou, 2004; Nowotschin *et al.*, 2009b). The latter approach allows monitoring throughout the cell cycle: while cytoplasmic GFP is diluted through cell division, histone H2B–FPs are stably expressed throughout the cell cycle (discussed in Hadjantonakis and Papaioannou, 2004; used for primitive erythroid and endothelial lineages in Fraser *et al.*, 2005; Isern *et al.*, 2008). The inner leaf of the cell membrane has also been targeted for specific fluorescent labeling by combining farnesyl (Heck *et al.*, 2003), myristoyl, or GPI moieties with the FP (discussed in Rhee *et al.*, 2006). Recently, transgenic mice expressing a lipid-modified GFP in combination with a nuclear-mCherry within the same cell have been used to simultaneously evaluate membrane dynamics and nuclear behavior (Nowotschin *et al.*, 2009a).

5. Photomodulatable FPs

   Photoactivatable and photoconvertible FPs allow the specific tagging and tracking of individual cells (Nowotschin and Hadjantonakis, 2009; Nowotschin *et al.*, 2009b). They can be activated to either change from a nonfluorescent to fluorescent state (i.e., nongreen-fluorescent in the case of PA-GFP), or emit at a distinct wavelength (such as cyan-green for PS-CFP or green-red for EosFP, Kaede, and KikGR; Nowotschin and Hadjantonakis, 2009) upon exposure to very brief but intense light. Individual cells or cohorts of photomodulated cells can be monitored within a larger population of cells that are either nonfluorescent or fluoresce at a different wavelength. Cells expressing the photomodulated FP can, therefore, be studied throughout the lifetime of that particular FP. This feature is particularly useful for monitoring protein dynamics and for tagging and tracking the development of specific cell types during embryogenesis.

1.3. Methodologies for monitoring hematopoietic development using fluorescent reporter mice

Transgenic mouse lines in which a fluorescent reporter is expressed in specific hematopoietic lineages are listed in Table 22.1. Below, we discuss a number of methodologies used in our own work that are useful for analyzing the development of the hematopoietic system. These methods are based upon our experience using *epsilon-globin::FP* transgenic mouse lines but could be applied to other hematopoietic lineages and include embryo dissection, flow cytometry, immunostaining, and coculture with macrophages (Dyer *et al.*, 2001; Fraser *et al.*, 2007; Isern *et al.*, 2008).

2. Materials

2.1. Dissecting tools

- Dissecting scissors (Roboz Surgical Instrument Inc., Gaithersburg, MD; Cat. # RS-6702 and RS-5882) and forceps (Sigma-Aldrich, St. Louis, MO; Cat. # F4267)
- Watchmaker's forceps, Dumont #5 and #55 (Roboz, FST)
- 3 ml sterile plastic transfer pipettes (VWR; Cat. # 414004-037)
- Stereomicroscope with transmitted and reflected light sources (Zeiss, Leica, or Nikon)

2.2. Glassware and plasticware

- BD Falcon 40 μm cell strainer (BD Biosciences, San Jose, CA; Cat. # 352340)
- 24-well Nunclon tissue culture plates (Nunc, Thermo Fisher Scientific; Cat. # 142475)
- 12 mm circular coverslips no. 1 (Thermo Fisher Scientific; Cat. # 12-545-80)
- 3 ml syringe (BD Biosciences; Cat. # 309585)
- 20-gauge needle (BD Biosciences; Cat. # 305176)
- 15 and 50 ml polypropylene tubes (Corning, Lowell, MA; Cat. # 430766 and 430291)
- 5, 10, and 25 ml Costar serological pipettes (Corning; Cat. # 7543R00, 7543R02 and 7543R04)
- 3.5-cm bacterial culture Petri dishes (BD Falcon, Franklin Lakes, NJ; Cat. # 351008)
- 15-cm Petri dishes (BD Falcon; Cat. # 351013)
- 16-gauge blunt-end needles (Stem Cell Technologies, Vancouver, BC; Cat. # 28110)

2.3. Embryo dissection and cell preparation

- Phosphate-buffered saline (PBS), pH 7.4 (GIBCO Invitrogen, Carlsbad, CA; Cat. # 10010-023)

- Iscove's modified Dulbecco's medium (IMDM) (GIBCO Invitrogen; Cat. # 12440-079)
- Fetal bovine serum (FBS) (Hyclone, Thermo Fisher Scientific)
- Dissection medium: IMDM + 10% FBS
- Heparin (Sigma-Aldrich; Cat. # H3149): dissolve in PBS to 12.5 mg/ml stock solution (100×)
- Cell strainers, 40 μm Nylon (BD Falcon; Cat. # 35243)
- Cell dissociation buffer (GIBCO Invitrogen; Cat. # 13150-016)
- Collagenase (Sigma-Aldrich; Cat. # C2674): stock solution prepared at 100 mg/ml in sterile water. Aliquots stored frozen at −20 °C.
- DNase I (Sigma-Aldrich; Cat. # D5025)
- RPMI-1640 medium (GIBCO Invitrogen; Cat. # A10491-01)
- EDTA (Sigma-Aldrich; Cat. # E5134). Stock solution is prepared at 0.5 *M*, pH 8.0.
- Bovine serum albumin (BSA) (Sigma-Aldrich; Cat. # B4287)
- α-Monothioglycerol (α-MTG) (Sigma-Aldrich; Cat. # M1753)
- Fetal liver macrophage (FLM) culture media: RPMI-1640 medium/10% FBS/2 m*M* L-glutamine/0.5 m*M* α-MTG/penicillin–streptomycin (Pen/Strep, see #16)
- L-Glutamine (200 m*M* (100×), liquid) (GIBCO Invitrogen; Cat. # 25030-164)
- Penicillin–streptomycin, liquid (10,000 units penicillin; 10,000 μg streptomycin) (GIBCO Invitrogen; Cat. # 15140-163)
- Erythroblastic island (EBI) culture medium: RPMI-1640 medium/10% FBS/2 m*M* L-glutamine/2 U/ml erythropoietin

2.4. Cytospin centrifugation and Giemsa staining

- Cytospin cytocentrifuge: Shandon Cytospin 3 (Pittsburgh, PA)
- Methanol (Fisher Chemicals; Cat. # 67-56-1)
- Giemsa stain (Sigma-Aldrich; Cat. # GS-500)

2.5. Flow cytometry

- FACS buffer: PBS containing 10% heat-inactivated FBS; see Section 4
- DAPI (4′,6-diamidino-2-phenylindole dihydrochloride; Sigma-Aldrich; Cat. # D9542); see Section 4
- DRAQ5 (eBioscience, San Diego, CA; Cat. # 65-0880); see Section 4
- Propidium iodide (Sigma-Aldrich; Cat. # P4170). Dilute powder in distilled water to prepare 1000× stock solution.

2.6. Immunostaining and microscopy

- 4% paraformaldehyde (Sigma-Aldrich; Cat. # 158127) in PBS. Mix 4 g paraformaldehyde in PBS at 65 °C until thoroughly dissolved. Cool before use.
- Washing buffers: PBS with 0.05% Tween-20 (v/v) (Sigma-Aldrich; Cat. # P1379) (PBST); PBST with 0.05% no-fat skim milk powder (Carnation) (PBSMT)
- Vectashield with DAPI (Vector Labs, Burlingame, CA; Cat. # H-1200)
- Primary antibodies:
  - Anti-mouse Forssman glycosphingolipid antigen (rat IgM, clone FOM-1, BMA Biomedicals AG, Switzerland, Cat. # T-2113)
  - Anti-mouse F4/80 monoclonal (clone CI:A3-1, Abd Serotec, Oxford, UK; Cat. # MCA497GA)
- Secondary antibodies:
  - AlexaFluor 594-conjugated goat anti-rat IgM (Molecular Probes, Eugene, OR; Cat. # A21213)
  - AlexaFluor 568-conjugated goat anti-rat (Molecular Probes; Cat. # A-11077)
- Triton X-100 (Sigma-Aldrich; Cat. # T8787)
- Blocking buffer: 2% BSA, 0.1% Triton X-100 in PBS

2.7. Primitive erythroid colony (progenitor) assay

- Plasma-derived serum (PDS) (Animal Technologies, Tyler, TX; Cat. # FBP-186); see Section 4
- Erythropoietin (Epogen 10,000 U/ml; Amgen Ltd., Thousand Oaks, CA); see Section 4
- Methylcellulose: prepare as a 1.5% (w/v) stock as described (Baron and Mohn, 2005)
- Penicillin/streptomycin (GIBCO Invitrogen; Cat. # 15140-163): 100×, liquid
- Ascorbic acid (Sigma-Aldrich; Cat. # A4544): prepare 5 mg/ml stock in distilled water. Aliquots can be stored at −20 °C and thawed when needed.
- L-Glutamine (GIBCO Invitrogen; Cat. # 25030-081): 200 m*M* (100×), liquid
- Protein-free hybridoma medium (GIBCO Invitrogen; Cat. # 12040-077): PFHM-II (1×)
- IMDM (GIBCO Invitrogen; Cat. # 12440-079): IMDM (1X), liquid

3. Methods

3.1. Isolation of hematopoietic cells from the developing mouse embryo

3.1.1. Embryo dissection

The hematopoietic system is initiated in the mouse shortly after gastrulation. The embryos typically dissected for analysis of hematopoietic development range from stages E7.5 to E14.5. Mouse embryos are dissected as described in detail (Baron and Mohn, 2005; Nagy *et al.*, 2003). The brief protocol presented below is representative of that used for dissecting embryos from E9.5 to E14.5. For dissection of earlier stage embryos please refer to Baron and Mohn (2005) and Nagy *et al.* (2003).

1. Euthanize pregnant mother by CO_2 asphyxiation and cervical dislocation.
2. Wet the fur of the euthanized mouse with 70% ethanol to flatten the dander. Make a midline incision to open the abdominal cavity. The two uterine horns will be seen extending laterally.
3. Using #5 watchmaker's forceps, lift up the end of each of the horns. Carefully trim away any connective tissue or fat that may remain attached and carefully remove both uterine horns. Place into a 10 cm Petri dish containing PBS to wash off maternal blood.
4. Carefully cut between each conceptus, releasing each ones into a 10 cm Petri dish containing PBS. Rinse briefly three times in PBS to remove maternal blood, then transfer to a Petri dish containing dissection medium.
5. Remove the uterine tissue using one #5 or #55 watchmaker's forceps to stabilize the uterus and a second forceps to widen the hole created in step 4 and then to peel away the tissue.
6. Remove the Reichert's membrane using a #5 or #55 watchmaker's forceps, stabilizing the embryo with one forceps and separating off the membrane using a second forceps.
7. The embryo can now be visualized inside the YS, with the placenta attached. The placenta is removed by stabilizing the embryo with a #5 forceps in one hand and dissecting the placenta away using a second #5 forceps in the other hand. Significant amounts of blood will be released into the medium once the placental vessels are ruptured. The placenta can then be dissociated as described below. The YS can be removed from the embryo essentially as described for the removal of the placenta.

3.1.2. Circulating embryonic blood

Peripheral blood can be obtained only from embryos at E9.5 or later (Fraser *et al.*, 2007), after the embryonic and extraembryonic vascular systems have become connected and the heart has begun to beat (McGrath *et al.*, 2003).

1. Individual embryos should be transferred separately to the wells of a 24-well dish containing dissection medium with 0.5% heparin. The fluorescence of each embryo can be easily examined under a microscope.
2. Once the fluorescence of each embryo in the litter has been evaluated, the embryonic blood can be isolated. Using #5 watchmaker's forceps, separate the region above the heart from the trunk/tail of the embryo. Sever the umbilical and vitelline vessels at the point where they enter the embryo. This step will release large numbers of peripheral blood cells into the medium.
3. Allow the embryos to exsanguinate on ice for 10 min. Blood will collect in a pool at the bottom of the dish.
4. Remove the pieces of embryo from each well and carefully resuspend the blood in dissection medium using a P1000 Gilson Pipetman.
5. Filter the blood through a cell strainer, then collect the cells by centrifugation at 1200 rpm (100×*g*) in an Eppendorf Centrifuge 5415D.
6. Dilute the cells to an appropriate concentration for cytocentrifugation (see Section 3.2) or for flow cytometry (see Section 3.3).

3.1.3. Fetal liver dissection

Hematopoietic cells can be isolated relatively easily from the FL through E16.5, as this tissue has not yet developed a tightly adherent epithelial structure.

1. Remove the FL from the embryo using #5 watchmaker's forceps.
2. Transfer the FL to a 1.5 ml Eppendorf tube containing 0.5 ml dissection medium.
3. Disperse the FL by pipetting the tissue in dissection medium using a P1000 Gilson Pipetman. Pipet the FL 5–10 times until the tissue is clearly dispersed into a suspension.
4. Filter through a cell strainer, collect the cells by centrifugation at 1200 rpm (100×*g*) in an Eppendorf Microcentrifuge 5415D and resuspend in FACS buffer for flow cytometry or in PBS for cytocentrifugation.

3.1.4. Dissection of yolk sac and placenta

Isolation of hematopoietic cells from YS and placenta, tissues that contain endothelial cells with adherens junctions and endodermal cells with tight junctions, requires more vigorous dissociation steps than for FL.

1. Remove the YS and placenta from each embryo.
2. Place each tissue into an individual 1.5 ml Eppendorf tube containing 0.3–0.4 ml collagenase. If numerous YS or placentae from a single embryonic stage are being dissociated, they can be pooled in a 15 ml Corning tube containing 4 ml collagenase. Dissociation to single cells is more efficient if narrow dissection scissors (Roboz RS-6702) are inserted into a 1.5 ml Eppendorf tube and used to macerate the YS or placental tissue. Vortex briefly.
3. Incubate the suspension at 37 °C for at least 20 min. Check every 5 min and shake vigorously (and, if necessary, vortex briefly) to help break up the

tissues. Shaking helps to break up the tissue pieces. Vortexing, which can kill some of the cells, is used only sparingly to complete the cell dispersion. Hold the tube up to the light to determine when a uniform cell suspension has been obtained (no large clumps remain). Filter the sample through a cell strainer, collect by centrifugation in an Eppendorf Microcentrifuge 5415D at 1200 rpm (100 × *g*) and resuspend in FACS buffer or PBS, as described above.

3.2. Cytological analysis of hematopoietic cells

3.2.1. Cytospin centrifugation

The morphology of individual hematopoietic cell types can be clearly delineated by microscopic examination after cytocentrifugation onto glass slides and cytological staining.

1. Place up to 40,000 cells into the chamber of a Shandon Cytospin cytocentrifuge.
2. Centrifuge the samples for 5 min at 300 × *g*.
3. Air-dry for at least 15 min.
4. Store under moisture-free conditions (e.g., in a slide box containing desiccant).

3.2.2. Giemsa staining

Giemsa stain is one of the best known histological stains, coloring the nuclei dark blue and the cytoplasm blue to pink, according to the acidity of the cytoplasmic contents.

1. Fix air-dried samples in methanol for 10 min.
2. Air-dry until all methanol has evaporated.
3. Stain in coplin jar containing 5% Giemsa stain (diluted in tap water) for 20 min.
4. Wash sample in large beaker filled with tap water until excess Giemsa stain is removed.
5. Air-dry and examine under microscope.

3.3. Flow cytometry of embryonic hematopoietic cells

Hematopoietic lineages lend themselves particularly well to flow cytometric analysis, as they can be isolated from tissues as single cell suspensions relatively easily. Flow cytometry is a fundamental method for assessing fluorescently labeled hematopoietic lineages. Once tissues have been dispersed into single cell suspensions, staining with hematopoietic markers can be performed.

3.3.1. FACS analysis of embryonic hematopoietic cells

1. Isolate single embryonic cell suspensions from peripheral blood or from embryonic tissues as described above. Count cells using a hemocytometer. Dispense 1×10^6 cells into a 1.5 ml Eppendorf tube. Collect by centrifugation in an Eppendorf Microcentrifuge 5415D at 1200 rpm ($100\times g$). Aspirate supernatant.
2. Dilute commercially available fluorescently conjugated antibodies in FACS buffer according to manufacturer's suggestions (we generally use 2 μg antibody per 100,000 cells; however, the optimal amount of each antibody needs to be determined in each case). We routinely combine antibodies conjugated with different fluorochromes in the same "cocktail" to reduce the number of incubation and washing steps.
3. As EryP do not express Fc receptors, they exhibit low background binding to antibodies. We have not found that treatment of cells with normal mouse serum or FcBlock is necessary.
4. Add 100 μl antibody cocktail to cell pellet and resuspend thoroughly to ensure uniform labeling of cells.
5. Incubate on ice for 20 min. Wash with 1 ml FACS buffer per sample. Collect by centrifugation in an Eppendorf Microcentrifuge 5415D at 1200 rpm ($100\times g$).
6. If primary antibodies are biotin-conjugated, prepare a cocktail of diluted streptavidin conjugated to a fluorochrome. Resuspend the cell pellet in 100 μl diluted streptavidin. Incubate on ice for 20 min. Wash with 1 ml FACS buffer per sample. Collect by centrifugation at 1200 rpm ($100\times g$).
7. Resuspend pellet in 0.4 ml FACS buffer/DAPI if ultraviolet (UV) laser available or in FACS buffer/PI if UV laser is not available. Analyze using flow cytometer.

3.3.2. Sorting of embryonic hematopoietic populations

1. Prepare antibody-stained single cell suspensions for flow cytometry, as described above.
2. Prepare collection tubes for sorted cells. For sorting of live cells, collect the cells into 15 ml Corning tubes containing 4 ml sterile dissection medium. For isolation of embryonic RNA, we recommend sorting directly into 1.5 ml Eppendorf tubes containing the RTL lysis buffer from the Qiagen RNAeasy kit. In our case, this approach has yielded high quality RNA from low numbers of cells.
3. Sorting is typically performed by, or at least in the presence of, a trained flow cytometry operator. To obtain large numbers of cells from embryos at E12.5 onwards, we allow several hours of sorting time. Sorting from dispersed tissues such as the YS, placenta, and FL also requires more time than sorting cells from peripheral blood. The higher rate of cell death

observed after lengthy tissue dissociation results in a higher abortion rate (the rate at which droplets containing single cells are not selected by the cytometer to be sorted) and, therefore, lower recovery of viable sorted cells. This may lead to a less efficiency of recovery of viable, sorted cells compared to that obtained from the circulating, peripheral blood. Sorted cells are kept on ice, then collected by centrifugation in an Eppendorf Microcentrifuge 5415D at 1200 rpm ($100 \times g$) and cultured or used for isolation of RNA as soon as possible.

3.4. Primitive erythroid colony assay

Primitive erythroid progenitor activity is measured using a colony assay. The frequency of primitive erythroid colony forming cells (EryP-CFC) can be assessed by plating cells from dissociated YS tissue (E7.5 and E8.5) in methylcellulose and scoring colonies of 50–100 cells after 3–5 days of culture (Palis *et al.*, 1999; Wong *et al.*, 1986). Only erythroid progenitors, but not more differentiated erythroid cells, will grow under these growth conditions. This assay, and photographs illustrating the morphology of EryP-CFC, has been described in detail previously (Baron and Mohn, 2005; Palis and Koniski, 2005; Palis *et al.*, 1999) and is described more briefly below.

1. Prepare single cell suspension from dispersed embryos (E7.0–E9.25) for plating directly in methylcellulose. Alternatively, cells can be sorted prior to plating.
2. Prepare methylcellulose plating mixture (for 10 ml total) as follows:

Methylcellulose/IMDM	5.5 ml
PDS	1.0 ml
PFHM-II	500 μl
L-Glutamine (200 nm stock)	100 μl
Ascorbic acid (5 mg/ml stock)	50 μl
α-MTG (dilute 26 μl/2 ml IMDM)	30 μl
Erythropoietin (Epogen 2000 U/ml)	20 μl
1× IMDM + Pen/Strep	to 10 ml

This mixture will be quite viscous and must be vortexed thoroughly.

3. Resuspend cells in IMDM to final volume of 1 ml. Add to methylcellulose mix prepared above. Vortex to ensure even distribution of cells. Allow to sit at room temperature until all bubbles have disappeared from the mixture.
4. Plate 1 ml of methylcellulose + cells in a 3.5 cm Petri dish using a 3 ml syringe fitted with a 16-gauge blunt-end needle.

5. Place up to six of the 3.5 cm dishes in a 15 cm Petri dish along with one open 3.5 cm dish containing autoclaved water to humidify the cultures. Place lid on the 15 cm dish and incubate at 37 °C, 5% CO_2 for 4–5 days. EryP colonies will appear within 2.5 days.

3.5. Monitoring the nucleation status of erythroid populations

Enucleation is crucial for the production of erythrocytes during terminal erythroid differentiation. During this process, the nucleus becomes condensed to a fraction of its original volume and is expelled. The nucleation status of erythroid cells can be monitored using a transgenic mouse line in which a fluorescent reporter is targeted to the nucleus (Isern *et al.*, 2008) or using fluorescent dyes that bind to nucleic acids (Fraser *et al.*, 2007; Isern *et al.*, 2008; McGrath *et al.*, 2008).

3.5.1. Assessment of nucleation status using DRAQ5 and flow cytometry

1. Prepare working solution of DRAQ5 (25 μg/ml) by diluting 4 μl of 5 m*M* DRAQ5 stock to 1000 μl of FACS buffer.
2. Prepare a suspension of 1×10^6 erythroid cells (from peripheral blood, bone marrow, fetal liver, or other embryonic tissue) in 10 μl of FACS buffer.
3. Thoroughly resuspend pellet in 100 μl of diluted DRAQ5. Incubate at room temperature in dark for 1 min.
4. Add 1 ml FACS buffer. Collect cells by centrifugation in an Eppendorf Microcentrifuge 5415D at 1200 rpm ($100\times g$) and resuspend in FACS buffer/DAPI.

3.5.2. Monitoring erythroid enucleation utilizing a nuclear-localized FP

Nuclei can be labeled by transgenic expression of a histone H2B–FP fusion (Hadjantonakis and Papaioannou, 2004). We have used human *epsilon-globin* regulatory elements drive expression of H2B–GFP within the primitive erythroid lineage, to assess nuclear condensation, enucleation, and nuclear engulfment by FLMs (Isern *et al.*, 2008). Following culture of EryP with FLMs (see below) or flow cytometry of a FL suspension in which macrophages were labeled using anti-F4/80 antibody and H2B–GFP(+) nuclei were identified with the cytoplasm of the macrophages (Isern *et al.*, 2008). Free nuclei could also be detected using flow cytometry (Isern *et al.*, 2008). Single cell suspensions of peripheral blood or fetal liver were analyzed by flow cytometry using a combination of low side scatter (SSC, indicating granularity) and GFP fluorescence (Isern *et al.*, 2008). FACS-sorted SSC(low);GFP(+) structures were identified by microscopy as intact nuclei surrounded by a thin membrane (Isern *et al.*, 2008). The SSC (medium/high); GFP(+) cells were nucleated EryP (Isern *et al.*, 2008).

3.6. Assessing macrophage–erythroid interactions

Erythroblastic islands (EBIs) are the specialized microenvironment within fetal liver and adult bone marrow in which erythroid precursors develop and terminally differentiate, in close proximity to specialized macrophages (Bessis, 1958; Chasis and Mohandas, 2008). These hematopoietic islands comprise three-dimensional clusters of maturing erythroblasts that surround a central macrophage. The membrane of the macrophage extends around and makes intimate contact with the erythroblasts. It is thought that local cell–cell interactions within the EBI, in concert with cytokines, play critical roles in regulating erythroid development and apoptosis (Chasis and Mohandas, 2008). Intact EBIs can be isolated and maintained in culture for a short time, for time-lapse imaging or other studies. The central macrophages of the EBI can be isolated on the basis of their adhesion to glass and then reconstituted with erythroblasts, to examine adhesive interactions between these cells under different conditions (Iavarone *et al.*, 2004; Lee *et al.*, 2006; Morris *et al.*, 1988). The protocol for isolating EBIs is summarized in Fig. 22.1. All procedures described here should be carried out under sterile conditions.

3.6.1. Preparation of erythroblastic islands

3.6.1.1. Bone marrow EBIs

1. Sacrifice adult (8–24-week-old) mice, dissect the femora and flush out the marrow by gently inserting a 20-gauge needle fitted with a 3 ml syringe into the opening of the bone. To flush, use cold IMDM supplemented with 10% FBS.
2. Dissociate the marrow cells by passing flushed cells through a 3 ml syringe twice.
3. Remove any remaining cell clumps by filtering the dispersed marrow through a 40 μm cell strainer.
4. Collect the cells by centrifugation for 5 min at 1200 rpm/100$\times g$, then wash the cell pellet with 10 ml of dissection medium.
5. Proceed to step 5, below.

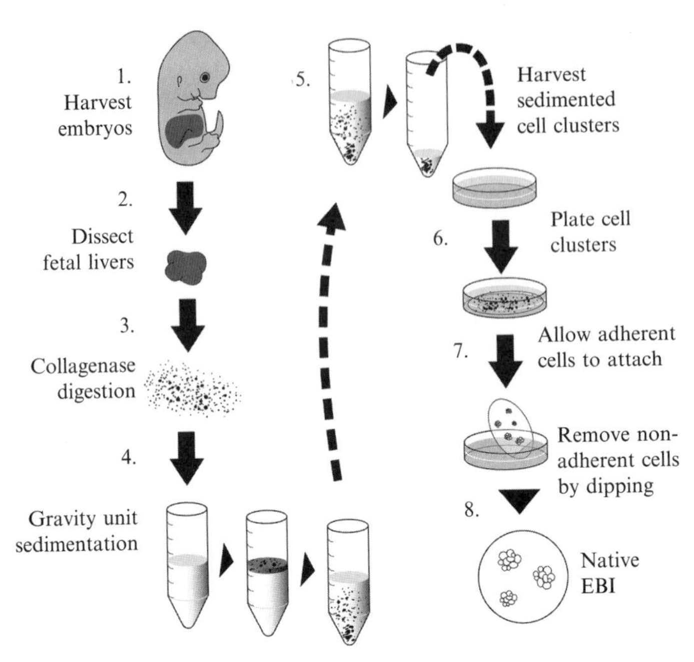

Figure 22.1 Schematic representation of the experimental approach used for the isolation of erythroblastic islands (EBIs) from the fetal liver, as described in this chapter. Fetal livers are dissected from E13.5–E14.5 mouse embryos and enzymatically digested using collagenase. A fraction enriched in EBIs is isolated by unit gravity sedimentation and then plated onto glass coverslips. The EBIs are allowed to attach to the glass surface. Nonadherent cells are removed by gentle rinsing. The coverslips bound with native EBIs are then analyzed using immunofluorescence microscopy.

3.6.1.2. Fetal liver EBIs (FL EBIs)

1. Dissect embryos (E13.5–E15.5) in cold dissection medium and carefully subdissect FL, removing any nonhepatic tissue. Rinse FL twice in dissection medium. (Any transfers of FL should be performed using a plastic transfer pipette with the narrow tip cut back to prevent the tissue from breaking up.)
2. Transfer the FLs to a Falcon tube containing 10 ml of medium and disaggregate by gently pipetting several times using a 25 ml pipette. Allow the liver fragments to settle for 5 min and remove the supernatant.
3. Add 5 ml of prewarmed digestion buffer (0.05% (w/v) Collagenase, 0.002% (w/v) DNase I in RPMI). Digest for 1 h at 37 °C, with gentle rotation.
4. Disperse the sample by pipetting several times with a 5 ml pipette, collect by centrifugation in an Eppendorf Microcentrifuge 5415D for 5 min at 1200 rpm (100×*g*). Resuspend the pellet in 10 ml RPMI/2% FBS. Wash twice with this medium.

5. Resuspend the pellet in a small volume of RPMI/2% FBS (three to four times the original volume of the cell pellet).
6. Erythroblastic clusters may be enriched by unit gravity sedimentation. Carefully layer the resupended sample over 30 ml RPMI/30% FBS in a 50 ml Falcon tube. Allow the clusters to sediment at RT for 15 min. Aspirate the supernatant very slowly, taking care to avoid disrupting the settled clusters and leaving 3–5 ml, containing EBIs, at the bottom of the tube.
7. Transfer the EBIs to glass coverslips inserted into the wells of a 24-well plate. A volume of 200–300 μl is sufficient to cover the surface of the coverslip.
8. Allow the clusters to adhere to the glass for 2 h at 37 °C in a tissue culture incubator.
9. Using #5 watchmaker's forceps, carefully dip the coverslips serially into each of two 100 ml beakers filled with 80 ml of PBS to wash off unattached cells. Transfer the coverslips to a new 24-well plate and add 600 μl of culture medium to each well. Cells can be cultured for several days at 37 °C.

3.6.2. Preparation of fetal liver macrophages

1. Dissect out the FL (see above). For a good yield, we routinely pool the FLs ($\geq$6 FL, E13.5; $\geq$5 FL, E14.5).
2. Transfer FLs to a 15 ml Falcon tube containing 2 ml FACS buffer.
3. Disperse the pooled FL by pipetting until a homogeneous suspension is obtained (start by using a 5 ml pipette to disrupt the livers and then use a P1000 Gilson Pipetman to dissociate the tissue further). If suspension becomes viscous, further dilution with FACS buffer can be helpful.
4. Filter the suspension through a 40 μm cell strainer to remove any remaining clumps.
5. Add 10 ml of PBS/2 m*M* EDTA and mix well to wash the cells.
6. Collect the cells by centrifugation in an Eppendorf Microcentrifuge 5415D for 5 min at 1200 rpm (100$\times$*g*). Aspirate the supernatant.
7. Repeat steps 7–9 once more, making sure to resuspend the pellet well.
8. Resuspend the final pellet in 10–20 ml of FLM culture medium.
9. Plate the cells on 12 mm circular coverslips in the wells of a 24-well plate. Use 400–500 μl/well (see Section 4, Note 1).
10. Culture cells in the incubator at 37 °C, 5% CO_2 for 2–3 h. Wash by dipping serially into each of two 100 ml beakers containing 80 ml PBS to remove nonadherent cells. Transfer each coverslip to a new well containing fresh medium.
11. The FLM-enriched adherent cells remaining attached to the coverslip can be cultured for several days. Media should be replaced every 2 days.

3.6.3. EBI reconstitution (rosetting) assays

Wash the FLMs prepared above.

1. Strip off the remaining nonadherent cells by dipping the coverslips in a 100 ml glass beaker filled with 80 ml of PBS and transfer the coverslip to a new well containing binding medium (see Section 4, Note 2).
2. Pipet isolated erythroid cells (1–5 × 10^6 erythroid cells in 300 μl in binding medium) onto the macrophage-bound coverslip and incubate 45 min at 37 °C. Erythroid cells are isolated either from the peripheral blood or the fetal liver (as described in Sections 3.1.2 or 3.1.3, respectively).
3. Rinse coverslips by gently dipping serially into three 100 ml beakers, each containing 80 ml PBS, to remove nonadherent cells. Fix immediately in 4% PFA for 15 min at 4 °C.
4. Wash the fixed coverslips three times in PBS and proceed to staining or scoring of fluorescent cells. Fixed cells can be air-dried and stored at 4 °C in the dark in the 24-well dish, prior to microscopic analysis.

3.6.4. Coculture of macrophages and EryP

1. Prepare FLMs as described above.
2. Add 1–2 × 10^6 erythroid cells/well containing a FLM-coated coverslip in 500 μl of culture medium.
3. Cocultures can be maintained for up to 48 h. Coverslips are processed as described in steps 4 and 5 of "rosetting assays" (see Section 3.6.3). Nonadherent cells that remain in the well after washing can be analyzed by FACS.

3.6.5. Immunostaining of EBIs

1. Permeabilize EBIs in 0.5% Triton X-100/PBS for 5 min and wash twice for 5 min in PBS.
2. Block coverslips for 15–30 min in blocking buffer.
3. Dilute primary anti-F4/80 antibody (1:100) in blocking buffer and add 100 μl per coverslip. Incubate for 1 h at room temperature.
4. Wash three times for 5 min with PBS + 0.01 % (v/v) Triton X-100.
5. Dilute secondary antibody (1:300) in blocking buffer and add 100 μl per coverslip. Incubate for 45 min.
6. Wash three times for 5 min with PBS + 0.01 % (v/v) Triton X-100. Rinse in PBS.
7. Mount with Vectashield mounting medium containing DAPI.

4. Notes

FACS buffer: FACS buffer is PBS containing protein that prevents cells from adhering to the plastic of the FACS tubes or the tubing of the fluidics system of the flow cytometer. We routinely use PBS containing 10% FBS that has been heat-inactivated to eliminate complement protein. If cells appear to agglutinate, as is often observed with erythroid cells, PBS containing 0.2–1% (w/v) BSA can be used.

DAPI: To exclude dead cells, which may bind nonspecifically to antibodies and produce a false positive signal, we routinely resuspend the final cell pellet in FACS buffer containing 0.2 mg/ml DAPI. DAPI is a poorly cell-permeable DNA binding dye that is excited by the violet laser in the flow cytometer. DAPI is soluble in water but not in PBS. Hence, we prepare a 1000× stock solution in deionized water. This stock is then diluted in FACS buffer.

DRAQ5: DRAQ5 is a synthetic anthraquinone dye that binds strongly to DNA (Smith *et al.*, 1999, 2000). DRAQ5 has two main advantages for use in flow cytometry compared to other dyes. First, it is highly cell permeable and can cross the cell membrane and stain nucleic acids within a minute of incubation. Second, it is excited by the 488 nm laser but fluoresces in the far-red region (665 to >800 nm) of the spectrum. Therefore, DRAQ5 can be used in combination with GFP or with antibodies or FPs conjugated to fluorochromes that emit at lower wavelengths. The manufacturer's instructions indicate that samples need not be washed following staining with DRAQ5. However, we have found that prolonged exposure to DRAQ5 leads to widespread fluorescent signal, even in enucleated cells.

Assessing erythroid–macrophage interactions

1. Coverslips are cleaned and sterilized by acid-washing in a solution of 10% HCl/70% ethanol, rinsed twice in water and then in 95% ethanol and then air-dried.
2. Binding medium is prepared using a lower serum concentration (2%) than used for culture medium, to prevent cells from sticking nonspecifically. Coverslips are handled carefully using #5 watchmaker's forceps.

REFERENCES

Baron, M. H., and Fraser, S. T. (2005). The specification of early hematopoiesis in the mammal. *Curr. Opin. Hematol.* **12,** 217–221.

Baron, M. H., and Mohn, D. (2005). Mouse embryonic explant culture system for analysis of hematopoietic and vascular development. *Methods Mol. Med.* **105,** 231–256.

Bessis, M. (1958). Erythroblastic island, functional unity of bone marrow. *Rev. Hématol.* **13,** 8–11.

Cairns, L. A., Moroni, E., Levantini, E., Giorgetti, A., Klinger, F. G., Ronzoni, S., Tatangelo, L., Tiveron, C., De Felici, M., Dolci, S., Magli, M. C., Giglioni, B., and Ottolenghi, S. (2003). Kit regulatory elements required for expression in developing hematopoietic and germ cell lineages. *Blood* **102,** 3954–3962.

Chasis, J. A., and Mohandas, N. (2008). Erythroblastic islands: Niches for erythropoiesis. *Blood* **112,** 470–478.

Dyer, M. A., Farrington, S. M., Mohn, D., Munday, J. R., and Baron, M. H. (2001). Indian hedgehog activates hematopoiesis and vasculogenesis and can respecify prospective neurectodermal cell fate in the mouse embryo. *Development* **128,** 1717–1730.

Dzierzak, E., and Speck, N. A. (2008). Of lineage and legacy: The development of mammalian hematopoietic stem cells. *Nat. Immunol.* **9,** 129–136.

Faust, N., *et al.* (2000). Insertion of enhanced green fluorescent protein into the lysozyme gene creates mice with green fluorescent granulocytes and macrophages. *Blood* **96,** 719–726.

Fontenot, J. D., Dooley, J. L., Farr, A. G., and Rudensky, A. Y. (2005). Developmental regulation of Foxp3 expression during ontogeny. *J. Exp. Med.* **202,** 901–906.

Fraser, S. T., Hadjantonakis, A. K., Sahr, K. E., Willey, S., Kelly, O. G., Jones, E. A., Dickinson, M. E., and Baron, M. H. (2005). Using a histone yellow fluorescent protein fusion for tagging and tracking endothelial cells in ES cells and mice. *Genesis* **42,** 162–171.

Fraser, S. T., Isern, J., and Baron, M. H. (2007). Maturation and enucleation of primitive erythroblasts during mouse embryogenesis is accompanied by changes in cell-surface antigen expression. *Blood* **109,** 343–352.

Fuxa, M., and Busslinger, M. (2007). Reporter gene insertions reveal a strictly B lymphoid-specific expression pattern of Pax5 in support of its B cell identity function. *J. Immunol.* **178,** 8222–8228.

Garrick, D., Fiering, S., Martin, D. I., and Whitelaw, E. (1998). Repeat-induced gene silencing in mammals. *Nat. Genet.* **18,** 56–59.

Giepmans, B. N., Adams, S. R., Ellisman, M. H., and Tsien, R. Y. (2006). The fluorescent toolbox for assessing protein location and function. *Science* **312,** 217–224.

Hadjantonakis, A. K., and Papaioannou, V. E. (2004). Dynamic in vivo imaging and cell tracking using a histone fluorescent protein fusion in mice. *BMC Biotechnol.* **4,** 33.

Hamada, M., Moriguchi, T., Yokomizo, T., Morito, N., Zhang, C., and Takahashi, S. (2003). The mouse mafB 5′-upstream fragment directs gene expression in myelomonocytic cells, differentiated macrophages and the ventral spinal cord in transgenic mice. *J. Biochem.* **134,** 203–210.

Heck, S., Ermakova, O., Iwasaki, H., Akashi, K., Sun, C. W., Ryan, T. M., Townes, T., and Graf, T. (2003). Distinguishable live erythroid and myeloid cells in beta-globin ECFP x lysozyme EGFP mice. *Blood* **101,** 903–906.

Heinrich, A. C., *et al.* (2004). A mouse model for visualization and conditional mutations in the erythroid lineage. *Blood* **104,** 659–666.

Hosen, N., Yamane, T., Muijtjens, M., Pham, K., Clarke, M. F., and Weissman, I. L. (2007). Bmi-1-green fluorescent protein-knock-in mice reveal the dynamic regulation of bmi-1 expression in normal and leukemic hematopoietic cells. *Stem Cells* **25,** 1635–1644.

Iavarone, A., King, E. R., Dai, X. M., Leone, G., Stanley, E. R., and Lasorella, A. (2004). Retinoblastoma promotes definitive erythropoiesis by repressing Id2 in fetal liver macrophages. *Nature* **432,** 1040–1045.

Isern, J., Fraser, S. T., He, Z., and Baron, M. H. (2008). The fetal liver is a niche for maturation of primitive erythroid cells. *Proc. Natl. Acad. Sci. USA* **105,** 6662–6667.

Joyner, A. L. (ed.), (1995). Gene Targeting: A Practical Approach, IRL Press, New York.

Jung, S., Aliberti, J., Graemmel, P., Sunshine, M. J., Kreutzberg, G. W., Sher, A., and Littman, D. R. (2000). Analysis of fractalkine receptor CX(3)CR1 function by targeted deletion and green fluorescent protein reporter gene insertion. *Mol. Cell. Biol.* **20,** 4106–4114.

Kallies, A., Hasbold, J., Tarlinton, D. M., Dietrich, W., Corcoran, L. M., Hodgkin, P. D., and Nutt, S. L. (2004). Plasma cell ontogeny defined by quantitative changes in blimp-1 expression. *J. Exp. Med.* **200,** 967–977.

Kissenpfennig, A., Henri, S., Dubois, B., Laplace-Builhe, C., Perrin, P., Romani, N., Tripp, C. H., Douillard, P., Leserman, L., Kaiserlian, D., Saeland, S., Davoust, J., and Malissen, B. (2005). Dynamics and function of Langerhans cells in vivo: Dermal dendritic cells colonize lymph node areas distinct from slower migrating Langerhans cells. *Immunity* **22,** 643–654.

Kuwata, N., Igarashi, H., Ohmura, T., Aizawa, S., and Sakaguchi, N. (1999). Cutting edge: Absence of expression of RAG1 in peritoneal B-1 cells detected by knocking into RAG1 locus with green fluorescent protein gene. *J. Immunol.* **163,** 6355–6359.

Laiosa, C. V., Stadtfeld, M., and Graf, T. (2006). Determinants of lymphoid-myeloid lineage diversification. *Annu. Rev. Immunol.* **24,** 705–738.

Lee, G., Lo, A., Short, S. A., Mankelow, T. J., Spring, F., Parsons, S. F., Yazdanbakhsh, K., Mohandas, N., Anstee, D. J., and Chasis, J. A. (2006). Targeted gene deletion demonstrates that the cell adhesion molecule ICAM-4 is critical for erythroblastic island formation. *Blood* **108,** 2064–2071.

Lochner, M., Peduto, L., Cherrier, M., Sawa, S., Langa, F., Varona, R., Riethmacher, D., Si-Tahar, M., Di Santo, J. P., and Eberl, G. (2008). In vivo equilibrium of proinflammatory IL-17+ and regulatory IL-10+ Foxp3+ RORgamma t+ T cells. *J. Exp. Med.* **205,** 1381–1393.

Lohmann, F., and Bieker, J. J. (2008). Activation of Eklf expression during hematopoiesis by Gata2 and Smad5 prior to erythroid commitment. *Development* **135,** 2071–2082.

Lorsbach, R. B., Moore, J., Ang, S. O., Sun, W., Lenny, N., and Downing, J. R. (2004). Role of RUNX1 in adult hematopoiesis: Analysis of RUNX1-IRES-GFP knock-in mice reveals differential lineage expression. *Blood* **103,** 2522–2529.

Luc, S., Buza-Vidas, N., and Jacobsen, S. E. (2008). Delineating the cellular pathways of hematopoietic lineage commitment. *Semin. Immunol.* **20,** 213–220.

Ma, X., Robin, C., Ottersbach, K., and Dzierzak, E. (2002). The Ly-6A (Sca-1) GFP transgene is expressed in all adult mouse hematopoietic stem cells. *Stem Cells* **20,** 514–521.

McGrath, K., and Palis, J. (2008). Ontogeny of erythropoiesis in the mammalian embryo. *Curr. Top. Dev. Biol.* **82,** 1–22.

McGrath, K. E., Koniski, A. D., Malik, J., and Palis, J. (2003). Circulation is established in a stepwise pattern in the mammalian embryo. *Blood* **101,** 1669–1676.

McGrath, K. E., Kingsley, P. D., Koniski, A. D., Porter, R. L., Bushnell, T. P., and Palis, J. (2008). Enucleation of primitive erythroid cells generates a transient population of "pyrenocytes" in the mammalian fetus. *Blood* **111,** 2409–2417.

Monroe, R. J., Seidl, K. J., Gaertner, F., Han, S., Chen, F., Sekiguchi, J., Wang, J., Ferrini, R., Davidson, L., Kelsoe, G., and Alt, F. W. (1999). RAG2:GFP knockin mice reveal novel aspects of RAG2 expression in primary and peripheral lymphoid tissues. *Immunity* **11,** 201–212.

Morris, L., Crocker, P. R., and Gordon, S. (1988). Murine fetal liver macrophages bind developing erythroblasts by a divalent cation-dependent hemagglutinin. *J. Cell Biol.* **106,** 649–656.

Nagy, A., Gertsenstein, M., Vintersten, K., and Behringer, R. (2003). Manipulating the Mouse Embryo: A Laboratory Manual. Cold Spring Harbor Laboratory Press, Cold Spring Harbor, NY.

Nishimura, S., Takahashi, S., Kuroha, T., Suwabe, N., Nagasawa, T., Trainor, C., and Yamamoto, M. (2000). A GATA box in the GATA-1 gene hematopoietic enhancer is a critical element in the network of GATA factors and sites that regulate this gene. *Mol. Cell. Biol.* **20,** 713–723.

Norris, H. H., *et al.* (2007). TCRbeta enhancer activation in early and late lymphoid progenitors. *Cell. Immunol.* **247,** 59–71.

North, T. E., de Bruijn, M. F., Stacy, T., Talebian, L., Lind, E., Robin, C., Binder, M., Dzierzak, E., and Speck, N. A. (2002). Runx1 expression marks long-term repopulating hematopoietic stem cells in the midgestation mouse embryo. *Immunity* **16,** 661–672.

Nowotschin, S., and Hadjantonakis, A. K. (2009). Use of KikGR a photoconvertible green-to-red fluorescent protein for cell labeling and lineage analysis in ES cells and mouse embryos. *BMC Dev. Biol.* **9,** 49.

Nowotschin, S., Eakin, G. S., and Hadjantonakis, A. K. (2009a). Dual transgene strategy for live visualization of chromatin and plasma membrane dynamics in murine embryonic stem cells and embryonic tissues. *Genesis* **47,** 330–336.

Nowotschin, S., Eakin, G. S., and Hadjantonakis, A. K. (2009b). Live-imaging fluorescent proteins in mouse embryos: Multi-dimensional, multi-spectral perspectives. *Trends Biotechnol.* **27,** 266–276.

Nutt, S. L., Metcalf, D., D'Amico, A., Polli, M., and Wu, L. (2005). Dynamic regulation of PU.1 expression in multipotent hematopoietic progenitors. *J. Exp. Med.* **201,** 221–231.

Orkin, S. H., and Zon, L. I. (2008). Hematopoiesis: An evolving paradigm for stem cell biology. *Cell* **132,** 631–644.

Palis, J., and Koniski, A. (2005). Analysis of hematopoietic progenitors in the mouse embryo. *Methods Mol. Med.* **105,** 289–302.

Palis, J., Robertson, S., Kennedy, M., Wall, C., and Keller, G. (1999). Development of erythroid and myeloid progenitors in the yolk sac and embryo proper of the mouse. *Development* **126,** 5073–5084.

Pirity, M., Hadjantonakis, A. K., and Nagy, A. (1998). Embryonic stem cells, creating transgenic animals. *Methods Cell Biol.* **57,** 279–293.

Preis, J. I., Downes, M., Oates, N. A., Rasko, J. E., and Whitelaw, E. (2003). Sensitive flow cytometric analysis reveals a novel type of parent-of-origin effect in the mouse genome. *Curr. Biol.* **13,** 955–959.

Pruitt, S. C., Mielnicki, L. M., and Stewart, C. C. (2004). Analysis of fluorescent protein expressing cells by flow cytometry. *Methods Mol. Biol.* **263,** 239–258.

Rhee, J. M., Pirity, M. K., Lackan, C. S., Long, J. Z., Kondoh, G., Takeda, J., and Hadjantonakis, A. K. (2006). In vivo imaging and differential localization of lipid-modified GFP-variant fusions in embryonic stem cells and mice. *Genesis* **44,** 202–218.

Sasmono, R. T., Oceandy, D., Pollard, J. W., Tong, W., Pavli, P., Wainwright, B. J., Ostrowski, M. C., Himes, S. R., and Hume, D. A. (2003). A macrophage colony-stimulating factor receptor-green fluorescent protein transgene is expressed throughout the mononuclear phagocyte system of the mouse. *Blood* **101,** 1155–1163.

Schedl, A., Ross, A., Lee, M., Engelkamp, D., Rashbass, P., van Heyningen, V., and Hastie, N. D. (1996). Influence of PAX6 gene dosage on development: Overexpression causes severe eye abnormalities. *Cell* **86,** 71–82.

Shaner, N. C., Steinbach, P. A., and Tsien, R. Y. (2005). A guide to choosing fluorescent proteins. *Nat. Methods* **2,** 905–909.

Shu, X., Royant, A., Lin, M. Z., Aguilera, T. A., Lev-Ram, V., Steinbach, P. A., and Tsien, R. Y. (2009). Mammalian expression of infrared fluorescent proteins engineered from a bacterial phytochrome. *Science* **324,** 804–807.

Singbartl, K., Thatte, J., Smith, M. L., Wethmar, K., Day, K., and Ley, K. (2001). A CD2-green fluorescence protein-transgenic mouse reveals very late antigen-4-dependent CD8+ lymphocyte rolling in inflamed venules. *J. Immunol.* **166,** 7520–7526.

Smith, P. J., Wiltshire, M., Davies, S., Patterson, L. H., and Hoy, T. (1999). A novel cell permeant and far red-fluorescing DNA probe, DRAQ5, for blood cell discrimination by flow cytometry. *J. Immunol. Methods* **229,** 131–139.

Smith, P. J., Blunt, N., Wiltshire, M., Hoy, T., Teesdale-Spittle, P., Craven, M. R., Watson, J. V., Amos, W. B., Errington, R. J., and Patterson, L. H. (2000). Characteristics of a novel deep red/infrared fluorescent cell-permeant DNA probe, DRAQ5, in intact human cells analyzed by flow cytometry, confocal and multiphoton microscopy. *Cytometry* **40,** 280–291.

Stadtfeld, M., Varas, F., and Graf, T. (2005). Fluorescent protein-cell labeling and its application in time-lapse analysis of hematopoietic differentiation. *Methods Mol. Med.* **105,** 395–412.

Suzuki, N., *et al.* (2006). Combinatorial Gata2 and Sca1 expression defines hematopoietic stem cells in the bone marrow niche. *Proc. Natl. Acad. Sci. USA* **103,** 2202–2207.

Tadjali, M., *et al.* (2006). Prospective isolation of murine hematopoietic stem cells by expression of an Abcg2/GFP allele. *Stem Cells* **24,** 1556–1563.

Vassen, L., Okayama, T., and Moroy, T. (2007). Gfi1b:green fluorescent protein knock-in mice reveal a dynamic expression pattern of Gfi1b during hematopoiesis that is largely complementary to Gfi1. *Blood* **109,** 2356–2364.

Wong, P. M., Chung, S. W., Chui, D. H., and Eaves, C. J. (1986). Properties of the earliest clonogenic hemopoietic precursors to appear in the developing murine yolk sac. *Proc. Natl. Acad. Sci. USA* **83,** 3851–3854.

Yang, X., Model, P., and Heintz, N. (1997). Homologous recombination based modification in *Escherichia coli* and germline transmission in transgenic mice of a bacterial artificial chromosome. *Nat. Biotechnol.* **15,** 859–865.

Yang, J., Hills, D., Taylor, E., Pfeffer, K., Ure, J., and Medvinsky, A. (2008). Transgenic tools for analysis of the haematopoietic system: Knock-in CD45 reporter and deletor mice. *J. Immunol. Methods* **337,** 81–87.

Zhang, Y., Buchholz, F., Muyrers, J. P. P., and Stewart, A. F. (1998). A new logic for DNA engineering using recombination in *Escherichia coli. Nat. Genet.* **20,** 123–128.

Zhang, J., *et al.* (2007). CD41-YFP mice allow in vivo labeling of megakaryocytic cells and reveal a subset of platelets hyperreactive to thrombin stimulation. *Exp. Hematol.* **35,** 490–499.

CHAPTER TWENTY-THREE

Identification and *In Vivo* Analysis of Murine Hematopoietic Stem Cells

Serine Avagyan,[*,†] Yacine M. Amrani,[*,†] *and* Hans-Willem Snoeck[*,†]

Contents

Abstract

Hematopoietic stem cells (HSCs) can self-renew and give rise to all the cells of the blood and the immune system. As they differentiate, HSCs progressively lose their self-renewal capacity and generate lineage-restricted multipotential progenitor cells that in turn give rise to mature cells. The development of rigorous quantitative *in vivo* assays for HSC activity combined with multicolor flow cytometry and high-speed sorting have resulted in the phenotypic definition of HSCs to virtual purity. Here, we describe the isolation and identification of HSCs by flow cytometry and the use of competitive repopulation to assess HSC number and function.

1. Hematopoietic Stem Cells

Hematopoiesis maintains the lifelong production of billions of blood cells every day. This process is highly responsive to stress, such as blood loss, infection, or irradiation and is maintained by hematopoietic stem cells (HSCs). HSCs mostly reside in the bone marrow microenvironment or niche, which provides signaling by cytokines and by membrane-bound ligands and adhesion molecules (Moore and Lemischka, 2006; Schofield, 1978).

* Department of Gene and Cell Medicine, Mount Sinai of School of Medicine, New York, USA
† Black Family Stem Cell Institute, Mount Sinai of School of Medicine, New York, USA

Methods in Enzymology, Volume 476
ISSN 0076-6879, DOI: 10.1016/S0076-6879(10)76023-7

The nature of that microenvironment is still controversial. Although most studies focus on the role of the endosteal lining of the bone marrow (Calvi *et al.*, 2003; Zhang *et al.*, 2003), bone marrow sinusoidal endothelial cells and CXCL12-expressing reticular cells have also been implicated as part of the HSC niche (Kiel *et al.*, 2005; Sugiyama *et al.*, 2006). HSC can undergo self-renewal, differentiation, apoptosis, or can remain quiescent (Orkin and Zon, 2008). In addition, HSCs can move to the blood stream and home to other bone marrow sites (Wright *et al.*, 2001). The latter characteristic allows HSCs to be delivered intravenously to reconstitute the hematopoietic system of conditioned hosts. Although they are the best characterized stem cells in mouse and human, it is still unclear how the balance between these fate choices is maintained *in vivo*, both in steady state and in conditions of stress.

Long-term HSCs (LT-HSCs) can fully reconstitute all the hematopoietic lineages of lethally irradiated host for the life span of that host, and represent $1/10^4$ to $1/10^5$ cells in the bone marrow (Chao *et al.*, 2008; Luc *et al.*, 2008; Orkin and Zon, 2008). In addition, these cells can be serially transplanted, although, for reasons that are unclear, repopulation capacity is entirely lost after four to five passages through irradiated hosts, even when telomere length is maintained by overexpression of telomerase (Allsopp *et al.*, 2003a,b; Spangrude *et al.*, 1995). Capacity for serial reconstitution *in vivo* is generally taken as evidence for self-renewal, as self-renewal of HSCs *in vitro* has not been achieved thus far. LT-HSCs develop into short-term (ST)-HSCs. These are capable of reconstituting all the lineages of the hematopoietic system, but have limited self-renewal capacity, and cannot sustain serial repopulation (Yang *et al.*, 2005). Multipotential progenitor (MPP) cells have lost all self-renewal capacity, but can still reconstitute most or all lineages of the hematopoietic system (Adolfsson *et al.*, 2001; Christensen and Weissman, 2001). As MPPs proliferate, they differentiate and become progressively committed to a specific lineage. According the classical scheme of hematopoiesis, MPPs differentiate into either common lymphoid progenitor (CLP) or common myeloid progenitors (CMPs). The former generate T, B, and NK cells, while the latter develop into megakaryocyte/erythroid (MEP) and granulocyte/macrophage (GMP) progenitors. These progenitors become further lineage restricted and finally give rise to mature blood cells (Chao *et al.*, 2008). While attractive for its relative simplicity and appealing logic, in reality, phenotypical and functional transitions between different classes of progenitors are neither synchronous nor abrupt. Furthermore, it is still a matter of debate to what extent megakaryocytic/erythroid potential is lost prior to differentiation into MPPs, instead of after the CMP stage (Adolfsson *et al.*, 2005; Chao *et al.*, 2008; Forsberg *et al.*, 2006; Luc *et al.*, 2008), and whether CLPs are physiological precursors of T cells in the thymus (Bell and Bhandoola, 2008; Bhandoola *et al.*, 2007; Bhattacharya *et al.*, 2008; Wada *et al.*, 2008).

2. Assays of Hematopoietic Stem Cells

A fundamental observation that led to the rigorous efforts to identify HSC was that death due to irradiation was caused primarily by hematopoietic failure. This was experimentally shown in mice where shielding of the spleen during irradiation or injection of spleen or bone marrow cells after irradiation prevented death of the animals (Jacobson *et al.*, 1950; Lorenz *et al.*, 1951). A decade later, Till and McCulloch (1961) demonstrated that the bone marrow of the mouse contained highly proliferative cells that were able to form macroscopic colonies in the spleens of irradiated hosts. These colonies contained cells from all hematopoietic lineages and could also form more colonies when transplanted into a secondary host. Analysis of retroviral integration sites provided evidence that a cell capable of clonally reconstituting multiple lineages does exist as it was observed that a single retroviral insertion could be detected in multiple lineages after injection of transduced bone marrow into a lethally irradiated host (Lemischka *et al.*, 1986).

Properties that have been assigned to stem and progenitor cells include radioprotection and the generation of spleen colony-forming unit at day 12 (d12 CFU-S).

In a radioprotection assay, bone marrow cells are injected into lethally irradiated syngeneic hosts, and survival is assessed. Survival after lethal irradiation relies on more committed progenitor cells that can rapidly generate mature cells, in particular platelets and neutrophils. At later time points, remaining endogenous HSCs that survived irradiation can reconstitute the hematopoietic system. The radioprotection assay therefore does not measure the function of transferred stem cells (Bhattacharya *et al.*, 2008; Kondo *et al.*, 2003; Pallavicini *et al.*, 1997). In the CFU-S assay, cells are injected into lethally irradiated mice, and progenitor cell-derived colonies are counted in the spleen. The later the colonies appear, the less differentiated the cell is they are derived from. While late-appearing CFU-S (d12 CFU-S) were originally considered to be derived from HSC, physical separation of d12 CFU-S from functional long-term repopulating cells indicated that the CFU-S assays did not measure stem cell activity (Hodgson and Bradley, 1979; Jones *et al.*, 1990; Ploemacher and Brons, 1989).

The gold standard for the detection of HSC activity is the long-term multilineage reconstitution of the hematopoietic system of a lethally irradiated host (Bhattacharya *et al.*, 2008; Harrison, 1980; Harrison *et al.*, 1993; Kondo *et al.*, 2003; Orkin and Zon, 2008). This assay measures a complex process that starts with the injection of the HSC, followed by its homing to a niche where it engrafts and proliferates to restore the HSC compartment and to produce the various multilineage progenitors that eventually will give rise to the mature blood cells. Once the HSC has engrafted in the primary recipient, it not only differentiates but also produces more HSCs that would repopulate the secondary recipient in a similar manner (Spangrude *et al.*, 1995). Serial repopulation capacity of the transplanted cell is therefore considered to be a measure of its self-renewal capacity.

To engraft, HSC need "space," a term that appears intuitively appropriate, but for which the nature of the physical reality is unclear. Most likely, the number of available physical niches for HSCs is limited in steady state (Bhattacharya *et al.*, 2006, 2008). Space has to be created by cytoreductive agents or irradiation to achieve significant engraftment of transplanted HSCs. Such assays may not reflect steady-state hematopoiesis because they measure the capacity to rescue a lethally irradiated host (Metcalf, 2007). Ideally, the function of HSC should be measured in steady-state conditions. In unconditioned mice, massive numbers of bone marrow cells need to be injected to achieve detectable engraftment (Sykes *et al.*, 1997). The only feasible ways to avoid these treatments involve the use of immunodeficient mouse recipients ($Rag2^{-/-}\gamma_c^{-/-}$) that are either treated with a c-Kit receptor blocking antibody (Czechowicz *et al.*, 2007) or are HSC-deficient due to a mutant c-Kit receptor (Waskow *et al.*, 2009). Space also needs to be created in the periphery, as after engraftment HSC will preferentially reconstitute those lineages where the peripheral demand is the highest (Bhattacharya *et al.*, 2006; Lemischka, 1991; Lemischka *et al.*, 1986). For example, in $Rag1^{-/-}Il2rg^{-/-}$ mice, rare HSCs can even engraft without conditioning, likely by homing to a few niches made available by the constant flux of HSCs into the peripheral blood, but will almost exclusively generate lymphoid cells, which are absent in the host (Bhattacharya *et al.*, 2006). Therefore, conditioning using cytotoxic agents or irradiation is also required to deplete mature cells. Irradiation is the mostly used conditioning. The appropriate dose is highly mouse strain dependent, and split dose irradiation is preferred. A typical dose in C57BL/6 mice is 750 cG followed by 500 cG 3 h later. Injection of hematopoietic cells has to occur preferably within 24 h after irradiation.

Since stem cell engraftment is a competitive process, relative quantification of stem cell number and/or function can be achieved by competing populations with a sample that serves as an internal control. This analysis can be done by repeated bleeding of the recipients to quantify mature cells of various lineages of donor origin weeks and months after transplantation. The most commonly used mouse strain is C57BL/6. This technique utilizes congenic mice that have distinguishable alleles of the tyrosine phosphatase, CD45/Ly-5, expressed on all hematopoietic cells except mature erythrocytes and committed erythroid progenitors (Shiku *et al.*, 1975; Thomas, 1989), termed CD45.1 and CD45.2 (or Ly5.1 and Ly5.2). C57BL/6 mice carry the CD45.2 allele. The CD45.1 allele has been backcrossed onto the

C57BL/6 background from SLJ mice (B6.SJL-*Ptprca*$^{Pep3b/BoyJ}$ mice). Antibodies that distinguish both allelic variants are used to track donor, competitor, and recipient hematopoiesis after transplantation (Spangrude *et al.*, 1988). It is not always possible to use C57LB/6, as knockout mice are mostly made on mixed 129.B6 background. In that case, the mice need to be backcrossed onto the C57LB/6 background for at least six or, preferably, 10 generations. Alternatively, 129.B6F1 mice can be used as recipients. However, in a competitive setting with CD45.1^{+} C57BL/6 bone marrow as a competitor, the donor sample has to be depleted of T and NK cells, as these cells contaminating the graft can cause graft versus host disease. Another and probably better option is to use *Rag*$^{-/-}$ *Il2rg*$^{-/-}$ mice, which lack T, B, and NK cells, and are therefore receptive to engraftment with bone marrow of a different MHC haplotype. Here too, however, T and NK cell depletion of the grafted cells is required.

In a typical experiment, equal numbers of bone marrow cells from, for example, a particular knockout mouse and from wild-type (wt) controls (the donors) are mixed with equal numbers of competitor bone marrow cells and injected into lethally irradiated C57BL/6 recipients (Harrison, 1980; Harrison *et al.*, 1993). The higher the cell number injected, the lower the variability of the data will be, as when HSC numbers in the samples gets lower, the standard deviation on the repopulation data will increase (Harrison *et al.*, 1993). Typically, 1–2 $\times$ 10^{6} cells of each population are injected when unseparated bone marrow is used. However, single purified HSCs can be detected using this assay (Osawa *et al.*, 1996). To distinguish donor-derived from recipient-derived and competitor cells, congenic C57BL/6.SJL-*Ptprca*$^{Pep3b/BoyJ}$ (CD45.1^{+}, B6-CD45.1 mice) mice are used as competitors, and heterozygous B6.B6.SJL-*Ptprca*$^{Pep3b/BoyJ}$ F1 mice (CD45.1^{+}CD45.2^{+}) as recipients or vice versa. Alternatively, recipient and competitor can be of the same genotype. However, in that case, host-derived, radioresistant, long-lived lymphoid cells cannot be distinguished from competitor-derived cells, and may complicate the analysis of donor/competitor ratios within the lymphoid lineage (Fig. 23.1). It is then recommended to analyze T cell development in the thymus, instead of peripheral T cells. Assuming similar stem cell function, the ratio of donor and competitor cells in the peripheral blood of the recipients is a reflection of the ratio of the number of repopulating stem cells in the two competing sources of bone marrow cells (Harrison *et al.*, 1993). This analysis can be refined by gating on specific lineages (T cells, B cell, or myeloid cells) to examine lineage-specific variation in the repopulation potential (Fig. 23.1). As lymphocytes are long-lived, the reconstitution of neutrophils, which have a very short half-life, is generally considered the best measure of HSC activity (Wright *et al.*, 2001). Indeed, as shown in the representative example in Fig. 23.1, even 16 weeks after transplantation, a significant fraction of T cells are still of host origin, whereas all myeloid cells are of donor or

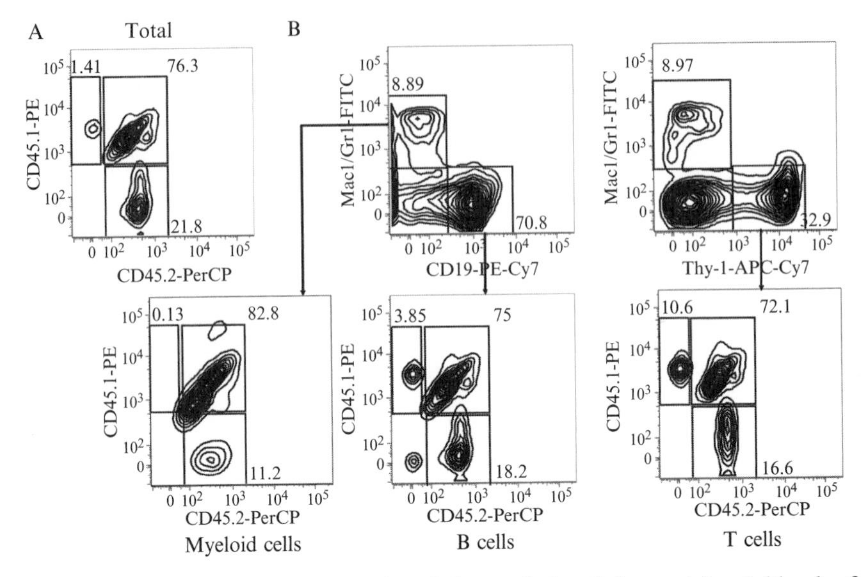

Figure 23.1 Representative example of the analysis of the peripheral blood of $CD45.1^+$ B6.SLJ congenic mice transplanted with $CD45.2^+$ B6 and $CD45.1^+CD45.2^+$ F1 mice 16 weeks earlier. (A) Analysis of the total white blood cell population. (B) Analysis gated on myeloid cells ($Mac1/Gr1^+$), B cells ($CD19^+$), and T cells ($Thy1^+$). Host contribution is the highest among the long-lived T cell population and is absent in the myeloid fraction.

competitor origin. It is important to note that in this experimental design, it is essential that both donor populations under study (e.g., cells from a knockout and from wt controls) be competed with the same competitor cells, and not against each other, and are transplanted into recipients of the same CD45 allotype. The reason for this strategy is that HSC function in $CD45.2^+$ mice might be slightly different than in the backcrossed $CD45.1^+$ B6.SJL-*Ptprca*$^{Pep3b/BoyJ}$ mice. In fact, recent evidence suggests that several alleles in the remaining interval surrounding the CD45 gene in B6.SJL-*Ptprca*$^{Pep3b/BoyJ}$ mice that are still of SLJ origin do affect hematopoiesis (Waterstrat *et al.*, 2010). The difference between the CD45 alleles may also be immunogenic (Bhattacharya *et al.*, 2006; van Os *et al.*, 2001). While the host is rendered immune deficient by lethal irradiation, contaminating T cells in the donor and competitor populations can still respond to each other's antigenic differences when donor and competitor differ in CD45 allotype.

HSCs cannot be expanded in culture using cytokines (Metcalf, 2007, 2008). Hence *in vitro* culture will typically compromise repopulation capacity. This is particularly relevant for experiments involving retroviral transduction, where preculture in a mixture of cytokines enhances transduction of repopulating HSCs (Luskey *et al.*, 1992; Robbins *et al.*, 1998). To test the effect of transduced genes on HSC function, transduced bone marrow cells should be competed either with a much smaller number of fresh bone marrow cells, with bone marrow cells which have undergone serial transplantation, or, preferably, with an equal number of cells of a different CD45 allotype transduced with a control vector. The appropriate control in this type of experiments is then mice where both donor populations differ in CD45 allele, but have been transduced with the same control vector.

When injected in limiting numbers, extensive HSC renewal is required to sustain long-term multilineage hematopoiesis. However, the most rigorous test for self-renewal is serial transplantation. Here, bone marrow cells from reconstituted animals are transferred to lethally irradiated secondary recipients. Populations or samples containing HSCs with limited renewal capacity will have a decreased contribution in the secondary hosts compared to the primary hosts. The simplest way to perform serial transplantation is to inject 10^6 cells from the primary, competitively repopulated host into secondary hosts. Alternatively, specific populations can be purified prior to serial transplantation. Serial transplantation is best performed 4 months after reconstitution of the primary host, as at that time, hematopoiesis has returned to steady state (Cheshier *et al.*, 1999). Typically, several secondary hosts are injected with cells from one primary host. After 2–4 months, the contribution of the donor cells to the secondary host is measured in the peripheral blood, spleen, and thymus, and is compared to the contribution of the same donor cells in the primary host. A difference between the contributions of two donor cell populations compared to the competitor cells reflects a difference in the capacity of HSC to renew in the primary recipients, or to withstand proliferative stress of serial transplantation. A caveat in the interpretation of these data is that the change in the percentage contribution of donor cells in primary compared to secondary hosts is not necessarily a good measure of a shift in reconstitution capacity in secondary recipients. When the contribution of the donors in the primary host is very skewed, data comparing reconstitution in primary and secondary recipients are best log transformed. A 5% change around 50% does not reflect a major change in HSC function, whereas 5% change around 95% (in case repopulation was already very skewed in the primary hosts) actually represents a large change HSC function caused by serial transplantation. A better measure is the ratio between donor/competitor ratio pre- (input) and post (output)-secondary reconstitution. By analogy with the way ratiometric data are handled in the analysis of cDNA microarrays, the difference between log donor/competitor ratios in primary and secondary recipients can be used. An advantage of log transformation is that a ratio smaller than 1 will give a negative value, and negative ratios will extend over the same numerical ranges as positive ones (e.g., a ratio of 0.01 gives a log ratio of −2, a ratio of 100 gives a log ratio of +2), thus normalizing the data. Accordingly, data are best presented as the difference between input log donor/competitor (i.e., the log ratio in the primary recipient) and the output log donor/competitor (i.e., the log ratio 3 months after reconstitution of the secondary recipients), and is referred to as Δ(log ratio) (Langer *et al.*, 2004).

To more rigorously quantify the number of repopulating HSCs, competitive transplantation can be performed in a limited dilution setting. Alternatively, candidate HSC populations can be transplanted as single cells. In the latter approach, single candidate HSCs are mixed with 2×10^5 bone marrow cells expressing a different CD45 allelic variant (Ema *et al.*, 2006; Osawa *et al.*, 1996). The helper bone marrow cells provide short-term radioprotection, allowing the single injected HSC to expand and reconstitute long-term multilineage hematopoiesis. The fraction of mice with detectable multilineage reconstitution is a reflection of the HSC frequency (also termed the frequency of competitive repopulation units, CRU). The limiting dilution approach is similar to that used to determine the frequency of antigen-specific T cells *in vitro*, and is based on the assumption of single hit kinetics and on a Poisson distribution for the frequency of "hits" (Hu and Smyth, 2009; Szilvassy *et al.*, 1990). Here, decreasing numbers of bone marrow cells or of enriched HSC populations are competed against the same competitor cells, for example, 2×10^5 unselected bone marrow cells. The log of frequency of mice without donor repopulation in each cohort is plotted against the number of cells injected. Interpolation of best fitting line at 37% of negative mice indicates the HSC frequency. This analysis is implemented in a Web-based application tailored specifically to the calculation of stem cell frequencies (http://bioinf.wehi.edu.au/software/elda/index.html). While informative, the data generated from these assays are often interpreted too literally. In a typical experiment, a 1% contribution to myeloid and lymphoid lineages is the cutoff to consider a recipient positive (Ema *et al.*, 2006). This is a rather arbitrary criterium. In the original publication describing limit dilution analysis of HSCs, for example, a 5% cutoff was used, but the cells were competed against 2×10^5 bone marrow cells that had undergone serial transplantation, and therefore had compromised HSC function (Szilvassy *et al.*, 1990). These discrepancies do not pose a problem when the difference in relative HSC content between two populations is compared, but it does matter when the absolute content of a phenotypically defined population is determined, and when data are compared between different investigators. A further consideration in this type of experiment is the fact that individual HSC can display an intrinsic myeloid or lymphoid bias. While in fetal and young mice, lymphoid-biased HSCs appear predominant, with age, myeloid-biased HSC are enriched (Cho *et al.*, 2008; Rossi *et al.*, 2005; Sudo *et al.*, 2000). HSCs that are intrinsically strongly biased toward one or the other lineage may not easily satisfy the criterium of 1% contribution to all lineages, yet they are multipotent and are capable of long-term maintenance of hematopoiesis and of self-renewal.

In the interpretation of functional assays of HSC, it is important to take into consideration the fact that the microenvironment, phenotype, and function of HSCs change throughout development and aging. Hematopoiesis in the early embryo begins around embryonic day 7 (E7) in two independent sites, the yolk sac and an intraembryonic region called the para-aortic splanchnopleura (P-Sp) which later develops into the aorta, gonads, and mesonephros (AGM) (Dzierzak and Speck, 2008; Orkin and Zon, 2008). The developmental programs in these two regions are quite different. The yolk sac generates primitive and later definitive (or "adult") erythrocytes, definitive macrophages, and mast cells (Palis *et al.*, 1999). However, cells capable of reconstituting the hematopoietic system of a lethally irradiated adult host, the defining characteristic of HSCs, are not generated in the yolk sac (Dzierzak and Speck, 2008). In contrast, the P-Sp/AGM mostly produces myeloid, lymphoid, definitive erythroid lineages, and HSCs, which can be fully detected around E11, but not primitive erythrocytes. Significant numbers of HSCs can also be detected between E10 and E13 in the placenta (Gekas *et al.*, 2005). The exact origin of HSC in the AGM region has been controversial for a long time. Recently, however, three papers provided convincing evidence that HSCs arise from the endothelium of the dorsal aorta of the mouse embryo (Chen *et al.*, 2009; Eilken *et al.*, 2009; Lancrin *et al.*, 2009). Of note is the fact that at these early stages of hematopoiesis, hematopoietic express CD41, and not CD45 (Mikkola *et al.*, 2003a). The next and dominant site of hematopoeisis in the embryo is the fetal liver. HSC activity can be detected here as early as E9 (Dzierzak and Speck, 2008; Orkin and Zon, 2008). Fetal liver hematopoiesis differs dramatically from adult hematopoiesis in the bone marrow, the final destination of most HSCs. These differences include higher cycling activity, higher competitive repopulation capacity (Rebel *et al.*, 1996), and production of different types of lymphoid cells such as $CD5^+$ B cells (B1) (Kantor and Herzenberg, 1993) and $V\gamma3/V\gamma4$ T cells (Ikuta *et al.*, 1990). Interestingly, even surface markers used to detect fetal HSCs are different from those in the adults (see below). A further functional change in hematopoiesis occurs postnatally. Between weeks 3 and 4 of postnatal life, most HSC become quiescent (Bowie *et al.*, 2007). The temporal and stage-specific regulation of hematopoiesis from the yolk sac to bone marrow is controlled at least in part by intrinsic developmental changes in HSCs. For example, while *Runx1* and *Scl1* are required for the establishment of fetal hematopoiesis in the AGM, they are dispensable for the function of adult HSCs (Chen *et al.*, 2009; Mikkola *et al.*, 2003b). On the other hand, *Sox17* is essential for fetal liver, but not adult HSCs (Kim *et al.*, 2007), while *Bmi1* and *Gfi-1* (Hock *et al.*, 2004; Zeng *et al.*, 2004) are critical for adult hematopoiesis.

3. Isolation of Hematopoietic Stem Cells

The development of fluorescence-activated cell sorting (FACS) made it possible to home in on the likely phenotype of HSCs and to allow their prospective isolation, based on the capacity of single cells from immunophenotypically defined populations to competitively repopulate a lethally irradiated recipient. Currently available high-speed cell sorters with up to five lasers allow the isolation of rare cells based on multiple intracellular and surface membrane markers. High sorting speeds (25,000 s^{-1}) are achieved by using high sheath pressures. At these high pressures, shear stress at the nozzle unit causes significant cell death. We find that we get the best recovery of stem cells, as measured in competitive repopulation assays using lower pressures (<35 p.s.i.) and consequently sorting speeds (12,000–15,000 s^{-1}) as well as wider nozzle opening (100 μ instead of 70 μ).

Murine HSCs are enriched in a population characterized by the absence of lineage-specific markers for mature lymphoid, myeloid, and erythroid cells ($lineage^{-}$), and the high expression of stem cell antigen-1 (Sca1) and CD117 (c-Kit), referred to as $Lin^{-}Sca1^{++}Kit^{+}$ or LSK cells (Ikuta and Weissman, 1992; Morrison and Weissman, 1994; Ogawa *et al.*, 1991; Spangrude *et al.*, 1988) (Fig. 23.2). A typical cocktail of lineage antigens should cover all hematopoietic lineages, and would include CD2, CD3, CD4, CD8 (T cell, NK cell), NK1.1 (NK cells), CD19, B220 (B cells), Gr1, Mac1 (myeloid cells), Ter119 (erythroid cells). Of note, HSCs in fetal liver have been reported to express Mac1 however (Morrison *et al.*, 1995). An important caveat is that while in the prototypical mouse strain used in hematopoiesis research, C57BL/6, in addition to DBA/2, 129, and AKR, all HSCs express Sca1, in other mouse strains, such as Balb/c, C3H, and CBA, only a minor fraction of HSCs express Sca1 (Spangrude and Brooks, 1993). The LSK population also includes short-term HSCs and MPPs (Kondo *et al.*, 2003). It was then discovered that the expression of Flt3 on differentiating HSCs was associated with a loss of renewal potential. Furthermore, mouse HSCs were shown to be negative for CD34. Combining these two markers within the LSK fraction leads three distinct populations (Fig. 23.2): LT-HSCs are $CD34^{lo}Flt3^{lo/-}$, ST-HSCs are $CD34^{hi}Flt3^{lo/+}$, and MPPs are $CD34^{hi}Flt3^{hi}$ (Adolfsson *et al.*, 2001; Christensen and Weissman, 2001; Osawa *et al.*, 1996; Yang *et al.*, 2005). It is important to note, however, that before the age of 5 weeks, when HSCs are more actively cycling, all HSCs do express CD34, while by the age of 10 weeks, most HSCs are $CD34^{-}$ (Ito *et al.*, 2000). More recently, Kiel *et al.* (2005) identified a group of SLAM family receptors that are differentially expressed on HSC and progenitor cells. HSC are highly enriched in the $CD150^{+}CD48^{-}$ population, which constitutes less than 0.01% of the total

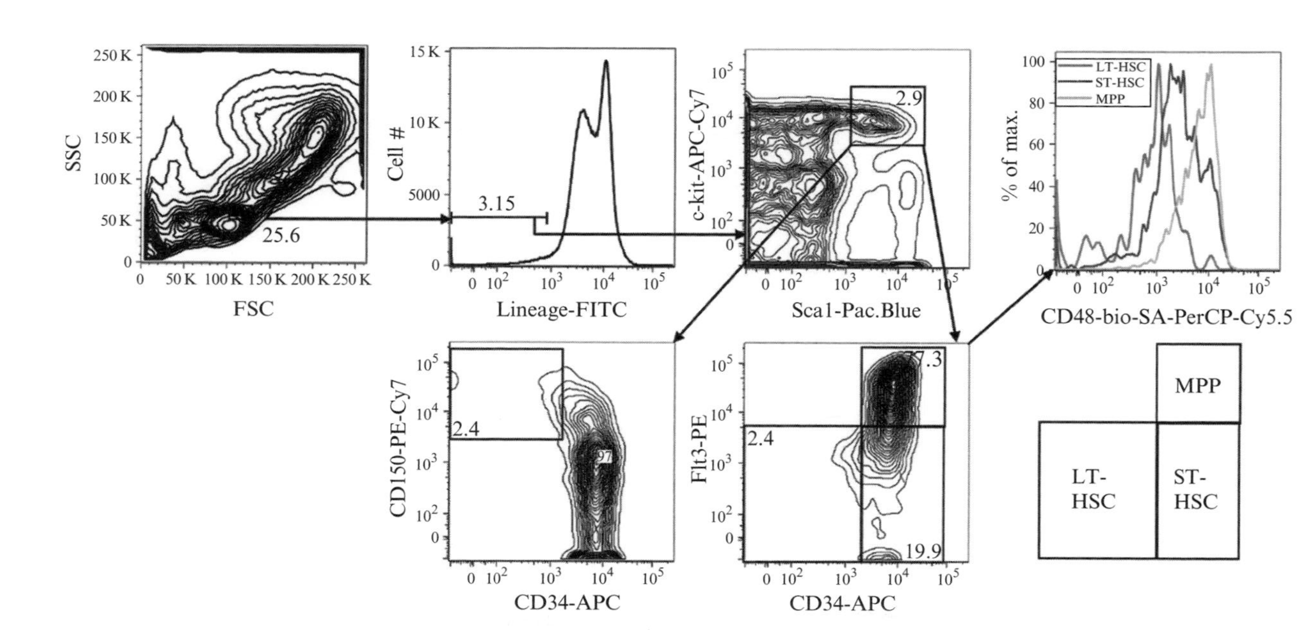

Figure 23.2 Representative example of the identification of HSC in mouse bone marrow. Within the LSK fraction, the $CD34^{-}$ phenotype identifies $Flt3^{-}CD150^{+}CD48^{lo}$ HSCs.

bone marrow cells. Twenty-one percent of $CD150^+CD48^-$ cells injected intravenously gave long-term reconstitution in lethally irradiated mice. As this phenotype also includes megakaryocytes, negativity for CD41 is often included as a criterium. SLAM markers have since been used to isolate HSC-enriched populations, but not without lineage exclusion by mature cell markers (Kim *et al.*, 2007; Pronk *et al.*, 2007). In a population defined by the combination of SLAM and LSK markers, $CD150^+CD48^-$LSK cells, 47% of cells were long-term repopulating HSC, while single-cell reconstitution efficiencies of 20–40% have been reported for $CD34^-$LSK cells (Ema *et al.*, 2006; Osawa *et al.*, 1996). A head-to-head comparison of the repopulation efficiencies of cells defined by these phenotypes has not been reported yet. However, virtually all $CD34^-$LSK cells are $CD150^+$, $Flt3^-$ and express very low levels of CD48 (Fig. 23.2), suggesting that the $CD150^+CD48^-$LSK and the $CD34^-$LSK identify largely the same population. Therefore, it can be argued that a stain for CD34, lineage markers, Sca1 and c-Kit is sufficient to define the most homogenous HSC population currently possible (Ema *et al.*, 2006; Wilson *et al.*, 2008) (see Fig. 23.2). It is interesting to note that this state-of-the-art for HSC identification and isolation was already attained more than a decade ago by Osawa *et al.* (1996). Nonetheless, there is still functional heterogeneity within the $LSKCD34^-CD150^+CD48^-$ fraction. Wilson *et al.* (2008) have shown that this population consists of a minor, largely dormant fraction, and a larger fraction that divides, though only rarely, and expresses CD34 mRNA. The only way to detect this fraction, which appears to contain most LT-HSC activity, is to use histone2B-GFP reporter mice, originally developed by Tumbar *et al.* (2004). When H2B-GFP is expressed in a doxycyclin-dependent fashion and the tTA is expressed from a hematopoietic-specific promoter, such as *Scl*, the expression of H2B-GFP is shut down by treatment with doxycyclin. After a long chase period, only cells that have not divided retain the label. These cells appear to be the most primitive, dormant HSCs. There is currently no surface marker available to distinguish these two subpopulations within the $LSKCD34^-CD150^+CD48^-$ fraction however.

Less commonly used methods of HSC activity enrichment include isolation of cells by their unique ability to efflux the DNA-binding dye Hoechst 33342 (Goodell *et al.*, 1996; reviewed in Challen *et al.*, 2009). When bone marrow cells are incubated with this membrane permeable, DNA-binding dye at 37 °C, it was found that the more primitive cells actively effluxed the dye, a process that can be blocked by the Ca^{2+} channel antagonist verapamil. This efflux results in a so-called side population (SP), when blue versus red Hoechst fluorescence is plotted. Furthermore, cells in the tip of the SP tend to be more undifferentiated than those closer to the main population (MP). However, one study subsequently showed that while the SP is highly enriched for HSCs, approximately half of the most primitive HSCs, defined by the CD34$^-$LSK phenotype were still within the MP (Morita *et al.*, 2006). Furthermore, slight variations in staining conditions or different batches of Hoechst may profoundly affect the data (Pearce *et al.*, 2004).

Other characteristics of HSCs used for their flow cytometric identification include low staining with the mitochondrial dye Rho123 (Visser *et al.*, 1984), and expression of Tie2 (Arai *et al.*, 2004), CD105 (Chen *et al.*, 2003), endothelial protein C receptor (EPCR) (Balazs *et al.*, 2006; Kent *et al.*, 2009), and aldehyde dehydrogenase 1a1 (Armstrong *et al.*, 2004). The latter enzymatic activity can be detected using a commercially available fluorometric enzymatic assay.

Quiescence of HSCs is critical for their long-term maintenance as in several models increased HSC cycling was associated with premature exhaustion of their function. Therefore, decreased competitive serial repopulation capacity can be explained by either an intrinsically lower cell cycle activity, or by enhanced cycling followed by exhaustion. Example of the latter situation include deletion of p21 (Cheng *et al.*, 2000), Gfi1 (Hock *et al.*, 2004; Zeng *et al.*, 2004), and c-mpl (Qian *et al.*, 2007; Yoshihara *et al.*, 2007). Tumor suppressors, on the other hand, tend to limit expansion of HSCs. Simultaneous deletion of p16Ink4, p19Arf, and p53 leads to expansion of MPPs, which adopt functional characteristics of HSCs in that they provide long-term reconstitution activity (Akala *et al.*, 2008). The polycomb family member, Bmi-1, represses these tumor suppressors and, consequently, deletion of Bmi1 leads to a severe decrease in HSCs (Iwama *et al.*, 2004; Park *et al.*, 2003). Assessment of the cell cycle status of HSCs is therefore important for the interpretation of serial reconstitution experiments. *In vivo* analyses of the stem and progenitor cell compartment using labeling with BrdU showed that no more than 9% of the long-term repopulating cells actively cycle in steady-state conditions (Cheshier *et al.*, 1999). However, it has been shown that myelotoxicity of BrdU may recruit HSCs into cell cycle, leading to an overestimation of the proliferating HSCs fraction when using continuous BrdU labeling (Wilson *et al.*, 2008). Currently, the best way to estimate the proliferating fraction is a combination of staining with Pyronin Y and Hoecht3342. PY stains only RNA after DNA has been blocked in fixed cells with the DNA-binding dye Hoechst 33342 (Shapiro, 1981). In this way, it is possible to plot a cell cycle histogram based on Hoechst staining against RNA content. Cells with 2n DNA that have low PY staining are considered to be in G0. The G0 fraction decreases as cells differentiate from LT-HSCs over ST-HSCs to MPP (Fig. 23.3). A similar approach is staining for the nuclear antigen Ki67 (Gerdes *et al.*, 1983). 2n cells that are negative for Ki67 are in the G0 phase of the cell's cycle.

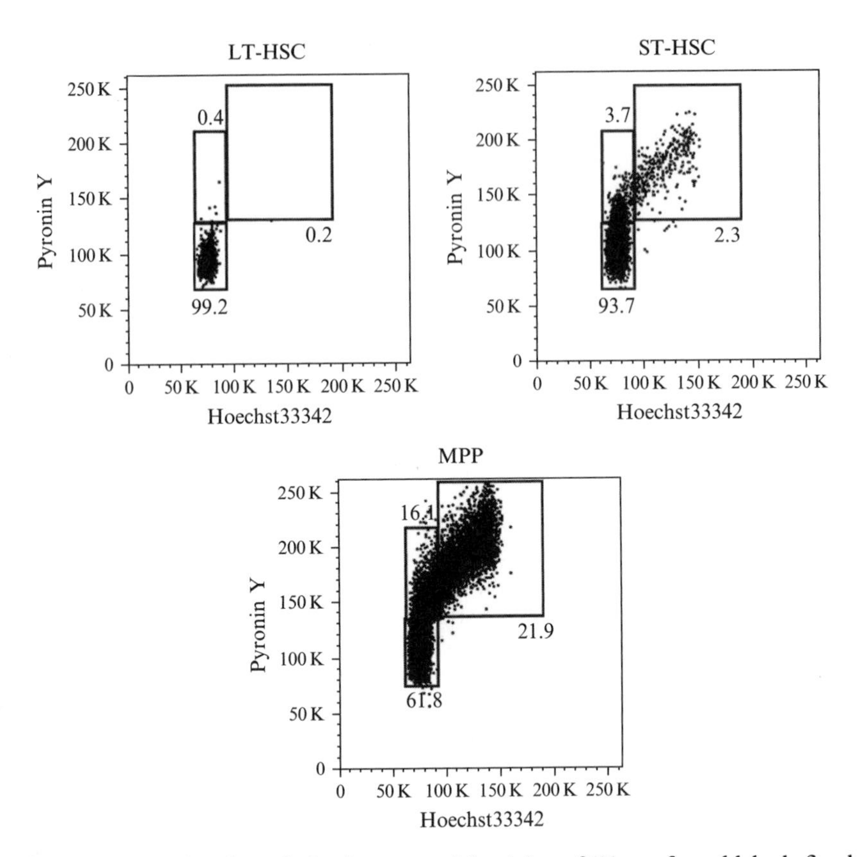

Figure 23.3 Cell cycle analysis after sequential staining of 4% paraformaldehyde fixed with Hoechst 33342 and Pyronin Y, gated on LT-HSCs, ST-HSCs, and MPP (see Fig. 23.2 for representative gates defining these populations).

REFERENCES

Adolfsson, J., Borge, O. J., Bryder, D., Theilgaard-Monch, K., Astrand-Grundstrom, I., Sitnicka, E., Sasaki, Y., and Jacobsen, S. E. (2001). Upregulation of Flt3 expression within the bone marrow Lin(−)Sca1(+)c-kit(+) stem cell compartment is accompanied by loss of self-renewal capacity. *Immunity* **15,** 659–669.

Adolfsson, J., Månsson, R., Buza-Vidas, N., Hultquist, A., Liuba, K., Jensen, C. T., Bryder, D., Yang, L., Borge, O. J., Thoren, L. A., Anderson, K., Sitnicka, E., *et al.* (2005). Identification of Flt3+ lympho-myeloid stem cells lacking erythro-megakaryocytic potential a revised road map for adult blood lineage commitment. *Cell* **121,** 295–306.

Akala, O. O., Park, I. K., Qian, D., Pihalja, M., Becker, M. W., and Clarke, M. F. (2008). Long-term haematopoietic reconstitution by Trp53-/-p16Ink4a-/-p19Arf-/- multipotent progenitors. *Nature* **453,** 228–232.

Allsopp, R. C., Morin, G. B., DePinho, R., Harley, C. B., and Weissman, I. L. (2003a). Telomerase is required to slow telomere shortening and extend replicative lifespan of HSCs during serial transplantation. *Blood* **102,** 517–520.

Allsopp, R. C., Morin, G. B., Horner, J. W., DePinho, R., Harley, C. B., and Weissman, I. L. (2003b). Effect of TERT over-expression on the long-term transplantation capacity of hematopoietic stem cells. *Nat. Med.* **9,** 369–371.

Arai, F., Hirao, A., Ohmura, M., Sato, H., Matsuoka, S., Takubo, K., Ito, K., Koh, G. Y., and Suda, T. (2004). Tie2/angiopoietin-1 signaling regulates hematopoietic stem cell quiescence in the bone marrow niche. *Cell* **118,** 149–161.

Armstrong, L., Stojkovic, M., Dimmick, I., Ahmad, S., Stojkovic, P., Hole, N., and Lako, M. (2004). Phenotypic characterization of murine primitive hematopoietic progenitor cells isolated on basis of aldehyde dehydrogenase activity. *Stem Cells* **22,** 1142–1151.

Balazs, A. B., Fabian, A. J., Esmon, C. T., and Mulligan, R. C. (2006). Endothelial protein C receptor (CD201) explicitly identifies hematopoietic stem cells in murine bone marrow. *Blood* **107,** 2317–2321.

Bell, J. J., and Bhandoola, A. (2008). The earliest thymic progenitors for T cells possess myeloid lineage potential. *Nature* **452,** 764–767.

Bhandoola, A., von Boehmer, H., Petrie, H. T., and Zúñiga-Pflücker, J. C. (2007). Commitment and developmental potential of extrathymic and intrathymic T cell precursors: Plenty to choose from. *Immunity* **26,** 678–689.

Bhattacharya, D., Rossi, D. J., Bryder, D., and Weissman, I. L. (2006). Purified hematopoietic stem cell engraftment of rare niches corrects severe lymphoid deficiencies without host conditioning. *J. Exp. Med.* **203,** 73–85.

Bhattacharya, D., Ehrlich, L. I., and Weissman, I. L. (2008). Space-time considerations for hematopoietic stem cell transplantation. *Eur. J. Immunol.* **38,** 2060–2067.

Bowie, M. B., Kent, D. G., Dykstra, B., McKnight, K. D., McCaffrey, L., Hoodless, P. A., and Eaves, C. J. (2007). Identification of a new intrinsically timed developmental checkpoint that reprograms key hematopoietic stem cell properties. *Proc. Natl. Acad. Sci. USA* **104,** 5878–5882.

Calvi, L. M., Adams, G. B., Weibrecht, K. W., Weber, J. M., Olson, D. P., Knight, M. C., Martin, R. P., Schipani, E., Divieti, P., Bringhurst, F. R., *et al.* (2003). Osteoblastic cells regulate the haematopoietic stem cell niche. *Nature* **425,** 841–846.

Challen, G. A., Boles, N., Lin, K. K., and Goodell, M. A. (2009). Mouse hematopoietic stem cell identification and analysis. *Cytometry A* **75,** 14–24.

Chao, M. P., Seita, J., and Weissman, I. L. (2008). Establishment of a normal hematopoietic and leukemia stem cell hierarchy. *Cold Spring Harb. Symp. Quant. Biol.* **73,** 439–449.

Chen, C. Z., Li, L., Li, M., and Lodish, H. F. (2003). The endoglin(positive) sca-1(positive) rhodamine(low) phenotype defines a near-homogeneous population of long-term repopulating hematopoietic stem cells. *Immunity* **19,** 525–533.

Chen, M. J., Yokomizo, T., Zeigler, B. M., Dzierzak, E., and Speck, N. A. (2009). Runx1 is required for the endothelial to haematopoietic cell transition but not thereafter. *Nature* **457,** 887–891.

Cheng, T., Rodrigues, N., Shen, H., Yang, Y., Dombkowski, D., Sykes, M., and Scadden, D. T. (2000). Hematopoietic stem cell quiescence maintained by p21cip1/waf1. *Science* **287,** 1804–1808.

Cheshier, S. H., Morrison, S. J., Liao, X., and Weissman, I. L. (1999). In vivo proliferation and cell cycle kinetics of long-term self-renewing hematopoietic stem cells. *Proc. Natl. Acad. Sci. USA* **96,** 3120–3125.

Cho, R. H., Sieburg, H. B., and Muller-Sieburg, C. E. (2008). A new mechanism for the aging of hematopoietic stem cells: Aging changes the clonal composition of the stem cell compartment but not individual stem cells. *Blood* **111,** 5553–5561.

Christensen, J. L., and Weissman, I. L. (2001). Flk-2 is a marker in hematopoietic stem cell differentiation: A simple method to isolate long-term stem cells. *Proc. Natl. Acad. Sci. USA* **98,** 14541–14546.

Czechowicz, A., Kraft, D., Weissman, I. L., and Bhattacharya, D. (2007). Efficient transplantation via antibody-based clearance of hematopoietic stem cell niches. *Science* **318,** 1296–1299.

Dzierzak, E., and Speck, N. A. (2008). Of lineage and legacy: The development of mammalian hematopoietic stem cells. *Nat. Immunol.* **9,** 129–136.

Eilken, H. M., Nishikawa, S., and Schroeder, T. (2009). Continuous single-cell imaging of blood generation from haemogenic endothelium. *Nature* **457,** 896–900.

Ema, H., Morita, Y., Yamazaki, S., Matsubara, A., Seita, J., Tadokoro, Y., Kondo, H., Takano, H., and Nakauchi, H. (2006). Adult mouse hematopoietic stem cells: Purification and single-cell assays. *Nat. Protoc.* **1,** 2979–2987.

Forsberg, E. C., Serwold, T., Kogan, S., Weissman, I. L., and Passegué, E. (2006). New evidence supporting megakaryocyte-erythrocyte potential of flk2/flt3+ multipotent hematopoietic progenitors. *Cell* **126,** 415–426.

Gekas, C., Dieterlen-Lievre, F., Orkin, S. H., and Mikkola, H. K. (2005). The placenta is a niche for hematopoietic stem cells. *Dev. Cell* **8,** 365–375.

Gerdes, J., Schwab, U., Lemke, H., and Stein, H. (1983). Production of a mouse monoclonal antibody reactive with a human nuclear antigen associated with cell proliferation. *Int. J. Cancer* **31,** 13–20.

Goodell, M. A., Brose, K., Paradis, G., Conner, A. S., and Mulligan, R. C. (1996). Isolation and functional properties of murine hematopoietic stem cells that are replicating in vivo. *J. Exp. Med.* **183,** 1797–1806.

Harrison, D. E. (1980). Competitive repopulation: A new assay for long-term stem cell functional capacity. *Blood* **55,** 77–81.

Harrison, D. E., Jordan, C. T., Zhong, R. K., and Astle, C. M. (1993). Primitive hemopoietic stem cells: Direct assay of most productive populations by competitive repopulation with simple binomial, correlation and covariance calculations. *Exp. Hematol.* **21,** 206–219.

Hock, H., Hamblen, M. J., Rooke, H. M., Schindler, J. W., Saleque, S., Fujiwara, Y., and Orkin, S. H. (2004). Gfi-1 restricts proliferation and preserves functional integrity of haematopoietic stem cells. *Nature* **431,** 1002–1007.

Hodgson, G. S., and Bradley, T. R. (1979). Properties of haematopoietic stem cells surviving 5-fluorouracil treatment: Evidence for a pre-CFU-S cell? *Nature* **281,** 381–382.

Hu, Y., and Smyth, G. K. (2009). ELDA: Extreme limiting dilution analysis for comparing depleted and enriched populations in stem cell and other assays. *J. Immunol. Methods* **347,** 70–78.

Ikuta, K., and Weissman, I. L. (1992). Evidence that hematopoietic stem cells express mouse c-kit but do not depend on steel factor for their generation. *Proc. Natl. Acad. Sci. USA* **89,** 1502–1506.

Ikuta, K., Kina, T., MacNeil, I., Uchida, N., Peault, B., Chien, Y. H., and Weissman, I. L. (1990). A developmental switch in thymic lymphocyte maturation potential occurs at the level of hematopoietic stem cells. *Cell* **62,** 863–874.

Ito, T., Tajima, F., and Ogawa, M. (2000). Developmental changes of CD34 expression by murine hematopoietic stem cells. *Exp. Hematol.* **28,** 1269–1273.

Iwama, A., Oguro, H., Negishi, M., Kato, Y., Morita, Y., Tsukui, H., Ema, H., Kamijo, T., Katoh-Fukui, Y., Koseki, H., van Lohuizen, M., and Nakauchi, H. (2004). Enhanced self-renewal of hematopoietic stem cells mediated by the polycomb gene product Bmi-1. *Immunity* **21,** 843–851.

Jacobson, L. O., Simmons, E. L., Marks, E. K., Robson, M. J., Bethard, W. F., and Gaston, E. O. (1950). The role of the spleen in radiation injury and recovery. *J. Lab. Clin. Med.* **35,** 746–770.

Jones, R. J., Wagner, J. E., Celano, P., Zicha, M. S., and Sharkis, S. J. (1990). Separation of pluripotent haematopoietic stem cells from spleen colony-forming cells. *Nature* **347,** 188–189.

Kantor, A. B., and Herzenberg, L. A. (1993). Origin of murine B cell lineages. *Annu. Rev. Immunol.* **11,** 501–538.

Kiel, M. J., Yilmaz, O. H., Iwashita, T., Terhorst, C., and Morrison, S. J. (2005). SLAM family receptors distinguish hematopoietic stem and progenitor cells and reveal endothelial niches for stem cells. *Cell* **121,** 1109–1121.

Kim, I., Saunders, T. L., and Morrison, S. J. (2007). Sox17 dependence distinguishes the transcriptional regulation of fetal from adult hematopoietic stem cells. *Cell* **130,** 470–483.

Kondo, M., Wagers, A. J., Manz, M. G., Prohaska, S. S., Scherer, D. C., Beilhack, G. F., Shizuru, J. A., and Weismann, I. L. (2003). Biology of hematopoietic stem cells and progenitors: Implications for clinical application. *Annu. Rev. Immunol.* **21,** 759–806.

Lancrin, C., Sroczynska, P., Stephenson, C., Allen, T., Kouskoff, V., and Lacaud, G. (2009). The haemangioblast generates haematopoietic cells through a haemogenic endothelium stage. *Nature* **457,** 892–895.

Langer, J. C., Henckaerts, E., Orenstein, J., and Snoeck, H. W. (2004). Quantitative trait analysis reveals transforming growth factor-beta2 as a positive regulator of early hematopoietic progenitor and stem cell function. *J. Exp. Med.* **199,** 5–14.

Lemischka, I. R. (1991). Clonal, in vivo behavior of the totipotent hematopoietic stem cell. *Semin. Immunol.* **3,** 349–355.

Lemischka, I. R., Raulet, D. H., and Mulligan, R. C. (1986). Developmental potential and dynamic behavior of hematopoietic stem cells. *Cell* **45,** 917–927.

Lorenz, E., Uphoff, D., Reid, T. R., and Shelton, E. (1951). Modification of irradiation injury in mice and guinea pigs by bone marrow injections. *J. Natl. Cancer Inst.* **12,** 197–201.

Luc, S., Buza-Vidas, N., and Jacobsen, S. E. (2008). Delineating the cellular pathways of hematopoietic lineage commitment. *Semin. Immunol.* **20,** 213–220.

Luskey, B. D., Rosenblatt, M., Zsebo, K., and Williams, D. A. (1992). Stem cell factor, interleukin-3, and interleukin-6 promote retroviral-mediated gene transfer into murine hematopoietic stem cells. *Blood* **80,** 396–402.

Metcalf, D. (2007). On hematopoietic stem cell fate. *Immunity* **26,** 669–673.

Metcalf, D. (2008). Hematopoietic cytokines. *Blood* **111,** 485–491.

Mikkola, H. K., Fujiwara, Y., Schlaeger, T. M., Traver, D., and Orkin, S. H. (2003a). Expression of CD41 marks the initiation of definitive hematopoiesis in the mouse embryo. *Blood* **101,** 508–516.

Mikkola, H. K., Klintman, J., Yang, H., Hock, H., Schlaeger, T. M., Fujiwara, Y., and Orkin, S. H. (2003b). Haematopoietic stem cells retain long-term repopulating activity and multipotency in the absence of stem-cell leukaemia SCL/tal-1 gene. *Nature* **421,** 547–551.

Moore, K. A., and Lemischka, I. R. (2006). Stem cells and their niches. *Science* **311,** 1880–1885.

Morita, Y., Ema, H., Yamazaki, S., and Nakauchi, H. (2006). Non-side-population hematopoietic stem cells in mouse bone marrow. *Blood* **108,** 2850–2856.

Morrison, S. J., and Weissman, I. L. (1994). The long-term repopulating subset of hematopoietic stem cells is deterministic and isolatable by phenotype. *Immunity* **1,** 661–673.

Morrison, S. J., Hemmati, H. D., Wandycz, A. M., and Weissman, I. L. (1995). The purification and characterization of fetal liver hematopoietic stem cells. *Proc. Natl. Acad. Sci. USA* **92,** 10302–10306.

Ogawa, M., Matsuzaki, Y., Nishikawa, S., Hayashi, S., Kunisada, T., Sudo, T., Kina, T., Nakauchi, H., and Nishikawa, S. (1991). Expression and function of c-kit in hemopoietic progenitor cells. *J. Exp. Med.* **174,** 63–71.

Orkin, S. H., and Zon, L. I. (2008). Hematopoiesis: An evolving paradigm for stem cell biology. *Cell* **132,** 631–644.

Osawa, M., Hanada, K., Hamada, H., and Nakauchi, H. (1996). Long-term lymphohematopoietic reconstitution by a single CD34-low/negative hematopoietic stem cell. *Science* **273,** 242–245.

Palis, J., Robertson, S., Kennedy, M., Wall, C., and Keller, G. (1999). Development of erythroid and myeloid progenitors in the yolk sac and embryo proper of the mouse. *Development* **126,** 5073–5084.

Pallavicini, M. G., Redfearn, W., Necas, E., and Brecher, G. (1997). Rescue from lethal irradiation correlates with transplantation of 10–20 CFU-S-day 12. *Blood Cells Mol. Dis.* **23,** 157–168.

Park, I. K., Qian, D., Kiel, M., Becker, M. W., Pihalja, M., Weissman, I. L., Morrison, S. J., and Clarke, M. F. (2003). Bmi-1 is required for maintenance of adult self-renewing haematopoietic stem cells. *Nature* **423,** 302–305.

Pearce, D. J., Ridler, C. M., Simpson, C., and Bonnet, D. (2004). Multiparameter analysis of murine bone marrow side population cells. *Blood* **103,** 2541–2546.

Ploemacher, R. E., and Brons, R. H. (1989). Separation of CFU-S from primitive cells responsible for reconstitution of the bone marrow hemopoietic stem cell compartment following irradiation: Evidence for a pre-CFU-S cell. *Exp. Hematol.* **17,** 263–266.

Pronk, C. J., Rossi, D. J., Mansson, R., Attema, J. L., Norddahl, G. L., Chan, C. K., Sigvardsson, M., Weissman, I. L., and Bryder, D. (2007). Elucidation of the phenotypic, functional, and molecular topography of a myeloerythroid progenitor cell hierarchy. *Cell Stem Cell* **1,** 428–442.

Qian, H., Buza-Vidas, N., Hyland, C. D., Jensen, C. T., Antonchuk, J., Mansson, R., Thoren, L. A., Ekblom, M., Alexander, W. S., and Jacobsen, S. E. (2007). Critical role of thrombopoietin in maintaining adult quiescent hematopoietic stem cells. *Cell Stem Cell* **1,** 671–684.

Rebel, V. I., Miller, C. L., Eaves, C. J., and Lansdorp, P. M. (1996). The repopulation potential of fetal liver hematopoietic stem cells in mice exceeds that of their liver adult bone marrow counterparts. *Blood* **87,** 3500–3507.

Robbins, P. B., Skelton, D. C., Yu, X. J., Halene, S., Leonard, E. H., and Kohn, D. B. (1998). Consistent, persistent expression from modified retroviral vectors in murine hematopoietic stem cells. *Proc. Natl. Acad. Sci. USA* **95,** 10182–10187.

Rossi, D. J., Bryder, D., Zahn, J. M., Ahlenius, H., Sonu, R., Wagers, A. J., and Weissman, I. L. (2005). Cell intrinsic alterations underlie hematopoietic stem cell aging. *Proc. Natl. Acad. Sci. USA* **102,** 9194–9199.

Schofield, R. (1978). The relationship between the spleen colony-forming cell and the haemopoietic stem cell. *Blood Cells* **4,** 7–25.

Shapiro, H. M. (1981). Flow cytometric estimation of DNA and RNA content in intact cells stained with Hoechst 33342 and Pyronin Y. *Cytometry* **2**(143–150), 1981.

Shiku, H., Kisielow, P., Bean, M. A., Takahashi, T., Boyse, E. A., Oettgen, H. F., and Old, L. J. (1975). Expression of T-cell differentiation antigens on effector cells in cell-mediated cytotoxicity in vitro. Evidence for functional heterogeneity related to the surface phenotype of T cells. *J. Exp. Med.* **141,** 227–241.

Spangrude, G. J., and Brooks, D. M. (1993). Mouse strain variability in the expression of the hematopoietic stem cell antigen Ly-6A/E by bone marrow cells. *Blood* **82,** 3327–3332.

Spangrude, G. J., Heimfeld, S., and Weissman, I. L. (1988). Purification and characterization of mouse hematopoietic stem cells. *Science* **241,** 58–62.

Spangrude, G. J., Brooks, D. M., and Tumas, D. B. (1995). Long-term repopulation of irradiated mice with limiting numbers of purified hematopoietic stem cells: In vivo expansion of stem cell phenotype but not function. *Blood* **85,** 1006–1016.

Sudo, K., Ema, H., Morita, Y., and Nakauchi, H. (2000). Age-associated characteristics of murine hematopoietic stem cells. *J. Exp. Med.* **192,** 1273–1280.

Sugiyama, T., Kohara, H., Noda, M., and Nagasawa, T. (2006). Maintenance of the hematopoietic stem cell pool by CXCL12-CXCR4 chemokine signaling in bone marrow stromal cell niches. *Immunity* **25,** 977–988.

Sykes, M., Szot, G. L., Swenson, K. A., and Pearson, D. A. (1997). Induction of high levels of allogeneic hematopoietic reconstitution and donor-specific tolerance without myelosuppressive conditioning. *Nat. Med.* **3,** 783–787.

Szilvassy, S. J., Humphries, R. K., Lansdorp, P. M., Eaves, A. C., and Eaves, C. J. (1990). Quantitative assay for totipotent reconstituting hematopoietic stem cells by a competitive repopulation strategy. *Proc. Natl. Acad. Sci. USA* **87,** 8736–8740.

Thomas, M. L. (1989). The leukocyte common antigen family. *Annu. Rev. Immunol.* **7,** 339–369.

Till, J. E., and McCulloch, E. A. (1961). A direct measurement of the radiation sensitivity of normal mouse bone marrow cells. *Radiat. Res.* **14,** 213–222.

Tumbar, T., Guasch, G., Greco, V., Blanpain, C., Lowry, W. E., Rendl, M., and Fuchs, E. (2004). Defining the epithelial stem cell niche in the skin. *Science* **303,** 359–363.

van Os, R., Sheridan, T. M., Robinson, S., Drukteinis, D., Ferrara, J. L., and Mauch, P. M. (2004). Immunogenicity of Ly5 (CD45)-antigens hampers long-term engraftment following minimal conditioning in a murine bone marrow transplantation model. *Stem Cells* **19,** 80–87.

Visser, J. W., Bauman, J. G., Mulder, A. H., Eliason, J. F., and de Leeuw, A. M. (1984). Isolation of murine pluripotent hemopoietic stem cells. *J. Exp. Med.* **159,** 1576–1590.

Wada, H., Masuda, K., Satoh, R., Kakugawa, K., Ikawa, T., Katsura, Y., and Kawamoto, H. (2008). Adult T-cell progenitors retain myeloid potential. *Nature* **452,** 768–772.

Waskow, C., Madan, V., Bartels, S., Costa, C., Blasig, R., and Rodewald, H. R. (2009). Hematopoietic stem cell transplantation without irradiation. *Nat. Methods* **6,** 267–269.

Waterstrat, A., Liang, Y., Swiderski, C. F., Shelton, B. J., and Van Zant, G. (2010). Congenic interval of CD45/Ly-5 congenic mice contains multiple genes that may influence hematopoietic stem cell engraftment. *Blood* **115,** 408–417.

Wilson, A., Laurenti, E., Oser, G., van der Wath, R. C., Blanco-Bose, W., Jaworski, M., Offner, S., Dunant, C. F., Eshkind, L., Bockamp, E., *et al.* (2008). Hematopoietic stem cells reversibly switch from dormancy to self-renewal during homeostasis and repair. *Cell* **135,** 1118–1129.

Wright, D. E., Wagers, A. J., Gulati, A. P., Johnson, F. L., and Weissman, I. L. (2001). Physiological migration of hematopoietic stem and progenitor cells. *Science* **294,** 1933–1936.

Yang, L., Bryder, D., Adolfsson, J., Nygren, J., Mansson, R., Sigvardsson, M., and Jacobsen, S. E. (2005). Identification of Lin(-)Sca1(+)kit(+)CD34(+)Flt3- short-term hematopoietic stem cells capable of rapidly reconstituting and rescuing myeloablated transplant recipients. *Blood* **105,** 2717–2723.

Yoshihara, H., Arai, F., Hosokawa, K., Hagiwara, T., Takubo, K., Nakamura, Y., Gomei, Y., Iwasaki, H., Matsuoka, S., Miyamoto, K., *et al.* (2007). Thrombopoietin/MPL signaling regulates hematopoietic stem cell quiescence and interaction with the osteoblastic niche. *Cell Stem Cell* **1,** 685–697.

Zeng, H., Yucel, R., Kosan, C., Klein-Hitpass, L., and Moroy, T. (2004). Transcription factor Gfi1 regulates self-renewal and engraftment of hematopoietic stem cells. *EMBO J.* **23,** 4116–4125.

Zhang, J., Niu, C., Ye, L., Huang, H., He, X., Tong, W. G., Ross, J., Haug, J., Johnson, T., Feng, J. Q., Harris, S., Wiedemann, L. M., *et al.* (2003). Identification of the haematopoietic stem cell niche and control of the niche size. *Nature* **425,** 836–841.

Author Index

C

D

E

F

G

H

I

J

M

N

Q

R

S

T

Subject Index

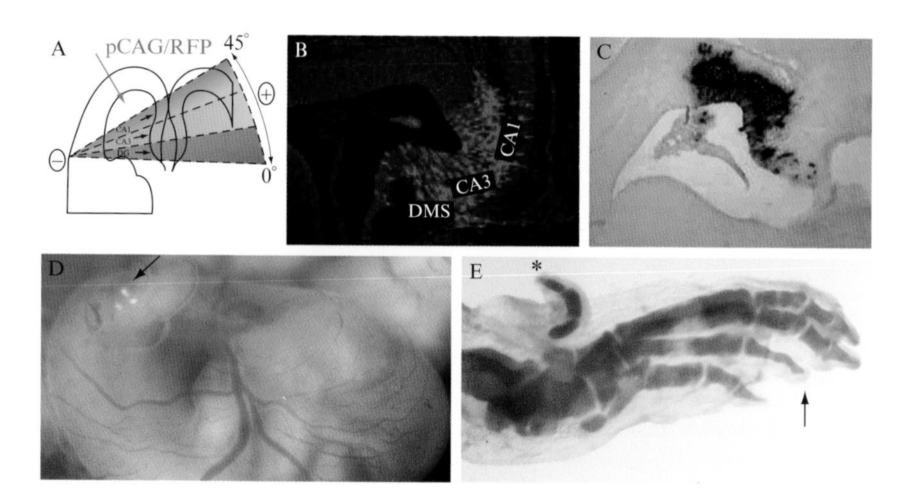

Valérie Ngô-Muller and Ken Muneoka, Figure 12.3 *In utero* and *exo utero* manipulation of the mouse embryo. (A–C) *In utero* electroporation of the E14.5 hippocampus with a plasmid encoding red fluorescent protein under the control of the chicken β-actin promoter (pCAG/RFP) as a lineage tracer. (A) Efficient transfection of specific areas of the hippocampal anlage (CA1, CA3, and dentate gyrus (DG)) is accomplished by manually changing the orientation of the electric field. (B, C) Two days after electroporation, widespread expression of RFP (B) and RFP mRNA (C) are present in the ventricular zone, the initial differentiating fields of the CA1–CA3 regions, and the dentate migratory stream (DMS). (D) Targeted injection of labeled 3T3 cells (arrow) into the zeugopodial region of the E12.5 hindlimb was used to explore regional growth dynamics during development. Cells were coinjected with fluorescent latex beads to identify the injection site. (E) Digit amputations at E13.5 combined with grafts of the amputated digit to the dorsal surface of the limb was used to characterize the extent of the regenerative response (arrow) in relation to the level of the amputation (*). (A–C) From Navarro-Quiroga *et al.* (2007), (D) from Trevino *et al.* (1992), (E) from Reginelli *et al.* (1995).

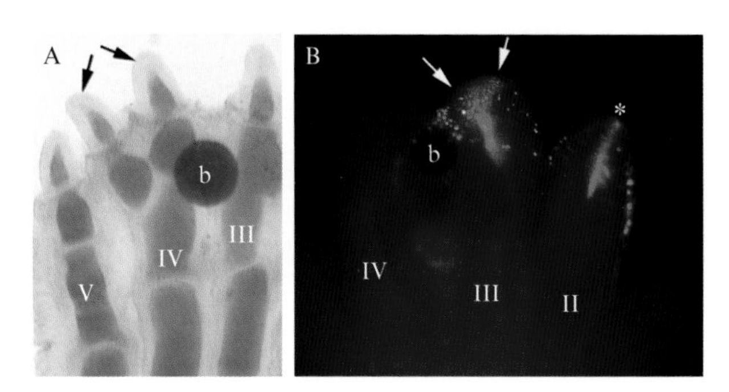

Valérie Ngô-Muller and Ken Muneoka, Figure 12.5 Targeted application of growth factors: (A) microcarrier beads (b) soaked in a high concentration of purified FGF4 implanted between digits III and IV of a E12.5 hindlimb induces the bifurcation of the distal region of digit IV (arrows). (B) Microcarrier bead implantation into the region between digits III and IV and in conjunction with targeted injections of DiI into digits II and III demonstrates a localized effect on the movements of digit III cells (arrows) with the cells of digit II (*) completely unaffected. From Ngo-Muller and Muneoka (2000a,b).

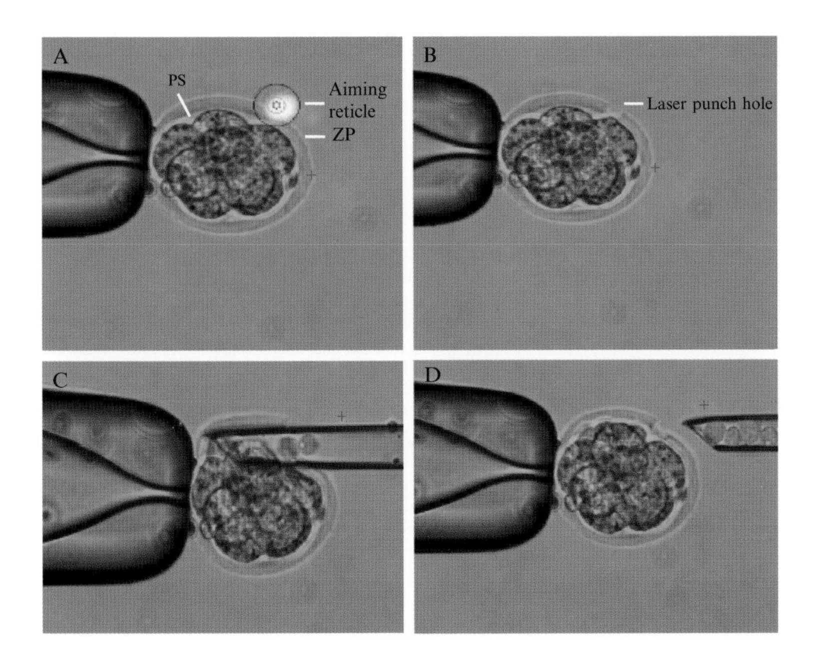

Thomas M. DeChiara *et al.*, Figure 16.1 The eight-cell embryo injection process. (A) An uncompacted eight-cell embryo is held on the holding pipette with the laser aiming reticle positioned over the target site on the zona pellucida (ZP) at the "1 o'clock" position. ES cells are deposited in the perivitelline space (PS) between the blastomeres and the ZP. (B) Following one pulse of the laser, a hole in the ZP is easily visualized. (C) The injection pipette containing ES cells is gently pushed through the hole in the ZP and into the ps to deposit seven to nine ES cells at the "10 o'clock" position. (D) When the injection pipette is removed the ES cells remain in the ps. The injection of more than nine ES cells will normally result in them leaking out of the laser punch hole.

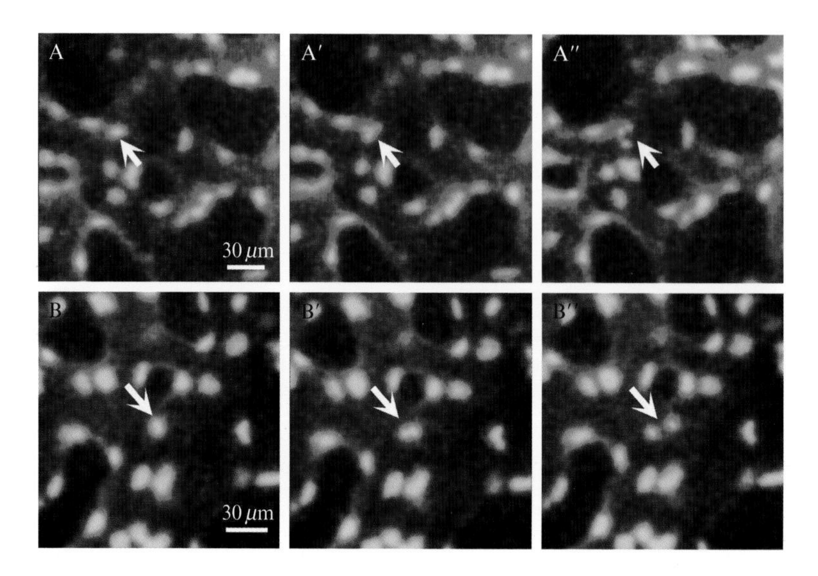

Ryan S. Udan and Mary E. Dickinson, Figure 19.3 Time-lapse movie of the developing Yolk sac vessels (cultures starting at E8.5). Yolk sac vessels of intact Flk1-myr::mCherry$^{tg/tg}$ (to visualize endothelial cell membrane); Flk1-H2B::eYFP$^{tg/tg}$ (to visualize endothelial cell nuclei) embryos are imaged on the LSM 5 LIVE confocal with a 25× objective (NA of 0.45). Images are taken every 6 min at three *Z*-planes (~30 *μM*) each with the 488 nm (0.3% power) and 561 nm (2.0% power) lasers to image eYFP and mCherry, respectively. Endothelial cell apoptosis (panels A-A″) and mitosis (panels B-B″) can be captured in movies by observing fragmentation of nuclei and division of a single nucleus into two nuclei, respectively. Beginning of chromatin condensation is observed in A′ (6 minutes after A), and nuclear fragmentation is observed in A″ (1 hour after A). Beginning of mitosis is observed in B′ (12 minutes after B) and formation of two separate nuclei is apparent in B″ (18 minutes after B).

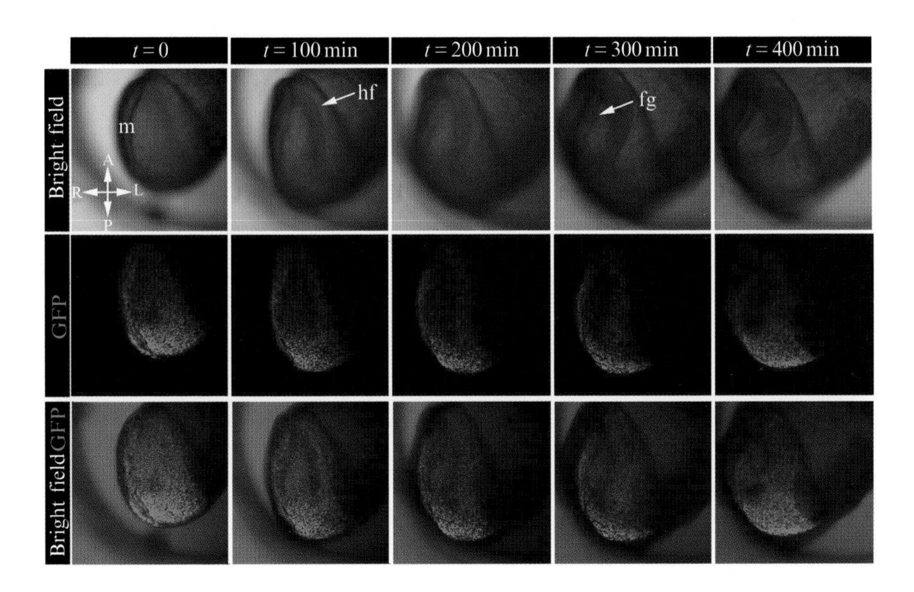

Sonja Nowotschin *et al.*, Figure 20.7 Live imaging of postimplantation mouse development, visualization of mesoderm cell dynamics during foregut invagination. Sequence of stills from a laser scanning confocal time-lapse experiment of an embryo expressing a histone H2B–YFP reporter in all mesoderm cells. Normal morphogenesis of the headfolds (hf) and invagination of the foregut (fg) are evident in the bright field panels of the time series. Embryo was imaged for a total of 8.5 h. Optical sections were taken every 4 μm, and z-stacks were acquired at 20 min intervals. Upper row shows a time series of bright field images. Middle row shows 3D rendered z-stack of the GFP channel from individual time points of the same time series. Bottom row shows the overlay of the GFP channel onto the bright field images. m, midline.

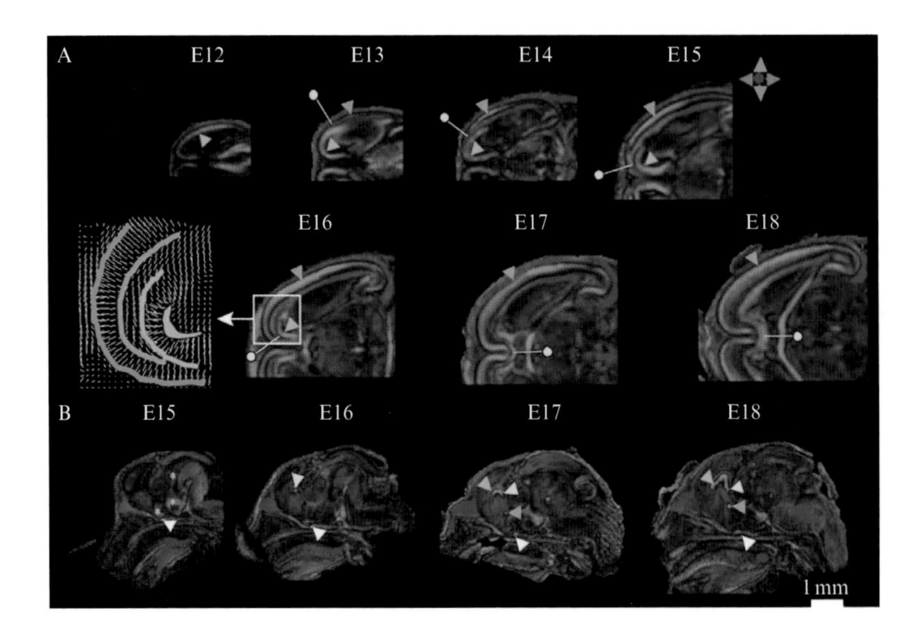

Brian J. Nieman and Daniel H. Turnbull, Figure 21.6 Diffusion tensor imaging in the developing mouse brain. Diffusion tensor data is shown with a color map, where red, green, and blue each represent diffusion in an orthogonal direction as indicated by the arrows. In panel (A), horizontal images from E12 to E18 are provided. Pink and blue arrowheads represent the cortical plate and neuroepithelium, respectively. Yellow pins indicate the leading edge of the growing intermediate zone. A white box is expanded at E16 to show the fiber orientation in a vector picture. In panel (B), the equivalent sagittal images are provided for E15–E18. In this case, white, yellow, pink, and blue arrowheads indicate the optic chiasm, hippocampal commissure, anterior commissure, and corpus callosum, respectively. Asterisks (*) mark the neuroepithelium around the third ventricle. All data are from full three-dimensional data sets. Reprinted from Zhang *et al.* (2003) with permission.

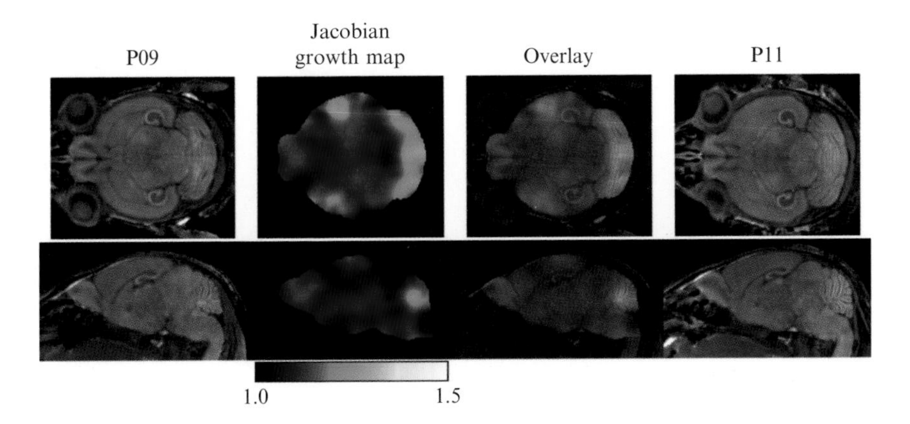

Brian J. Nieman and Daniel H. Turnbull, Figure 21.8 Computational mapping of growth in the developing mouse brain. *In vivo* Mn-enhanced MRI images at days 9 and 11 after birth (left and rightmost columns, respectively) provide a visual representation of growth-related changes. Computational image processing can provide a quantitative growth map (second column). An overlay of the growth map on the outline of the P09 mouse shows that the most significant growth is occurring in the cerebellum and regions of the cortex.

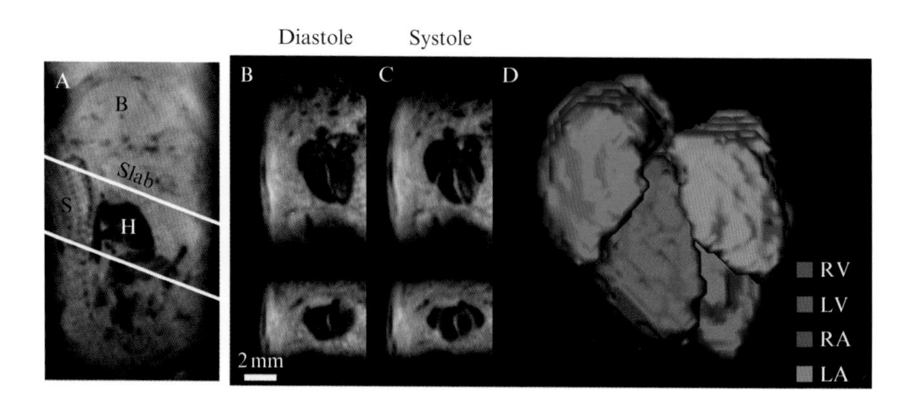

Brian J. Nieman and Daniel H. Turnbull, Figure 21.9 *In utero* MRI of the beating embryonic heart. With appropriate motion-correction, it is possible to image a volume (depicted in as an imaging slab in (A)) including the embryonic heart (day E17). Orthogonal long- and short-axis views are shown at diastole in (B) and (C). A volumetric rendering of the four heart chambers in (D) emphasizes the three-dimensional character of the data and shows the similarity in the four chamber volumes. Reprinted from Nieman *et al.* (2009) with permission.